Biodiversity Enrichment in a Changing World

Biodiversity Enrichment in a Changing World

Editor: Neil Griffin

R CALLISTO REFERENCE

www.callistoreference.com

Callisto Reference,
118-35 Queens Blvd., Suite 400,
Forest Hills, NY 11375, USA

Visit us on the World Wide Web at:
www.callistoreference.com

ISBN: 978-1-63239-829-1 (Hardback)

The publisher's policy is to use permanent paper from mills that operate a sustainable forestry policy. Furthermore, the publisher ensures that the text paper and cover boards used have met acceptable environmental accreditation standards.

Trademark Notice: Registered trademark of products or corporate names are used only for explanation and identification without intent to infringe.

Printed in the United States of America.

Cataloging-in-publication Data

Biodiversity enrichment in a changing world / edited by Neil Griffin.
    p. cm.
Includes bibliographical references and index.
ISBN 978-1-63239-829-1
1. Biodiversity. 2. Ecological heterogeneity. 3. Ecology. 4. Species diversity. I. Griffin, Neil.
QH541.15.B56 B56 2017
577--dc23

Table of Contents

Preface

The emphasis on biodiversity has increased in the past few decades. Biodiverse species occupy a vast portion of the earth's ecosystem. This book on biodiversity enrichment in a changing world discusses the strategies and programs adopted by organizations and projects around the world for diversity enhancement, wildlife conservation and forest management. Studies enumerating the ecological, rather than economic aspects emphasizes the changing interaction between the wildlife-urban continuums. A number of latest researches have been included to keep the readers up-to-date with the global concepts in this area of study. The book will serve as a reference text to ecologists, environmentalists, conservationists, researchers, academicians and students engaged in this field.

The information contained in this book is the result of intensive hard work done by researchers in this field. All due efforts have been made to make this book serve as a complete guiding source for students and researchers. The topics in this book have been comprehensively explained to help readers understand the growing trends in the field.

I would like to thank the entire group of writers who made sincere efforts in this book and my family who supported me in my efforts of working on this book. I take this opportunity to thank all those who have been a guiding force throughout my life.

Editor

New Insights into the Consequences of Post-Windthrow Salvage Logging Revealed by Functional Structure of Saproxylic Beetles Assemblages

Simon Thorn[1]*, **Claus Bässler**[1], **Thomas Gottschalk**[2], **Torsten Hothorn**[3], **Heinz Bussler**[4], **Kenneth Raffa**[5], **Jörg Müller**[1,6]

1 Sachgebiet Forschung und Dokumentation, Nationalparkverwaltung Bayerischer Wald, Grafenau, Germany, 2 Hochschule für Forstwirtschaft Rottenburg, Rottenburg am Neckar, Germany, 3 Abteilung Biostatistik, Universität Zürich, Zürich, Switzerland, 4 Bavarian State Institute for Forestry, Freising, Germany, 5 Department of Entomology, University of Wisconsin-Madison, Madison, United States of America, 6 Chair for Terrestrial Ecology, Department of Ecology and Ecosystem Management, Technische Universität München, Freising, Germany

Abstract

Windstorms, bark beetle outbreaks and fires are important natural disturbances in coniferous forests worldwide. Windthrown trees promote biodiversity and restoration within production forests, but also cause large economic losses due to bark beetle infestation and accelerated fungal decomposition. Such damaged trees are often removed by salvage logging, which leads to decreased biodiversity and thus increasingly evokes discussions between economists and ecologists about appropriate strategies. To reveal the reasons behind species loss after salvage logging, we used a functional approach based on four habitat-related ecological traits and focused on saproxylic beetles. We predicted that salvage logging would decrease functional diversity (measured as effect sizes of mean pairwise distances using null models) as well as mean values of beetle body size, wood diameter niche and canopy cover niche, but would increase decay stage niche. As expected, salvage logging caused a decrease in species richness, but led to an increase in functional diversity by altering the species composition from habitat-filtered assemblages toward random assemblages. Even though salvage logging removes tree trunks, the most negative effects were found for small and heliophilous species and for species specialized on wood of small diameter. Our results suggested that salvage logging disrupts the natural assembly process on windthrown trees and that negative ecological impacts are caused more by microclimate alteration of the dead-wood objects than by loss of resource amount. These insights underline the power of functional approaches to detect ecosystem responses to anthropogenic disturbance and form a basis for management decisions in conservation. To mitigate negative effects on saproxylic beetle diversity after windthrows, we recommend preserving single windthrown trees or at least their tops with exposed branches during salvage logging. Such an extension of the green-tree retention approach to windthrown trees will preserve natural succession and associated communities of disturbed spruce forests.

Editor: Marc Hanewinkel, Swiss Federal Institute for Forest, Switzerland

Funding: ST was funded by the Scholarship Programme of the German Federal Environmental Foundation. The funders had no role in study design, data collection and analysis, decision to publish, or preparation of the manuscript.

Competing Interests: The authors have declared that no competing interests exist.

* Email: simon@thornonline.de

Introduction

Forest ecosystems worldwide are periodically affected by natural disturbances, such as wind storms, fires, avalanches and insects [1,2]. Since the 1990s, disturbance events in forests of the northern hemisphere have increased, particularly in mature conifer stands, owing to both an increase in growing stock and global climate change [3–5]. After such disturbances, forest managers try to limit the economic loss by focusing on saving downed wood from fungal infestation and avoiding an increase of pest species populations [6,7]. Even if such salvage logging is broadly publicly accepted [8], both ecologists and conservationists are increasingly aware that natural disturbances conserve biodiversity in forests moulded by anthropogenic impacts [9,10]. In addition, whether salvage logging should be conducted and how it would be best conducted, particularly in coniferous forests, is highly controversial [11–13].

Between 1950 and 2000 in Europe, wind storms annually damaged an estimated average of 18.7 million m^3 of wood [3]. Such events often are followed by outbreaks of the European spruce bark beetle *Ips typographus* (Linnaeus, 1758), which damages an additional 2.9 million m^3 wood annually [3]. In contrast to our knowledge of the high value of windthrows for biodiversity [14,15], our knowledge about why particular species are affected or not by salvage logging is limited; most recent studies have focused on post-fire salvage logging [16–19] or economic consequences and bark beetles [20,21]. The relatively few studies on post-windthrow salvage logging focus mainly on the decrease in species numbers [14,22,23].

Recent studies using, for example, guild-specific analysis of bird assemblages demonstrate that species richness poorly reflects the effect of human intervention [24]. A quantification of species loss is not sufficient to guide conservation efforts and resource management directly [25,26]. Therefore, functional approaches have become increasingly important throughout broad areas of ecological research [26–28]. It has recently been proposed that the species position in a functional space can be used as a tool to reveal advanced warnings for changes in disturbed ecosystems [29].

Saproxylic beetles are highly diverse, play important roles in the decomposition of wood [30] and are sensitive to forest management, and are thereby an ideal model group to study the impact of salvage logging on biodiversity [31,32]. Here we studied the impact of salvage logging and focused on recently published ecological traits of saproxylic beetles [33]: mean body size, diameter and decay stage of wood in which larvae develop and canopy cover of forests in which the species is known to occur.

The body size of saproxylic beetles is positively correlated to the diameter of the substrate used for larval development [34]. Assemblage values should therefore decrease with the removal of tree sections of large diameter [35,36]. Similarly, a shift in assemblages toward species with a preference for wood of smaller diameters might follow the removal of major tree trunks. Since the cutting of uprooted trees in salvage-logging operations removes the trunk and the remaining branches rapidly decompose to advanced decomposition stages [37], we therefore expect that colonizers of earlier successional stages of wood decomposition also decline [35]. Finally, the removal of dead wood might decrease shady conditions provided by the cross-laminated arrangement of trees after windthrows, thereby promoting heliophilous species [38].

We calculated effect sizes of saproxylic beetle functional diversity, and mean assemblage values and diversity values of each single trait to test the predictions that salvage logging results in 1) a lower overall functional diversity, 2) lower mean body size and body size diversity, 3) a minor occurrence of species preferring dead wood of large diameter, 4) a decrease in species preferring early successional stages of decomposition and 5) an increase in species preferring open canopies.

Methods

Study area

The study was carried out in the high montane spruce forest in the Bavarian Forest National Park in south-eastern Germany. Forest stands in this area, at an elevation above 1,100 m, are naturally dominated by Norway spruce (*Picea abies*). Annual precipitation ranges from 1,300 to 1,800 mm, and annual mean air temperature ranges from 3.0 to 4.0°C [39].

On January 16, 2007, an area of approximately 1,000 ha of spruce forests was felled to various extents by the windstorm Kyrill, ranging from single trees to stands covering several hundreds of hectares. From a total affected amount of wood of about 160,000 m³, 50,000 m³ are concentrated on four larger windthrow areas (~170 ha). These centres were partially excluded from the overall salvage-logging operation. Such operations basically remove the main trunk to preserve it from fungal and pest species infestation. The branches are cut off the trunk and remain on the ground surface, which is covered by a grass layer of *Calamagrostis villosa* (32±11 cm height on logged plots and 32±9 cm height on non-logged plots, measured by relevés [39]). Standardized measurements of dead-wood objects per plot revealed that on salvaged-logged plots, 90% (n = 29) of dead-wood objects had direct contact with the ground surface, in contrast to only 10% (n = 35) on non-logged plots. Wind-felled trees on non-logged plots remained mostly alive after the storm in spring 2007 until they were colonized by *Ips typographus* in 2008. In contrast, salvage logging typically kill trees immediately. Salvage logging in the major windthrow areas removed about 255 m³/ha and was completed in autumn 2007 (Fig. 1).

Beetle sampling

To reflect the emerging beetle fauna of surrounding dead wood, we used flight-interception traps [40]. Traps were established throughout windthrow centres and surrounding salvage-logged areas: 22 in logged areas and 22 in non-logged areas in spring 2008. Traps in logged areas were surrounded by at least a 50 m radius of completely salvage-logged windthrows (all trees removed); traps in non-logged areas were surrounded by at least a 50 m radius of completely non-logged windthrow (all trees of the previous stand were wind felled). Each trap consisted of a crossed pair of transparent plastic shields (40×60 cm) and contained a 3.0% copper-vitriol solution to preserve trapped specimens [41]. The shortest distance between two traps was 50 m, and the largest distance between traps was 6,500 m. Sampling was conducted during the entire growing season between May after the snow melted until September over four consecutive years until 2011. Traps were emptied monthly. All sampled beetles were identified to the species level, but only saproxylic beetles were considered in the analysis [42,43].

Trait characterization

We used four ecological traits that enabled us to link species habitat selection [33] directly to forest management: mean body size, diameter of wood in which the larvae of the species was recorded, decay stage of the wood, and canopy cover of forests in which the larvae of the species is known to occur. The single classes of niche positions were classified as follows: *wood diameter class*: 1, <15 cm; 2, 15–35 cm; 3, 35–70 cm; and 4, >70 cm; *wood decay stage*: 0, alive; 1, freshly dead (up to two years); 2, initiated decomposition with loose bark and tough sapwood; 3, advanced decomposition with soft sapwood and partly tough hardwood; and 4, extremely decomposed and mouldered; *canopy cover*: 1, open; 2, semi-open; and 3, closed (for mean niche positions of species, see Table S1).

Data analysis

Prior to the main analysis, trapped specimens were grouped according to the trap level in each year, and subsequent analyses were conducted at the trap per year level. We calculated the mean trait value of the assemblage for each of the four single traits as an arithmetic mean, weighted by the number of trapped individuals (e.g. log-transformed number of individuals) per species. The results of abundance-weighted data, abundance-weighted data based on the log-transformed number of individuals, and presence/absence data were similar; therefore, we present only abundance-weighted results, which do not overestimate singletons and represents the majority of trapped beetles [40]. To test the impact of salvage logging on saproxylic beetles, we selected these mean assemblage trait values of every single trait and their related trait diversity as target variable and year, and we selected logged/non-logged as predictor variables.

A challenge in functional diversity measures is to compare dispersion of functional traits in the functional space independent of species numbers [29]. Hence, we calculated a distance matrix based on the pairwise Euclidean distance between functional traits of all possible species pairs within an assemblage per plot for each single trait [44]. Similarly, we used the function *dist* to calculate an

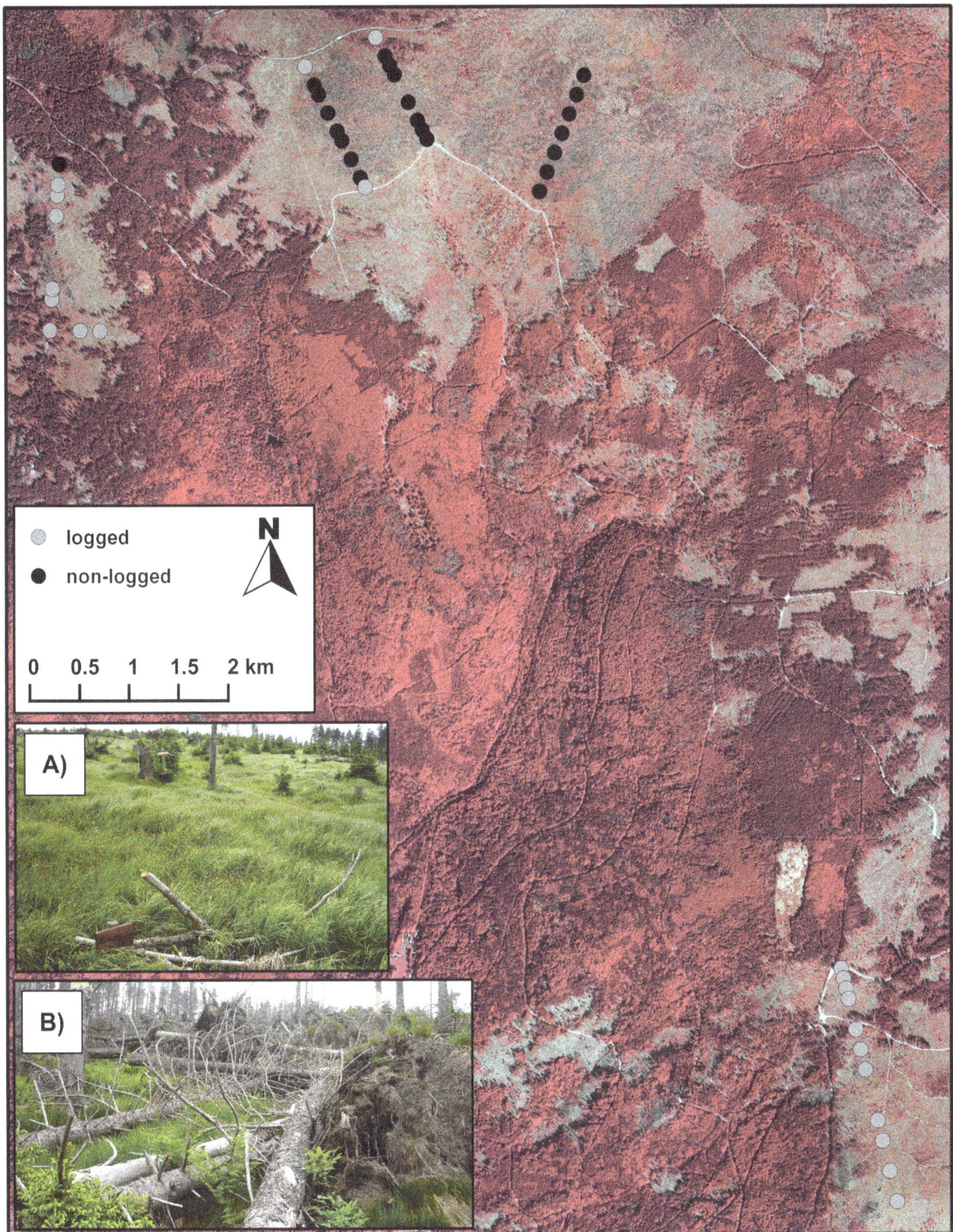

Figure 1. Locations of flight-interception traps within our study area in salvage-logged (A) and non-salvage-logged windthrows, sampled from 2008 to 2011 (B).

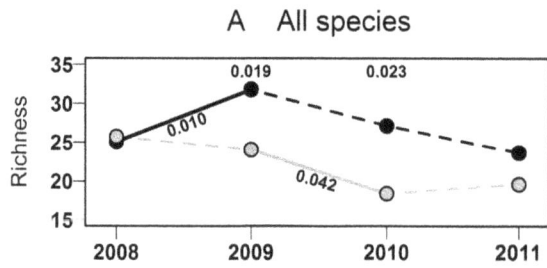

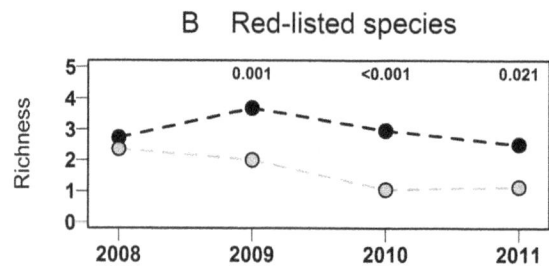

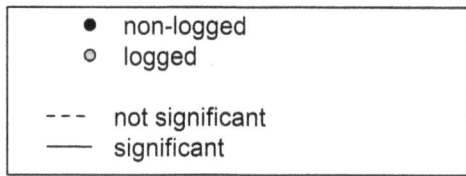

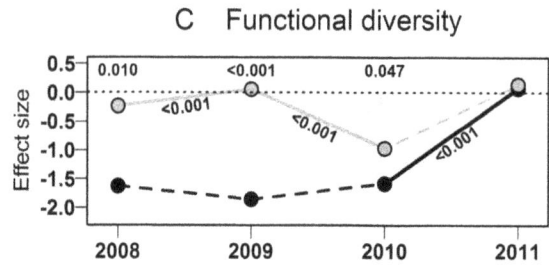

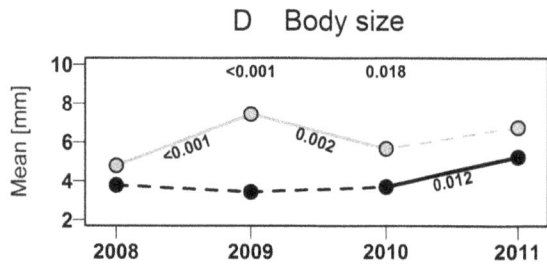

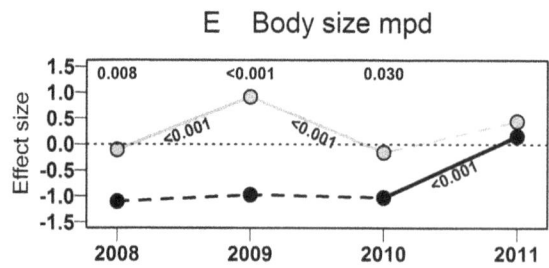

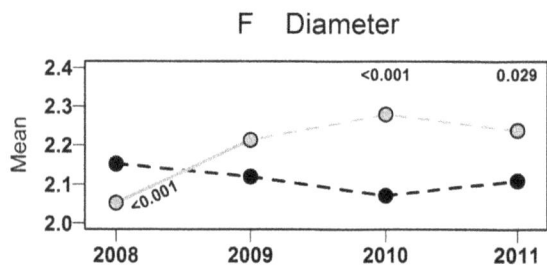

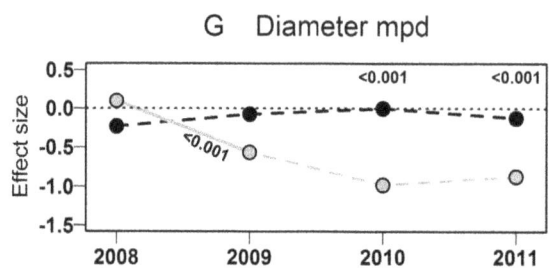

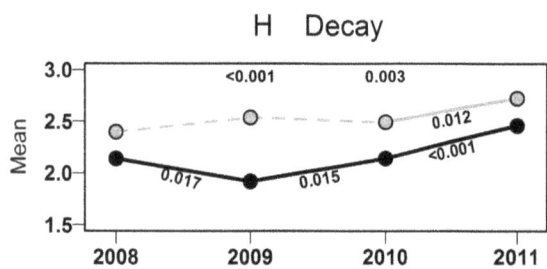

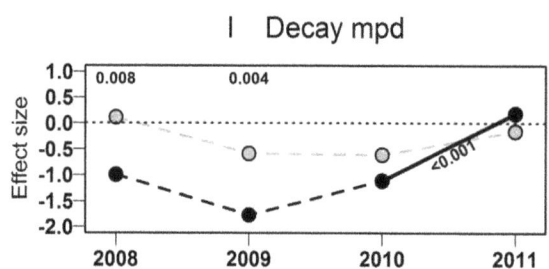

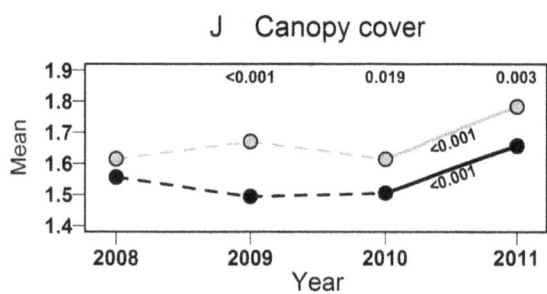

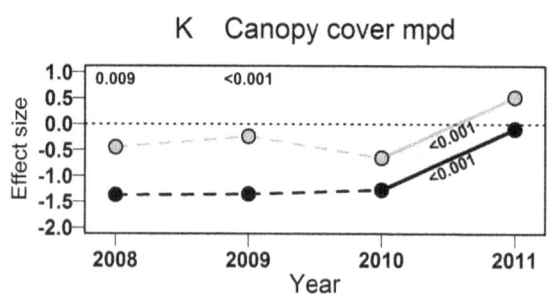

Figure 2. Effects of salvage logging on saproxylic beetles. Mean species richness of saproxylic beetles (A) and red-listed saproxylic beetles (B), abundance-weighted mean niche positions and standardized effect size (based on mean pairwise distance) of functional diversity (C) and body size (D, E), niche diameter (F, G), niche decay (H, I) and niche canopy cover (J, K) in logged and non-logged windthrow areas in a spruce mountain forest based on a GLMM with treatment and year as fixed factors, and space and plot as random factors. For multiple comparisons between treatments, the adjusted p-values are drawn above the respective points; for comparisons between different years within the same treatment, p-values are drawn below respective lines. Full details on p-values and model estimators can be found in Tables S2 and S3.

overall distance matrix based on Euclidean distance between among four traits. We compared the observed value of mean niche position for each trait in each assemblage to an artificial assemblage with equal number of randomly selected species from the regional species pool (all species recorded on plots pooled). To create these random assemblages, we used null models with tip shuffling and abundance weighting in 999 randomizations using the function *ses.mpd* (abundance.weighted = TRUE) in the add-on package *picante* of R version 2.15 [45,46]. The effect size, which is provided by the null model, indicates the dispersion of a specific trait and compares it to the trait dispersion in an artificial assemblage. Values >0 indicate over-dispersion of a trait. In turn, values <0 indicate clustering and a trait dispersion in observed assemblages smaller than expected from a random assemblage. In general, over-dispersion is a result of competition or facilitation, and clustering is a result of environmental filtering [47].

The natural occurrence of windthrow patches necessitates that logged and non-logged plots are near each other in the same forest site (Fig. 1). This might reduce the independence of observations. Furthermore, we cannot treat the measurement on each plot as independent as the measurements were conducted in consecutive years. To address these issues, we analysed the data using a linear mixed random effects model with sampling plot and coordinates of sampling plot as random factors. The plot coordinates enable us to account for possible spatial autocorrelation within the arrangement of our plots, as a second-order trend surface (for the R code, see Text S1; for the method, see [48]).

To illustrate the impact of salvage logging on the primary target species, we modelled the mean number of *I. typographus* individuals per trap by using an observation-specific random factor to control the mixed model with Poisson distribution for potential over-dispersion [49]. To consider the resulting problem of testing hypothesis families, we applied a multiple post-hoc comparison with adjusted p-values using the function *glht* in the add-on package *multcomp* and constructed a matrix of coefficients [50]. This enabled us to compare the impact of salvage logging within each year and to compare changes within consecutive years on logged or non-logged plots.

Results

Effect of pest control

In total, 33,796 specimens belonging to 179 species of 37 saproxylic beetles families were sampled. Of these, 29 (89 individuals) species were found exclusively on non-logged plots, and 34 species (85 individuals) were found only on salvage-logged plots. Forty-two species (19 on non-logged plots) were caught as singletons. The most abundant species in our data on non-logged plots were *Ips typographus* (7,785 individuals), *Pityogenes chalcographus* (6,697 individuals) and *Xyloterus lineatus* (1,579 individuals), all of subfamily *Scolytinae* (Curculionidae). The most frequent species on logged plots were *P. chalcographus* (3,147 individuals), *Hylastes cunicularius* (1,101 individuals) and *I. typographus* (796 individuals).

Salvage logging significantly decreased the mean number of the primary target species *I. typographus* per trap in 2008, 2009 and 2010. The mean number of *I. typographus* individuals significantly increased from 2008 to 2009 and significantly decreased from 2009 to 2010, reflecting the colonization of the windthrow by this pest species. The species richness of saproxylic beetles was significantly lower on salvage-logged plots than on non-logged plots in 2009 and 2010 (Fig. 2A). On non-logged plots, species richness significantly increased from 2008 to 2009, whereas on logged plots, richness significantly decreased from 2009 to 2010 (Fig. 2A). A similar trend was found only for red-listed species, which displayed significantly lower species richness on salvage-logged plots in 2009, 2010 and 2011 (Fig. 2B; for complete species list, see Table S1).

Assemblage functional response

Beetle assemblages of logged and non-logged windthrows differed in their functional trait dispersion. The total functional diversity was lower on non-logged plots and revealed a clustered pattern, but was random on logged plots (Fig. 2C). From 2010 to 2011, the functional assemblage composition of non-logged plots developed significantly toward more random assemblages. The mean body sizes and the corresponding niche diversity were consistently and significantly higher on logged plots than on non-logged plots in 2009 and 2010 (Fig.2 D, E). We also found significantly higher mean diameter niche values on logged plots in 2010 and 2011, but not in 2008 and 2009 (Fig. 2F). The corresponding niche diversity of assemblages was significantly more randomly distributed on non-logged plots in 2010 and 2011 (Fig. 2G). In 2009 and 2010, the mean decay niche, which reflects the process of decomposition, was significantly higher on logged plots (Fig. 2H). The corresponding decay niche diversity (Fig. 2I) showed a significantly more clustered pattern on non-logged plots than on logged plots in 2008 and 2009. The mean canopy niche value of the local assemblages of logged and non-logged plots differed in 2009–2011. Significantly more species preferring open-canopy conditions (sunny habitats) were found on non-logged plots, as indicated by a significantly higher mean niche value of the canopy niche in 2009, 2010 and 2011 (Fig. 2J). A similar trend was found in the corresponding niche diversity, which displayed a significantly clustered pattern in the canopy cover niche in 2008 and 2009 (Fig. 2K). Full details on p-values and model estimators can be found in Tables S2 and S3.

Discussion

Our study confirmed long-held empirical and scientific findings that salvage logging after windthrows reduces populations of pest bark beetles [7,12] and has accompanying negative effects on the species richness of saproxylic beetles [51]. Surprisingly, our examination of the functional traits did not support most of our predictions: single-trait analyses revealed species on logged plots that were on average larger, preferred dead wood of larger diameter, and were adapted to shady habitats, compared to species on non-logged plots. These unexpected results underline the high sensitivity of an analysis based on species functional traits to detect complex and subtle changes in species assemblages caused by anthropogenic impact [26,52–54].

Salvage logging aims at controlling populations of one or a few pest species but reduces biodiversity per se as collateral damage

[6,55,56]. Accordingly, our results demonstrated that salvage logging dramatically decreases species richness of saproxylic beetles, including red-listed species and not only the target species *I. typographus* (Fig. 2A, B). These reductions seem to be mainly caused by the loss of species directly associated with *I. typographus* (compare [57]), e.g. by the decrease in predators of *I. typographus*, such as *Thanasimus* sp.; by the decrease in species that exploit bark beetle galleries, such as *Crypturgus cinereus*; and by the loss of species associated with a similar early decay stage of wood. Saint-Germain *et al.* [58] demonstrated that the majority of saproxylic beetle species on the conifer black spruce (*Picea mariana*) colonize the early decay stages, while saproxylic beetle species on the broadleaf tree aspen (*Populus tremula*) occur mostly on wood of later decay stages — a commonly observed pattern [41,59,60]. Our study, which focuses on the first four years of the decay process, therefore reflects the most important stage of succession within windthrown stands of coniferous trees [61]. In particular, because colonization patterns of saproxylic organisms this stage determines subsequent saproxylic communities, like.g. the early-arriving bark beetle *Hylurgops palliates* and the wood-decaying fungus *Fomitopsis pinicola*, enabled a higher colonization success of the endangered beetle *Peltis grossa* after 10 years [62].

Larger beetles on logged plots

In contrast to our prediction, our results showed a consistent separation of beetle assemblages, from on average large species on logged plots toward small species on non-logged plots. One explanation for this pattern may be found behind the functional structure of the assemblages, i.e. in the habitat specificity of the species in our study: the majority of the larger species identified are widely distributed habitat generalists (except *Ampedus auripes*; [63]). One of the largest species, *Hylobius abietis*, which was more frequent in logged stands (see Table S1), is a known pest species attracted by the odour of resin in tree stumps [64]. Such species breed well in stumps and in logging residuals. Hence, an increase in the harvesting of stumps for bioenergy might expand the negative impacts of salvage logging to include currently less-affected species [65–67].

Species preferring wood of large diameters on logged plots

After regular clear-cutting, the remaining dead wood amounts to approximately 10 m^3 ha^{-1} [68]. Post-windthrow salvage-logged sites offer much more dead wood, e.g. from 45 to >70 m^3 ha^{-1} on 90 sites in Switzerland [37]. Such high amounts of remaining dead wood on salvage-logged areas regularly surpass the critical thresholds of dead-wood amount for diversity in boreal forests [32]. Accordingly, the mean diameter niche position of beetles in our study was significantly higher in 2010 and 2011. Thereby, simply the amount of dead wood does not seem to be the limiting factor for saproxylic beetles species richness in salvage-logged plots. The limiting factor is rather the loss of small branches, as indicated by the loss of species preferring wood of small diameter (Fig. 2J). Accordingly, an alteration of the remaining dead-wood resources appears to be more crucial than the simple removal of the main wood volume by salvage logging.

This assumption is strongly supported by our findings on functional trait diversity (Fig. 2C): we found a clustering of functional assemblage structure on non-logged plots, which indicated a strong habitat-filtering (dead-wood resources) effect on dead-wood communities. Using co-occurrence null-model approaches, Azeria *et al.* [69] also found a strong habitat-filtering effect on saproxylic assemblages on burned trees. In accordance, Ding *et al.* [70] proposed that disturbance in forest ecosystems generates communities by abiotic filtering. Hence, our data suggested that anthropogenic intervention, i.e. salvage logging, of natural disturbances can disrupt the natural habitat filtering in the assembly process.

Accelerated decomposition on salvage-logged plots

The amount and diversity of dead-wood resources of salvage-logged areas does not significantly differ between salvage-logged and non-salvage-logged windthrows [37]. But, in contrast to natural windthrows, the remaining dead wood on salvage-logged windthrows in our study area is mostly scattered on the ground surface (e.g. Fig. 1). Owing to the stronger attraction of wood-inhabiting fungi, salvage-logged sites tend to harbour more advanced decomposition stages than non-salvage-logged sites [37,71,72]. This shift within decay stages of available dead-wood resources was well reflected by our finding of a mean decay niche with significantly higher mean decay niche values on salvage-logged plots in 2009 and 2010. Furthermore, the corresponding niche diversity indicated a strong habitat filtering effect towards species of early decay stages on non-salvage-logged plots.

Decrease of heliophilous species through salvage logging

Sun exposure increases the probability of the presence of red-listed species in aspen retention trees in clear-cuts in Norway [38]. Similarly, the endangered longhorn beetle *Rosalia alpina* prefers trees with a lower percentage of canopy closure and higher sun exposure than the average tree [73]. Hence, sun exposure is a major predictor determining saproxylic beetle communities in dead-wood resources [74,75]. In our study, we demonstrated a shift of assemblages comprising heliophilous species on non-logged plots toward species preferring shady habitats on logged plots. This result is contrasts our prediction that the removal of dead wood might decrease the shady conditions provided by the cross-laminated arrangement of trees after windthrow. However, the majority of logging residuals on logged plots lie on the ground surface and are covered by an extensive grass layer, both of which create a moist microclimate, which decreases the availability of sun-exposed dead wood. An accelerated growth of natural regeneration on salvage-logged plots [76] does not appear to be of great importance, since natural regeneration in our study area is still poor and not able to shade the complete surroundings of a flight-interception trap (see Fig. 1 insets). Furthermore, Priewasser [77] demonstrated for a comparable study area that local factor, such as soil pH, are the main predictors for the growth of natural regeneration and not salvage logging per se. Based on our findings, it seems necessary to experimentally test our assumptions of the importance of microclimate and to estimate the amount of retention trees in windthrows sufficient for conserving saproxylic biodiversity.

Conclusion

Spruce forests in Europe will be affected heavily by increasing storm damages in the future, which will lead to increasingly heated debates between economists and ecologists on the appropriate means to limit the negative effects of salvage logging on biodiversity [3,4]. Our analysis based on functional traits revealed an unexpected response of saproxylic beetles to salvage logging and suggested that microclimate conditions are more crucial for the use of dead-wood resources by saproxylic beetles than dead-wood diameter or diversity [32,37]. The direct relationship between species traits and logging-affected structures enables us to derive new guidelines for conservationists and managers to optimize salvage logging with a consideration of biodiversity conservation: downed tree tops unaffected by salvage logging operations and complete single windthrown trees should be preserved and allowed to naturally decay on salvage-logged areas to help sustain heliophilous species, colonizers of early decay stages and species that prefer wood of small diameter. Such a preservation of some windthrown trees in salvage-logged areas is an extension of the "green-tree retention approach" to downed trees in forests worldwide.

Acknowledgments

We are grateful to all those who contributed to this study in the field. We thank Karen A. Brune for linguistic revision of the manuscript and Ulrich Bense for beetle determination.

Author Contributions

Conceived and designed the experiments: JM CB. Performed the experiments: JM CB. Analyzed the data: ST TH JM. Contributed to the writing of the manuscript: ST TG HB CB JM TH KR.

References

1. Attiwill PM (1994) The disturbance of forest ecosystems - the ecological basis for conservative management. Forest Ecology and Management 63: 247–300.
2. Shorohova E, Kuuluvainen T, Kangur A, Jogiste K (2009) Natural stand structures, disturbance regimes and successional dynamics in the Eurasian boreal forests: a review with special reference to Russian studies. Ann For Sci 66: 20.
3. Schelhaas MJ, Nabuurs GJ, Schuck A (2003) Natural disturbances in the European forests in the 19th and 20th centuries. Glob Change Biol 9: 1620–1633.
4. Seidl R, Schelhaas M-J, Lexer MJ (2011) Unraveling the drivers of intensifying forest disturbance regimes in Europe. Glob Change Biol 17: 2842–2852.
5. Raffa KF, Aukema BH, Bentz BJ, Carroll AL, Hicke JA, et al. (2008) Cross-scale drivers of natural disturbances prone to anthropogenic amplification: the dynamics of bark beetle eruptions. Bioscience 58: 501–517.
6. Overbeck M, Schmidt M (2012) Modelling infestation risk of Norway spruce by Ips typographus (L.) in the Lower Saxon Harz Mountains (Germany). Forest Ecology and Management 266: 115–125.
7. Schröder LM (2011) Colonization of storm gaps by the spruce bark beetle: influence of gap and landscape characteristics. Agric For Entomol 12: 29–39.
8. Lindenmayer DB, Foster DR, Franklin JF, Hunter ML, Noss RF, et al. (2004) Ecology - Salvage harvesting policies after natural disturbance. Science 303: 1303–1303.
9. Lindenmayer D (2006) Salvage harvesting - past lessons and future issues. Forestry Chronicle 82: 48–53.
10. Noss RF, Lindenmayer DB (2006) The ecological effects of salvage logging after natural disturbance. Conservation Biology 20: 946–948.
11. Black SH (2005) Logging to control insects: the science and myths behind managing forest insect "pests". A synthesis of independently reviewed research. Portland: The Xerces Society for Invertebrate Conservation.
12. Fettig CJ, Klepzig KD (2006) The effectiveness of vegetation management practices for mitigating the impacts of insects on forest ecosystems: a science synthesis. U. S. Department of Agriculture Forest Service.
13. Stokstad E (2006) Ecology: salvage logging research continues to generate sparks. Science 311: 761–761.
14. Bouget C, Duelli P (2004) The effects of windthrow on forest insect communities: a literature review. Biological Conservation 118: 281–299.
15. Chambers JQ, Robertson AL, Carneiro VMC, Lima AJN, Smith ML, et al. (2009) Hyperspectral remote detection of niche partitioning among canopy trees driven by blowdown gap disturbances in the Central Amazon. Oecologia 160: 107–117.
16. Castro J, Moreno-Rueda G, Hodar JA (2010) Experimental Test of Postfire Management in Pine Forests: Impact of Salvage Logging versus Partial Cutting and Nonintervention on Bird-Species Assemblages. Conservation Biology 24: 810–819.
17. Hayes JP (2009) Post-fire Salvage Logging in Central Oregon: Short-term Response in Bats, Birds and Small Mammals. Fire Science Brief: 1–6.
18. Hebblewhite M, Munro RH, Merrill EH (2009) Trophic consequences of postfire logging in a wolf-ungulate system. Forest Ecol Manage 257: 1053–1062.
19. Morissette JL, Cobb TP, Brigham RM, James PC (2002) The response of boreal forest songbird communities to fire and post-fire harvesting. Can J For Res-Rev Can Rech For 32: 2169–2183.
20. Gautam S, Pulkki R, Shahi C, Leitch M (2010) Economic and energy efficiency of salvaging biomass from wildfire burnt areas for bioenergy production in northwestern Ontario: a case study. Biomass Bioenerg 34: 1562–1572.
21. Jakus R, Edwards-Jonasova M, Cudlin P, Blazenec M, Jezik M, et al. (2011) Characteristics of Norway spruce trees (Picea abies) surviving a spruce bark beetle (Ips typographus L.) outbreak. Trees Struct Funct 25: 965–973.
22. Lain EJ, Haney A, Burris JM, Burton J (2008) Response of vegetation and birds to severe wind disturbance and salvage logging in a southern boreal forest. Forest Ecology and Management 256: 863–871.
23. Zmihorski M (2010) The effect of windthrow and its management on breeding bird communities in a managed forest. Biodivers Conserv 19: 1871–1882.
24. Kroll AJ, Giovanini J, Jones JE, Arnett EB, Altman B (2012) Effects of salvage logging of beetle-killed forests on avian species and foraging guild abundance. J Wildl Manage 76: 1188–1196.
25. Cadotte MW, Carscadden K, Mirotchnick N (2011) Beyond species: functional diversity and the maintenance of ecological processes and services. Journal of Applied Ecology 48: 1079–1087.
26. Stuart-Smith RD, Bates AE, Lefcheck JS, Duffy JE, Baker SC, et al. (2013) Integrating abundance and functional traits reveals new global hotspots of fish diversity. Nature 501: 539–542.
27. Reich PB, Tilman D, Isbell F, Mueller K, Hobbie SE, et al. (2012) Impacts of biodiversity loss escalate through time as redundancy fades. Science 336: 589–592.
28. Yang Q, Wang X, Shen Y, Philp JNM (2013) Functional diversity of soil microbial communities in response to tillage and crop residue retention in an eroded loess soil. Soil Science and Plant Nutrition 59: 311–321.
29. Mouillot D, Graham NAJ, Villéger S, Mason NWH, Bellwood DR (2013) A functional approach reveals community responses to disturbances. Trends Ecol Evol 28: 167–177.
30. Alexander KNA (2008) Tree biology and saproxylic coleoptera: issues of definitions and conservation language. Rev Ecol-Terre Vie: 9–13.
31. Lassauce A, Paillet Y, Jactel H, Bouget C (2011) Deadwood as a surrogate for forest biodiversity: Meta-analysis of correlations between deadwood volume and species richness of saproxylic organisms. Ecol Indic 11: 1027–1039.
32. Müller J, Bütler R (2010) A review of habitat thresholds for dead wood: a baseline for management recommendations in European forests. Eur J For Res 129: 981–992.
33. Gossner MM, Lachat T, Brunet J, Isacsson G, Bouget C, et al. (2013) Current near-to-nature forest management effects on functional trait composition of saproxylic beetles in beech forests. Conservation Biology 27: 605–614.
34. Brin A, Bouget C, Brustel H, Jactel H (2011) Diameter of downed woody debris does matter for saproxylic beetle assemblages in temperate oak and pine forests. J Insect Conserv 15: 653–669.
35. Bussler H, Bouget C, Brustel H, Brandle M, Riedinger V, et al. (2011) Abundance and pest classification of scolytid species (Coleoptera: Curculionidae, Scolytinae) follow different patterns. Forest Ecology and Management 262: 1887–1894.
36. Foit J (2010) Distribution of early-arriving saproxylic beetles on standing dead Scots pine trees. Agric For Entomol 12: 133–141.
37. Priewasser K, Brang P, Bachofen H, Bugmann H, Wohlgemuth T (2013) Impacts of salvage-logging on the status of deadwood after windthrow in Swiss forests. Eur J For Res 132: 231–240.
38. Sverdrup-Thygeson A, Ims RA (2002) The effect of forest clearcutting in Norway on the community of saproxylic beetles on aspen. Biological Conservation 106: 347–357.
39. Bässler C, Muller J, Dziock F (2010) Detection of climate-sensitive zones and identification of climate change indicators: a case study from the Bavarian Forest National Park. Folia Geobotanica 45: 163–182.
40. Sverdrup-Thygeson A, Birkemoe T (2009) What window traps can tell us: effect of placement, forest openness and beetle reproduction in retention trees. J Insect Conserv 13: 183–191.
41. Hyvärinen E, Kouki J, Martikainen P (2006) Fire and green-tree retention in conservation of red-listed and rare deadwood-dependent beetles in Finnish boreal forests. Conservation Biology 20: 1711–1719.

42. Schmidl J, Bußler H (2004) Ökologische Gilden xylobionter Käfer Deutschlands. Naturschutz und Landschaftsplanung 36: 202–218.

43. Freude H, Harde K, Lohse GA (1964-1983) Die Käfer Mitteleuropas. Goecke und Evers, Krefeld.

44. Laliberte E, Legendre P (2010) A distance-based framework for measuring functional diversity from multiple traits. Ecology 91: 299–305.

45. Purves DW, Turnbull LA (2010) Different but equal: the implausible assumption at the heart of neutral theory. J Anim Ecol 79: 1215–1225.

46. Webb CO, Ackerly DD, McPeek MA, Donoghue MJ (2002) Phylogenies and community ecology. Annu Rev Ecol Syst 33: 475–505.

47. Pausas JG, Verdu M (2010) The jungle of methods for evaluating phenotypic and phylogenetic structure of communities. Bioscience 60: 614–625.

48. Hothorn T, Mueller J, Schröder B, Kneib T, Brandl R (2011) Decomposing environmental, spatial, and spatiotemporal components of species distributions. Ecological Monographs 81: 329–347.

49. Elston DA, Moss R, Boulinier T, Arrowsmith C, Lambin X (2001) Analysis of aggregation, a worked example: numbers of ticks on red grouse chicks. Parasitology 122: 563–569.

50. Hothorn T, Bretz F, Westfall P (2008) Simultaneous inference in general parametric models. Biometrical Journal 50: 346–363.

51. Cobb TP, Morissette JL, Jacobs JM, Koivula MJ, Spence JR, et al. (2011) Effects of postfire salvage logging on deadwood-associated beetles. Conservation Biology 25: 94–104.

52. Ernst R, Linsenmair KE, Rodel MO (2006) Diversity erosion beyond the species level: dramatic loss of functional diversity after selective logging in two tropical amphibian communities. Biological Conservation 133: 143–155.

53. Flynn DFB, Gogol-Prokurat M, Nogeire T, Molinari N, Richers BT, et al. (2009) Loss of functional diversity under land use intensification across multiple taxa. Ecol Lett 12: 22–33.

54. Winter M, Devictor V, Schweiger O (2013) Phylogenetic diversity and nature conservation: where are we? Trends Ecol Evol 28: 199–204.

55. Grodzki W, Jakus R, Lajzova E, Sitkova Z, Maczka T, et al. (2006) Effects of intensive versus no management strategies during an outbreak of the bark beetle Ips typographus (L.) (Col.: Curculionidae, Scolytinae) in the Tatra Mts. in Poland and Slovakia. Ann For Sci 63: 55–61.

56. McFarlane BL, Parkins JR, Watson DOT (2012) Risk, knowledge, and trust in managing forest insect disturbance. Can J For Res-Rev Can Rech For 42: 710–719.

57. Weslien J (1992) The arthropod complex associated with Ips typographus (L) (Coleoptera, Scolytidae), species composition, phenology, and impact on bark beetle productivity. Entomol Fenn 3: 205–213.

58. Saint-Germain M, Drapeau P, Buddle CM (2007) Host-use patterns of saproxylic phloeophagous and xylophagous Coleoptera adults and larvae along the decay gradient in standing dead black spruce and aspen. Ecography 30: 737–748.

59. Müller J, Noss RF, Bussler H, Brandl R (2010) Learning from a "benign neglect strategy" in a national park: Response of saproxylic beetles to dead wood accumulation. Biological Conservation 143: 2559–2569.

60. Kouki J, Hyvärinen E, Lappalainen H, Martikainen P, Simila M (2012) Landscape context affects the success of habitat restoration: large-scale colonization patterns of saproxylic and fire-associated species in boreal forests. Divers Distrib 18: 348–355.

61. Stokland JN, Siitonen J, Jonsson BG (2012) Biodiversity in Dead Wood. Octavo: Cambridge University Press.

62. Weslien J, Djupstrom LB, Schroeder M, Widenfalk O (2011) Long-term priority effects among insects and fungi colonizing decaying wood. J Anim Ecol 80: 1155–1162.

63. Jarzabek-Müller A, Müller J (2008) On the distinction between Ampedus auripes (Reitter, 1895) and Ampedus nigrinus (Herbst, 1784) (Coleoptera: Elateridae). Elateridarium 2: 199–212.

64. Pitkdnen A, Kouki J, Viiri H, Martikainen P (2008) Effects of controlled forest burning and intensity of timber harvesting on the occurrence of pine weevils, Hylobius spp., in regeneration areas. Forest Ecology and Management 255: 522–529.

65. Lassauce A, Lieutier F, Bouget C (2012) Woodfuel harvesting and biodiversity conservation in temperate forests: Effects of logging residue characteristics on saproxylic beetle assemblages. Biological Conservation 147: 204–212.

66. Victorsson J, Jonsell M (2012) Effects of stump extraction on saproxylic beetle diversity in Swedish clear-cuts. Insect Conservation and Diversity 6: 483–493.

67. Brin A, Bouget C, Valladares L, Brustel H (2012) Are stumps important for the conservation of saproxylic beetles in managed forests? Insights from a comparison of assemblages on logs and stumps in oak-dominated forests and pine plantations. Insect Conservation and Diversity 6: 255–264.

68. Gibb H, Ball JP, Johansson T, Atlegrim O, Hjalten J, et al. (2005) Effects of management on coarse woody debris volume and composition in boreal forests in northern Sweden. Scand J Forest Res 20: 213–222.

69. Azeria ET, Ibarzabal J, Hebert C (2012) Effects of habitat characteristics and interspecific interactions on co-occurrence patterns of saproxylic beetles breeding in tree boles after forest fire: null model analyses. Oecologia 168: 1123–1135.

70. Ding Y, Zang RG, Letcher SG, Liu SR, He FL (2012) Disturbance regime changes the trait distribution, phylogenetic structure and community assembly of tropical rain forests. Oikos 121: 1263–1270.

71. Olsson J, Jonsson BG, Hjalten J, Ericson L (2011) Addition of coarse woody debris: the early fungal succession on Picea abies logs in managed forests and reserves. Biological Conservation 144: 1100–1110.

72. Jacobs JM, Work TT (2012) Linking deadwood-associated beetles and fungi with wood decomposition rates in managed black spruce forests. Can J For Res-Rev Can Rech For 42: 1477–1490.

73. Russo D, Cistrone L, Garonna AP (2011) Habitat selection by the highly endangered long-horned beetle Rosalia alpina in southern Europe: a multiple spatial scale assessment. J Insect Conserv 15: 685–693.

74. Buse J, Schroder B, Assmann T (2007) Modelling habitat and spatial distribution of an endangered longhorn beetle - A case study for saproxylic insect conservation. Biological Conservation 137: 372–381.

75. Jonsell M, Nordlander G, Ehnström B (2001) Substrate Associations of Insects Breeding in Fruiting Bodies of Wood-Decaying Fungi. Ecological Bulletins 49: 173–194.

76. Fischer A, Fischer HS (2012) Individual-based analysis of tree establishment and forest stand development within 25 years after wind throw. Eur J For Res 131: 493–501.

77. Priewasser K (2013) Factors influencing tree regeneration after windthrow in Swiss forests. PhD ETH Zurich 157.

2

Political Systems Affect Mobile and Sessile Species Diversity – A Legacy from the Post-WWII Period

Sara A. O. Cousins[1]*, **Mitja Kaligarič**[2,3], **Branko Bakan**[2], **Regina Lindborg**[1]

1 Landscape Ecology, Department of Physical Geography and Quaternary Geology, Stockholm University, Stockholm, Sweden, **2** University of Maribor, Biology Department, Faculty of Natural Sciences and Mathematics, Maribor, Slovenia, **3** Faculty of Agriculture and Life Sciences, University of Maribor, Pivola 10, Hoče, Slovenia

Abstract

Political ideologies, policies and economy affect land use which in turn may affect biodiversity patterns and future conservation targets. However, few studies have investigated biodiversity in landscapes with similar physical properties but governed by different political systems. Here we investigate land use and biodiversity patterns, and number and composition of birds and plants, in the borderland of Austria, Slovenia and Hungary. It is a physically uniform landscape but managed differently during the last 70 years as a consequence of the political "map" of Europe after World War I and II. We used a historical map from 1910 and satellite data to delineate land use within three 10-kilometre transects starting from the point where the three countries meet. There was a clear difference between countries detectable in current biodiversity patterns, which relates to land use history. Mobile species richness was associated with current land use whereas diversity of sessile species was more associated with past land use. Heterogeneous landscapes were positively and forest cover was negatively correlated to bird species richness. Our results provide insights into why landscape history is important to understand present and future biodiversity patterns, which is crucial for designing policies and conservation strategies across the world.

Editor: David L. Roberts, University of Kent, United Kingdom

Funding: Funding was provided by the program group "Biodiversity" (P1-0078) and granted by Slovenian Research Agency to MK and by FORMAS to SC and RL. MK was supported by program group P1-0164, funded by Slovenian Research Agency. The funders had no role in study design, data collection and analysis, decision to publish, or preparation of the manuscript.

* Email: sara.cousins@natgeo.su.se

Introduction

It is increasingly recognized that conservation biology should have a "landscape perspective" [1–4]. This is generally understood in a spatial context when considering targets for conservation, but a temporal dimension of the landscape is also necessary to understand effects of delayed species responses. This is however rarely considered. Land use change, either by intensification or abandonment, is one of the main drivers causing deterioration of species-richness across the world [5]. Land use and vegetation structure and composition, commonly used as explanatory factor for biodiversity patterns, is to a large extent outcomes of political and socio-economic decisions or constraints. However, the same driving forces may lead to different effects depending on the physical landscape [6,7], e.g. differences in soil fertility, topography or water availability. Land use effects on biodiversity are highly debated topics, especially in conservation research [8,9], because a lack of spatially explicit historical biodiversity data. However, there is a consensus that the decline in traditional agriculture often has negative effects on biodiversity, as the low intensive utilisation of grasslands and forests in the past has been a prerequisite for much of the high small-scale species richness found in the rural landscape of Europe today [10–12]. Species' richness, abundance and composition may respond directly to land use changes but a delayed response has been detected in several studies [13,14]. Such responses often differ depending on organism group, where many mobile organisms respond more quickly to landscape change compared to long-lived sessile organism [15,16].

Despite the increased awareness of social-ecological linkages [17] in conservation, few studies have used large scale *in situ* experimental designs to analyze direct or indirect effects of non-ecological drivers on biodiversity patterns. One reason is the difficulty to find suitable study systems as the divisions into countries or regions often are a result of underlying physical landscape differences [18]. Furthermore, magnitude and timing of intensifications or abandonment is also constrained by physical properties at local or regional scales. For example, Cousins [19] found that areas with a larger proportion of clayey soils changed towards intensive crop-production earlier than areas with smaller proportions of clayey soil. Landscapes on more marginal soils or locations have shown a tendency to be abandoned and afforested [20,21]. However, because of geopolitical reasons, during the last 100 years, there are regions all over the world that have been divided without considering physical landscape divisions or uniformities. Beside former colonies in Africa and Asia some recent examples are Korea and New Guinea. Considerable differences between Eastern and Western Europe, regarding bird and plant diversity, and forestry, have previously been highlighted by several authors [22–24], but are rarely addressed in the design and interpretation of research or policy, but see [25,26]. In Europe, the division between socialist and non-socialist states since WWII, and the recent agenda to privatize or re-privatize the former socialist economies, has especially affected agricultural systems. During the socialist era, there was a widespread collectivization resulting in few but large farming units, contrasted by many small co-existing non-industrial farms outside the industrial food production [27]. Also in Western Europe, driven by market economy, the majority of traditional small scale farms have disappeared due to intensified and specialized agriculture [28,29] or abandonment and afforestation [19].

In this study, we compare current biodiversity patterns in a physically uniform area in the borderland where three countries meet, Austria, Hungary and Slovenia. The historical management in this part of Europe was dominated by small-scale traditional farming 100 years ago when it was part of the same state; the Austrian-Hungarian Empire. Although the area was divided into three countries after 1921, the economic system (market oriented traditional agriculture) remained more or less the same until 1945. After 1945 the difference between countries became prominent, due to changed political systems. In Austria almost all land were privately owned, driven by a free market Western economic policy, whereupon fields became more intensively used and larger in general. Slovenia (part of Yugoslavia) became a pronounced socialist market economy. Large farms on the lowlands were confiscated and collectivised, but in the hilly remote region near the borders to Hungary and Austria the majority of the land remained private and fairly small-scaled. In contrast, the command economy in Hungary eroded the traditional farming system through collectivisation of farms, where former landowners became employed as workers in big state-owned farms. Thus, both free-market economy and collectivization resulted in larger and more intensively used farms where the physical landscape made it possible. Along the Iron Curtain (i.e. here the border to Austria and Slovenia) there was a policy of depopulation. Since the fall of former Yugoslavia and the Eastern Block in the 1990s, Hungary and Slovenia has moved to a free market economy and are both part of the European Union since 2004, whereas Austria became a member state in 1995. Today all nations are part of EU and the "borderless" Schengen region. It is important to note that after the socialist era, no quick, abrupt or radical changes in land use or land ownership occurred in the studied region.

Our primary focus is on landscape matrix effects (as a consequence of land use change) on biodiversity. The rationale behind this study is that the different political systems during the last 70 years will be reflected in the land use and hence also in biodiversity patterns [13], here analysed by using mobile (birds) and sessile (plants) species. We hypothesize that traditionally managed landscape has highest biodiversity and intensified agricultural landscape has the lowest biodiversity [29–31] but see [32]. Furthermore, mobile organisms will be more associated with current land use structure than less mobile organisms [15], exposing a legacy from the land use prior to the post-war era.

Methods

Land cover

Because mobile organisms are expected to respond more quickly to landscape change compared to long-lived sessile organism we chose to investigate both birds and plants. Methods were chosen to best fit the mobility of the organisms i.e. transect mapping for birds and plot inventory for plants. Birds and plants were inventoried in 2-kilometers long cross-transects along three 10 kilometre long main transects radiating out from the point (46,869° N; 16,114° E) where Austria, Slovenia and Hungary meet (Fig. 1). Each main transect ends in Austria at 46,556 N; 16,116 E, in Hungary at 46,532 N; 16,143 E and in Slovenia at 46,465 N;

16,637 E. The climate is moderate continental or sub-Pannonic, with relatively dry winters and with an average annual rainfall of 900 mm. Mean temperature in January is −2°C and in July 19°C. Geologic substrates are mainly tertiary sediments, which forms a soft hilly landscape of sandy-acid soils with networks of running fresh water. The landscape is a mixture of forest and open areas with small farms scattered along hilltops. The investigated landscapes belong to the Trilateral Park: Raab (Austria), Goričko (Slovenia) and Őrség (Hungary). Raab is a nature park (established 1997) aiming to preserve traditional landscapes, Goričko is a Natura 2000 area (established 2002) with the goal to keep traditional and extensive small-scale farming, and Őrség National Park and Natura 2000 area was established 2004 to promote wild-life and tourism and preserve the unique Oak-Pine forests.

Hereafter these different landscapes will be referred to as Raab (Austria), Goričko (Slovenia) and Őrség (Hungary) although it should be noted that the study does not encompass the whole of each park. We used a historical map from 1910 (Fig. 1) from the Austrian-Hungarian Empire (3rd Military Mapping Survey of Austria-Hungary, sheet "Szombathely"), to estimate the relationship between open and forest land in the past. Unfortunately it was not possible to carry out any detailed analyses on land cover composition because of low thematic resolution. To link current biodiversity to land use, we calculated current land use in the study region using CORINE land-cover data from 2000 with a resolution of 50×50 m. We used each 10-km transect (Fig. 1) with sample cross-section width, i.e. 10×2 km, to calculate percentages of different land-cover classes from CORINE in a geographical information system (GIS). Maps of topography and soils were cross-checked to detect dissimilarities between the different transects.

Field survey

A field inventory of birds and plants was designed to capture the differences in land use as well as biodiversity patterns within the countries (File S1). Bird diversity was investigated by slowly walking along the cross-section transect and noting all birds seen or heard following the transect mapping method [33]. All cross-transects were visited twice during the breeding season (spring and late summer), which resulted in 5 samples for each country. Bird classifications follows the nomenclature by Geister [34] and Svensson & Mullarney [35].

The plant inventory was conducted during the field-season 2010. Along the cross-section transects 21 sampling points were placed evenly every 50 meters, in total 105 sampling points in each country. First, all vascular plants found within a 2×2 m square were noted, and then all additional plants found in a circle with radius of 10 m around the plot were added. In addition, we noted the main land-cover type for each 10 m radius plot: grassland (grazed or mown), forest, field (for crop-production) and ruderal or urban surfaces land (for example house, road, playground), which is hereafter referred to as ruderal. For each sampling point in forest the age of trees were categorized as: >30 years old, between 15–30 years or <15 years. Plant nomenclature followed Martinčič et al. [36]. No specific permissions were required for any of the field studies. Only observational studies were performed without interference of plants or birds.

Statistical analysis

Number of bird and plant species in relation to nation and land use type was analysed in separate ANOVAs. To examine how the number of bird species was affected by land use type we used the proportion of present day forest-cover in an ANCOVA, using forest cover as explanatory variable and nationality as covariate.

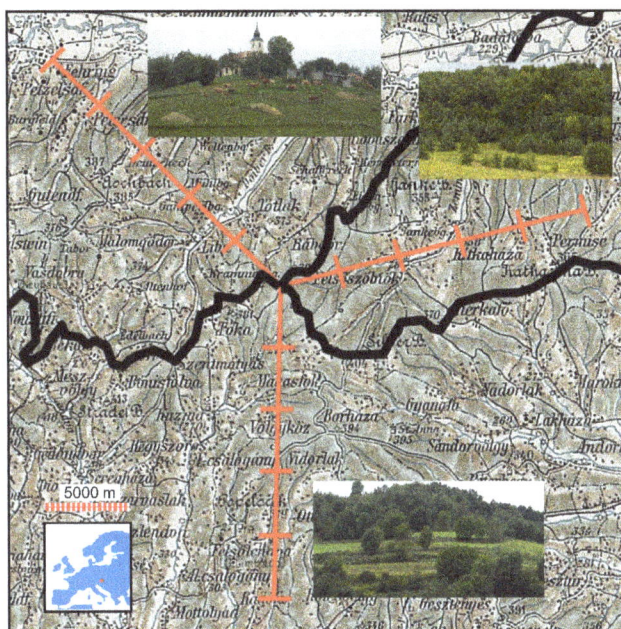

Figure 1. Sampling design for investigating bird and plant diversity in three bordering landscapes in Austria, Slovenia and Hungary. Birds and plants were inventoried in 2-kilometers long cross-transects along three 10 kilometer long main transects radiating out from the point where Austria, Slovenia and Hungary meet today. The investigated landscapes belong to the Trilateral Park: Raab (Austria), Goričko (Slovenia) and Őrség (Hungary). The historical land cover map is from 1910, under the Austrian-Hungarian Empire, showing the study area with the present-day borders (black) superimposed. Green areas are forested land and pink areas are arable land.

"Nationality" (Austria, Hungary, and Slovenia) is arbitrarily used here to reflect past land use history i.e. past political system. To investigate differences in plant composition, nationality and each cross transect distance from the border were predictor variables. We used the model Generalized Estimating Equations (GEE) [37] which allows for terms specifying autocorrelation and are well suited for evaluating landscape processes. First we did a Principal Coordinates Analysis (PCoA), based on dissimilarity (Bray-Curtis) of plant species at the 10 m scale, using the first three axes. We would expect similarity to decrease with distance from where the transects meet, both due to distance in itself and because of differences in land use. Models including both predictor variables transect and nationality, as well as a model with nationality only, were tested against the null model (only including random effects). The significance of each model was tested with likelihood ratio test and post hoc test to separate each variable within the predictors. Data deviating from a normal distribution was log10 transformation and all numbers were increased by one before analysis. The statistical software R 2.13.0 was used for the analyses using the *geepack* package for GEE modeling [37].

Results

Land use change

In 1910, the landscape along each transect was dominated by open agricultural land with a forest cover between 30–39%; a landscape composition that today is inverted with forests covering between 57–74% of the landscapes (Fig 2). Based on the plot surveys, Raab (Austria) has the highest current percentage (23%) of forest older than 30 years, whereas 65% of the forests in Goričko (Slovenia) and 67% in Őrség (Hungary) are between 15 and 30 years old. Young forests (<30 years) are primarily on former arable fields or grasslands. In the investigated landscapes grasslands are few and arable fields even more rare: 11% and 4%

Raab (Austria), 19% and 1% Őrség (Hungary) and 15% and 9% Goričko (Slovenia) for grassland and fields, respectively.

Bird diversity patterns

We found 53 different birds species in total (Table 1; File S1). The bird species composition was 40% in total overlap between the three countries, and 55% to 58% when comparing countries pair wise. There was a significant difference between countries ($F_{2,12} = 6.747$, $p = 0.0108$) (Fig. 2), where Goričko (Slovenia) had significantly higher diversity of birds than Raab (Austria) (Tukey HSD, $p < 0.0087$), but there were no significant difference between Őrség (Hungary) and Raab (Austria) or Goričko (Slovenia) (Tukey HSD, $p = 0.37$ and $p = 0.1$ respectively). When associating different bird species to habitat most birds were classified as forest species; Goričko (Slovenia) had 28% (29 species), Őrség (Hungary) 25% (27 sp.) and Raab (Austria) 24% (28 sp.). Only a few percent of the bird species were associated to open grassland habitats: Raab (5%), Goričko (15%) and Őrség (7%). Number of birds was clearly related to forest cover in each transect, with a significant difference between frequencies of birds found in transects depending on forest cover and land use history (i.e. nationality) as bird diversity declined with an increase in forest cover (Fig. 2). Although the trend is similar for all countries it is only significant for Slovenia ($F_3 = 4.7$, $p = 0.02$, adjusted R2 = 0.45, ANCOVA).

Plant diversity patterns

In total, we found 407 vascular plant species with relatively few endangered plant species, according to (separately considered) national Red Lists: Raab (Austria) 7% (18 sp.), Goričko (Slovenia) 1% (3 sp.) and Őrség (Hungary) 5% (14 sp.) (File S1). Only 180 out of 407 species occurred in all the three countries. There was a significant difference in plant species richness among countries at the larger (314 m^2) scale. Őrség (Hungary) had more plant species

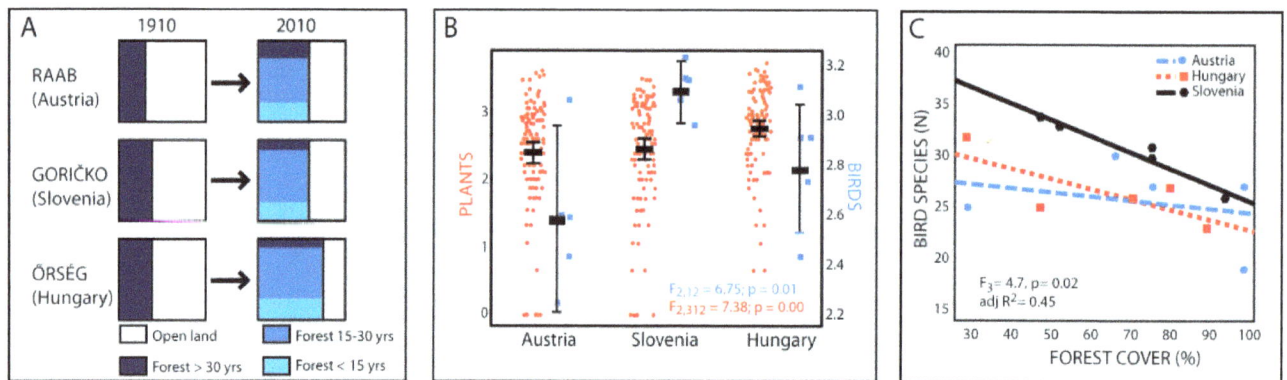

Figure 2. 100 year of land cover change and present-day plant and bird diversity in the border landscapes in Austria, Slovenia and Hungary. (A) Open and forest land cover within the transects between 1910 to 2010. The map from 1910 is thematically coarser so it is not possible to separate the difference in forest age. (B) Mean numbers of plants (red circles) and birds (blue squares) from plots along transects in each country. (C) The percentage of forest cover along each cross-transect and the relationship to number of bird species. There were 5 cross-transects investigated in each country.

than Goričko (Slovenia) and Raab (Austria) ($F_{2,312} = 7.38$; p = 0.0007) (Fig. 2). At the smaller scale ($4 \, m^2$) no significant difference could be detected for plant richness between countries (p>0.11; Table 1). Plots in grassland habitats had significantly higher plant richness compared to all other habitats, irrespective of country ($F_{3,310} = 14.9$; p<0.0001).

In the GEE analyses, using distance of cross-transects and country as predictive variables on plant composition, we found that distance from the border clearly affected plant community structure were there also was a clear effect of nationality (likelihood ratio test: $\chi^2 = 19.1$, DF = 4, p = 0.0008). The post hoc test showed that the species composition in first cross-transect (closest to the border, Fig. 3b) is separated from the second which are both separate from the third to fifth. Furthermore the species composition in Raab (Austria) is different compared to both Őrség (Hungary) and Goričko (Slovenia).

Discussion

To understand the processes behind, and to predict, biodiversity patterns we need to analyse landscape history [38]. Unfortunately there are hardly any records of historical species diversity patterns, making it impossible to analyse direct effects of landscape changes on species richness and composition but see [21,39]. However, historical maps together with landscape change trajectories have been used to indirectly analyse how biodiversity patterns are

affected [10,40,41]. Here we show, by using 100 year old landscape data, that the difference in historical political systems during the last 70 years can be detected on present-day species diversity patterns. Hundred years ago the landscape along the investigated transects was dominated by open agricultural land with a forest cover between 30–40%, whereas it is today inverted with forests covering between 60–70%. Many arable fields and grasslands have become afforested, particularly in Hungary. The current landscape composition is fairly similar in the three landscapes, considering the percentage of forest cover to open land (Fig. 2a) although land use changes and different conservation strategies have resulted in clear differences in species composition (Fig. 2b). In the Hungarian Őrség, rewilding [42,43] has led to increasing forest area, whereas Goričko (Slovenia) is more similar to the traditional landscape before the changes after 1945. Despite Austria and Slovenia having similarities in conservation goals, higher biodiversity of both plants and birds was noted in Slovenia compared to Austria. These biodiversity patterns could be indirectly linked to differences in political system and economic drivers, where the market oriented agriculture in Austria, compared to the more subsistence agriculture in the area in Slovenia during the whole period, has resulted in an intensification of agriculture through the EU Common Agricultural Policy (CAP) funding structures.

The distribution pattern of mobile and sessile organisms varied depending on country (i.e. nationality). The more traditional

Table 1. Species richness of plants and birds based on surveys along three 10 km long transects in Hungary, Slovenia and Austria.

	Austria		Slovenia		Hungary	
	plants	**birds**	**plants**	**birds**	**plants**	**birds**
Total (n)	256	43	281	47	299	42
2×2 plot (SD)	7.2 (5.3)	-	8.5 (5.8)	-	9.3 (6.6)	-
10 m radius (SD)	14.8 (9.7)	-	15.6 (9.4)	-	19.7 (10.4)	-
Cross-transect (SD)	-	13.4 (4.6)	-	21.8 (2.3)	-	16.0 (3.1)

Total (n) is the total number of different species found in each country. Richness of plants is recorded at two different spatial scales, 2×2 m and 10×10 m (105 plots in each country), and bird richness is recorded per cross-transect (5 transects in each country). Grassland/forest species is the number of specialists species for grassland and forest habitats found. (-) indicate not applicable.

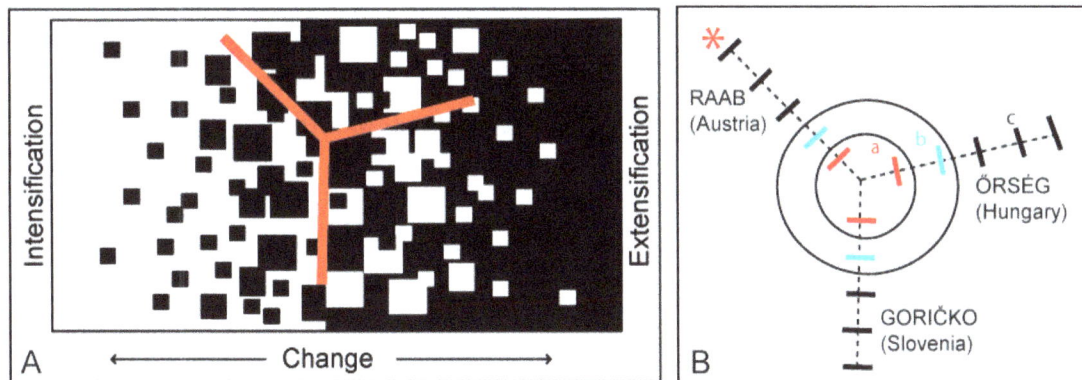

Figure 3. A conceptual model (A) shows how the three areas have either changed because of intensification (Austria) or extensification (Hungary) or remained more or less status quo (Slovenia). Similarity in plant species composition along cross-transects (B) spanning out from the point where the three countries meet. Significant difference in composition is marked with different letters a, b or c. An * marks the significant difference in species composition between countries.

landscape in Goričko (Slovenia) had significantly higher diversity of birds compared to the more intensively managed Raab (Austria), but there were no difference between the Hungarian landscape and the other two. The bird composition similarity among areas was low compared to other studies [44,45], especially considering the high mobility of birds that can move across national borders for nesting and feeding. The frequencies of birds found in transects depended on forest cover, where bird diversity declined with increasing forest cover. Small-scale heterogeneity, i.e. structurally more complex landscapes, in contemporary landscapes favour bird species richness, whilst a denser forest-cover has a negative effect on bird diversity. Several models suggest that more wildlife-friendly farming and heterogeneous landscapes, with many small natural or semi-natural habitats, help to support a relative high diversity compared to large scale farming and commercial forestry which is negative for biodiversity [46,47].

The highest plant species richness was found in Őrség (Hungary), with many typical grassland plant species, despite being primarily conserved and managed as forest. Prior to the Eastern Bloc policy to depopulate and reforest the area it was managed as a traditional agricultural landscape with many orchards and grasslands, and today remnant grassland communities intermingle with colonizing forest species. Particularly long-lived organisms, like plants, may survive as remnant populations for a long time after management has ceased [48], creating a so-called extinction debt [14,15,40,49]. Cousins [50] estimated a threshold for extinction debt in plant communities in grasslands to be settled after around 70 years, in Northern Europe. Many rural ecosystems have a long history of co-evolution with human management and today the survival of many species depends on the maintenance of low intensity farmland practices [51–53]. A historical dimension is hence a necessary complement to the spatial conservation perspective, particularly in landscapes where biodiversity is associated to traditional management. Here, we expect that many plant species associated to the remnant grassland habitats will disappear in Őrség (Hungary). However, other organisms might benefit and those associated to forest habitats should increase with succession of young forests [54]. We envisage based on the past and current trajectories that in the future the rewilding in Hungary will lead to that the legacy from past grassland composition will disappear in favour of forest biodiversity. The traditional landscape in Austria will probably remain

fairly stable, but for the traditional landscape in Slovenia to remain, subsidies are needed and a functioning infrastructure to increase retailing of farm products. Thus, the long-term legacies from the pre-war landscape will disappear slowly and differences in plant diversity patterns become even more pronounced in the future also close to the border at a local scale, despite that there formally are no borders any more.

Although we cannot explain the direct causes for the differences in biodiversity patterns, due to possible co-variation of unknown environmental variables, the study area is relatively small and the abiotic conditions similar, i.e. bedrock and soil types, topography, climate; which strongly infers that the effects are a consequence of land use change, linked to past political systems. Both intensification and abandonment are clear results of political systems during the last 70 years. Similar effects have been observed also within the countries that used to lie within the Eastern Bloc where the different political systems established after the collapse of Soviet Union. For example, a comparison between e.g. Poland, as a EU country, to Russia and Belarus showed different trajectories of landscape transformations caused by agricultural abandonment [55]. Studies from areas, not confounded by underlying abiotic landscape differences, but driven by different political and economical policies can further disentangle how biodiversity patterns may change in the future [25,26]. There are several other comparable political divisions outside Europe that potentially can be used as experimental sites for investigating landscape history's effect on biodiversity and conservation.

In this study we give an *in situ* example of how national political priorities for social structure and economy may drive regional changes in land use that affects species diversity and composition. As expected, traditional agricultural landscape in Slovenia had the highest diversity but only for birds. Heterogeneity (here traditional agricultural landscape) at a landscape scale is expected to favour also plants, but we found that the heterogeneity in time (new land uses superimposed on former land use) created higher plant richness (Fig. 3a and 3b). Thus the hypothesis that mobile species richness is more associated to current land use and many sessile species are more associated to past land use is confirmed, as shown for plant diversity and composition in the Hungarian transect. We stress that awareness of how political and economical decisions directly or indirectly affect land use and biodiversity is crucial information, not only for the managers of these particular conservation areas, but also for designing sustainable policies and conservation strategies across the world.

Supporting Information

File S1 Species data. Plants and birds species found in Austria (AUS), Slovenia (SLO) and Hungary (HUN). + indicates occurrence during the field survey in respectively country. If a plant specie occurs on the National Red List (Anonymous [56] for Slovenia, Gergely [57] for Hungary and Niklfeld [58] for Austria) it is indicated with a 1. The typical habitats for plants were classified as forest (F), ruderal (R) and grassland (G) on the basis of the local flora monograph [59]. The birds were classified; [43] into species typical for forest (F), open and grassland habitats (O), settlements (S) or mixed habitats (M) including both open and forest landscape habitats. Bird abundance was classified as very common (vC), common (C) or rare (R) and breeding status as resident (rB),

migratory (mB) and possible breeding species (rB?). Total numbers of plant and bird taxa found were 407 and 53 respectively.

Acknowledgments

This research is part of the EkoKlim-project at Stockholm University. J. Plue and R. Schmucki gave valuable advice on statistical analyses and M. Hjernqvist on bird inventories. We also thank M. Stenseke, and P. Batáry for valuable comments.

Author Contributions

Conceived and designed the experiments: SAOC RL MK. Performed the experiments: BB. Analyzed the data: SAOC RL. Contributed reagents/materials/analysis tools: SAOC BB MK RL. Wrote the paper: SAOC BB MK RL. Provided expert knowledge on the flora and bird fauna of the region: BB.

References

1. Tscharntke T, Tylianakis JM, Rand TA, Didham RK, Fahrig L, et al. (2012) Landscape moderation of biodiversity patterns and processes - eight hypotheses. Biological Reviews 87: 661–685.
2. Ibbe M, Milberg P, Tuner A, Bergman KO (2011) History matters: Impact of historical land use on butterfly diversity in clear-cuts in a boreal landscape. Forest Ecology and Management 261: 1885–1891.
3. Eriksson O, Cousins SAO (2014) Historical landscape perspectives on grasslands in Sweden and the Baltic Region. Land 3: 300–321.
4. Lindborg R, Bengtsson J, Berg Å, Cousins SAO, Eriksson O, et al. (2008) A landscape perspective on conservation of semi-natural grasslands. Agriculture Ecosystems & Environment 125: 213–222.
5. Sala OE, Chapin FS, Armesto JJ, Berlow E, Bloomfield J, et al. (2000) Global biodiversity scenarios for the year 2100. Science 287: 1770–1774.
6. Lambin EF, Turner BL, Geist HJ, Agbola SB, Angelsen A, et al. (2001) The causes of land-use and land-cover change: moving beyond the myths. Global Environmental Change-Human and Policy Dimensions 11: 261–269.
7. Bürgi M, Hersperger AM, Schneeberger N (2004) Driving forces of landscape change - current and new directions. Landscape Ecology 19: 857–868.
8. Balmford A, Green RE, Scharlemann JPW (2005) Sparing land for nature: exploring the potential impact of changes in agricultural yield on the area needed for crop production. Global Change Biology 11: 1594–1605.
9. Vandermeer J, Perfecto I (2007) The agricultural matrix and a future paradigm for conservation. Conservation Biology 21: 274–277.
10. Kull K, Zobel M (1991) High species richness in an Estonian wooded meadow. Journal of Vegetation Science 2: 715–718.
11. Cousins SAO, Eriksson O (2008) After the hotspots are gone: Land use history and grassland plant species diversity in a strongly transformed agricultural landscape. Applied Vegetation Science 11: 365–374.
12. Wilson JB, Peet RK, Dengler J, Pärtel M (2012) Plant species richness: the world records. Journal of Vegetation Science 23: 796–802.
13. Kuussaari M, Bommarco R, Heikkinen RK, Helm A, Krauss J, et al. (2009) Extinction debt: a challenge for biodiversity conservation. Trends in Ecology & Evolution 24: 564–571.
14. Tilman D, May RM, Lehman CL, Nowak MA (1994) Habitat destruction and the extinction debt. Nature 371: 65–66.
15. Krauss J, Bommarco R, Guardiola M, Heikkinen RK, Helm A, et al. (2010) Habitat fragmentation causes immediate and time-delayed biodiversity loss at different trophic levels. Ecology Letters 13: 597–605.
16. Bommarco R, Lindborg R, Marini L, Öckinger E (2014) Extinction debt for plants and flower-visiting insects in landscapes with contrasting land use history. Diversity and Distributions 20: 591–599.
17. Folke C (2006) Resilience: The emergence of a perspective for social-ecological systems analyses. Global Environmental Change-Human and Policy Dimensions 16: 253–267.
18. Serra P, Pons X, Sauri D (2008) Land-cover and land-use change in a Mediterranean landscape: A spatial analysis of driving forces integrating biophysical and human factors. Applied Geography 28: 189–209.
19. Cousins SAO (2009) Landscape history and soil properties affect grassland decline and plant species richness in rural landscapes. Biological Conservation 142: 2752–2758.
20. Bender O, Boehmer HJ, Jens D, Schumacher KP (2005) Analysis of land-use change in a sector of Upper Franconia (Bavaria, Germany) since 1850 using land register records. Landscape Ecology 20: 149–163.
21. Hooftman DAP, Bullock JM (2012) Mapping to inform conservation: A case study of changes in semi-natural habitats and their connectivity over 70 years. Biological Conservation 145: 30–38.
22. Baldi A, Batary P (2011) Spatial heterogeneity and farmland birds: different perspectives in Western and Eastern Europe. Ibis 153: 875–876.

23. Liira J, Schmidt T, Aavik T, Arens P, Augenstein I, et al. (2008) Plant functional group composition and large-scale species richness in European agricultural landscapes. Journal of Vegetation Science 19: 3–14.
24. Mikusinski G, Angelstam P (1998) Economic geography, forest distribution, and woodpecker diversity in central Europe. Conservation Biology 12: 200–208.
25. Kuemmerle T, Hostert P, Radeloff VC, van der Linden S, Perzanowski K, et al. (2008) Cross-border comparison of post-socialist farmland abandonment in the Carpathians. Ecosystems 11: 614–628.
26. Kuemmerle T, Muller D, Griffiths P, Rusu M (2009) Land use change in Southern Romania after the collapse of socialism. Regional Environmental Change 9: 1–12.
27. Primdahl J, Swaffield SR (2010) Globalisation and agricultural landscapes: change patterns and policy trends in developed countries. Cambridge, UK, New York: Cambridge University Press. xv, 275 p.
28. Sutherland WJ (2002) Restoring a sustainable countryside. Trends in Ecology & Evolution 17: 148–150.
29. Stoate C, Baldi A, Beja P, Boatman ND, Herzon I, et al. (2009) Ecological impacts of early 21st century agricultural change in Europe - A review. Journal of Environmental Management 91: 22–46.
30. Benton TG, Vickery JA, Wilson JD (2003) Farmland biodiversity: is habitat heterogeneity the key? Trends in Ecology & Evolution 18: 182–188.
31. Fahrig L, Baudry J, Brotons L, Burel FG, Crist TO, et al. (2011) Functional landscape heterogeneity and animal biodiversity in agricultural landscapes. Ecology Letters 14: 101–112.
32. Batáry P, Fischer J, Baldi A, Crist TO, Tscharntke T (2011) Does habitat heterogeneity increase farmland biodiversity? Frontiers in Ecology and the Environment 9: 152–153.
33. Robbins CS (1970) An international standard for a mapping method in bird census work. International Bird Census Committee. Audubon Field Notes 24: 722–726.
34. Geister I (1995) Ornitološki atlas Slovenije. Ljubljana: DZS.
35. Svensson L, Mullarney K (2009) Birds of Europe. Princeton, NJ [u.a.]: Princeton Univ. Press. 448 S. p.
36. Martinčič A, Wraber T, Jogan N, Podobnik A, Turk B, et al. (2007) Mala flora Slovenije. Ključ za določanje praprotnic in semenk. Četrta, dopolnjena in spremenjena izdaja. Ljubljana: Tehniška založba Slovenije. 968 p.
37. Højsgaard S, Halekoh U, Yan J (2006) The R Package geepack for Generalized Estimating Equations. Journal of Statistical Software 15: 1–11.
38. Ewers RM, Didham RK, Pearse WD, Lefebvre V, Rosa IMD, et al. (2013) Using landscape history to predict biodiversity patterns in fragmented landscapes. Ecology Letters 16: 1221–1233.
39. Aggemyr E, Cousins SAO (2012) Landscape structure and land use history influence changes in island plant composition after 100 years. Journal of Biogeography 39: 1645–1656.
40. Lindborg R, Eriksson O (2004) Historical landscape connectivity affects present plant species diversity. Ecology 85: 1840–1845.
41. Cousins SAO, Ohlson H, Eriksson O (2007) Effects of historical and present fragmentation on plant species diversity in semi-natural grasslands in Swedish rural landscapes. Landscape Ecology 22: 723–730.
42. Donlan CJ, Berger J, Bock CE, Bock JH, Burney DA, et al. (2006) Pleistocene rewilding: An optimistic agenda for twenty-first century conservation. American Naturalist 168: 660–681.
43. Navarro LM, Pereira HM (2012) Rewilding Abandoned Landscapes in Europe. Ecosystems 15: 900–912.
44. Franklin JF, Lindenmayer DB (2009) Importance of matrix habitats in maintaining biological diversity. Proceedings of the National Academy of Sciences of the United States of America 106: 349–350.

45. Zurita GA, Bellocq MI (2010) Spatial patterns of bird community similarity: bird responses to landscape composition and configuration in the Atlantic forest. Landscape Ecology 25: 147–158.

46. Tscharntke T, Klein AM, Kruess A, Steffan-Dewenter I, Thies C (2005) Landscape perspectives on agricultural intensification and biodiversity - ecosystem service management. Ecology Letters 8: 857–874.

47. Fischer J, Lindenmayer DB, Montague-Drake R (2008) The role of landscape texture in conservation biogeography: a case study on birds in south-eastern Australia. Diversity and Distributions 14: 38–46.

48. Eriksson O (1996) Regional dynamics of plants: a review of evidence for remnant, source-sink and metapopulations. Oikos 77: 248–258.

49. Vellend M, Verheyen K, Jacquemyn H, Kolb A, Van Calster H, et al. (2006) Extinction debt of forest plants persists for more than a century following habitat fragmentation. Ecology 87: 542–548.

50. Cousins SAO (2009) Extinction debt in fragmented grasslands: paid or not? Journal of Vegetation Science 20: 3–7.

51. Bignal EM, McCracken DI (1996) Low-intensity farming systems in the conservation of the countryside. Journal of Applied Ecology 33: 413–424.

52. Plieninger T, Hochtl F, Spek T (2006) Traditional land-use and nature conservation in European rural landscapes. Environmental Science & Policy 9: 317–321.

53. Katoh K, Sakai S, Takahashi T (2009) Factors maintaining species diversity in satoyama, a traditional agricultural landscape of Japan. Biological Conservation 142: 1930–1936.

54. Öckinger E, Lindborg R, Sjödin NE, Bommarco R (2012) Landscape matrix modifies richness of plants and insects in grassland fragments. Ecography 35: 259–267.

55. Gobster PH, Nassauer JI, Daniel TC, Fry G (2007) The shared landscape: what does aesthetics have to do with ecology? Landscape Ecology 22: 959–972.

56. Anonymous (2002) Pravilnik o uvrstitvi rastlinskih in živalskih vrst v rdeči seznam: Official Gazette of the Republic of Slovenia.

57. Gergely K (2007) Vörös Lista. A magyarorszagi edenyes flora veszelyeztett fajai. Red list of the vascular flora of Hungary. Sopron.

58. Niklfeld H (1986) Rote Listen gefährdeter Pflanzen Oesterreichs: 1. Fassung. Wien: Bundesministerium für Gesundheit und Umweltschutz. 202 S. p.

59. Bakan B (2006) Slikovni pregled višjih rastlin Prekmurja. Prispevek k poznavanju flore Prekmurja. Lendava: RC.

3

How Does Conversion of Natural Tropical Rainforest Ecosystems Affect Soil Bacterial and Fungal Communities in the Nile River Watershed of Uganda?

Peter O. Alele[1,2,3,7]*, **Douglas Sheil**[4,5,6,7], **Yann Surget-Groba**[1], **Shi Lingling**[1,2], **Charles H. Cannon**[1,8]

1 Key Laboratory of Tropical Forest Ecology, Xishuangbanna Tropical Botanical Garden (XTBG), Chinese Academy of Sciences, Kunming, Yunnan, P. R. China, **2** University of the Chinese Academy of Sciences, Beijing, P. R. China, **3** Great Nile Conservation Centre (GNCC), Lira, Uganda, **4** Department of Ecology and Natural Resource Management, Norwegian University of Life Sciences, Ås, Norway, **5** Center for International Forestry Research (CIFOR), Bogor, Indonesia, **6** Department of Ecology and Natural Resource Management, School of Environment, Science and Engineering, Southern Cross University, Lismore, New South Wales, Australia, **7** Institute of Tropical Forest Conservation (ITFC), Mbarara University of Science and Technology (MUST), Kabale, Uganda, **8** Texas Tech University, Lubbock, Texas, United States of America

Abstract

Uganda's forests are globally important for their conservation values but are under pressure from increasing human population and consumption. In this study, we examine how conversion of natural forest affects soil bacterial and fungal communities. Comparisons in paired natural forest and human-converted sites among four locations indicated that natural forest soils consistently had higher pH, organic carbon, nitrogen, and calcium, although variation among sites was large. Despite these differences, no effect on the diversity of dominant taxa for either bacterial or fungal communities was detected, using polymerase chain reaction-denaturing gradient gel electrophoresis (PCR-DGGE). Composition of fungal communities did generally appear different in converted sites, but surprisingly, we did not observe a consistent pattern among sites. The spatial distribution of some taxa and community composition was associated with soil pH, organic carbon, phosphorus and sodium, suggesting that changes in soil communities were nuanced and require more robust metagenomic methods to understand the various components of the community. Given the close geographic proximity of the paired sampling sites, the similarity between natural and converted sites might be due to continued dispersal between treatments. Fungal communities showed greater environmental differentiation than bacterial communities, particularly according to soil pH. We detected biotic homogenization in converted ecosystems and substantial contribution of β-diversity to total diversity, indicating considerable geographic structure in soil biota in these forest communities. Overall, our results suggest that soil microbial communities are relatively resilient to forest conversion and despite a substantial and consistent change in the soil environment, the effects of conversion differed widely among sites. The substantial difference in soil chemistry, with generally lower nutrient quantity in converted sites, does bring into question, how long this resilience will last.

Editor: Morag McDonald, Bangor University, United Kingdom

Funding: This work was funded by Xishuangbanna Tropical Botanical Garden (XTBG) and the Chinese Academy of Sciences. The funders had no role in study design, data collection and analysis, decision to publish, or preparation of the manuscript.

Competing Interests: The authors have declared that no competing interests exist.

* Email: alelepeter@gmail.com

Introduction

Tropical rainforests (TRF) possess most of the world's terrestrial biodiversity and deforestation is the leading cause of biodiversity loss [1,2]. Due to their high biodiversity and endemism, the tropical rainforests in Uganda's Nile river watershed are among the world's most important for their conservation values. But these areas are under pressure. The United Nations Population Division [3] predicts that the population of the Nile Basin states will increase by 57% from 2010 to 2030, reaching 647 million people. This rapid population growth, high levels of poverty and prevalent civil insecurity continue to exert severe pressure on natural resources in the region. Uganda in particular has one of the world's highest population growth rates (3.2% per year) [4]. Most of this growing population (nearly 80%) is dependent on

agriculture leading to large scale and continuing conversion of natural habitats [5].

Soil communities form the foundation of any ecosystem, in terms of nutrient cycling and availability, so understanding how land conversion affects these communities is an important first step. The effect of land use change on soil microbial communities has been studied in South American and Southeast Asian forests [6,7], but not in the biodiversity hotspots of the Nile river watershed. There is considerable global concern about the loss of biodiversity and the consequences for human well-being [8]. Microorganisms in particular play a vital role in many ecological processes and environmental services [9]: these roles are not always apparent or well characterized but if all microbes died the world would rapidly become buried in undecomposed dead material. Due to their significance in maintaining ecosystem function and productivity [9,10], our study offers a vital exploratory appraisal of microbial community dynamics in natural TRF and human-converted sites. We don't know if there are reasons to be concerned unless we look. Developing such knowledge is critical at this point, because populations in the Nile river watershed are highly dependent on forests for basic requirements such as food and fuel wood, with the environment contributing between 40–60% of the gross domestic product (GDP) of the Nile riparian states [11].

Because of widespread loss of biodiversity, focus from species conservation within particular habitats has been shifted to conservation of communities [12,13]. It is therefore important to explore and understand how composition and diversity changes across spatial scales in a given context [14–16]. Changes in ecosystems caused by conversion to intensive management can lead to biotic homogenization, the increase in community similarity over time and/or space and an implied loss of rare and vulnerable taxa when examined at larger scales [17–19]. Because microorganisms are the most diverse organisms on earth with most taxa and respective functions and behaviors as yet unknown, determining their sensitivities and biogeography remains a major challenge. But in the longer term such knowledge will help us better understand the sustainability of land-use systems and associated environmental values.

This study was therefore necessary as a first step in exploring these relationships, and to enhance understanding so as to contribute to the informed and appropriate stewardship of the region's natural resources. Our objective was to establish how forest conversion and soil factors affect soil bacterial and fungal diversity and community composition in the tropical rain-forests in the Nile river watershed of Uganda. We chose four forest sites found within protected areas, with paired treatments within each forest; (1) natural and (2) converted ecosystem sites. The natural forest ecosystem at each site had suffered minimal human disturbance, while converted areas had been transformed to cropland. These matched sites found in different locations and environmental conditions each experienced different land use histories, conservation circumstances and individual challenges for management.

In each matched set of natural and converted sites, we compared soil physical and chemical properties and microbial community diversity and composition using standard PCR-based genotyping techniques. We then calculated community similarity indices between sites. This approach would allow us to examine both environmental and biotic changes in the soil community associated with conversion. Disturbances of sufficient magnitude or duration may alter an ecosystem and force a different regime of predominant processes and structures that favor some populations over others [20].

We tested the null hypotheses that there was no difference in soil properties, band-types, and diversity between treatments. The influence of soil properties on microbial community diversity was measured by discriminant analysis and canonical correspondence analysis (CCA), [21,22]. Because additive partitioning of diversity provides a useful framework for quantifying the spatial patterns of diversity across hierarchical spatial scales [23,24], we partitioned total diversity (γ) in each ecosystem type (natural and converted) into additive components representing within-community diversity (α) and between-community diversity (β). Our objective was to identify the most important sources of total diversity so as to propose conservation measures for microbial communities in the TRF ecosystems of the Nile river watershed of Uganda.

Methods

Site description

We selected four tropical rainforest (TRF) sites because of their relative size, biodiversity, socio-economic and scientific importance (Fig 1). Mabira forest is located between the highly populated and urbanized Kampala city on the western side; the extensive and mechanized Lugazi sugar and tea plantations on the Eastern; and Lake Victoria on the southern side. Budongo forest is located next to the extensive Kinyala sugar plantations on one side and a densely populated mainly subsistence population scattered around it. Maramagambo and Kaniyo Pabidi are located within Queen Elizabeth and Murchison Falls national parks (NP) respectively. These two NP forests had perhaps the best protection due to presence of Uganda Wildlife Authority (UWA) personnel. However, Maramagambo's location starting on the steep slopes of the rift valley subjected it to frequent storms with strong runoff flow that swept away most of its top soil (Table 1).

Soil sampling design

We collected 400 core soil samples within 40 plots (1000 m^2 each) in four TRF sites (Fig 1). We sampled five plots from each site of the natural TRF and five plots from the converted TRF. We established the plots at least 100 m from the ecosystem edge and 500 m apart and collected 10 evenly placed core subsamples of top soil (0–15 cm) from each plot and homogenized them into one sample per plot. We then derived a 500 g composite sample from the mixture, sieved and packed it for physical and chemical analyses and DNA extraction.

Sample preparation

We sieved 100 g of the soil on-site through a 4 mm mesh, transported it to the laboratory on ice, and stored in a freezer at − 40°C prior to nucleic acid extraction and analysis. We kept the rest of the soil for drying and physical and chemical analysis. We performed DNA extractions from 1 g of soil using the Ultra Clean soil DNA kit (Mo Bio Labs, Solana Beach, CA, USA) following the manufacturer's protocol. The purified DNA was detected by agarose gel electrophoresis, and the DNA was amplified by polymerase chain reaction (PCR).

Soil property analyses

We measured the soil pH in 2.5:1 water to soil suspension using a pH meter (10 g soil+ 25 ml of distilled water, shaken for 30 min and read on a calibrated pH meter). We then used the Walkley and Black method [25] to analyze soil organic carbon (SOC) and the Kjeldahl method [26] to determine soil nitrogen. We measured the soil phosphorus by the Bray and Kurtz no. 1 method [27]. The photoelectric flame photometer was used to determine the soil potassium, sodium and calcium after extraction with neutral ammonium acetate. We used the atomic absorption spectrometer to measure the soil magnesium after extraction with neutral ammonium acetate. The Bouyoucos hydrometer method adopted from Gee and Bauder [28] was used to determine soil texture. The soil copper and iron were then determined using the atomic absorption spectrometer after extraction with EDTA.

PCR amplification and DGGE analysis

Polymerase chain reaction-denaturing gradient gel electrophoresis (PCR-DGGE) method has been used extensively in microbial ecology and is a robust and cost effective method for exploratory classification of microbial communities [29]. Following soil DNA extraction, we performed a PCR for each DNA extraction to

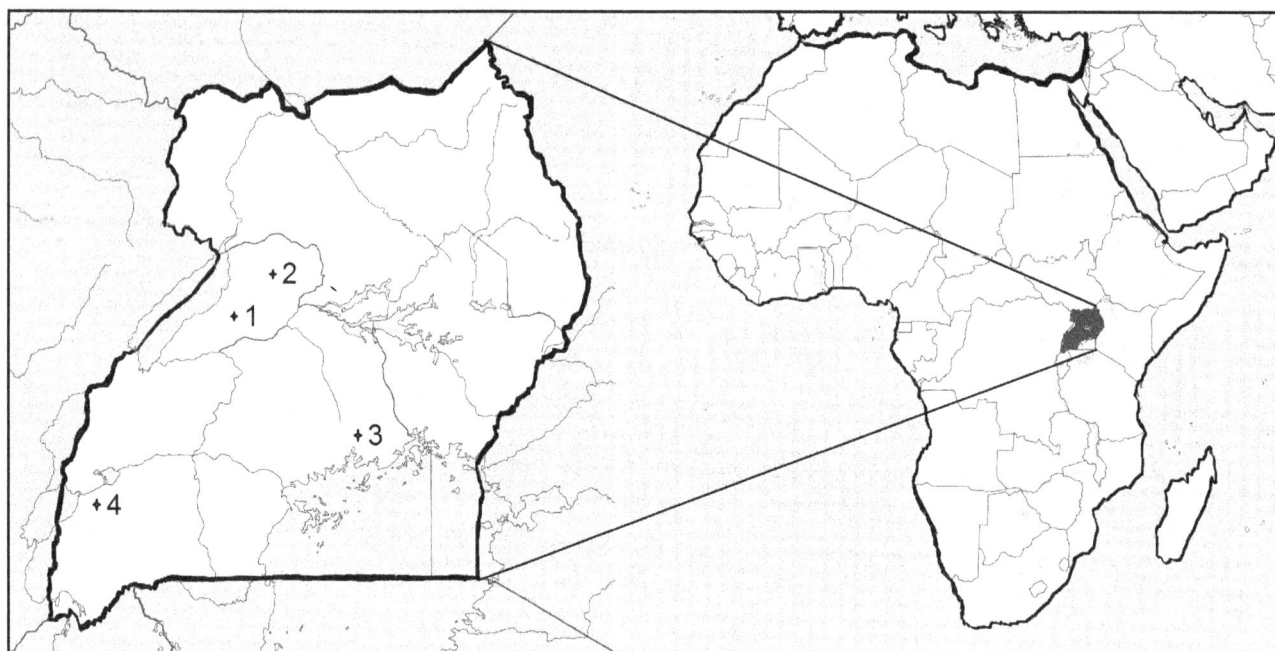

Figure 1. Map of Uganda showing the distribution of sampling sites; Budongo forest (1), Kaniyo Pabidi (2), Mabira forest (3), and Maramagambo forest (4).

amplify the 16S rRNA genes for bacteria and 18S rRNA genes for fungi using universal primers (Table 2).

PCR reactions had a final volume of 25 µl containing a final concentration of 1× TaKaRa ExTaq PCR buffer with $MgCl_2$, 300 pM of primers for bacteria. We then added 200 µM dNTPs, 2.5 U ExTaq DNA polymerase (TaKaRa Bio, Otsu, Japan) and milliQ H_2O to complete the volume, BSA was also added for the fungal community analysis. We performed PCR cycles with an initial denaturing temperature of 95 °C for 5 min, followed by 35 cycles of 95 °C for 30 sec, annealing temperature of 50 °C for 30 sec, extension of 72 °C for 1 min; and a final extension of 72 °C for 10 min. We checked the product of the PCR-rounds and quantified by agarose gel-electrophoresis.

We then performed 16S rRNA and 18S rRNA-DGGE analysis using a universal mutation detection system (Dcode Bio-Rad, Richmond, CA, USA) with a 6% and 8% acrylamide gel for bacteria and fungi respectively containing a gradient of 40–60% denaturant (100% denaturant contains 7 mol urea and 40% formamide). We applied 100 ng of PCR samples to the DGGE gel. DGGE was performed in 1 × TAE Buffer (40 mol Tris/ acetate, pH 8; 1 mol ethylene diaminetetra acetic acid) at 60 °C and a constant voltage of 60 V for 16 hours. After staining with SYBR Green1, we recorded the DGGE gels as digital images and analyzed the DNA band numbers using image-processing software after subtracting background noise.

Data analysis

We used the Rolling disk method with Quantity One (Bio-Rad laboratories Inc.), which normalizes the band pattern from electrophoresis for identification of each band. We then converted the band patterns into binary data based on the presence or absence of each band for part of our analysis. The DGGE fingerprints were interpreted in terms of band richness (number of predominant DGGE bands/population). The pixel intensity of each band was detected by Quantity One software and is

expressed as relative abundance (P_i) [30]. Shannon Index (H') and Simpson index (D), the most widely used diversity indices were then calculated using the richness and relative abundance data following the equations:

$$H' = -\sum_{i=1}^{R} P_i \ln P_i \qquad (1)$$

$$D = \sum_{i=1}^{R} (P_i)^2 \qquad (2)$$

Where R, the richness, is number of different bands each data set contains, $P_i = \dfrac{n_i}{N}$ and n_i is the abundance of the ith band and N the total abundance of all bands in the sample.

Band-type data of the DGGE fingerprints was then used to derive the alpha diversity (bands per sample and ecosystem type), beta diversity (total bands per site) [31]. Jaccard's similarity indices [32] between converted and natural TRF sites were determined using the equation:

Jaccard's Similarity Index $= {}^A\!/\!(A+B+C)$

Where,

A = Total number of bands present in both converted (C) and natural (N) ecosystem samples (plots) (also β-diversity)

B = Number of bands present in C but not in N

C = Number of bands present in N but not in C

We determined the influence of site factors as revealed by soil physicochemical properties on the variation of soil microbial communities by applying discriminant analysis using Statistical Package for the Social Sciences (SPSS). This was done to assess the relative importance of each predictor variable (pH, SOC, N, P, K, Na, Ca, Mg, and soil texture). We also used the Mann-Whitney

Table 1. Summary of study site description.

Forest Site	Location	Size (km²)	Altitude (masl)	Geology	Forest type	Habitat type	Ecosystem description
Budongo	31°N 35° E 1°S 45° N	793	700–1,270	Weathered pre-cambium with ferrallitic sandy clay loams	Ironwood forest (*Cynometra alexandri*); Mixed forest (*Maesopis*), and colonizing forest (*Entandrophragma*)	Primary forest	Consists of a medium altitude moist semi-deciduous forest with areas of savanna and woodland. Converted areas consisted of deforested agricultural land being cultivated and planted with maize, beans, sweet potatoes and cassava. This land has existed as agricultural land for at least 15 years.
Mabira	33° 0.00' E 0° 30.00' N	300	1,070–1,340	Ferrallitic soils with mainly sandy clay loams	Mixed forest	Secondary forest, heavily influenced by humans	The forest is surrounded by a densely populated area. Converted areas were actively cultivated and used to grow maize, groundnuts, beans, yams, cassava, sweet potatoes and a few scattered plants of coffee and sugarcane. The converted land had existed as agricultural land for at least 10 years.
Maramagambo	00° 33' 00" S and 29° 53' 00" E	1,978 (QENP)	910–1390	Ferrallitic soils with undifferentiated dark horizons	Medium altitude, moist, semi-deciduous forest	Secondary forest influenced by wildlife and humans	Forms part of the Queen Elizabeth NP (QENP) which is 1,978 km2. Converted areas consisted of cultivated and grazed areas with gardens of sweet potatoes, beans, maize, and sorghum with areas commonly grazed by cattle and goats.
Kaniyo Pabidi	Lat 1.916667 Long 31.666667	N/A	700–1,270	Freely drained ferruginous tropical soils	Mixed forest	Primary forest	Located north of Budongo forest and part of Murchison Falls N.P. Converted areas consisted of cultivated areas with crops like maize, beans, cassava and sweet potatoes.

[5,61,70,71].

Table 2. Sequences of primers used in study.

Microorganism	Primer	Sequence (5'–3')	Reference
Bacteria	F357	CGC CCG CCG CGC GCG GCG GGC GGG GCG GGG GCA CGG GGG GCC TAC GGG AGG CAG CAG	[72]
	907R	CCG TCA ATT CMT TTG AGT TT	
Fungi	FF390	CGA TAA CGA ACG AGA CCT	[73]
	FR1GC	AIC CAT TCA ATC GGT AIT	

test to examine differences between soil properties in natural and converted ecosystems, and microbial communities in natural and converted ecosystems.

We tested the null hypothesis that diversity is uniform at all spatial scales by additive partitioning of total diversity (γ diversity). To determine contributions of α and β diversity to overall diversity across a range of spatial scales [14,23,33], an additive relationship between diversity components (i.e., $\beta = \gamma - \alpha$) was derived (Fig 2.). The scale at which diversity is maximized was therefore identified [23,34], to facilitate planning processes and management strategies to conserve natural levels of diversity accordingly [35–38].

We used PARTITION 3.0 software [39] to calculate average diversity at each scale and diversity was measured as band richness. Individual-based randomization procedure in the software was used to test whether the observed partitions of diversity within the ecosystem could have been obtained by a random allocation of lower-level samples nested among higher-level samples [34]. Null values of β_i obtained from 1,000 randomizations were used to obtain a p-value for the observed β_i at each hierarchical scale. Deviations of the observed diversity from the null expectation indicated a nonrandom spatial distribution of fungi or bacteria at a given scale.

Results

Soil property variations

Soil pH comparisons using a Mann-Whitney U test of significance, between five plots of natural and five plots of converted TRF ecosystems in each of the four forest sites, found significantly higher (less acidic) pH in three of the four sites at Budongo ($p = 0.0107$), Kaniyo Pabidi ($p = 0.0112$), and Mabira ($p = 0.0269$); and non-significant difference at Maramagambo ($p = 0.1706$). Percentage soil organic carbon (SOC) was signifi-

cantly higher in natural than converted ecosystems in all four sites i.e. Budongo ($p = 0.0119$), Kaniyo Pabidi ($p = 0.0212$), Mabira ($p = 0.0122$) and Maramagambo ($p = 0.0119$) with combined %SOC in natural sites more than double of that in converted sites; whereas %soil nitrogen was only significantly higher in natural forests at Budongo ($p = 0.0112$) and Kaniyo Pabidi ($p = 0.0119$), and non-significant at Mabira ($p = 0.6015$) and Maramagambo ($p = 0.0947$) (Table 3).

Ecosystem and site comparisons of microbial community diversity

Bacterial (B) communities were significantly richer ($p = 0.0304$; Mann-Whitney U) in detectable bands than fungal (F) communities in both converted (C) and natural (N) ecosystems (converted: medians; F = 36, B = 61.5; natural: medians; F = 39.5, B = 60.5). While total band richness (B+F) did not differ between natural and converted forests we observed greater fungal richness in natural than in converted forests (medians: C = 36, N = 39.5; test stat = 18.5) and more bacterial bands in converted than in natural ecosystems (medians: C = 61.5, N = 60.5; test stat = 18.5). Kaniyo Pabidi was the most diverse site overall with the highest number of bacterial and fungal bands, while Maramagambo had the least band richness (Fig 3).

Natural sites harbored more bands unique to one site than converted sites for bacteria at Kaniyo Pabidi and Maramagambo and for fungi at Maramagambo and Budongo. Mabira and Kaniyo Pabidi had higher numbers of unique bacterial bands than at Maramagambo and Budongo. There were also more unique fungal bands at Mabira and Budongo than at Maramagambo and Kaniyo Pabidi (Fig 4).

We also found that Mabira and Maramagambo had the lowest bacterial Jaccard's community similarity indices [32] between natural and converted ecosystems, whereas Budongo and Mabira had the lowest fungal community similarity between natural and converted ecosystems (Table 4). Dissimilarity between natural and converted ecosystems was nonetheless non-significant in all sites for both fungal and bacterial communities. Also, there was generally greater dissimilarity between sites of fungal communities than in bacterial communities suggesting a higher susceptibility to habitat change among fungi than bacteria (Table 4).

Ecosystem classification and importance of predictor variables

The CCA showed that, despite the relatively small amount of difference between sites, soil pH, average phosphorus, and texture (%sand) had strong influence on bacterial diversity in the TRF ecosystem (Fig 5); whereas organic carbon, sodium, pH and average phosphorus were strongly associated with fungal community variation in both natural and converted TRF ecosystems (Fig 5). The CCA also showed that bacterial communities in both Kaniyo Pabidi and Mabira were unique to bacterial communities

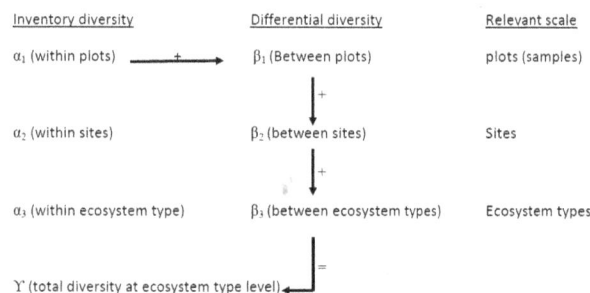

Inventory diversity   Differential diversity   Relevant scale

α_1 (within plots) ——+——→ β_1 (Between plots)   plots (samples)

α_2 (within sites) β_2 (between sites)   Sites

α_3 (within ecosystem type) β_3 (between ecosystem types)   Ecosystem types

Υ (total diversity at ecosystem type level)

Figure 2. Illustration of hierarchical spatial scales in our additive partitioning model. The α scale is the within-level, and β scale, the between-level components. Because a diversity at a given scale is the sum of the α and β diversity at the next lower scale, the total diversity (γ) can be described by the following formula: $\alpha_1 + \beta_1 + \beta_2 + \beta_3$ [14,22].

Table 3. Mean (Standard deviation) for diversity indices of Bacterial (B) and Fungal (F) communities and soil properties in natural (N) and converted (C) ecosystems.

	Budongo		Kaniyo Pabidi		Mabira		Maramagambo	
	Natural	Converted	Natural	Converted	Natural	Converted	Natural	Converted
Shannon (B)	2.49(0.68)	2.73(0.10)	3.25(0.21)	3.12(0.25)	3.06(0.18)	3.19(0.18)	3.04(0.16)	2.93(0.07)
Simpson (B)	0.89(0.10)	0.93(0.01)	0.95(0.13)	0.95(0.02)	0.95(0.01)	0.95(0.01)	0.95(0.01)	0.94(0.01)
Shannon (F)	2.44 (0.30)	2.49 (0.17)	2.83(0.19)	2.77(0.20)	2.57(0.62)	2.64(0.30)	2.10(0.50)	1.93(0.67)
Simpson (F)	0.90(0.29)	0.90(0.03)	0.93(0.01)	0.93(0.02)	0.90(0.06)	0.92(0.03)	0.84(0.08)	0.82(0.12)
pH	5.88(0.11)*	5.08(0.20)*	6.24(0.23)*	5.38(0.27)*	6.46(0.54)*	6.18(0.42)*	6.18(0.45)	5.80(0.24)
OC (%)	6.14(0.67)*	1.59(0.17)*	5.69(1.20)*	3.65(0.54)*	5.87(2.03)*	3.53(1.19)*	9.08(0.79)*	3.98(1.09)*
N (%)	0.43(0.05)*	0.11(0.01)*	0.43(0.08)*	0.25(0.03)*	0.21(0.10)	0.22(0.06)	0.28(0.04)	0.19(0.08)
Ca (Cmoles/kg)	8.00(2.56)*	4.00(0.58)*	13.54(3.36)*	6.74(0.98)*	8.14(1.19)*	6.20(0.76)*	12.12(2.13)*	7.34(1.16)*

*significant differences (p<0.05) between natural and converted ecosystems.

in the other sites and there was high contrast between bacterial communities of converted and natural ecosystems at Kaniyo Pabidi. Fungal communities at Maramagambo and Mabira were also unique to those in other sites and there was high contrast between fungal communities at Mabira's natural and converted ecosystems. Furthermore, the CCA showed that fungal communities responded more to soil pH levels than bacterial communities (Fig 5), with site-specific patterns showing that bacteria and fungi were grouping according to sites.

A discriminant analysis to predict whether bacterial or fungal communities were from natural or converted ecosystems found that only OC, Ca, N, and pH for bacterial communities; and OC, N, Ca, and pH for fungal communities (all ranked from most important to least important) were found to be significant predictors of soil physicochemical properties. All other variables were poor predictors in this context (Table 5).

Hierarchical scaling

We found 58 and 56 fungal bands in natural and converted forests respectively, from 17 plots of natural ecosystems and 20 plots of converted forests. There were also 92 and 88 bacterial bands in natural and converted ecosystems respectively found in 20 plots of converted ecosystems and 17 plots of natural ecosystem. All these were within four sites. β-diversity varied more than α-diversity between natural and converted ecosystems for both bacteria and fungi. We found higher bacterial and fungal β-diversity in converted ecosystems than in natural ecosystems at lower hierarchical scales (β_1); higher β-diversity in natural than converted at between-site scale (β_2), and higher β-diversity in converted than in natural ecosystems at the between-ecosystem type scale (β_3) (Fig 6).

We also found substantial contribution of observed β-diversity (β_1, β_2, and β_3) to total band richness (γ-diversity), while α-diversity of both bacteria and fungi in converted and natural ecosystems were similar. Spatial partitioning of total diversity also consistently showed that the beta components (β_1 and β_2) were always greater than expected by chance, whereas the alpha component (α_1) was always lower than expected. For both fungal and bacterial communities in natural and converted ecosystems, observed within plot diversity were substantially less than values expected from individual-based randomizations (Fig 7).

Discussion

Soil property variations and site differences

Studies in both tropical and temperate zones show that soils in converted or cropped areas normally have reduced soil aggregation, structural stability and organic matter, and an increase in bulk density when compared to forests [40,41]. Habitat conversion may also alter soil properties such as nutrient levels, and abiotic conditions and may affect associations between organisms. In our study there are some local details that may influence our results.

Both Maramagambo and Kaniyo Pabidi are located within Queen Elizabeth NP and Murchison Falls NP respectively and are protected by Uganda Wildlife Authority (UWA) personnel. They are well protected and there is little evidence of recent encroachment. There is significant wildlife populations including elephants, buffaloes, zebras and the areas are frequented by tourists. Protection by UWA and presence of dangerous animals (such as buffalos and lions) reduce damaging human activity at Kaniyo Pabidi and Maramagambo which should enhance the difference between natural and converted ecosystems. Maramagambo's location, in contrast, means the

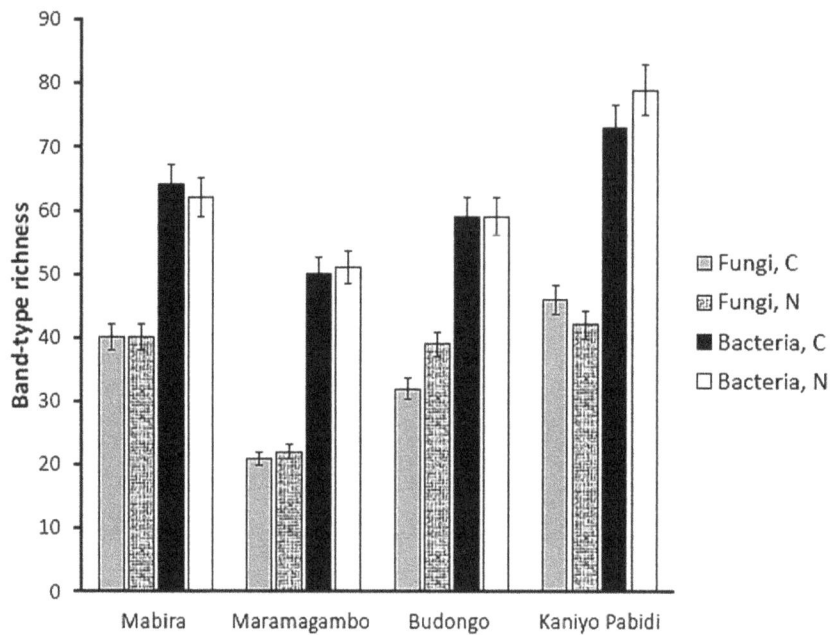

Figure 3. Band richness for fungal and bacterial communities in converted and natural ecosystems. All richness values are total bands present in five samples of each ecosystem treatment (error bars are 5% confidence interval).

forest is subjected to substantial natural disturbance from frequent storms and strong erosive runoff even within the natural forest, whereas tourist activity at Kaniyo Pabidi seemed to have little impact on soil properties. Converted areas at Kaniyo Pabidi were also sparsely populated with limited human impacts on the environment. Its sites were old and might have been cultivated for at least 20years.

In our study, Budongo is located next to a high, mainly subsistence population and resultant population activity. But even though encroachment, illegal hunting and logging in natural habitats in Budongo are not uncommon, there seems to be minimal impact of conversion on soil properties in our sample locations; whereas proximity of Mabira's natural forest to densely populated urban areas exposes it to increased human activities, likely reducing its difference with converted sites.

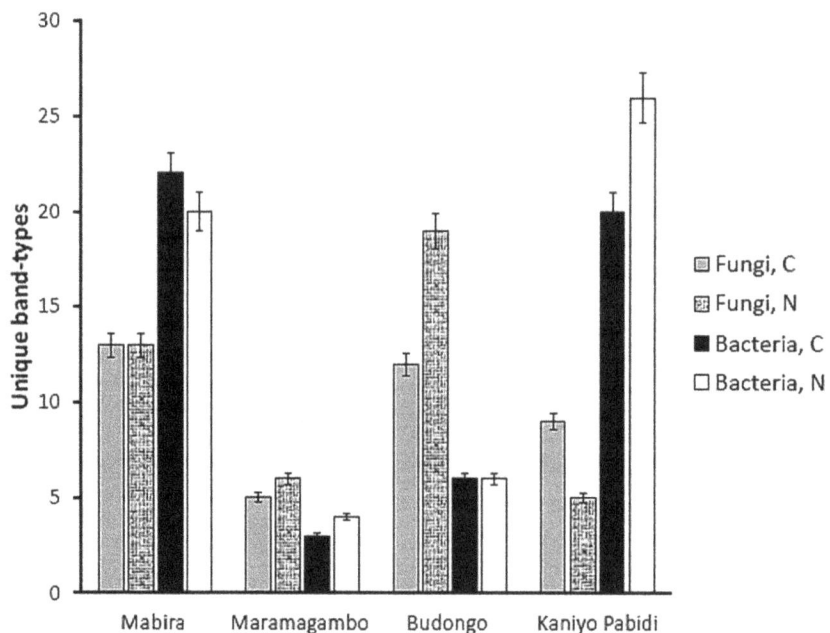

Figure 4. Bacterial and fungal bands unique to converted (C) and natural (N) ecosystems at each site (error bars are 5% confidence interval).

Table 4. Jaccard's similarity indices between bacterial and fungal communities in natural (N) and converted (C) sites of Mabira (Mb), Maramagambo (Mg), Budongo (Bd) and Kaniyo Pabidi (Kp).

			F Mb	Mg	U Bd	Kp	N Mb	Mg	G Bd	I F Kp
			N	N	N	N	C	C	C	C
B	Mb	N	1	0.713	0.769	0.732	0.671	0.685	0.696	0.768
A	Mg	N	0.841	1	0.733	0.666	0.666	0.711	0.646	0.666
C	Bd	N	0.803	0.840	1	0.804	0.808	0.693	0.622	0.827
T	Kp	N	0.731	0.765	0.698	1	0.774	0.670	0.794	0.785
E	Mb	C	0.667	0.788	0.739	0.741	1	0.663	0.710	0.761
R	Mg	C	0.866	0.885	0.846	0.755	0.809	1	0.684	0.698
I	Bd	C	0.825	0.857	0.844	0.696	0.731	0.856	1	0.786
A	Kp	C	0.805	0.869	0.716	0.683	0.759	0.776	0.733	1

Microbial community variations

Soil properties determine many aspects of soil microbial community structure [42–44]. Carbon availability [45–47], nitrogen availability [45,48,49] and soil pH [44,50,51] can all influence microbial community composition and diversity. In addition, correlation studies have shown that plant species [52–54] and soil type [45,54,55] are associated with variation in microbial communities. It has also been shown that land use indirectly affects bacterial community structure by modification of soil properties [56] but similarity between converted and natural ecosystem bacterial communities may also suggest a high number of generalists.

Nacke et al., [56] found that bacterial community composition in forests and grassland was largely determined by tree species and soil pH. Jesus et al. [6] also showed that bacterial community structure is influenced by changes linked to soil acidity and nutrient concentration. Other studies also suggest that soil pH is a major factor influencing microbial community composition [50,57–59]. This influence of soil pH has been recognized at different taxonomic levels [45,60] with most microorganisms thriving within a limited pH range. This is because acids can denature proteins and large pH changes may inhibit growth in microorganisms. Fierer and Jackson, 2006 [44] found, in contrast, that net carbon mineralization rate (an index of C availability) was the best predictor of phylum-level abundances of dominant bacterial groups, and Bisset et al., [42] found that soil microbial communities were consistent with disturbance gradients within different agricultural treatments and relatively undisturbed non-agricultural sites.

Because of widespread forest conversion in Uganda as a result of increasing population pressure, estimated at between 1.1% and 3.15% per year [61]; natural ecosystems and their inhabitant biodiversity are at risk [62]. Loss of diversity increases the likelihood of losing important functional roles and associated ecosystem processes. At landscape scale, spatial and temporal variations of microbial communities in forest soils are influenced by numerous biotic and abiotic factors. These factors may include climate, soil types, and vegetation associations [50,63,64]. Owing to this study design, many of these factors were assumed to be similar between natural and converted ecosystems. For instance, the proximity of natural and converted ecosystem sites meant that climate and geology were, we assume, similar in the two treatments. Even though there could still be a number of underlying causes of community differences, two likely influences were assumed to be soil properties [45,54,55] and vegetation types [52–54]. Clearly both of these sets of factors change when forest is converted for agriculture or range lands.

Despite substantial reductions in SOC, N, Ca and pH in converted sites in this study, differences in microbial communities were small meaning that converted sites still had sufficient SOC, N and Ca to sustain the same microbial populations. The close proximity of the matched pairs could also lead to a source-sink relationship between the natural and converted forests, with the presence of unique taxonomic groups a likely indication of habitat preference (endemism) for some taxonomic groups. It may also be an indicator of relative habitat dissimilarity. The high numbers of unique bacterial bands at Mabira and Kaniyo Pabidi and unique fungal bands at Mabira and Budongo (Fig 4) thus suggests that ecosystem alteration at these sites was sufficient to force a different regime of processes and structures enabling a new set of taxonomic groups to predominate. Mabira had high numbers of both unique bacterial and unique fungal bands that can be attributed to the extent of disturbance at its sites (Mabira

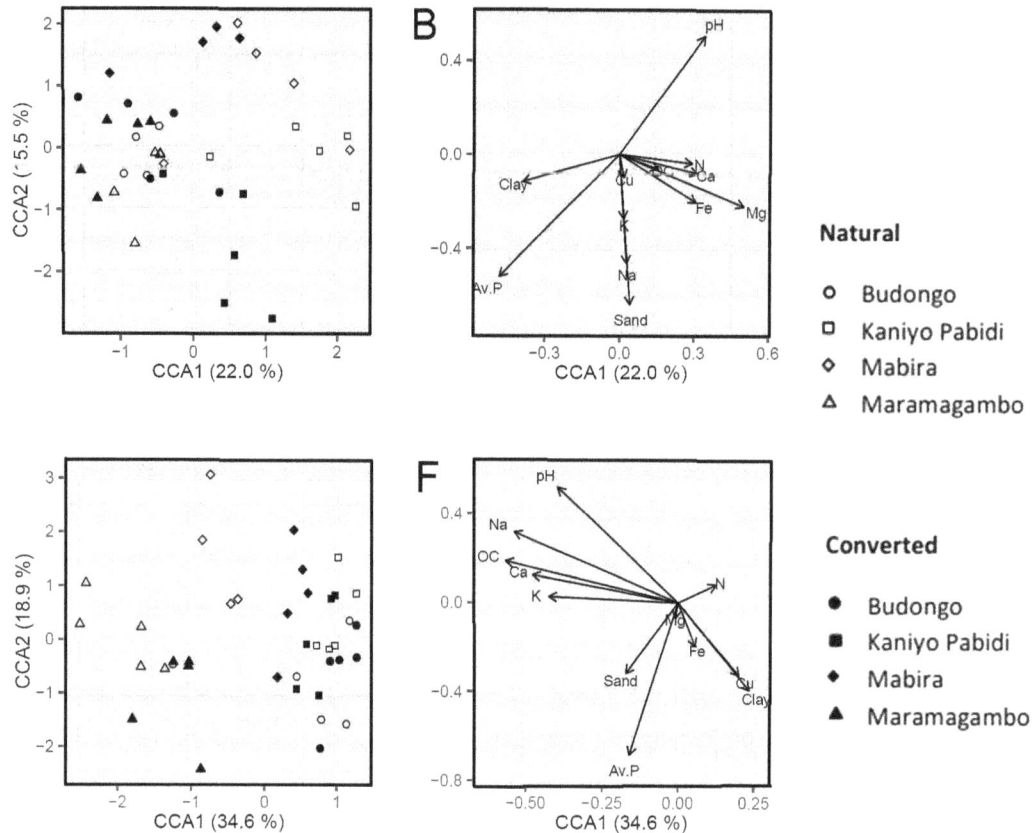

Figure 5. CCA for bacterial (B) and fungal (F) relationships using relative abundance of bands and soil physicochemical properties in natural and converted ecosystems. The symbols (left graphs) represent the similarity between each sample (plot) as defined by their diversity, and the vectors (right graphs) represent the structural matrix for soil properties and their influence on relative abundance of each band. The length of the vectors represents the relative strength of influence of the particular aspect of soil physicochemical property.

is the only peri-urban tropical rainforest site among the four selected sites).

The low numbers of unique bacterial and fungal bands at Maramagambo can be attributed to the high erosion at natural sites that reduced the contrast between the natural and converted sites. For the other sites, bacteria and fungi had different responses to ecosystem alterations. This could indicate separate influences on microbial distribution that exist when alteration is moderate. Similarity indices suggested that bacterial and fungal communities were determined by separate forces leading to distinct responses across the study locations.

Hierarchical scaling

Many studies have shown that specialist species are more negatively affected by current global changes than generalists [7,65]. The process of biotic homogenization can involve the replacement of native biota with non-natives or the introduction of generalist species [66]. In this study, the net decrease in β-diversity from natural to converted TRF ecosystem at the between-site scale (β_2) for both fungi and bacteria was an indication of biotic homogenization [18,66]. This can result from ecosystem alterations which can in-turn alter ecosystem function and reduce ecosystem resilience to disturbance [65,67].

We also showed that the β components of diversity (β_1 and β_2; the average diversity between the plots and sites, respectively) were consistently higher than those expected by chance, whereas the local α_1 diversity component (α_1, the average diversity within the plots) was consistently lower than that expected (Fig 7). Such scale-dependent deviations of the observed diversity from the expected can be generally explained by aggregation at a relatively small "local" scale and, spatial differentiation of diversity at a larger "landscape" scale [33,34,68,69].

Relatively lower diversity within converted ecosystems suggests that conversion of natural TRF ecosystems results in reduced diversity for both bacteria and fungi. This is compatible with recent studies that show that conversion of TRF ecosystems threatens microbial diversity [7] and because microorganisms, like all other organisms, have habitat preferences and may be affected by land-use changes [6,64]. While we cannot be certain that such decline in diversity has led to a decline in any particular ecosystem functions or services, this is a possibility that deserves further evaluation, and we speculate that such loss of diversity will at the very least cause a reduction in functional redundancy and associated resilience.

Higher β-diversity of both bacterial and fungal communities at the between-plot scale (β_1) in converted ecosystems indicates

Table 5. Structure matrix rank showing absolute size of correlation between discriminant analysis function from most important to least important predictor variable of site factors (soil physiochemical properties) and their influence on the variation of soil microbial communities.

Bacteria		Fungi	
Predictor Variables	Function 1	Predictor variables	Function 1
OC	0.625*	OC	0.625*
Ca	0.471*	N	0.493*
N	0.421*	Ca	0.473*
pH	0.355*	pH	0.340*
Mg	0.298	Mg	0.281
Cu	0.197	Cu	0.221
Na	0.181	Na	0.183
K	0.169	K	0.160
Av.P	−0.077	Sand	0.064
Sand	0.070	Av.P	−0.060
Simpson	−0.065	Fe	0.024
Fe	0.029	Shannon	0.012
Shannon	−0.023	Simpson	0.011

(* = important predictor variable, with 0.30 used as the threshold).

differentiation (reduced community similarity) in converted ecosystems at this hierarchical scale. Considering the multiple land-uses and cropping systems of converted areas, this was expected. There was also substantial contribution of β-diversity to total diversity (γ). This suggests the importance of nonrandom ecological processes at the between-plot and between-site scale in determining total richness and community composition [14,34]. Differences between the observed and expected diversity components could be due to ecological processes that lead to a nonrandom dispersion of individuals. These processes could include intra-specific aggregation, habitat selection, and limited dispersal capacity [33].

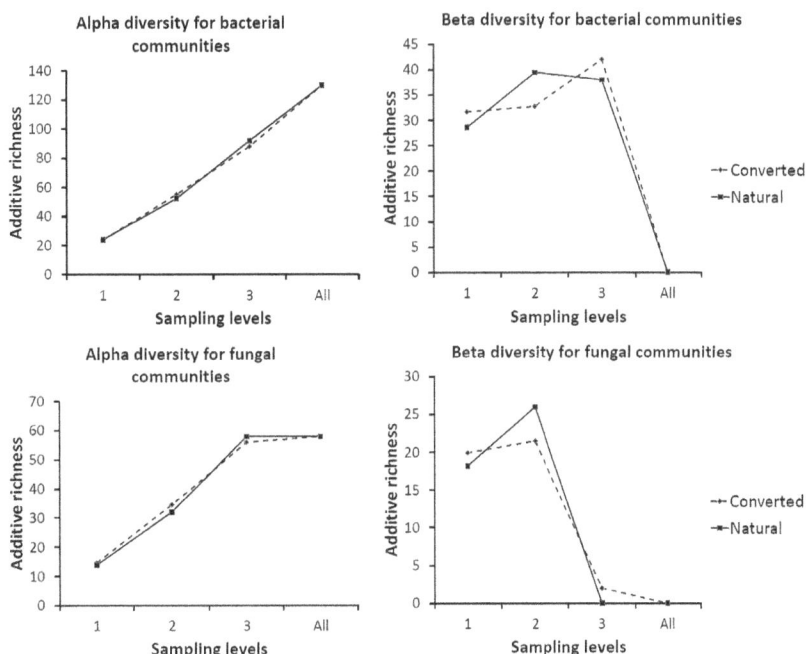

Figure 6. Additive partitioning of bacterial and fungal diversity (expressed as additive richness) across alpha and beta hierarchical spatial scales at three sampling levels (plot, site and ecosystem type) in natural and converted TRF ecosystems.

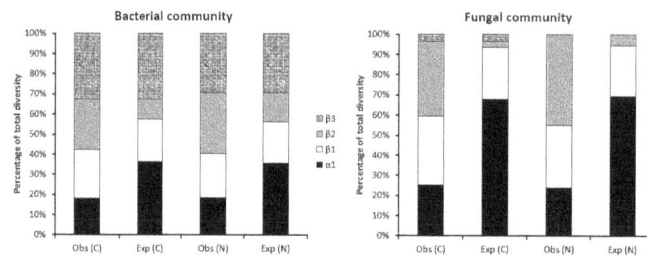

Figure 7. The additive partitioning of total bacterial and fungal community, γ -diversity into α and β components at three nested spatial scales, with each component expressing their relative contributions to γ -diversity; where γ -diversity is equal to $\alpha1+\beta1+\beta2+\beta3$. The observed (obs) partitions are compared with the expected (exp) values, as predicted by the null model based on 1000 iterations using individual-based randomization.

Conclusion

There is international concern about the threat to natural habitats in the Nile river watershed and the consequential loss of important biodiversity. Whereas aspects of microbial biogeography and influence of forest conversion in Uganda's Nile river watershed is largely unknown, this study offers an important first glimpse into indicators of spatial diversity patterns of soil fungal and bacterial communities in the Uganda's Nile river watershed. Our observations of reduced soil microbial diversity, both bacterial and fungal, in converted ecosystems though unsurprising in itself causes us some concern and would justify further work to determine the significance of the diversity lost and the wider implications.

By focusing on diversity patterns across multiple hierarchical spatial scales, we were able to identify the scale at which regional microbial diversity is maximized. We showed that there was substantial contribution of β-diversity to total ecosystem diversity (γ) which includes taxa at the between-plot, site and ecosystem scales and unique taxa, highlighting the necessity to conserve marginal habitats and ecotones. Soil microbial communities in Uganda's Nile river watershed exhibit considerable resilience to forest conversion even though SOC, N, Ca and pH were all significantly altered. This result is surprising given that these physical and chemical properties typically strongly influence microbial diversity. Additionally, the variation among sites was quite large, indicating that soil communities in this region vary considerably on a regional spatial scale. Our results do not explain this variation. Most studies suggest that biogeographic barriers play little role in the geographic structure of soil communities. Rather than a consistent general pattern of microbial community change following forest conversion we find that responses are largely site-specific and widely variable.

Acknowledgments

We confirm that our field work did not involve any endangered or protected species and that we were granted access to protected areas by Uganda Wildlife Authority (UWA) and National Forestry Authority (NFA). Our sample site coordinates included: (1) Mabira; Lat, N00°24.495'; Long, E033°02.464' (Required permission obtained from NFA); (2) Maramagambo; Lat, S00°04.421'; Long, E030°02.305' (Required permission obtained from UWA); (3) Budongo; Lat, N01°41.852'; Long, E031°29.363' (Required permission obtained from UWA/NFA); (4) Kaniyo Pabidi; Lat, N01°55.128'; Long, E031°43.116' (Required permission obtained from UWA).

We would like to thank the Institute of Tropical Forest Conservation (ITFC) of Mbarara University of Science and Technology for supporting and co-advising this research. We also thank the Genetics Lab at Faculty of Science and the soil department lab at Faculty of Agriculture both at Makerere University, Kampala, for their lab facilitation during extractions and analyses. And finally we acknowledge the ideas and insights of Masatoshi Katabuchi, Dossa Gbadamassi, and our field and lab technicians and assistants Solomon Echel, Francis Alele Ozirit and Boniface Balikuddembe.

Supporting Information

Data S1 Excel spreadsheet from Quantity One analysis of PCR-DGGE profiles of fungal and bacterial communities in natural and converted TRF ecosystems.

Author Contributions

Conceived and designed the experiments: POA CHC. Performed the experiments: POA SL. Analyzed the data: POA DS CHC. Contributed reagents/materials/analysis tools: DS YSG CHC. Contributed to the writing of the manuscript: POA DS YSG CHC.

References

1. Turner IM (1996) Species loss in fragments of tropical rain forest: a review of the evidence. J Appl Ecol 33: 200–209.
2. Joseph Wright S, Muller-Landau HC (2006) The Future of Tropical Forest Species. Biotropica 38: 287–301.
3. United Nations (2013) World Population Prospects: The 2012 Revision. Highlights and Advance Tables. New York.
4. United Nations D of E and SAPD (2007) World Population Prospects: The 2006 Revision.
5. Kayanja FIB, Byarugaba D (2001) Disappearing forests of Uganda: The way forward. Curr Sci 81: 936–947.

6. Jesus da CE, Marsh TL, Tiedje, M J, de S Moreira FM (2009) Changes in land use alter the structure of bacterial communities in Western Amazon soils. ISME J 3: 1004–1011. Available: http://www.ncbi.nlm.nih.gov/pubmed/19440233. Accessed 15 November 2012.
7. Rodrigues JLM, Pellizari VH, Mueller R, Baek K, Jesus EDC, et al. (2012) Conversion of the Amazon rainforest to agriculture results in biotic homogenization of soil bacterial communities. Proc Natl Acad Sci U S A. Available: http://www.ncbi.nlm.nih.gov/pubmed/23271810. Accessed 29 December 2012.

8. Cardinale BJ, Duffy JE, Gonzalez A, Hooper DU, Perrings C, et al. (2012) Biodiversity loss and its impact on humanity. Nature 486: 59–67. Available: http://www.ncbi.nlm.nih.gov/pubmed/22678280. Accessed 25 May 2014.

9. Venail PA, Vives MJ (2013) Positive effects of bacterial diversity on ecosystem functioning driven by complementarity effects in a bioremediation context. PLoS One 8: e72561. Available: http://www.pubmedcentral.nih.gov/articlerender.fcgi?artid = 3762786&tool = pmcentrez&rendertype = abstract. Accessed 24 June 2014.

10. Torsvik V, Øvreås L (2002) Microbial diversity and function in soil: from genes to ecosystems. Curr Opin Microbiol 5: 240–245. Available: http://www.ncbi.nlm.nih.gov/pubmed/12057676.

11. Nile Basin Initiative (2012) State of the River Nile Basin. Entebbe (Uganda): Nile Basin Initiative Secretariat.

12. Olson DM, Dinerstein E, Powell GVN, Wikramanayake ED (2002) Conservation Biology for the Biodiversity Crisis. Conserv Biol 16: 1–3.

13. Ricklefs RE (2004) A comprehensive framework for global patterns in biodiversity. Ecol Lett 7: 1–15. Available: http://doi.wiley.com/10.1046/j.1461-0248.2003.00554.x. Accessed 21 January 2014.

14. Gering JC, Crist TO, Veech J a. (2003) Additive Partitioning of Species Diversity across Multiple Spatial Scales: Implications for Regional Conservation of Biodiversity. Conserv Biol 17: 488–499. Available: http://doi.wiley.com/10.1046/j.1523-1739.2003.01465.x.

15. Whittaker RJ, Willis KJ, Field R (2001) Scale and species richness: towards a general, hierarchical theory of species diversity. J Biogeogr 28: 453–470. Available: http://doi.wiley.com/10.1046/j.1365-2699.2001.00563.x.

16. Demeny P (2003) Population Policy: A Concise Summary. Internatio. Demeny P, McNicoll G, editors New York: MacMillan Reference. doi:173.

17. Olden JD (2006) Biotic homogenization: a new research agenda for conservation biogeography. J Biogeogr 33: 2027–2039. Available: http://doi.wiley.com/10.1111/j.1365-2699.2006.01572.x. Accessed 5 May 2014.

18. Olden JD, Poff NL (2004) Ecological processes driving biotic homogenization: testing a mechanistic model using fish faunas. Ecology 85: 1867–1875.

19. Olden JD, Poff NL (2003) Toward a mechanistic understanding and prediction of biotic homogenization. Am Nat 162: 442–460. Available: http://www.ncbi.nlm.nih.gov/pubmed/14582007.

20. Folke C, Carpenter S, Walker B, Scheffer M, Elmqvist T, et al. (2004) Regime Shifts, Resilience, and Biodiversity in Ecosystem Management. Annu Rev Ecol Evol Syst 35: 557–581. Available: http://www.annualreviews.org/doi/abs/10.1146/annurev.ecolsys.35.021103.105711. Accessed 26 October 2012.

21. Ter Braak CJ F. (1986) Canonical Correspondence Analysis: A New Eigenvector Technique for Multivariate Direct Gradient Analysis. Ecol Soc Am 67: 1167–1179.

22. Legendre P, Legendre L (1998) Numerical Ecology. 2nd ed. Amsterdam: Elsevier.

23. Lande R (1996) Statistics and partitioning of species diversity, and similarity among multiple communities. Oikos, Nord Soc 76: 5–13.

24. Godfray HCJ, Lawton JH (2001) Scale and species numbers. Trends Ecol Evol 16: 400–404. Available: http://www.ncbi.nlm.nih.gov/pubmed/11403873.

25. Walkley A., Black I (1934) An examination of the Degtjareff method for determining organic carbon in soils: effect of variations in digestion conditions and of inorganic soil constituents. Soil Sci: 251–263.

26. Kjeldahl J (1883) A new method for the determination of nitrogen in organic matter. Zeitschrift fur Anal Chemie 22.

27. Bray RH, Kurtz LT (1945) Determination of total, organic, and available forms of phosphorus in soils. Soil Sci 59: 39–45.

28. Gee G., Bauder J. (1986) Particle-size analysis. p. 383–411. In A Klute (ed.) Methods of Soil Analysis, Part 1. Physical and Mineralogical Methods. Agronomy Monograph No. 9 (2ed). Am Soc Agron Sci Soc Am Madison, WI.

29. Cleary DFR, Smalla K, Mendonça-Hagler LCS, Gomes NCM (2012) Assessment of variation in bacterial composition among microhabitats in a mangrove environment using DGGE fingerprints and barcoded pyrosequencing. PLoS One 7: e29380. Available: http://www.pubmedcentral.nih.gov/articlerender.fcgi?artid = 3256149&tool = pmcentrez&rendertype = abstract. Accessed 27 October 2012.

30. Reche I, Pulido-Villena E, Morales-Baquero R, Casamayor EO (2005) Does ecosystem size determine aquatic bacterial richness? Ecology 86: 1715–1722. Available: http://www.ncbi.nlm.nih.gov/pubmed/17489473.

31. Whittaker RH (1972) Evolution and Measurement of Species Diversity. Int Assoc Plant Taxon (IAPT) 21: 213–251.

32. Jaccard P (1908) Nouvelles recherches sur la distribution florale. Bull Soc Vaud Sci Nat 44: 223–270.

33. Veech JA, Summerville KS, Crist TO, Gering JC (2002) The additive partitioning of species diversity: recent revival of an old idea. Nord Soc Oikos 99: 3–9.

34. Crist TO, Veech J a, Gering JC, Summerville KS (2003) Partitioning species diversity across landscapes and regions: a hierarchical analysis of alpha, beta, and gamma diversity. Am Nat 162: 734–743. Available: http://www.ncbi.nlm.nih.gov/pubmed/14737711.

35. Chandy S, Gibson DJ, Robertson P a. (2006) Additive partitioning of diversity across hierarchical spatial scales in a forested landscape. J Appl Ecol 43: 792–801. Available: http://doi.wiley.com/10.1111/j.1365-2664.2006.01178.x. Accessed 22 January 2014.

36. Ribeiro DB, Prado PI, Brown Jr KS, Freitas AVL (2008) Additive partitioning of butterfly diversity in a fragmented landscape: importance of scale and implications for conservation. Divers Distrib 14: 961–968. Available: http://doi.wiley.com/10.1111/j.1472-4642.2008.00505.x. Accessed 3 February 2014.

37. Wu F, Yang XJ, Yang JX (2010) Additive diversity partitioning as a guide to regional montane reserve design in Asia: an example from Yunnan Province, China. Divers Distrib 16: 1022–1033. Available: http://doi.wiley.com/10.1111/j.1472-4642.2010.00710.x. Accessed 6 February 2014.

38. Sasaki T, Katabuchi M, Kamiyama C, Shimazaki M, Nakashizuka T, et al. (2012) Diversity partitioning of moorland plant communities across hierarchical spatial scales. Biodivers Conserv. doi:10.1007/s10531-012-0265-7.

39. Veech JA, Crist TO (2009) Partition: software for hierarchical partitioning of species diversity, version 3.0. Available: http://www.users.muohio.edu/cristto/partition.htm.

40. Monkiedje A, Spiteller M, Fotio D, Sukul P (2006) The effect of land use on soil health indicators in peri-urban agriculture in the humid forest zone of southern cameroon. J Environ Qual 35: 2402–2409. Available: http://www.ncbi.nlm.nih.gov/pubmed/17071911. Accessed 9 November 2012.

41. Neris J, Jiménez C, Fuentes J, Morillas G, Tejedor M (2012) Vegetation and land-use effects on soil properties and water infiltration of Andisols in Tenerife (Canary Islands, Spain). Catena 98: 55–62. Available: http://linkinghub.elsevier.com/retrieve/pii/S0341816212001270. Accessed 28 December 2012.

42. Bissett A, Richardson AE, Baker G, Thrall PH (2011) Long-term land use effects on soil microbial community structure and function. Appl Soil Ecol 51: 66–78. Available: http://linkinghub.elsevier.com/retrieve/pii/S0929139311001922. Accessed 9 November 2012.

43. Garbisu C, Alkorta I, Epelde L (2011) Assessment of soil quality using microbial properties and attributes of ecological relevance. Appl Soil Ecol 49: 1–4. Available: http://linkinghub.elsevier.com/retrieve/pii/S0929139311001089. Accessed 9 November 2012.

44. Fierer N, Jackson RB (2006) The diversity and biogeography of soil bacterial communities. Proc Natl Acad Sci U S A 103: 626–631. Available: http://www.pubmedcentral.nih.gov/articlerender.fcgi?artid = 1334650&tool = pmcentrez&rendertype = abstract.

45. Shange RS, Ankumah RO, Ibekwe AM, Zabawa R, Dowd SE (2012) Distinct soil bacterial communities revealed under a diversely managed agroecosystem. PLoS One 7: e40338. Available: http://www.pubmedcentral.nih.gov/articlerender.fcgi?artid = 3402512&tool = pmcentrez&rendertype = abstract. Accessed 1 November 2012.

46. Wang Y, Boyd E, Crane S, Lu-Irving P, Krabbenhoft D, et al. (2011) Environmental conditions constrain the distribution and diversity of archaeal merA in Yellowstone National Park, Wyoming, U.S.A. Microb Ecol62: 739–752. Available: http://www.ncbi.nlm.nih.gov/pubmed/21713435. Accessed 9 November 2012.

47. Fierer N, Bradford MA, Jackson RB (2007) Toward an ecological classification of soil bacteria. Ecology 88: 1354–1364. Available: http://www.ncbi.nlm.nih.gov/pubmed/17601128.

48. Saiya-Cork K., Sinsabaugh R., Zak D. (2002) The effects of long term nitrogen deposition on extracellular enzyme activity in an Acer saccharum forest soil. Soil Biol Biochem 34: 1309–1315. Available: http://linkinghub.elsevier.com/retrieve/pii/S0038071702000743.

49. Steenwerth K, Jackson L, Calderon F, Scow K, Rolston D (2005) Response of microbial community composition and activity in agricultural and grassland soils after a simulated rainfall. Soil Biol Biochem 37: 2249–2262. Available: http://linkinghub.elsevier.com/retrieve/pii/S0038071705001549. Accessed 9 November 2012.

50. De Vries FT, Manning P, Tallowin JRB, Mortimer SR, Pilgrim ES, et al. (2012) Abiotic drivers and plant traits explain landscape-scale patterns in soil microbial communities. Ecol Lett 15: 1230–1239. Available: http://www.ncbi.nlm.nih.gov/pubmed/22882451. Accessed 2 November 2012.

51. Hartman WH, Richardson CJ, Vilgalys R, Bruland GL (2008) Environmental and anthropogenic controls over bacterial communities in wetland soils. Proc Natl Acad Sci U S A 105: 17842–17847. Available: http://www.pubmedcentral.nih.gov/articlerender.fcgi?artid = 2584698&tool = pmcentrez&rendertype = abstract.

52. Cadotte MW, Cardinale BJ, Oakley TH (2008) Evolutionary history and the effect of biodiversity on plant productivity. Proc Natl Acad Sci U S A 105: 17012–17017. Available: http://www.pubmedcentral.nih.gov/articlerender.fcgi?artid = 2579369&tool = pmcentrez&rendertype = abstract.

53. Garbeva P, Veen JA Van, Elsas JD Van (2004) Microbial Diversity in Soil: Selection of Microbial Populations by Plant and Soil Type and Implications for Disease Suppressiveness. Annu Rev Phytopathol 42: 243–270. doi:10.1146/annurev.phyto.42.012604.135455.

54. Schulz S, Giebler J, Chatzinotas A, Wick LY, Fetzer I, et al. (2012) Plant litter and soil type drive abundance, activity and community structure of alkB harbouring microbes in different soil compartments. ISME J doi:10.103: 1763–1774.

55. Kuramae EE, Yergeau E, Wong LC, Pijl AS, van Veen J a, et al. (2012) Soil characteristics more strongly influence soil bacterial communities than land-use type. FEMS Microbiol Ecol 79: 12–24. Available: http://www.ncbi.nlm.nih.gov/pubmed/22066695. Accessed 7 November 2012.

56. Nacke H, Thürmer A, Wollherr A, Will C, Hodac L, et al. (2011) Pyrosequencing-based assessment of bacterial community structure along different management types in German forest and grassland soils. PLoS One 6: e17000. Available: http://www.pubmedcentral.nih.gov/articlerender.

fcgi?artid = 3040199&tool = pmcentrez&rendertype = abstract. Accessed 5 November 2012.

57. Waldrop MP, Balser TC, Firestone MK (2001) Linking microbial community composition to function in a tropical soil. Soil Biol Biochem 32: 1837–1846.

58. Dinsdale E a, Edwards R a, Hall D, Angly F, Breitbart M, et al. (2008) Functional metagenomic profiling of nine biomes. Nature 452: 629–632. Available: http://www.ncbi.nlm.nih.gov/pubmed/18337718. Accessed 26 October 2012.

59. Lauber CL, Hamady M, Knight R, Fierer N (2009) Pyrosequencing-based assessment of soil pH as a predictor of soil bacterial community structure at the continental scale. Appl Environ Microbiol 75: 5111–5120. Available: http://www.pubmedcentral.nih.gov/articlerender.fcgi?artid = 2725504&tool = pmcentrez&rendertype = abstract. Accessed 19 March 2014.

60. Russo SE, Legge R, Weber K a., Brodie EL, Goldfarb KC, et al. (2012) Bacterial community structure of contrasting soils underlying Bornean rain forests: Inferences from microarray and next-generation sequencing methods. Soil Biol Biochem 55: 48–59. Available: http://linkinghub.elsevier.com/retrieve/pii/S0038071712002362. Accessed 8 November 2012.

61. Winterbottom B, Eilu G (2006) Uganda Biodiversity and Tropical Forest Assessment. Washington DC.

62. Laurance WF (1999) Reflections on the tropical deforestation crisis. Biol Conserv 91: 109–117.

63. Scheckenbach F, Hausmann K, Wylezich C, Weitere M, Arndt H (2010) Large-scale patterns in biodiversity of microbial eukaryotes from the abyssal sea floor. Proc Natl Acad Sci U S A 107: 115–120. Available: http://www.pubmedcentral.nih.gov/articlerender.fcgi?artid = 2806785&tool = pmcentrez&rendertype = abstract. Accessed 1 November 2012.

64. Martiny JBH, Bohannan BJM, Brown JH, Colwell RK, Fuhrman J a, et al. (2006) Microbial biogeography: putting microorganisms on the map. Nat Rev Microbiol 4: 102–112. Available: http://www.ncbi.nlm.nih.gov/pubmed/16415926. Accessed 27 October 2012.

65. McKinney M, Lockwood J (1999) Biotic homogenization: a few winners replacing many losers in the next mass extinction. Trends Ecol Evol 14: 450–453. Available: http://www.ncbi.nlm.nih.gov/pubmed/10511724.

66. Olden JD, Rooney TP (2006) On defining and quantifying biotic homogenization. Glob Ecol Biogeogr 15: 113–120. doi:10.1111/j.1466-822x.2006.00214.x.

67. Devictor V, Julliard R, Clavel J, Jiguet F, Lee A, et al. (2008) Functional biotic homogenization of bird communities in disturbed landscapes. Glob Ecol Biogeogr 17: 252–261. Available: http://doi.wiley.com/10.1111/j.1466-8238.2007.00364.x. Accessed 21 January 2014.

68. Summerville KS, Boulware MJ, Veech J a., Crist TO (2003) Spatial Variation in Species Diversity and Composition of Forest Lepidoptera in Eastern Deciduous Forests of North America. Conserv Biol 17: 1045–1057. Available: http://doi.wiley.com/10.1046/j.1523-1739.2003.02059.x.

69. Weiher E, Howe A (2003) Scale-dependence of environmental effects on species richness in oak savannas. J Veg Sci 14: 917–920.

70. Obua J, Agea JG, Ogwal JJ (2010) Status of forests in Uganda. Afr J Ecol 48: 853–859.

71. National Environment Management Authority (NEMA) (2008) State of the Environment Report for Uganda.

72. Amann RI, Ludwig W, Schleifer K-H (1995) Phylogenetic identification and in situ detection of individual microbial cells without cultivation. Microb Rev 59: 143–169.

73. Vainio EJ, Hantula J (2000) Direct analysis of wood-inhabiting fungi using denaturing gradient gel electrophoresis of amplified ribosomal DNA. Mycol Res 104: 927–936.

4

Low-Intensity Agricultural Landscapes in Transylvania Support High Butterfly Diversity: Implications for Conservation

Jacqueline Loos[1]*, Ine Dorresteijn[1], Jan Hanspach[1], Pascal Fust[2], László Rakosy[3], Joern Fischer[1]

1 Institute of Ecology, Leuphana University, Lueneburg, Germany, **2** Organic Agricultural Science Group, University Kassel, Witzenhausen, Germany, **3** Department Taxonomy and Ecology, Babes-Bolay University, Cluj-Napoca, Romania

Abstract

European farmland biodiversity is declining due to land use changes towards agricultural intensification or abandonment. Some Eastern European farming systems have sustained traditional forms of use, resulting in high levels of biodiversity. However, global markets and international policies now imply rapid and major changes to these systems. To effectively protect farmland biodiversity, understanding landscape features which underpin species diversity is crucial. Focusing on butterflies, we addressed this question for a cultural-historic landscape in Southern Transylvania, Romania. Following a natural experiment, we randomly selected 120 survey sites in farmland, 60 each in grassland and arable land. We surveyed butterfly species richness and abundance by walking transects with four repeats in summer 2012. We analysed species composition using Detrended Correspondence Analysis. We modelled species richness, richness of functional groups, and abundance of selected species in response to topography, woody vegetation cover and heterogeneity at three spatial scales, using generalised linear mixed effects models. Species composition widely overlapped in grassland and arable land. Composition changed along gradients of heterogeneity at local and context scales, and of woody vegetation cover at context and landscape scales. The effect of local heterogeneity on species richness was positive in arable land, but negative in grassland. Plant species richness, and structural and topographic conditions at multiple scales explained species richness, richness of functional groups and species abundances. Our study revealed high conservation value of both grassland and arable land in low-intensity Eastern European farmland. Besides grassland, also heterogeneous arable land provides important habitat for butterflies. While butterfly diversity in arable land benefits from heterogeneity by small-scale structures, grasslands should be protected from fragmentation to provide sufficiently large areas for butterflies. These findings have important implications for EU agricultural and conservation policy. Most importantly, conservation management needs to consider entire landscapes, and implement appropriate measures at multiple spatial scales.

Editor: Robert B. Srygley, USDA-Agricultural Research Service, United States of America

Funding: The study was funded through a Sofja Kovalevskaja Award by the Alexander von Humboldt Foundation to Joern Fischer, financed by the German Ministry for Research and Education. The funders had no role in study design, data collection and analysis, decision to publish, or preparation of the manuscript.

Competing Interests: The authors have declared that no competing interests exist.

* Email: loos@leuphana.de

Introduction

Almost half of Europe's terrestrial surface consists of farmland, and many species, including rare and endangered ones, depend on farmland as habitat [1,2]. The loss of cultural-historic landscapes through intensification or abandonment of farming practices is causing declines of farmland biodiversity [1,3,4,5,6]. To effectively design conservation strategies, knowledge is needed about which variables influence species richness and distribution at different spatial scales [7,8,9].

In Western Europe, species loss in farmland has been associated with an increase of agricultural productivity [9,10,11], most likely caused by the use of agrochemicals [12] and the loss and fragmentation of semi-natural patches, especially grasslands [8,13]. In Eastern Europe, socio-economic conditions and land use have been rapidly changing since the breakdown of communism and accession of new member states to the European Union (EU) [14,15,16]. Current changes involve a dual threat to biodiversity, with a trend towards structural simplification on the one hand and abandonment of low-intensity practices on the other hand [17,18]. The current situation in Eastern Europe thus differs in important ways from Western European countries [1,19,20], and a better understanding is needed of how organisms respond to landscape features within low-intensity farming areas of Eastern Europe.

Heterogeneous landscapes typically harbour greater species richness than homogenous landscapes [3,21,22], most likely because of their greater niche diversity, as well as spillover effects and habitat complementation [23]. Agricultural simplification and land abandonment typically lead to a loss of landscape connectivity, which may not only dissect the habitats for species, but also causes flow-on effects on the composition and configuration of the landscape as a whole [24,25].

A particularly interesting cultural-historic region in Eastern Europe is Transylvania, which supports extraordinarily high levels of farmland biodiversity [26,27]. Especially in its South, Transylvania is characterised by a small-scale mosaic of different low-intensity land-uses that provide many different, well-connected structures such as field margins and roadside vegetation. The historic management of the area has created heterogeneity at multiple spatial scales: within tens of metres (hereafter termed the local scale), in the immediate surroundings around any given location (the context scale), as well as over thousands of metres (the landscape scale) [28,29].

Here, we focus on butterflies as a taxonomic group that rapidly responds to environmental changes [30] and is known to be sensitive to land use change worldwide [4]. In Europe, many butterflies use anthropogenic landscape elements [31], but species with different traits are expected to respond differently to land use change [8,32]. For example, Öckinger & Smith [33] found that the effects of landscape composition differed between species of different mobility classes, and Börschig et al. [34] found that intensively used agricultural landscapes mostly support generalists. Yet, evidence on the responses of butterflies to gradients of spatial heterogeneity is sparse, and more thorough studies at multiple scales are needed [22,35].

We sought to understand the responses of butterfly diversity to key landscape gradients in Southern Transylvania, using a snapshot natural experiment [36,37] that spanned the full range of environmental conditions with respect to heterogeneity and woody vegetation cover across multiple scales. Our overarching aim was to understand drivers of species richness and composition. Specifically, we asked (i) how landscape structures affect the composition of butterfly communities; (ii) which landscape structures explain butterfly species richness at various spatial scales; and (iii) which landscape structures affect abundance patterns of selected species. We discuss our findings in the context of possible landscape changes that may take place in Transylvania.

Materials and Methods

Ethics Statement

We obtained the necessary permit for surveying butterflies within the EU *Natura 2000* network from Progresul Silvic, the organization officially entrusted with the custody of the protected area by the Romanian government. The survey procedure was approved beforehand by the ethics committee of Leuphana University Lueneburg.

Data Availability Statement

All data underlying the findings reported in this study are available from the Dryad Digital Repository (http://doi.org/10.5061/dryad.97s1k).

Study area and experimental design

The study area covered approximately 7,000 km^2 in the lowlands of Southern Transylvania, Romania (Figure 1). We followed the notion of a natural experiment [36], with randomised site selection in pre-defined strata at two levels: study villages and survey sites within villages.

To select study villages, we first allocated each raster pixel of the study area to different "village catchments". These were calculated using a cost-distance algorithm in ARCGIS with the village centre as the reference point and the slope and the distance to the next village as the cost variables. Information about village locations was extracted from CORINE land cover data 2006 (http://www.eea.europa.eu/data-and-maps/data#c12 = corine+land+cover+version

+13&b_start = 0&c17 = CLC2006), and slope was calculated from the digital elevation model ASTER (Advanced Spaceborne Thermal Emission and Reflection Radiometer). Topographically based village catchments were used instead of administrative boundaries because administrative boundaries were only available at the commune level (typically 3–5 villages). However, we found that the resulting polygons accurently reflected historical land use responsibilities. Second, we stratified village catchments along a gradient of terrain ruggedness and according to their protection status under the EU Birds and Habitats Directives. Terrain ruggedness was calculated as the standard deviation of the altitude of the catchment, and we used quantiles to classify ruggedness as either low, medium or high. Protection status of the catchments was either unprotected, SCI (Site of Community Importance) or SPA (Special Protection Area). Third, we randomly chose 30 villages, covering all combination of ruggedness and protection status (Table S1).

To select survey sites, we stratified the agricultural area within these 30 villages according to CORINE land cover as grassland or arable land and excluded other land cover classes. Within these strata, we spanned two gradients that we assumed sensitive to change in the future as a result of structural simplification, namely woody vegetation cover and heterogeneity. We estimated woody vegetation cover in a circular one hectare area based on classified 10 m SPOT data (CNES, ISIS programme). To assess heterogeneity, we used the standard deviation of 2.5 m panchromatic SPOT data within a one hectare circle. We assigned each hectare of the agricultural landscape to a combination of three classes of woody vegetation cover by three classes of heterogeneity. We distinguished low (0–5%), medium (>5–15%) and high (>15%) woody vegetation cover and used the lower, middle and upper third of percentiles to classify heterogeneity. Within these combinations, we randomly selected replicates for each cross-combination (except for the combination of high heterogeneity and low woody vegetation cover, which did not exist (Table S2)). In total, we selected 120 circular 1 ha survey sites, with 60 in grassland and 60 in arable land, and an average of four survey sites per village catchment. Notably, sites in arable land in this context were consciously placed not to represent only arable fields specifically, but rather to capture the whole range of conditions within the mosaic of arable land [38], including field margins and fallow land.

Data collection

Butterfly surveys (response data). We assessed species richness and abundance of butterflies (Rhopalocera) and diurnal burnet moths (Zygaenidae) by walking four transects of 50 m length per survey site [39]. We included burnet moths because they are comparable to butterflies in their ecology [33,40]. These transect pointed north, east, south and west, and started 6 m from the centre of a given site. In a given transect walk, each butterfly observed within 2.5 m of each side of the transect and 5 m in front of the observer was identified and counted. Species that we could not identify in the field were treated as compound species: *L. sinapis/juvernica*, *C. alfacariensis/hyale* and *Zygaena minos/purpuralis*. *Adscita*, *Jordanita* and *Carcharodus* occurred within the study region, and are represented by two, two and three species, respectively [41]. However, these species are difficult to distinguish and therefore were only identified to the genus level. Surveys were repeated on four occasions between May and August 2012 by four different, trained observers. Surveys were conducted under suitable weather conditions (no rain, <90% cloud cover, > 17°C, no strong wind), between 9 am and 5 pm.

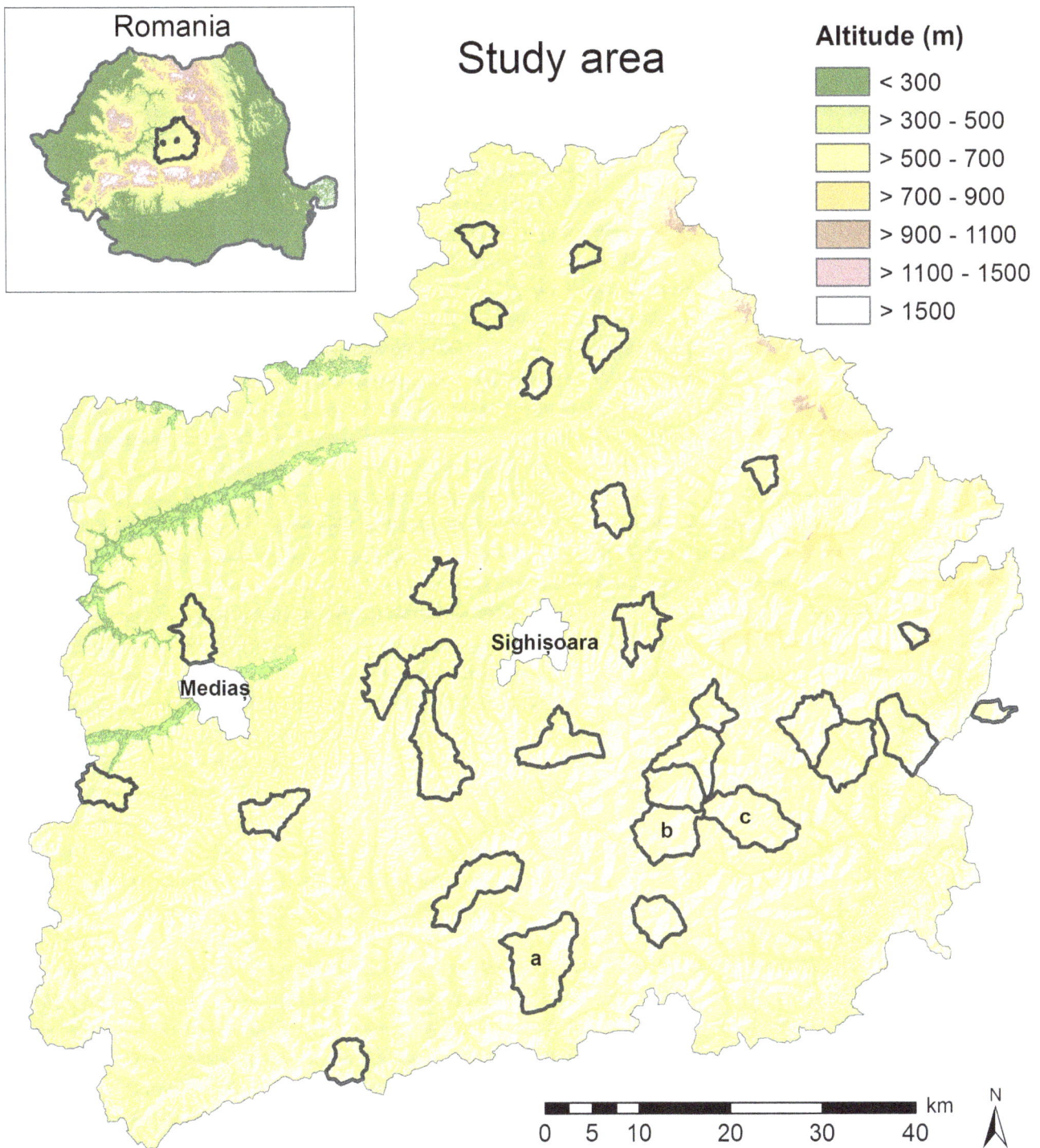

Figure 1. Location of the study area with investigated village catchments in Transylvania, Romania. The small letters indicate the village catchments illustrated for predictions in Figure 4 (a = Cincu, b = Granari, c = Viscri).

Environmental data (explanatory variables). We followed a multi-scale approach and included explanatory variables that could potentially explain butterfly distribution at the local (1 ha), context (50 ha) and landscape scale (i.e. village catchments, ranging from 430 to 4963 ha). An overview of all variables included in the analysis is presented in Table 1.

At the local scale, we collected data on vascular plant species richness in eight randomized quadrants (1×1 m). We used cumulative plant species richness per site as an explanatory variable. We also calculated indices for heatload (after [42]) and terrain wetness as a measure of potential soil wetness, and included heterogeneity assessed by the spectral variance of SPOT data (see Table 1 for details). We calculated percent woody vegetation cover at local and context scales, and used CORINE land cover to calculate percent forest at the landscape scale. For the context and landscape scales, we calculated the terrain ruggedness as the

Table 1. Definition of environmental variables used in the study at three different scales and method of obtaining those. Abbreviations are used in Figure 2 and Table 2.

Scale	Variable (abbreviation)	Definition and method
local (1 ha)	Number of plants species (NoPlant)	Vascular plant species richness assessed by eight randomly distributed quadrants of one by one meter
	Heterogeneity (het_1 ha)	Heterogeneity measured as the standard deviation of 2.5 m panchromatic SPOT picture (CNES, ISIS programme)
	Woody vegetation cover (woody_1 ha)	Proportion of woody vegetation cover based on classified 10 m SPOT satellite image (CNES, ISIS programme)
	Heat index (heatload)	Potential for ground heating calculated after Parker [42]: Heat index = cos (slope aspect − 225) * tan (slope angle)
	Terrain Wetness Index (TWI)	Measure of potential soil wetness, estimated as the position in the landscape and the slope from ASTER digital elevation model with 30 m resolution.
	Land Cover (LU_type)	Land use classification as arable land, grassland or forest based on CORINE land cover
context (50 ha)	Ruggedness (rugg_50 ha)	Terrain ruggedness, calculated as standard deviation of altitude
	Woody vegetation cover (woody_50 ha)	Proportion of woody vegetation cover based on classified 10 m SPOT satellite image
	Configurational heterogeneity (ED_50 ha)	Configuration of different land covers, calculated as the edge density with FRAGSTATS v4.2 based on CORINE land cover
landscape (village catchment)	Amount of pasture (past_catch)	Proportion of pasture, based on CORINE land cover
	Woody vegetation cover (woody_catch)	Proportion of forest cover based on CORINE land cover
	Ruggedness (catch_rugg)	Terrain ruggedness, calculated as the standard deviation of the altitude
	Compositional heterogeneity (SIDI)	Composition of different land covers, calculated as Simpson index of diversity with FRAGSTATS v4.2 based on CORINE land cover
	Configurational heterogeneity (ED)	Configuration of different land covers, calculated as edge density with FRAGSTATS v4.2 based on CORINE land cover
Random effects	Village catchment	Classification of the landscape into social-ecological units according to a cost distance algorithm of proximity to the nearest village as reference point and the slope of the terrain as cost factor
	Level	Observation level random effect

standard deviation of altitude. We also quantified compositional or configurational heterogeneity of the different land covers grassland, arable land and forest as provided by CORINE land cover data. At the context scale, our chosen heterogeneity measures (Simpson index of land cover diversity, edge density) were correlated ($r = 0.76$). Hence, we included only edge density as an explanatory variable (following [7]). At the landscape scale, we used both edge density and the Simpson index of diversity and added the amount of pasture and forest per village catchment, based on CORINE land cover data. Variables on compositional and configurational heterogeneity were calculated using FRAGSTATS v4.2 [43] and all other variables using ARCGIS 10.1 (ESRI Inc., Redland, CA).

Analysis

We pooled all observed butterfly species and individuals from the four survey rounds for each survey site. First, we tested for differences in species richness and abundance between different levels of official protection by using Analysis of Variance (ANOVA). Second, we conducted a detrended correspondence analysis (DCA) to describe species composition and its relation to environmental variables. We used a permutation test to fit and test the correlation of environmental variables with the ordination.

Third, we used generalized linear mixed effects models (GLMMs) to assess effects of environmental variables on butterfly species richness. Beforehand, we tested the explanatory variables for collinearity (all $r < 0.7$; Table S4; [44]). We log-transformed woody vegetation cover at local and context scales and heterogeneity at the local scale because these variables were highly skewed. All numerical explanatory variables were scaled to mean zero and unit variance. We included the variables listed in Table 1 to model species richness of butterflies. To test for a unimodal relationship in response to woody vegetation cover, we included a quadratic term of local woody vegetation cover. We furthermore expected that the effect of heterogeneity may differ between grassland and arable land and therefore included an interaction term between land cover type and heterogeneity. Grasslands are also interesting to look at separately because they are among the most species rich biotopes for butterflies in Europe [45]. We assessed the variance inflation factor (VIF) of the generalized linear model (GLM) and tested for spatial auto-correlation in the residuals. We included the village catchment as a random effect and corrected for overdispersion by adding an observation level random effect. We simplified the model by stepwise backwards selection retaining all variables with $p < 0.1$. For GLMMs, significance levels are only approximations, hence many statisticians suggest using a significance level of $p < 0.1$ [46].

Likewise, we modelled species richness of functional groups. To this end, we distinguished between species of low mobility (Bink's mobility classes 1 and 2) and high mobility (Bink's mobility classes 7, 8 and 9; [47]). Highly mobile species were *Colias crocea, Pieris brassicae, Vanessa atalanta* and *Vanessa cardui*. Low-mobility species were *Brenthis daphne, Brenthis ino, Coenonympha glycerion, Cupido minimus, Euphydryas aurinia, Hamaeris lucina, Heteropterus morpheus, Lopinga achine, Melitaea britomartis, Melitaea diamina* and *Satyrium acaciae*. As a third group we modelled the richness of grassland specialists, namely *Euphydryas aurinia, Polyommatus coridon, Cyaniris semiargus, Lysandra bellargus, Phengaris arion, Cupido minimus* and *Erynnis tages* [48].

We also modelled the abundance of individual species considered to be declining in Western and Northern Europe, but that are widespread or even increasing in Eastern Europe [48,49,50,51]. We only used species that were common enough in the study area to obtain reliable models, namely *Maniola jurtina, Coenonympha pamphilus, Polyommatus Icarus, Lycaena dispar* and *Glaucopsyche alexis*. We performed all statistical analyses in R [52], using the packages MASS, ade4, vegan, gdata and lme4.

Results

In total, we counted 19,878 individuals of 112 species of butterflies (Table S3). Site-level species richness varied between three and 45, and the number of individuals between seven and 452. Eighty-five percent of all individuals belonged to 12 species: *Colias alfacariensis/hyale, Minois dryas, Aphantopus hyperantus, Pieris rapae, Everes argiades, Coenonympha glycerion, Leptidea sinapis/juvernica, Melanargia galathea, Coenonympha pamphilus, Maniola jurtina, Polyommatus icarus,* and *Plebeius argus*. SCI, SPA and unprotected sites did not differ in species richness ($F = 0.54$, $p = 0.58$) but SCI sites appeared to have a slightly lower abundance of individuals than unprotected sites ($F = 2.37$, $p = 0.09$). Arable land and grassland did not differ in species richness ($F = 1.32$, $p = 0.25$) nor abundance of individuals ($F = 1.51$, $p = 0.22$).

Multivariate analysis showed substantial overlap in species composition between arable land and grassland (Figure 2), with less than one complete species turnover (length of first axis = 2.9). The first axis (Eigenvalue = 0.21) described a gradient from sites with a low terrain wetness index in homogenous landscapes to sites with a high terrain wetness index within highly heterogeneous landscapes. The second DCA axis (Eigenvalue = 0.18) described a gradient from low to high richness of vascular plants, ruggedness, woody vegetation cover and context-level heterogeneity and landscape-level woody vegetation cover.

Butterfly species richness was positively related to local plant species richness and local woody vegetation cover, but negatively to local heatload (Table 2). It increased in response to local heterogeneity in arable sites, but not in grasslands (Figure 3). Species richness furthermore increased with configurational heterogeneity and ruggedness at the context scale, but decreased with landscape woody vegetation cover. The models show suitable areas for species of conservation interest exist throughout village catchments, especially in large grassland areas and boundary areas of arable land (Figure 4).

Species richness of mobile butterflies was highest in arable land, and responded positively at the landscape scale to both compositional heterogeneity and ruggedness. By contrast, richness of low-mobility species was negatively related to landscape configurational heterogeneity, but responded positively to local-scale plant species richness and context heterogeneity (for additional details, see Table 2). Richness of specialist species was higher in grassland, in landscapes with high terrain ruggedness and at sites with high plant species richness.

For individual species, both *L. dispar* and *G. alexis* were more abundant in arable land, and were positively related to local plant species richness. *L. dispar* also responded positively to local woody vegetation cover, but negatively to local heatload, whereas *G. alexis* showed a positive response to context ruggedness and the amount of grassland in the landscape. The abundances of *P. icarus, M. jurtina* and *C. pamphilus* increased with heterogeneity in arable land, but not in grassland, and decreased with increasing heatload. Abundance of *M. jurtina* and *C. pamphilus* were positively related to local plant species richness, and negatively to landscape woody vegetation cover. *P. icarus* responded positively to the amount of grassland in the landscape. Abundance of *C. pamphilus* was unimodally related to local woody vegetation cover.

Discussion

We found a high diversity of butterflies in the cultural-historic landscape of Southern Transylvania. This is especially the case considering that we did not seek out sites expected a priori to harbour great diversity, but rather surveyed randomly selected sites within the agricultural matrix. An even greater diversity of butterflies, including rare and endangered species, would be expected to occur in dry grassland patches and traditionally managed hay-meadows, which occur within our study area but which we did not specifically target. Our findings suggest that some types of land use change could pose serious threats to butterfly diversity in Transylvania. Our findings can be summarised within four themes, which we discuss in the following: (i) both grassland and arable land have conservation value; (ii) low-intensity landscapes provide important resources for butterflies; (iii) heterogeneity has a different effect in arable land than in grassland; and (iv) it is important to consider multiple scales for effective butterfly conservation.

Both grassland and arable land have conservation value

Our findings revealed a high conservation value for butterflies of the small-scale farming system in the lowlands of Transylvania. Interestingly, butterfly species richness and abundance were similar in arable land and grassland. This is a surprising result and suggests a need to broaden the emphasis of conservation activities from grassland protection towards the maintenance of heterogeneous mosaic farmland, including cropland [38]. This is particularly important in the context of criticisms that the recent reform of the European Union's Common Agricultural Policy, for example, falls far short of what is needed in terms of biodiversity conservation [53]. Throughout Europe, grasslands are considered most important for butterfly conservation (e.g. [8,45]). Arable land, on the other hand, has received far less attention. In Western Europe, arable land has been found to support lower species richness and more homogenous butterfly communities than grassland [9,54]. Our results indicate that this situation may be different in Eastern Europe, and that certain types of arable land can in fact support similar levels of butterfly diversity as grasslands. A possible explanation for the similar species richness in arable land and grassland in Transylvania may be spillover effects [23], which may be more likely in small-scale mosaics of land covers. The mosaic character of the landscape also could explain the strong overlap in butterfly communities between arable land and grassland.

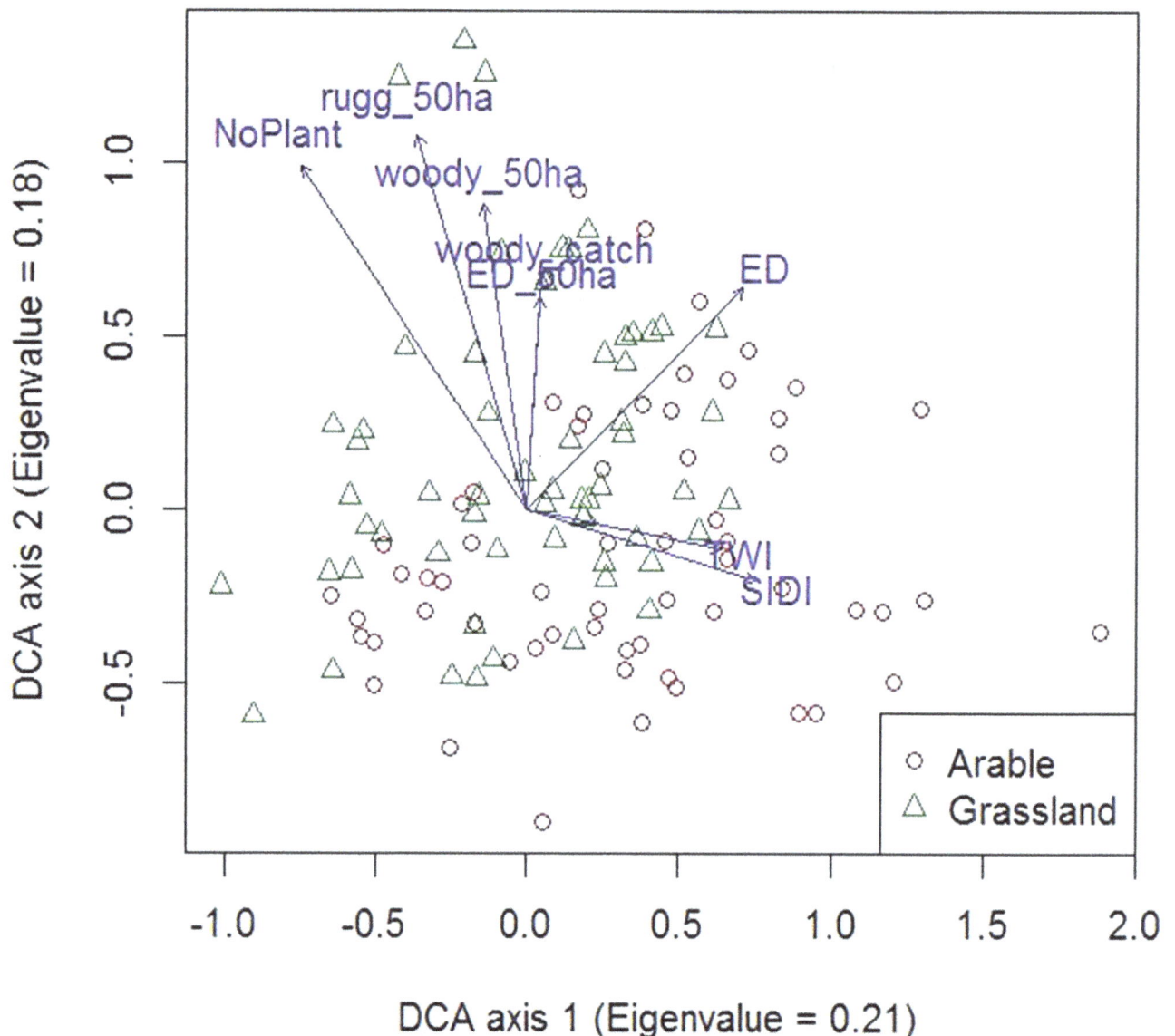

Figure 2. DCA ordination plot of butterfly species, with significant environmental variables superimposed (p<0.05) (Abbreviations: NoPlant = Local plant species richness; TWI = Local terrain wetness index; rugg_50 ha = context terrain ruggedness; woody_50 - ha = context woody vegetation cover; ED_50 ha = context edge density; woody_catch = landscape woody vegetation cover; SIDI = landscape compositional heterogeneity; Table 1).

Low-intensity landscapes provide important resources for butterflies

The fine-grained mosaic nature of arable land and the low-intensity nature of grassland in Southern Transylvania emphasize that low-intensity land use practices have major benefits for butterfly conservation. Semi-natural elements occur throughout the landscape, and are a likely reason why species richness is high throughout different land covers [22]. Furthermore, species richness of vascular plants can be high in field margins, which in turn may indicate high quality habitat for butterflies [55]. Consistent with the findings of Kumar, Simonson & Stohlgren [7], we found plant species richness strongly related to butterfly species richness. Currently, Transylvania contains some of the world's most species rich areas for plants [56], which is partly linked to the low use of fertilizers [57]. Agricultural intensification, by contrast, would likely lead to increased use of fertilizers and

hence reduced plant species richness [58,59,60]. Furthermore, intensification is typically associated with the use of fewer, high yielding crop varieties. Interestingly, many butterflies in Transylvania use the common crop *Medicago sativa ssp. sativa* (Alfalfa), a leguminous species that provides nectar and that we also observed to serve as a host plant for several butterfly species (e.g. *Glaucopsyche alexis*). Alfalfa is grown in small parcels, is primarily used as winter fodder for livestock, and may easily be lost as a result of intensification. However, high amounts of floral resources are critically important to maintain butterfly diversity. Similarly, woody vegetation offers important resources for butterflies, including shelter and space for thermoregulation [61]. At present, Transylvania contains many scattered trees and hedgerows, and we found that butterfly species richness responded positively to these structures at the local scale. By contrast, a large amount of

Table 2. Parameter estimates of the species distribution models with significance levels indicated by: †P<0.1; *P<0.005; **P<0.01; ***P<0.001.

	Species Richness	High mobile species	Low mobile species	Specialists	L. dispar	G. alexis	P. icarus	M. jurtina	C. pamphilus
Intercept	3.026	-0.739	0.681	-0.436	0.581	-1.172	2.914	2.637	1.959
NoPlant	0.261***		0.600***	0.685***	0.581*	0.941**		0.597***	0.313***
LU_type	-0.243**	-1.250***		0.052†	-2.059***	-1.554*	-0.040	0.046	0.210
het_1 ha	0.109*						0.278**	0.235*	0.224*
LU_type*het_1 ha	-0.140*						-0.326*	-0.415**	-0.387*
TWI									
woody_1 ha	0.072*		-0.054		0.443†				0.102
woody_1 ha^2			0.232†						0.177*
Heatload	-0.057*				-0.622†		-0.167*	-0.321***	-0.207**
rugg_50 ha	0.064†					0.511*			
woody_50 ha									
ED_50 ha	0.077*		0.261*						
woody_catch	-0.079*							-0.412***	-0.232†
past_catch						0.721**	0.256.		
rugg_catch		-0.423*	0.249*	0.051**					
ED			-0.374**						
SIDI		0.448*							

Arable land was used as the baseline land cover in all models. See Table 1 for abbreviations.

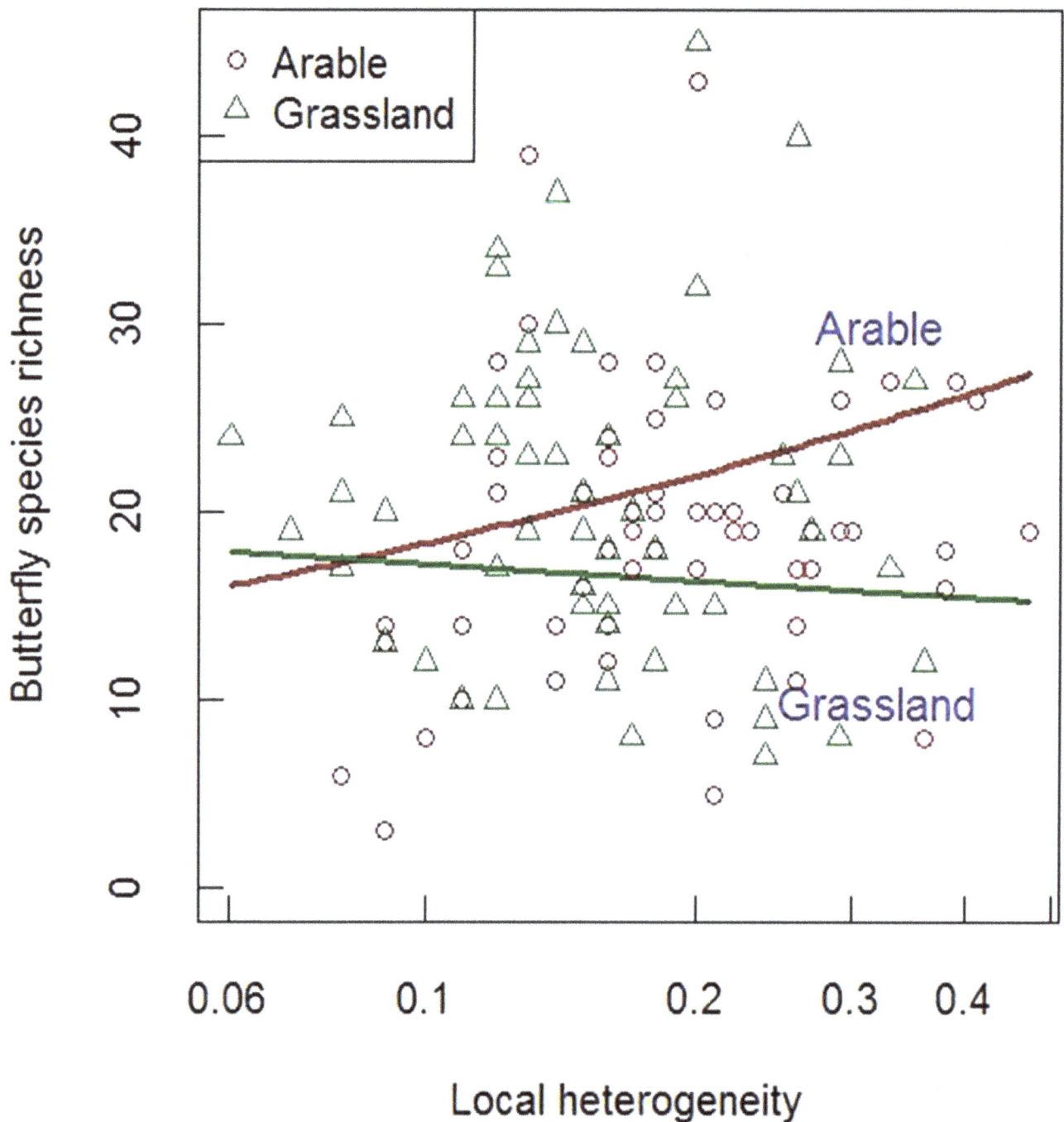

Figure 3. Predicted effect of local heterogeneity on species richness in arable land versus grassland, based on the simplified generalized linear mixed model (Table 2).

woody vegetation at the landscape scale may lead to decreased species richness, probably due to a lack of open habitat.

Heterogeneity has a different effect in arable land than in grassland

We considered heterogeneity and woody vegetation cover at the local scale as two potentially important gradients describing the structure of the landscape. Interestingly, our results showed that the effect of local heterogeneity on species richness depended on land cover. In arable land, species richness increased with

heterogeneity, supporting our hypothesis that small-scale farming benefits biodiversity by providing a range of different resources for butterflies. Notably, our land use class of "arable land" reflected the highly heterogeneous nature of traditional farmland, and included cropped areas as well as fallows and uncultivated field margins. These non-cropped areas are likely to be particularly important to maintain butterfly diversity in arable land. By contrast, in grassland, high heterogeneity was associated with reduced butterfly diversity. A possible explanation for this pattern is that heterogeneity of grassland may correspond to a higher

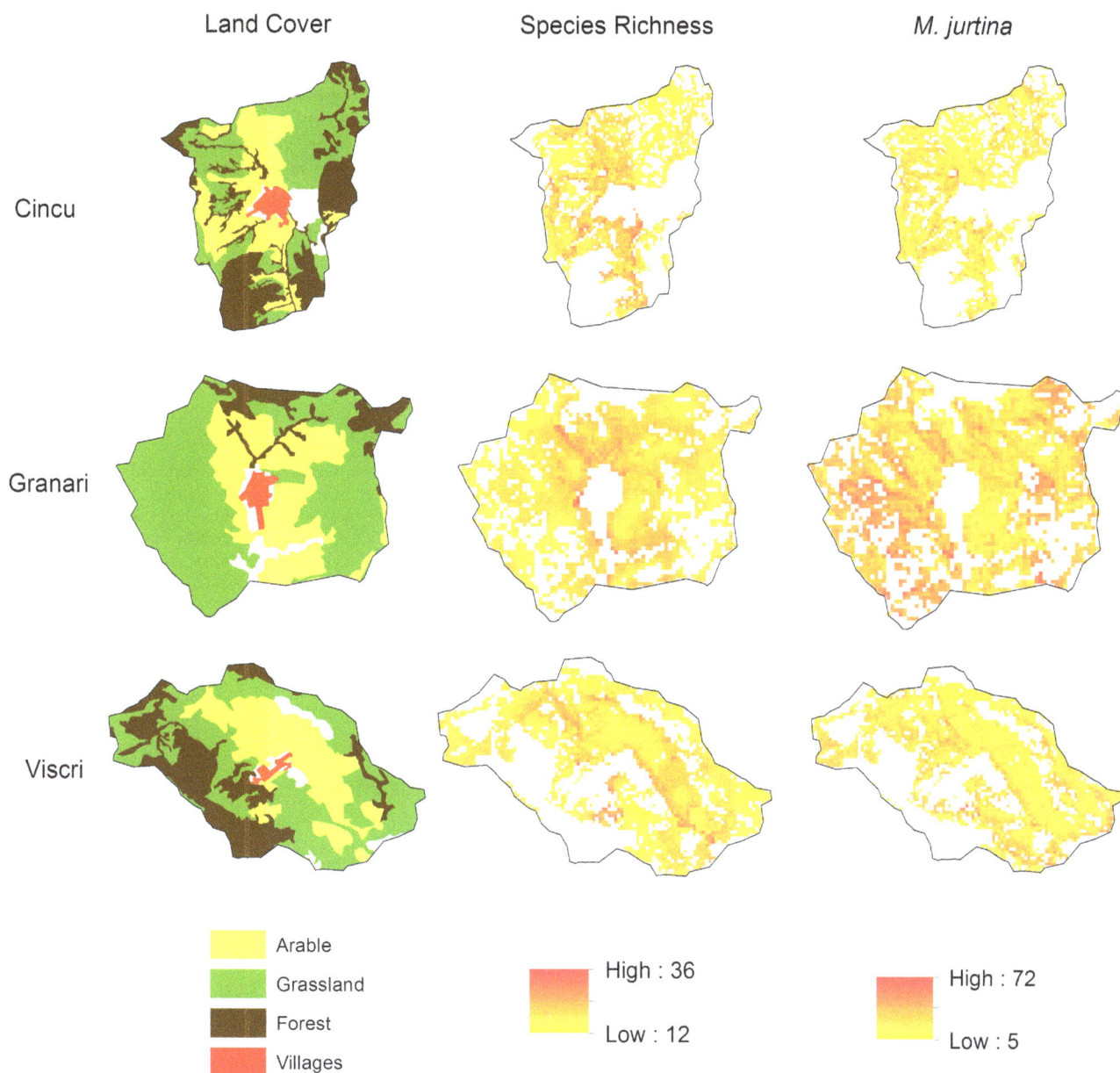

Figure 4. Maps of predicted butterfly distributions in three example villages. Left: Land cover map according to CORINE 2006; middle: predicted species richness for arable and grassland areas within each village catchment; right: predicted abundance of the Meadow Brown (*Maniola jurtina*).

degree of fragmentation of butterfly habitat, with likely negative consequences for species diversity [62]. Our study thus confirms that heterogeneity *per se* is not universally beneficial for species richness (see also [63]), although most work to date has focused on its positive effects (e.g. [64]).

The importance of considering multiple scales

To date, results from studies investigating multiple scales have been disparate and difficult to generalize [65]. We included three spatial scales in our study which we considered relevant for butterfly diversity and distribution. Our study revealed that all investigated scales affected butterfly community composition. Previous studies found local factors affecting butterfly species composition, with local heterogeneity in land cover being a good

predictor for species composition in Canada [54,64]. Butterfly species composition in Transylvania also showed a significant correlation with local factors, but was explained by heterogeneity and woody vegetation cover only at the two larger scales. Butterfly species richness also responded to variables at all different spatial scales, especially at the local scale, but also at the two larger scales (see also [9]). This suggests that local habitat conditions are particularly important, yet these cannot be considered in isolation from the surrounding landscape [33,66].

Our models also showed that the different functional groups of butterflies were affected by variables from different spatial scales. For example, landscape heterogeneity appeared to benefit mobile species but not low-mobility species. Furthermore, we found that woody vegetation cover was related to species richness. Land abandonment induces natural succession, whereas intensification leads to loss of scattered woody vegetation, and both have negative effects on butterfly richness in the long term [67]. Both processes decrease structural heterogeneity, which is important for viable butterfly populations in agricultural landscapes. In our study, only *Coenonympha pamphilus* showed a unimodal relationship to local woody vegetation cover. For such low-mobile species, presence of woody vegetation is crucial for wind shield and thermoregulation. *C. pamphilus* is abundant in Transylvania, however its population state in other European countries is declining [68]. Habitat heterogeneity from different spatial scales, including the presence of woody vegetation, should be further investigated as possible key elements in landscapes to halt biodiversity loss in farmland.

Conclusion

Collapse of communism and accession of Romania to the European Union have accelerated land use change in the rural areas of Transylvania, in particular towards land abandonment and agricultural intensification. The two key gradients considered in this study, namely woody vegetation cover and heterogeneity, would fundamentally change as a result of these two land use change processes. Along the gradients of woody vegetation cover and heterogeneity, we were able to show that butterfly abundance and distribution were affected by a range of different variables operating at multiple spatial scales. Not only local conditions, but the composition and configuration of the landscape as a whole need to be considered for effective conservation management of butterflies in low-intensively managed farming landscapes such as in Transylvania.

Our results showed that, unlike in Western Europe, species richness of butterflies was not only high in grassland, but also in arable land. This suggests that more emphasis needs to be placed on low-intensity farming practices and management of the landscape mosaic, and that arable land needs to be actively considered in butterfly conservation strategies. In our study area, butterfly richness would likely benefit from (1) the continuation of small-scale farming; (2) the production of a variety of crops, including legume species; and (3) the maintenance of broad field margins and uncultivated ruderal areas. New payment schemes under the Common Agricultural Policy have recently been criticised as grossly inadequate [53]. Our findings suggest that even measures considered adequate in Western Europe may not be directly transferable to Transylvania – in low-intensity

landscapes, it will be particularly important to consider the high nature value that entire agro-ecosystems provide, both inside and outside of protected areas (see also [69]). Ultimately, the continued existence of historic-cultural landscapes such as those in Transylvania hinges on the successful transfer of its appreciation and historic management to future generations of farmers. Substantial efforts are therefore needed in environmental education and in developing alternative ways for local people to make a living, for example through the development of cultural and ecological tourism.

Acknowledgments

We warmly thank Rémi Bigonneau, Paul Kirkland, Joerg Steiner, Kimberley Pope and Elek Telek for help in the field. Sincere thanks to the farmers, landowners, mayors and the Mihai Eminescu Trust for their support. We are grateful for analytical support by Henrik von Wehrden and Dave Abson. Comments by two anonymous reviewers greatly helped to improve an earlier version of this manuscript.

Author Contributions

Conceived and designed the experiments: JL JF JH ID LR. Performed the experiments: JL. Analyzed the data: JL JH ID. Contributed reagents/materials/analysis tools: PF. Wrote the paper: JL.

References

1. Stoate C, Baldi A, Beja P, Boatman ND, Herzon I, et al. (2009) Ecological impacts of early 21st century agricultural change in Europe - A review. Journal of Environmental Management 91: 22–46.
2. Kleijn D, Rundlöf M, Scheper J, Smith HG, Tscharntke T (2011) Does conservation on farmland contribute to halting the biodiversity decline? Trends in Ecology & Evolution 26: 474–481.
3. Benton TG, Vickery JA, Wilson JD (2003) Farmland biodiversity: is habitat heterogeneity the key? Trends in Ecology & Evolution 18: 182–188.
4. Thomas JA, Telfer MG, Roy DB, Preston CD, Greenwood JJD, et al. (2004) Comparative losses of British butterflies, birds, and plants and the global extinction crisis. Science 303: 1879–1881.
5. Foley JA, DeFries R, Asner GP, Barford C, Bonan G, et al. (2005) Global consequences of land use. Science 309: 570–574.
6. Cremene C, Groza G, Rakosy L, Schileyko AA, Baur A, et al. (2005) Alterations of steppe-like grasslands in Eastern Europe: a threat to regional biodiversity hotspots. Conservation Biology 19: 1606–1618.
7. Kumar S, Simonson SE, Stohlgren TJ (2009) Effects of spatial heterogeneity on butterfly species richness in Rocky Mountain National Park, CO, USA. Biodiversity and Conservation 18: 739–763.
8. Brückmann SV, Krauss J, Steffan-Dewenter I (2010) Butterfly and plant specialists suffer from reduced connectivity in fragmented landscapes. Journal of Applied Ecology 47: 799–809.
9. Weibull AC, Ostman O, Granqvist A (2003) Species richness in agroecosystems: the effect of landscape, habitat and farm management. Biodiversity and Conservation 12: 1335–1355.

10. Van Dyck H, Van Strien AJ, Maes D, Van Swaay CAM (2009) Declines in Common, Widespread Butterflies in a Landscape under Intense Human Use. Conservation Biology 23: 957–965.
11. Maes D, Van Dyck H (2001) Butterfly diversity loss in Flanders (north Belgium): Europe's worst case scenario? Biological Conservation 99: 263–276.
12. McLaughlin A, Mineau P (1995) The impact of agricultural practices on biodiversity. Agriculture, Ecosystems & Environment 55: 201–212.
13. Bergman KO, Askling J, Ekberg O, Ignell H, Wahlman H, et al. (2004) Landscape effects on butterfly assemblages in an agricultural region. Ecography 27: 619–628.
14. Kluvánková-Oravská T, Chobotová V, Banaszak I, Slavikova L, Trifunovova S (2009) From government to governance for biodiversity: the perspective of central and Eastern European transition countries. Environmental Policy & Governance 19: 186–196.
15. Kuemmerle T, Muller D, Griffiths P, Rusu M (2008) Land use change in Southern Romania after the collapse of socialism. Regional Environmental Change 9: 1–12.
16. Mikulcak F, Newig J, Milcu AI, Hartel T, Fischer J (2013) Integrating rural development and biodiversity conservation in Central Romania. Environmental Conservation 40: 129–137.
17. Schmitt T, Rákosy L (2007) Changes of traditional agrarian landscapes and their conservation implications: a case study of butterflies in Romania. Diversity and Distributions 13: 855–862.

18. Young J, Richards C, Fischer A, Halada L, Kull T, et al. (2007) Conflicts between Biodiversity conservation and human activities in the central and eastern European countries. Ambio 36: 545–550.

19. Pullin AS, Baldi A, Can OE, Dieterich M, Kati V, et al. (2009) Conservation Focus on Europe: Major Conservation Policy Issues That Need to Be Informed by Conservation Science. Conservation Biology 23: 818–824.

20. Tryjanowski P, Hartel T, Baldi A, Szymanski P, Tobolka M, et al. (2011) Conservation of farmland birds faces different challenges in Western and Central-Eastern Europe. Acta Ornithologica 46: 1–12.

21. Tscharntke T, Klein AM, Kruess A, Steffan-Dewenter I, Thies C (2005) Landscape perspectives on agricultural intensification and biodiversity - ecosystem service management. Ecology Letters 8: 857–874.

22. Ekroos J, Rundlöf M, Smith HG (2013) Trait-dependent responses of flower-visiting insects to distance to semi-natural grasslands and landscape heterogeneity. Landscape Ecology 28: 1283–1292.

23. Dunning JB, Danielson BJ, Pulliam HR (1992) Ecological Processes That Affect Populations in Complex Landscapes. Oikos 65: 169–175.

24. Fahrig L, Baudry J, Brotons L, Burel FG, Crist TO, et al. (2011) Functional landscape heterogeneity and animal biodiversity in agricultural landscapes. Ecology Letters 14: 101–112.

25. Persson AS, Olsson O, Rundlöf M, Smith HG (2010) Land use intensity and landscape complexity-Analysis of landscape characteristics in an agricultural region in Southern Sweden. Agriculture Ecosystems & Environment 136: 169–176.

26. Fischer J, Hartel T, Kuemmerle T (2012) Conservation policy in traditional farming landscapes. Conservation Letters 5: 167–175.

27. Page N, Bălan A, Popa SHR, Rákosy L, Sutcliffe L (2012) România/Romania. In: Oppermann R, Beaufoy GJ, G., editors. High Nature Value Farming in Europe. Ubstadt-Weiher: Verlag Regionalkultur. pp. 346–358.

28. Akeroyd JR, Page JN (2006) The Saxon villages of southern Transylvania: Conserving biodiversity in a historic landscape. In: Gafta D, Akeroyd JR, editors. Nature Conservation: Concepts and Practice. Heidelberg, Germany: Springer Verlag. pp. 199–210.

29. Hartel T, Moga CI, Öllerer K, Sas I, Demeter L, et al. (2008) A proposal towards the incorporation of spatial heterogeneity into animal distribution studies in Romanian landscapes. North-Western Journal of Zoology 4: 67–74.

30. Erhardt A (1985) Diurnal Lepidoptera - Sensitive Indicators of Cultivated and Abandoned Grassland. Journal of Applied Ecology 22: 849–861.

31. van Swaay C, Warren M, Lois G (2006) Biotope use and trends of European butterflies (vol 10, pg 189, 2006). Journal of Insect Conservation 10: 305–306.

32. Krauss J, Steffan-Dewenter I, Tscharntke T (2003) How does landscape context contribute to effects of habitat fragmentation on diversity and population density of butterflies? Journal of Biogeography 30: 889–900.

33. Öckinger E, Smith HG (2006) Landscape composition and habitat area affects butterfly species richness in semi-natural grasslands. Oecologia 149: 526–534.

34. Börschig C, Klein A-M, von Wehrden H, Krauss J (2013) Traits of butterfly communities change from specialist to generalist characteristics with increasing land-use intensity. Basic and Applied Ecology.

35. Öckinger E, Franzen M, Rundlof M, Smith HG (2009) Mobility-dependent effects on species richness in fragmented landscapes. Basic and Applied Ecology 10: 573–578.

36. Diamond JM (1986) Overview: laboratory experiments, field experiments, and natural experiments. In: Diamond JM, Case TJ, editors. Community Ecology. New York: Harper & Row. pp. 3–22.

37. Lindenmayer DB, Cunningham RB, MacGregor C, Crane M, Michael D, et al. (2008) Temporal changes in vertebrates during landscape transformation: a large-scale "natural experiment". Ecological Monographs 78: 567–590.

38. Bennett AF, Radford JQ, Haslem A (2006) Properties of land mosaics: Implications for nature conservation in agricultural environments. Biological Conservation 133: 250–264.

39. Pollard E, Yates TJ (1993) Monitoring butterflies for ecology and conservation: the British butterfly monitoring scheme; Institute of Terrestrial Ecology JNCC, editor. London: Chapman & Hall. 274 p.

40. Naumann CM, Tarmann GM, Tremewan WG (1999) The Western Palaearctic Zygaenidae (Lepidoptera). Stenstrup: Apollo Books. 304 p.

41. Rakosy L, Goia M, Kovács Z (2003) Catalogul Lepidopterelor Romaniei/ Verzeichnis der Schmetterlinge Rumäniens. In: Romana SL, editor. Cluj-Napoca: Romsver.

42. Parker KC (1991) Topography, Substrate, and Vegetation Patterns in the Northern Sonoran Desert. Journal of Biogeography 18: 151–163.

43. McGarigal K, Cushman S, Ene E (2012) FRAGSTATS v4: Spatial Pattern Analysis Program for Categorical and Continuous Maps. Computer software program produced by the authors at the University of Massachusetts, Amherst. Available: http://www.umass.edu/landeco/research/fragstats/fragstats.html.

44. Dormann CF, Elith J, Bacher S, Buchmann C, Carl G, et al. (2013) Collinearity: a review of methods to deal with it and a simulation study evaluating their performance. Ecography 36: 27–46.

45. van Swaay C, Maes D, Collins S, Munguira ML, Sasic M, et al. (2011) Applying IUCN criteria to invertebrates: How red is the Red List of European butterflies? Biological Conservation 144: 470–478.

46. Bolker BM, Brooks ME, Clark CJ, Geange SW, Poulsen JR, et al. (2009) Generalized linear mixed models: a practical guide for ecology and evolution. Trends in Ecology & Evolution 24: 127–135.

47. Bink FA (1992) Ecologische Atlas van de Dagvlinders van Noordwest-Europa. Haarlem: Schuyt & Co.

48. van Swaay C, Strien Av, Harpke A, Fontaine B, Stefanescu C, et al. (2013) The European Grassland Butterfly Indicator: 1990–2011. Copenhagen, Denmark: European Environment Agency. No 11/2013.

49. Franzén M, Ranius T (2004) Occurrence patterns of butterflies (Rhopalocera) in semi-natural pastures in southeastern Sweden. Journal for Nature Conservation 12: 121–135.

50. van Swaay C, Warren MS (1999) Red data book of European butterflies (Rhopalocera): Council of Europe.

51. Konvicka M, Maradova M, Benes J, Fric Z, Kepka P (2003) Uphill shifts in distribution of butterflies in the Czech Republic: effects of changing climate detected on a regional scale. Global Ecology and Biogeography 12: 403–410.

52. R Core Team (2013) R: A Language and Environment for Statistical Computing. Vienna, Austria: R Foundation for Statistical Computing.

53. Pe'er G, Dicks LV, Visconti P, Arlettaz R, Báldi A, et al. (2014) EU agricultural reform fails on biodiversity. Science 344: 1090–1092.

54. Weibull AC, Ostman O (2003) Species composition in agroecosystems: The effect of landscape, habitat, and farm management. Basic and Applied Ecology 4: 349–361.

55. Steffan-Dewenter I, Tscharntke T (2000) Butterfly community structure in fragmented habitats. Ecology Letters 3: 449–456.

56. Wilson JB, Peet RK, Dengler J, Partel M (2012) Plant species richness: the world records. Journal of Vegetation Science 23: 796–802.

57. Jones A (2009) Wildflower species indicators for lowland grassland habitat conservation in Transylvania (Romania). Contributii Botanice 44: 57–66.

58. Zechmeister HG, Schmitzberger I, Steurer B, Peterseil J, Wrbka T (2003) The influence of land-use practices and economics on plant species richness in meadows. Biological Conservation 114: 165–177.

59. Kleijn D, Kohler F, Baldi A, Batary P, Concepcion ED, et al. (2009) On the relationship between farmland biodiversity and land-use intensity in Europe. Proceedings of the Royal Society B-Biological Sciences 276: 903–909.

60. Van Landuyt W, Vanhecke L, Hoste I, Hendrickx F, Bauwens D (2008) Changes in the distribution area of vascular plants in Flanders (northern Belgium): eutrophication as a major driving force. Biodiversity and Conservation 17: 3045–3060.

61. Dover J, Sparks T, Greatorex-Davies J (1997) The importance of shelter for butterflies in open landscapes. Journal of Insect Conservation 1: 89–97.

62. Krauss J, Klein AM, Steffan-Dewenter I, Tscharntke T (2004) Effects of habitat area, isolation, and landscape diversity on plant species richness of calcareous grasslands. Biodiversity and Conservation 13: 1427–1439.

63. Ekroos J, Piha M, Tiainen J (2008) Role of organic and conventional field boundaries on boreal bumblebees and butterflies. Agriculture, Ecosystems & Environment 124: 155–159.

64. Kerr JT, Southwood TRE, Cihlar J (2001) Remotely sensed habitat diversity predicts butterfly species richness and community similarity in Canada. Proceedings of the National Academy of Sciences of the United States of America 98: 11365–11370.

65. Flick T, Feagan S, Fahrig L (2012) Effects of landscape structure on butterfly species richness and abundance in agricultural landscapes in eastern Ontario, Canada. Agriculture Ecosystems & Environment 156: 123–133.

66. Steffan-Dewenter I, Munzenberg U, Burger C, Thies C, Tscharntke T (2002) Scale-dependent effects of landscape context on three pollinator guilds. Ecology 83: 1421–1432.

67. Baur B, Cremene C, Groza G, Rakosy L, Schileyko AA, et al. (2006) Effects of abandonment of subalpine hay meadows on plant and invertebrate diversity in Transylvania, Romania. Biological Conservation 132: 261–273.

68. Conrad KF, Fox R, Woiwod IP (2007) Monitoring Biodiversity: Measuring Long-term changes in Insect Abundance. In: Stewart AJA, New TR, Lewis OT, editors. Insect Conservation Biology: Proceedings of the Royal Entomological Society's 23rd symposium. Oxford: The Royal Entomological Society. pp. 203–225.

69. González-Estébanez FJ, García-Tejero S, Mateo-Tomás P, Olea PP (2011) Effects of irrigation and landscape heterogeneity on butterfly diversity in Mediterranean farmlands. Agriculture, Ecosystems & Environment 144: 262–270.

5

Meta-Analysis of Gene Expression Signatures Reveals Hidden Links among Diverse Biological Processes in Arabidopsis

Liming Lai, Steven X. Ge*

Department of Mathematics and Statistics, South Dakota State University, Brookings, South Dakota, United States of America

Abstract

The model plant Arabidopsis has been well-studied using high-throughput genomics technologies, which usually generate lists of differentially expressed genes under various conditions. Our group recently collected 1065 gene lists from 397 gene expression studies as a knowledgebase for pathway analysis. Here we systematically analyzed these gene lists by computing overlaps in all-vs.-all comparisons. We identified 16,261 statistically significant overlaps, represented by an undirected network in which nodes correspond to gene lists and edges indicate significant overlaps. The network highlights the correlation across the gene expression signatures of the diverse biological processes. We also partitioned the main network into 20 sub-networks, representing groups of highly similar expression signatures. These are common sets of genes that were co-regulated under different treatments or conditions and are often related to specific biological themes. Overall, our result suggests that diverse gene expression signatures are highly interconnected in a modular fashion.

Editor: Francisco J. Esteban, University of Jaén, Spain

Funding: This work was supported by the National Institute of General Medical Studies at the National Institutes of Health (GM083226 to SXG, in part) and this material is based upon work supported partially by the National Science Foundation/EPSCoR Award No. IIA-1355423 and by the state of South Dakota. The funders had no role in study design, data collection and analysis, decision to publish, or preparation of the manuscript.

Competing Interests: The authors have declared that no competing interests exist.

* Email: xijin.ge@sdstate.edu

Introduction

Because of its small genome size, *Arabidopsis* thaliana has been a valuable model system for genetic mapping, sequencing and gene expression analysis [1]. Until March 2013, 1787 studies on gene expression of *Arabidopsis* were indexed in Gene Expression Omnibus (GEO) website in National Center for Biotechnology Information (NCBI) [2]. These studies investigated various biological processes by monitoring the gene expression level using the high-throughput genomics technologies such as DNA microarrays and RNA sequencing. The results were usually a set of genes associated with particular biological processes based on different experimental designs. Even though DNA microarrays suffer from noise and reproducibility issues [3], we believe that many of the noise could be filtered out by statistical analysis and that there are significant associations among these numerous results, or common modules in the transcriptional program.

Some studies have showed the relationships among gene lists in different species. Most researchers analyzed these gene lists using methodology of meta-analysis [4–7], which combines the results of studies that address a set of related research hypotheses, focusing on a special individual topic such as cancer or special treatment [8]. Several databases of gene lists have been created, such as L2L [9], LOLA [10], and MSigDB [11]. An network-based method was developed by Ge [12] to define associations among a large number of gene sets in human. Associations are defined as statistically significant overlaps between two gene lists. The method was applied successfully to a large number of human

gene lists [12], and identified molecular links among diverse biological processes.

In this study, we used the methodology in [12] to analyze a set of *Arabidopsis* gene lists identified by genome wide expression studies. These lists were collected for AraPath [13], an *Arabidopsis* gene lists database we created recently. The objective was to systematically evaluate relationships among the gene lists and interpret the relationships. This process provides not only a new tool to uncover hidden links among vast amounts of gene lists, but a quantitative measure to describe the global gene expression of the *Arabidopsis* system under diverse conditions.

Materials and Methods

Data in this study was extracted from the AraPath [13], which is a gene lists database in *Arabidopsis* we created (Availability: http://bioinformatics.sdstate.edu/arapath/). As part of the database, the data contains a total of 1,065 co-expression gene lists, which were manually retrieved from published papers linked to GEO [2] before February, 2011.

Methodology of the analysis includes four steps. Step 1 is to evaluate overlapping genes among the 1,065 gene lists. A Perl programs was written to evaluate overlapping genes between all 566,580 pairs of lists. An overlap refers to a pair of gene lists, which has at least two common genes. And overlaps from the same paper were considered trivial and were removed. Because there are too much overlaps and microarray experiments tends to produce noisy data, we selected significant overlaps using stringent threshold. Step 2 computes p-values and q-values to identify significant overlaps. Based on the Hypergeometric distribution, we first calculate the likelihood (p-value) of observing the number of overlapping genes if these two gene lists are randomly drawn without replacement from a collection of 28,024 unique genes in terms of R program [14] we compiled. Then, p-values were translated into q-values based on the false discovery rate (FDR) [15] to correct that for multiple testing. Overlaps with very small q-value were significant overlaps. In this case, significant overlaps were identified with a q-value $= 5.0E-9$ as a cutoff. In step 3, network of significant overlaps was constructed based on outputs of the step 2 using Cytoscape[16]. Because this network includes too many nodes and edges, we need to further break the big clusters into smaller subclusters. In step 4, There are many algorithms that could decompose large networks into small, densely connected subnetworks such as those in [17,18]. We chose a simply algorithm that is available as a plug-in to Cytoscape. MCODE [19] is used to identify interconnected sub-networks and their clusters within the network of the step 3. To generally find locally dense regions (or clusters) of a graph is based on the clustering coefficient [19], C_i, which measures "clique" of the neighborhood of a vertex: $C_i = 2n/k_i (k_i - 1)$, where k_i is the vertex size of the neighborhood of vertex i, n is the number of edges in the neighborhood. According to the MCODE algorithm [19], however, clustering the main network into sub-networks is by means of vertex weighting, which is to weight all vertices based on their local network density using the highest k-core of the vertex neighborhood rather than the clustering coefficient C_i. A k-core is a graph of minimal degree k. The highest k-core of a graph is the central most densely connected sub-graph. Given a highly connected vertex, in a dense region of a graph, v may be connected to many vertices of degree one. These low degree vertices do not interconnect within the neighborhood of v and thus would reduce the clustering coefficient, but not the core-clustering coefficient (for detailed information about the MCODE algorithms, see the paper [19]). Here we created the sub-networks and found the modules and clusters using MCODE algorithms based on the following parameters: Node Score Cutoff $= 0.15$; k-core $= 2$; Degree Cutoff $= 2$; Max. Depth $= 100$. The DAVID web site [20,21] was applied to analyze the most significant functions of most frequently shared genes in each of sub-networks.

Results

A total of 1,065 gene lists were analyzed in this study. They include 277,349 gene entries corresponding to 28,024 unique genes. The average size of these gene lists is 87 genes, ranging from 1 to 2,952 genes. Its distribution is close to normality on a log10 scale (Figure 1). The results of analysis of the data are as follows.

Significant overlaps

By comparing all pairs of 1,065 gene lists using the Perl program, 16,261 significant overlaps were identified from a total of 192,642 overlaps. Based on the Hyper-geometric distribution, the probabilities (p-values) of observing the number of overlapping genes or more were first calculated if these two gene lists were randomly drawn without the replacement from a collection of 28,024 unique genes. The p-values were translated into q-values according to the false discovery rate (FDR) [15] to correct for multiple testing. Overlaps from the same paper were considered trivial and were removed. With a q-value $= 5 \times 10^-$as a conservative cutoff, 16,261 significant overlaps were identified.

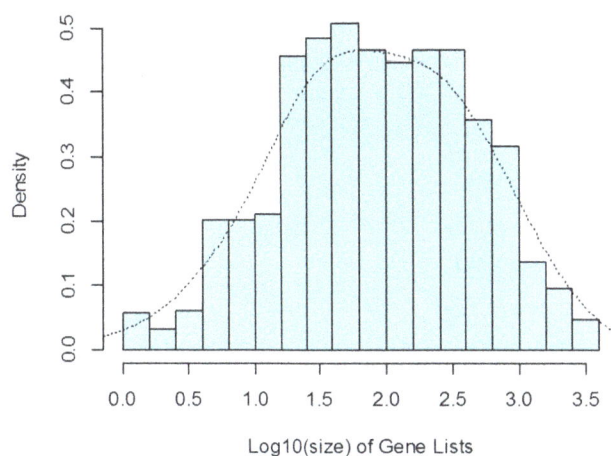

Figure 1. Histogram of log10 scale of size of gene lists.

Main network

The 16,261 significant overlaps are represented as an undirected network (Figure 2). In the network, nodes correspond to gene lists and edges indicate number of overlapping genes between two nodes within significant overlaps. This network highlights the correlation across gene expression signatures of diverse biological processes. It, thus, constitutes a "molecular signature map" in which the individual perturbations are placed in the context defined by others. This is a highly connected network with an average of 20.10 connections per gene list. The 809 nodes (75.96% of the 1,065 gene lists) and 16,261 edges are connected to a dominant main network. Most nodes are connected to a small number of gene lists. The network shows some different colors "cliques", which are some of the most connected graphs in terms of a vertex-weighting scheme based on the highest k-core of the vertex neighborhood. They are intuitively denser links within some neighborhoods.

Modules

To further explore these neighborhoods, we used the MCODE algorithm [19] to decompose the network into a total of 20 sub-networks. Of them, nine modules were further analyzed and their clustered information was shown in Table 1. The others were ignored because they are much lower score of density (less than 1.6) and have few nodes and edges (less than 6). The nine sub-networks are highly interconnected, suggesting that those genes are involved in common metabolic pathways or interact with each other under similar biological perturbations. The first three sub-networks are described in the following section. The remaining six sub-networks description and all the sub-networks figures and their composite outcomes tables are shown in Figures S1–S9 and Tables S1–S9 in File S1.

Sub-network 1

The sub-network 1 includes 46 nodes and 969 edges (Figure 3). The score of cluster density is 21.065, which is the highest score among nine sub-networks. This indicates it is the most densely connected. In Figure 3, the dark red nodes represent higher network density based on MCODE [19]. Dark green edges represent very small p-values. There are 31 nodes that represent up-regulated, eight down-regulated, and seven differently regulated. Most gene lists (67.39%) involving up-regulated nodes are related to seven biological themes and 25 treatments or conditions.

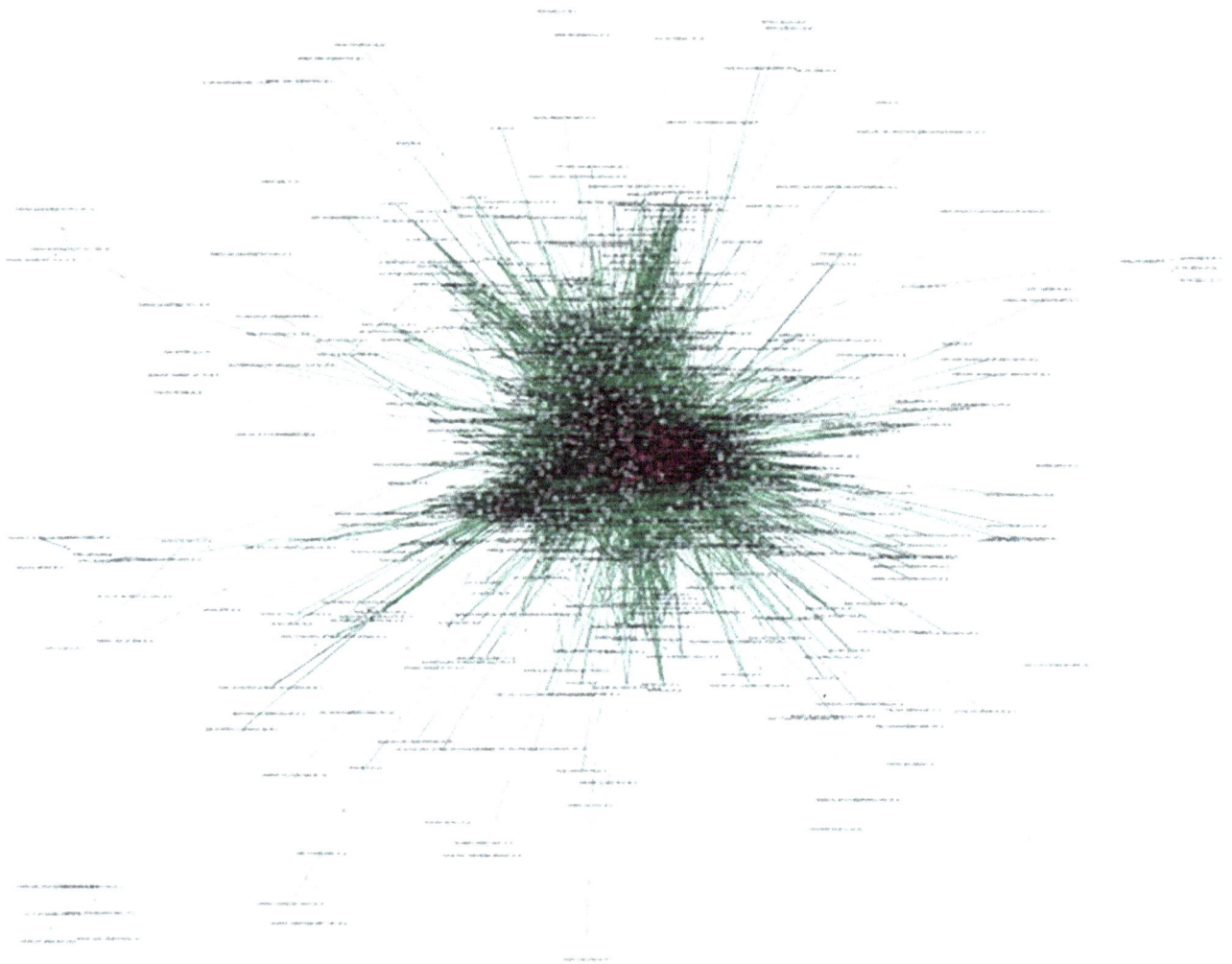

Figure 2. The main network created by Cytoscape. Node = name of gene list. Node Color = MCODE_Scores from small to large and corresponds to color from light green to dark red. Edge Color = p-values from large to small and corresponds to color from grey to dark green.

Table S1 in File S1 shows its composite outcomes. There are 46 gene lists and 256 frequently shared genes identified in this sub-network. They are regulated by 32 treatments or conditions from 32 publications related to 11 biological themes involving development, metabolism, disease, yield, function, genome analysis, immune, pathogen, mechanism, energy, virus, and photosynthesis in *Arabidopsis*. The top 10 most frequently shared genes with their gene descriptions were specifically listed corresponding to different gene lists, different biological themes, and treatments or conditions. For example, gene AT4G14365 ("putative E3 ubiquitin-protein ligase XBAT34") has the highest frequency of 35, which indicates it is the most active gene because it is regulated simultaneously under 35 gene lists in sub-network 1, namely, the gene connects directly 35 gene lists.

The most significant function of sub-network 1 is biological process in response to chitin (i.e. the most enriched term corresponding to the most frequently shared genes in sub-network 1 is "response to chitin") based on results of analysis of DAVID (Table 1). This indicates sub-network 1 is specifically associated with the chitin signaling pathway rather than by random chance. The other significant functions of sub-network 1 are responding to carbohydrate stimulus, organic substance, defense response, and bacterium based on the analysis of DAVID with a cutoff of

1.60×10^{-16} p-value. This suggests that sub-network 1 involves multiple signaling pathways.

Sub-network 2

Sub-network 2 is shown in Figure 4 and Table S2 in File S1. It includes 54 nodes (gene lists) and 168 frequently shared genes, which are regulated under 38 different treatments or conditions from 38 publications related to 10 biological themes. The score of cluster density is 9.907. There are 17 nodes to be up-regulated, 34 to be down-regulated, and three to be differently regulated. Most gene lists (62.96%) involving down-regulated nodes are related to nine biological themes and 23 treatments or conditions. Compared to sub-network 1, sub-network 2 has a lower cluster density score with even more treatments or conditions. Nine themes in sub-network 2 are common with sub-network 1: development, disease, function, genome analysis, mechanism, metabolism, photosynthesis, virus, and yield. This indicates the two sub-networks have relationships linked by same themes. No gene is common between the 256 frequently shared genes in sub-network 1 and the 168 frequently shared genes in sub-network 2. The two sub-networks have relatively independent functions. The most significant function of sub-network 2 is biological process of plastid thylakoid membrane based on results of DAVID (Table 1), suggesting that

Table 1. Summary of nine modules consisting of highly interconnected gene lists.

ID	Score Density†	#Nodes	#Edges	Unique genes	Shared genes*	Highest frequency	Most significantly enriched GO Term*	P-value*
1	21.065	46	969	10581	256	35	response to chitin	1.70E-38
2	9.907	54	535	12163	168	20	plastid thylakoid membrane	1.80E-59
3	9.545	33	315	9501	124	17	response to chitin	8.80E-13
4	9.062	48	435	13930	155	18	response to auxin stimulus	9.00E-10
5	2.559	34	87	9751	109	10	cell wall	5.30E-11
6	2.111	9	19	2565	34	6	External encapsulating structure organization	5.40E-07
7	2	15	30	6725	53	8	glycoside biosynthetic process	2.00E-05
8	2	6	12	5212	66	6	membrane-enclosed lumen	6.00E-14
9	1.846	13	24	3567	23	6	response to abiotic stimulus	6.80E-08
T	-	258	2426	-	988	-	-	-

Note: †Score Density = #Edges/#Nodes. *Most significantly enriched GO Term and p-value are the results of analysis of the. DAVID in terms of the most frequently shared genes (i.e. Shared genes in Table 1) in each sub-network.

sub-network 2 is specifically associated with plastid thylakoid membrane, i.e. the lipid bilayer membrane of any thylakoid within a plastid. The other significant functions of sub-network 2 are response to chloroplast thylakoid membrane, thylakoid membrane, plastid thylakoid, and chloroplast thylakoid based on DAVID with cut-off p-value of 6.8×10^{-58}. The top 10 most frequently shared genes with their gene descriptions corresponding to gene lists, biological themes, and treatments or conditions were specifically listed in Table S2 in File S1. Gene AT4G27030 (fatty acid desaturase A), for example, has the highest frequency of 20, indicating it is the most active gene in sub-network 2.

Sub-network 3

The sub-network 3 includes 33 nodes (gene lists) and 124 most frequently shared genes (Figure 5 and Table S3 in File S1). There are 28 nodes to be up-regulated, four to be down-regulated, and one to be differently regulated. Most gene lists (84.85%) involving up-regulated nodes are related to nine biological themes and 20 treatments or conditions. By contrast to sub-networks 1 and 2, sub-network 3 is smaller in size, has a lower cluster density score, and less treatments or conditions for gene lists. Sub-network 3 are regulated by 25 treatments or conditions from 25 publications associated with 11 biological themes. Nine themes in sub-network 3 are common with sub-network 1: development, disease, energy, function, immune, mechanism, metabolism, photosynthesis, and virus. There are seven common themes (development, disease, function, mechanism, metabolism, photosynthesis, and virus) between sub-networks 3 and 2. These indicate sub-networks 3 and 1 or 2 have relationships linked by the same themes.

The top 10 most frequently shared genes with their gene descriptions corresponding to each gene list in sub-network 3 are specifically listed in Table S3 in File S1. Gene AT2G18690 has the highest frequency at 17, indicating it is the most active genes in sub-network 3. Other shared genes in sub-network 3 have lower frequency, which means they are less active than those in sub-networks 1 and 2. There is no common gene between the 168 most frequently shared genes in sub-network 2 and the 124 most frequently shared genes in sub-network 3. However, there are 72 genes of intersection between sub-networks 1 and 3. This indicates sub-networks 2 and 3 have relatively independent functions and sub-networks 1 and 3 have dependent relationship linked by the common shared genes. The most significant function of sub-network 3 is biological process in response to chitin based on results of analysis of DAVID (Table 1). It is specifically associated with the chitin signaling pathway, which is the same as sub-network 1 but different from sub-networks 2.

Discussion

The gene lists in our data are highly connected. Out of the 1,065 gene lists, 75.96% are connected in the main network in which many seemingly unrelated stimuli/perturbation may activate or deactivate the same molecular pathways. All the gene lists within each sub-network are highly connected by the most frequently shared genes. For example, in sub-network 8 (Figure S8 and Table S8 in File S1), AT1G56110 ("homolog of nucleolar protein NOP56"), AT3G05060 ("putative SAR DNA-binding protein"), and AT3G44750 (HDA3 histone deacetylase HDT1) are regulated by six different treatments or conditions from five publications corresponding to five special biological themes, which involves reproduction, photosynthesis, metabolism, development, and yield in *Arabidopsis* (Table S8 in File S1). And the most enriched term in sub-network 8 is membrane-enclosed lumen based on functional analysis of DAVID (Table 1). Therefore, the

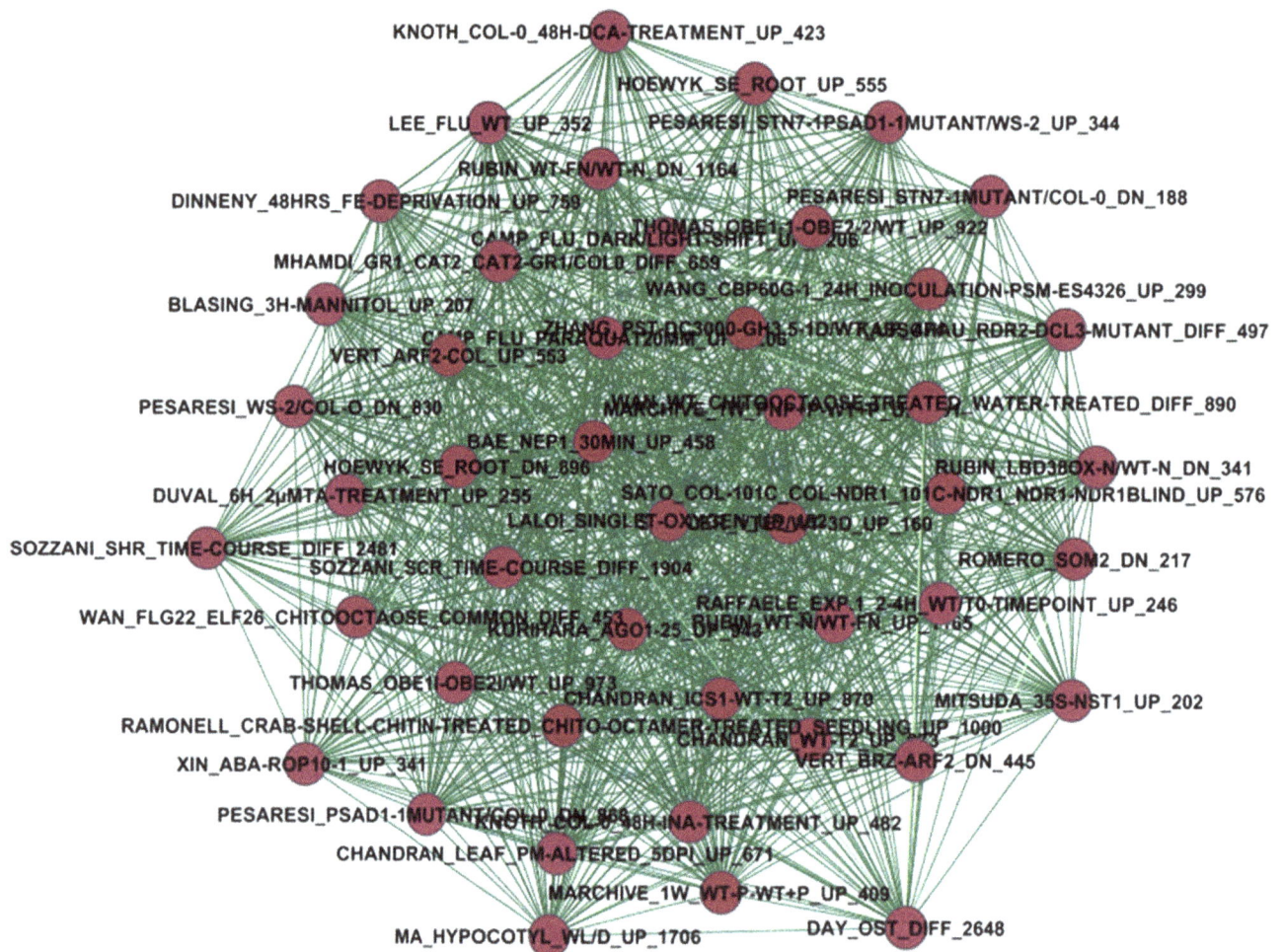

Figure 3. Sub-network 1 corresponding to the module 1/cluster 1 (see note in Figure 2 for meanings of node and edge and their color).

three genes not only connect the six gene lists, but have multiple functions by which we could propose a hypotheses that the three genes' interaction controls the reproduction, photosynthesis, metabolism etc. Furthermore, the three genes associated with the rapidly proliferating nature of the endosperm at 4 DA, with similar expression patterns to the early endosperm markers SUC5, PHE1, FWA, and FIS2 [22], were regulated by interploidy crosses, fis1X2x crosses at 5 DAP (two biological replicates of each), and unfertilized msi1 siliques at 7 DAF [23], kin10, starvation conditions, and sugar availability increase [24], sucrose [25], and 4h-carbon fixcation [26]. These all became the organized links among the six gene lists because they associated with the three genes.

However,there are significant differences among the nine sub-networks. Based on Table 1, sub-networks 1–5 have more nodes than sub-networks 6–9. This does not mean the genes in sub-network 6–9 are less important than those in sub-networks 1–5. Generally, the most significantly enriched GO terms in sub-networks 1–9 are different except for sub-networks 1 and 3, which have the same theme "response to chitin". The second, third, and fourth most significantly enriched GO Terms in sub-network 1 are response to carbohydrate stimulus, organic substance, and defense response, respectively, which are different from those in sub-network 3. This indicates there are significantly different functions

in sub-networks 1–9. Furthermore, "cliques" of all sub-networks are different in the whole network with 16,261 significant overlaps. These "cliques" are some of the most connected graphs in the NETWORK in terms of a vertex-weighting scheme based on the highest k-core of the vertex neighborhood. Therefore, they specify different meanings and information. Finally, all the most frequently shared genes of the sub-networks are different. For instance, sub-network 7 has the second smallest score of 2, with 15 nodes, and 30 edges, but possesses 53 most frequently shared genes, which are completely different from that of sub-network 1.

The most frequently shared genes are the strongest links within each of nine sub-networks and provide genomics research with important insights. They are the most important results we found in this study. For example, the gene AT5G39670 ("putative calcium-binding protein CML45", function as calcium ion binding) has the second highest frequency of 34 in the sub-network 1 (Table S1 in File S1). Its function as calcium ion binding indicates calcium-binding proteins participate in calcium cell signaling pathways by binding to Ca^{2+}. These proteins are expressed in many cell types during various growth stages in plants, and contribute to all aspects of the cell's functioning [24]. In the present study, we found that this gene responded to 24 treatments or conditions such as necrosis-ethylene, diurnal cycle, salicylic acid, iron deprivation, etc. according to various reports.

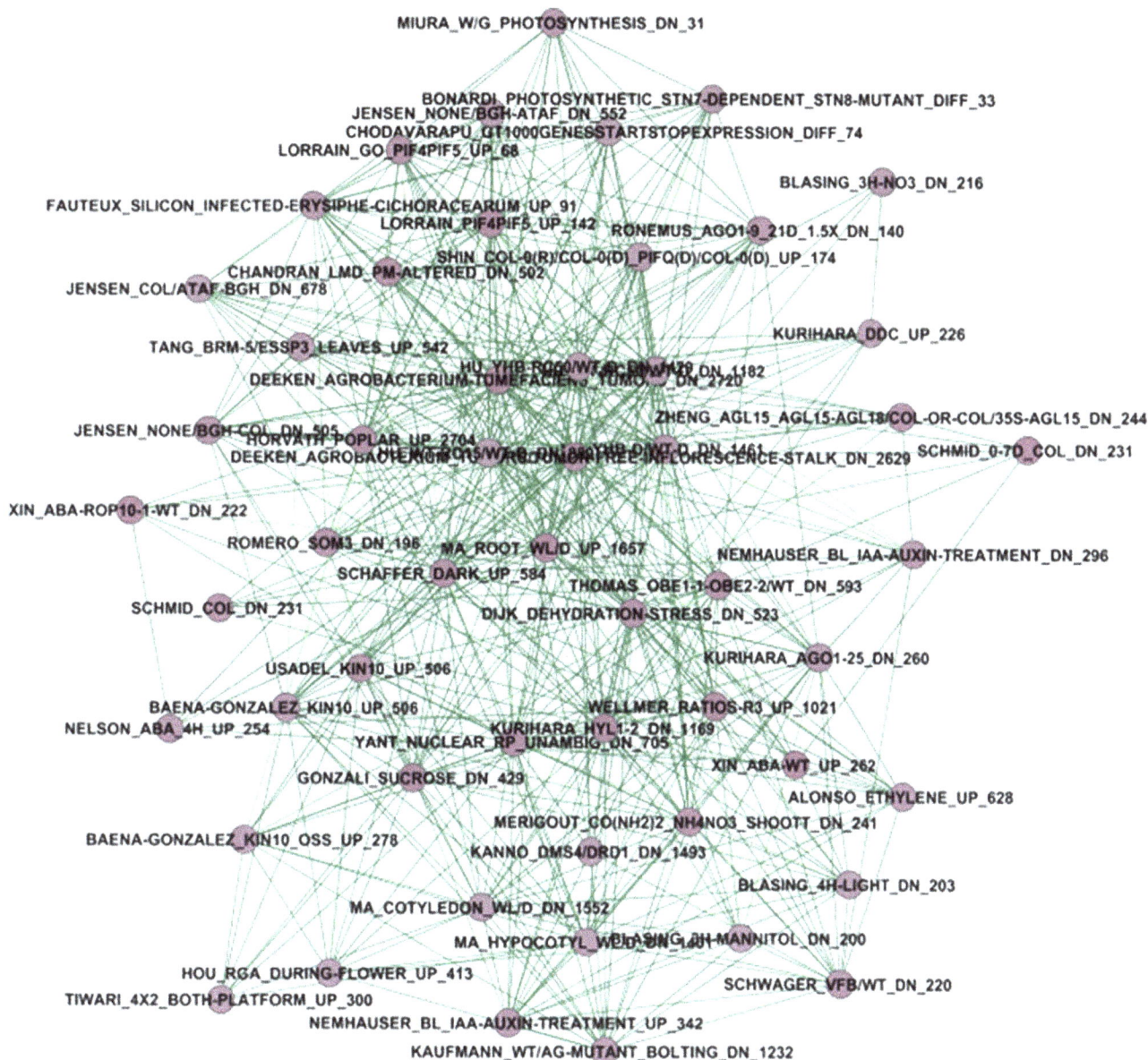

Figure 4. Sub-network 2 corresponding to the module 2/cluster 2 (see note in Figure 2 for meanings of node and edge and their color).

Furthermore, this gene mediates 10 crucial biological themes: immune, yield, disease, development, metabolism, function, photosynthesis, pathogen, energy, and virus. Also, the gene belongs to 24 up-regulated gene lists, seven down-regulated gene lists, and three different-regulated gene lists. In most cases, this gene is up-regulated, indicating that when the experimental treatments listed above were applied, there was an increased expression of the gene AT5G39670. However, some treatments or stimuli may cause a decreased expression of the above gene in order to protect its cells. Therefore, AT5G39670 is the second most active gene and second strongest link in sub-network 1. Similarly, gene AT4G14365 ("XBAT34", molecular functions: protein binding and zinc ion binding) is the strongest link as it is related to 35 gene lists in sub-network 1. Gene AT3G50930 ("BCS1", molecular functions: ATP binding and ATPase activity) is the third strongest link because it is associated with 33 gene lists

in sub-network 1. Also, the most frequently shared genes are significantly different in these nine sub-networks. For instance, there are 124 most frequently shared genes in sub-network 3 and 155 most frequently shared genes in sub-network 4. However, there are only two common genes between these two sub-networks. These important results and their biological mechanisms need to be further addressed.

The results from this study, summarized as a molecular signature map, provide key insights into the underlying connections of diverse perturbations. More importantly, compared to previous reports that focused on specific themes, this study explored and established the hidden links among the gene lists on a global scale in *Arabidopsis*. These sub-networks will provide new putative gene targets for future research. For example, sub-network 4 shows that the top three genes AT1G74670, AT1G04240, and AT1G69530 are down-regulated by bioactive

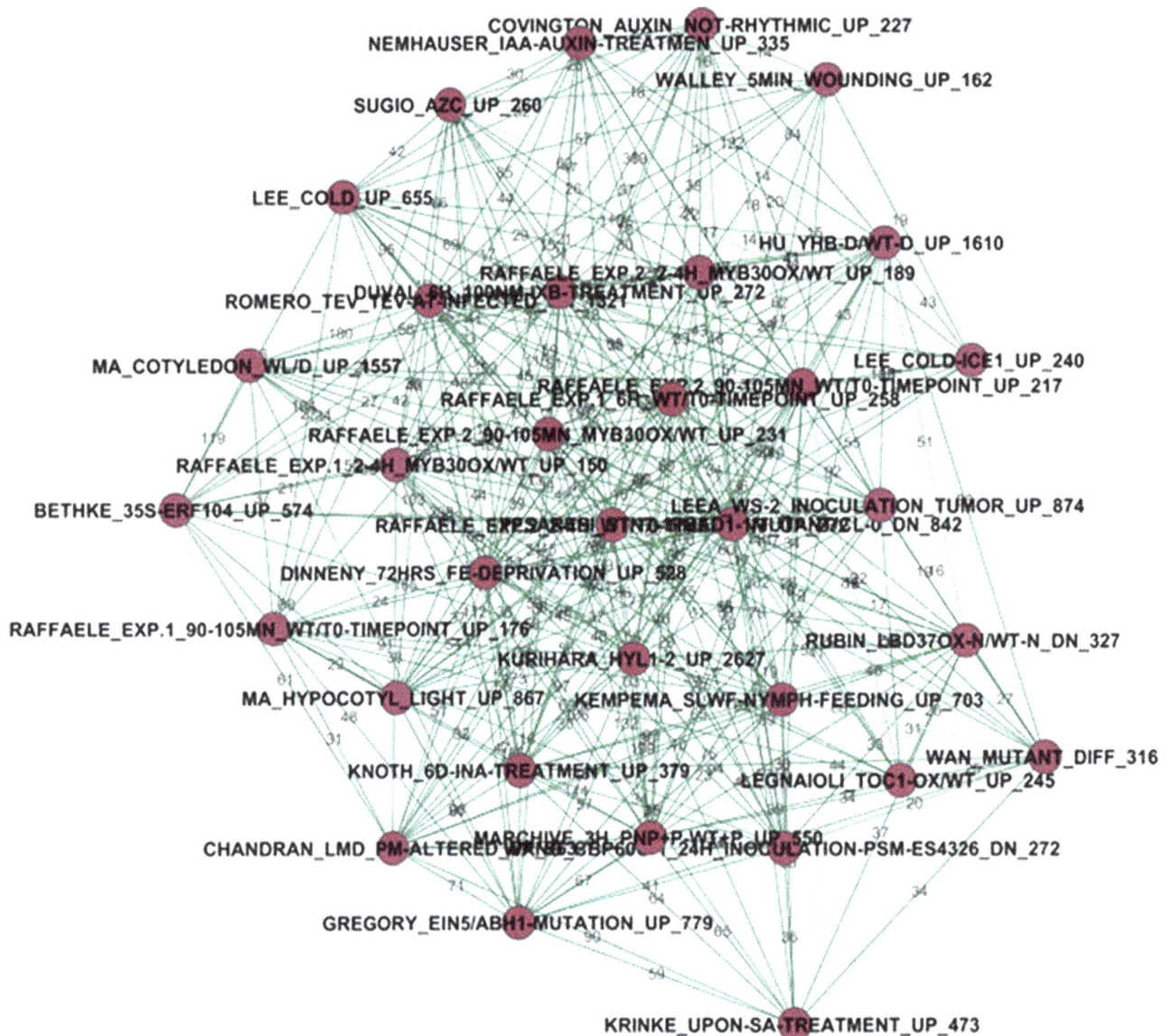

Figure 5. Sub-network 3 corresponding to the module 3/cluster 3 (see note in Figure 2 for meanings of node and edge and their color. edge label = number of overlapping genes between two nodes).

gibberellins corresponding to the gene list ZENTELLA_DEX_-VECTOR_DN_244 (Table S4 in File S1). This information provides some possible clues for future research regarding the mechanism of the regulation of plant growth by plant hormone gibberellins. Another example is gene AT4G27030, which has the highest frequency of 20 in sub-network 2. This gene is regulated by 14 treatments or conditions such as agrobacterium tumefaciens, kin10 and kin 11, dark, far-red light, etc. (Table S2 in File S1). This gene could be used to genetically modify crops for new and useful functions.

Conclusions

There are hidden links among the gene lists from the published papers concerning *Arabidopsis*. After performing systematic overlap analysis, we created 10 networks, including network and sub-networks 1–9 where there are a number of links among gene lists. Many seemingly unrelated stimuli/perturbation may activate or deactivate the same molecular pathways. These links are actually a set of overlapping genes. Of them, a total of 988 most frequently shared genes were identified from each sub-network. These genes are regulated by multiple treatments or conditions from different gene lists and related to different biological themes based on their sub-networks. They construct more active (stronger) links among the gene lists in our data.

Compared to previous reports focusing on specific themes, this study explored and established hidden links among the gene lists on a global scale in *Arabidopsis*. These results provide significant information about target genes or models for future research. In the future, it will be necessary for us to extend gene lists and develop more effective analysis methods to further explain the booming gene lists of microarray data.

Supporting Information Legends

File S1 Fig. S1. Sub-network 1 corresponding to cluster 1. Node = name of gene list. Node Color = MCODE_Scores from small to large and corresponds to color from light green to dark red. Edge Color = p-values from large to small and corresponds to color from grey to dark green. Fig. S2. Sub-network 2 corresponding to cluster 2. Fig. S3. Sub-network 3 corresponding to cluster 3. Fig. S4. The sub-network 4 correspoding to cluster 4. Fig. S5. The sub-network 5 corresponding to Cluster 5. Fig. S6. The sub-network 6 corresponding to Cluster 6. Fig. S7. The sub-network 7 corresponding to Cluster 7. Fig. S8. The sub-network 8 corresponding to Cluster 8. Fig. S9. The sub-network 9 corresponding to Cluster 9. Table S1. Results of sub-network 1 corresponding to cluster 1. Table S2. Results of sub-network 2 corresponding to cluster 2. Table S3. Results of sub-network 3 corresponding to cluster 3. Table S4. Results of sub-network 4 corresponding to cluster 4. Table S5. Results of sub-network 5 corresponding to cluster 5. Table S6. Results of sub-network 6 corresponding to cluster 6. Table S7. Results of sub-network 7 corresponding to cluster 7. Table S8. Results of sub-network 8 corresponding to cluster 8. Table S9. Results of sub-network 9 corresponding to cluster 9.

Author Contributions

Conceived and designed the experiments: SXG. Performed the experiments: LL. Analyzed the data: LL. Wrote the paper: LL SXG.

References

1. (2008) Arabidopsis thaliana. Wikipedia http://enwikipediaorg/wiki/Arabidopsis_thaliana: Wikipedia.
2. (2012). NCBI http://wwwncbinlmnihgov/geo/.
3. Ioannidis JP, Allison DB, Ball CA, Coulibaly I, Cui X, et al. (2009) Repeatability of published microarray gene expression analyses. Nat Genet 41: 149–155.
4. Miller BG, Stamatoyannopoulos JA (2010) Integrative meta-analysis of differential gene expression in acute myeloid leukemia. PLoS One 5: e9466.
5. Rogic S, Pavlidis P (2009) Meta-analysis of kindling-induced gene expression changes in the rat hippocampus. Front Neurosci 3: 53.
6. Yang XN, Sun X (2007) Meta-analysis of several gene lists for distinct types of cancer: A simple way to reveal common prognostic markers. BMC Bioinformatics 8.
7. Edwards YJK, Bryson K, Jones DT (2008) A Meta-Analysis of Microarray Gene Expression in Mouse Stem Cells: Redefining Stemness. PLoS One 3.
8. Glass GV Meta-analysis. Wikipedia http://enwikipediaorg/wiki/Meta-analysis. http://en.wikipedia.org/wiki/Meta-analysis: Wikipedia.
9. Newman JC, Weiner AM (2005) L2L: a simple tool for discovering the hidden significance in microarray expression data. Genome Biol 6: R81.
10. Cahan P, Ahmad AM, Burke H, Fu S, Lai YL, et al. (2005) List of lists-annotated (LOLA): A database for annotation and comparison of published microarray gene lists. Gene 360: 78–82.
11. Liberzon A, Subramanian A, Pinchback R, Thorvaldsdottir H, Tamayo P, et al. (2011) Molecular signatures database (MSigDB) 3.0. Bioinformatics 27: 1739–1740.
12. Ge SX (2011) Large-scale analysis of expression signatures reveals hidden links among diverse cellular processes. Bmc Systems Biology 5.
13. Lai LM, Liberzon A, Hennessey J, Jiang GX, Qi JL, et al. (2012) AraPath: a knowledgebase for pathway analysis in Arabidopsis. Bioinformatics 28: 2291–2292.
14. Team RC R: A language and environment for statistical computing. (2012) R Foundation for Statistical Computing, Vienna, Austria ISBN 3-900051-07-0, URL http://wwwR-projectorg/.
15. Benjamini Y, Hochberg Y (1995) Controlling the False Discovery Rate - a Practical and Powerful Approach to Multiple Testing. Journal of the Royal Statistical Society Series B-Methodological 57: 289–300.
16. Shannon P, Markiel A, Ozier O, Baliga NS, Wang JT, et al. (2003) Cytoscape: A software environment for integrated models of biomolecular interaction networks. Genome Research 13: 2498–2504.
17. Guimera R, Nunes Amaral LA (2005) Functional cartography of complex metabolic networks. Nature 433: 895–900.
18. Hsieh MH, Magee CL (2008) An algorithm and metric for network decomposition from similarity matrices: Application to positional analysis. Social Networks 30: 146–158.
19. Bader GD, Hogue CW (2003) An automated method for finding molecular complexes in large protein interaction networks. BMC Bioinformatics 4: 2.
20. Huang DW, Sherman BT, Lempicki RA (2009) Systematic and integrative analysis of large gene lists using DAVID bioinformatics resources. Nature Protocols 4: 44–57.
21. Huang DW, Sherman BT, Lempicki RA (2009) Bioinformatics enrichment tools: paths toward the comprehensive functional analysis of large gene lists. Nucleic Acids Research 37: 1–13.
22. Day RC, Herridge RP, Ambrose BA, Macknight RC (2008) Transcriptome analysis of proliferating Arabidopsis endosperm reveals biological implications for the control of syncytial division, cytokinin signaling, and gene expression regulation. Plant Physiology 148: 1964–1984.
23. Tiwari S, Spielman M, Schulz R, Oakey RJ, Kelsey G, et al. (2010) Transcriptional profiles underlying parent-of-origin effects in seeds of Arabidopsis thaliana. BMC Plant Biol 10: 72.
24. Baena-Gonzalez E, Rolland F, Thevelein JM, Sheen J (2007) A central integrator of transcription networks in plant stress and energy signalling. Nature 448: 938–942.
25. Gonzali S, Loreti E, Solfanelli C, Novi G, Alpi A, et al. (2006) Identification of sugar-modulated genes and evidence for in vivo sugar sensing in Arabidopsis. J Plant Res 119: 115–123.
26. Blasing OE, Gibon Y, Gunther M, Hohne M, Morcuende R, et al. (2005) Sugars and circadian regulation make major contributions to the global regulation of diurnal gene expression in Arabidopsis. Plant Cell 17: 3257–3281.

6

Incorporating Climate Change and Exotic Species into Forecasts of Riparian Forest Distribution

Dana H. Ikeda[1,2*]**, Kevin C. Grady**[2,3]**, Stephen M. Shuster**[1,2]**, Thomas G. Whitham**[1,2]

1 Department of Biological Science, Northern Arizona University, Flagstaff, Arizona, United States of America, **2** Merriam-Powell Center for Environmental Research, Northern Arizona University, Flagstaff, Arizona, United States of America, **3** School of Forestry, Northern Arizona University, Flagstaff, Arizona, United States of America

Abstract

We examined the impact climate change (CC) will have on the availability of climatically suitable habitat for three native and one exotic riparian species. Due to its increasing prevalence in arid regions throughout the western US, we predicted that an exotic species, *Tamarix*, would have the greatest increase in suitable habitat relative to native counterparts under CC. We used an ecological niche model to predict range shifts of *Populus fremontii*, *Salix gooddingii*, *Salix exigua* and *Tamarix*, from present day to 2080s, under five general circulation models and one climate change scenario (A1B). Four major findings emerged. 1) Contrary to our original hypothesis, *P. fremontii* is projected to have the greatest increase in suitable habitat under CC, followed closely by *Tamarix*. 2) Of the native species, *S. gooddingii* and *S. exigua* showed the greatest loss in predicted suitable habitat due to CC. 3) Nearly 80 percent of future *P. fremontii* and *Salix* habitat is predicted to be affected by either CC or *Tamarix* by the 2080s. 4) By the 2080s, 20 percent of *S. gooddingii* habitat is projected to be affected by both *Tamarix* and CC concurrently, followed by *S. exigua* (19 percent) and *P. fremontii* (13 percent). In summary, while climate change alone will negatively impact both native willow species, *Tamarix* is likely to affect a larger portion of all three native species' distributions. We discuss these and other results in the context of prioritizing restoration and conservation efforts to optimize future productivity and biodiversity. As we are accounting for only direct effects of CC and *Tamarix* on native habitat, we present a possible hierarchy of effects- from the direct to the indirect- and discuss the potential for the indirect to outweigh the direct effects. Our results highlight the need to account for simultaneous challenges in the face of CC.

Editor: Bazartseren Boldgiv, National University of Mongolia, Mongolia

Funding: This work was supported by the National Science Foundation (NSF) Integrative Graduate Education Research Traineeship (IGERT), Frontiers in Integrative Biological Research (FIBR) grant DEB-0425908, NSF Macrosystems grant DEB-1340852, the Northern Arizona University TRIF program, and Bureau of Land Management Challenge Cost Share grant L12AC20606. The funders had no role in study design, data collection and analysis, decision to publish, or preparation of the manuscript.

Competing Interests: The authors have declared that no competing interests exist.

* Email: Dana.Ikeda@nau.edu

Introduction

Climate change (CC) is predicted to be one of the leading causal agents of future extinctions, impacting biodiversity and ecosystem functioning worldwide [1], [2]. There is considerable evidence that shifts in species distributions both pole-ward and along elevation clines are already occurring [3–7], with regional die-off of species at the trailing edges of their range [8–11]. In the western U.S., recent climatic conditions have become more arid [12]; resulting in high mortality of numerous foundation species at the trailing edge of their distributions, in ecosystems that span from chaparral to alpine forests [10], [13]. Increasing aridity is expected to continue under CC, with models projecting higher temperatures and an increase in drought events in the next 80 years [12], [14], [15].

Projected increases in temperature and drought will negatively impact riparian ecosystems worldwide that have already experienced extensive modifications over the last century [16–18]. Changes in land use, particularly the alteration of flow regimes by damming [19] and extensive cattle grazing over the last century within riparian corridors, have resulted in a 97 percent decline of pre-20th century riparian habitat in the western U.S. [16]. In the southwestern U.S., riparian ecosystems are dominated by *Populus* and *Salix* species [20], which are considered to be foundation species that structure community composition across multiple trophic levels and influence ecosystem processes such as nutrient cycling [21], [22]. Because riparian tree fitness is influenced by a number of different processes (e.g., temperature, soil water availability, flooding regimes), CC will impact riparian species directly by altering growth, phenology and geographic distributions, and indirectly by altering flood regimes, such as the timing of spring runoff and the magnitude of floods (reviewed in [23]).

Rising temperatures, altered precipitation, and novel disturbances associated with land use also make these riparian ecosystems particularly vulnerable to invasive exotic plant species such as those within the *Tamarix* genus [19], [24], [25]. *Tamarix* spp. (*Tamarix chinensis*, *Tamarix ramosissima* and hybrids [26], hereafter referred to as *Tamarix*) have invaded much of the southwestern U.S., making them the third most frequently occurring woody riparian plant in the region [27]. The presence of *Tamarix* can detrimentally impact native riparian trees when combined with other stressors. In areas where the natural flood regime has been altered, such as along dammed river systems, *Tamarix* has invaded habitat once dominated by native *Populus* and *Salix* species [28], [29]. Once established, *Tamarix* has been shown to outcompete native species due to its ability to tolerate a wider range of salinity and soil moisture contents than its native counterparts [29]. In systems where *Tamarix* forms dense monocultures, *Tamarix* has been shown to alter soil salinity [30], [31], hydrology [32], [33], and change the surrounding floral and faunal communities including mycorrhizal mutualists [29], [34], [35]. As such, we also consider species within the *Tamarix* genus to be exotic foundation species [21] due to their role in redefining riparian communities.

An important step toward understanding the impact of CC on riparian ecosystems in the southwest is to assess how CC and a highly competitive species (*Tamarix*) may combine or act synergistically to further extirpate native species. We used the Maxent modeling algorithm [36] to identify areas of expansion and contraction of three native riparian species (*Populus fremontii, Salix exigua, Salix gooddingii*) and the exotic *Tamarix* under one future climate change scenario and multiple general circulation models (GCMs). Although other studies have created habitat suitability models for *Tamarix* (e.g., [37], [38]), no previous modeling efforts have identified regions where CC is projected to result in 1) native habitat loss and *Tamarix* gain, and 2) potential new range overlaps between natives and *Tamarix*.

Native riparian trees have largely been ignored in research using ecological niche models (ENMs) to assess the impact of CC, especially in respect to altered temperature and precipitation regimes. Projected increases in temperature may exceed the range of tolerance of many native riparian trees [39], [40], (e.g. record temperatures >50°C [39]) resulting in a decline of these stabilizing species, while these same conditions may promote the spread of invasive species [27]. Numerous studies have reported significant effects of temperature and drought on the productivity of riparian species. Grady *et al.* [39] found populations of the riparian trees *P. fremontii* and *S. gooddingii* from warm environments were more productive in a warm common garden than were populations that were from cooler locations, but populations of the riparian shrub *S. exigua* did not perform better based on home site temperature, suggesting local adaptation to temperature in riparian trees but not shrubs. There is conflicting evidence regarding the response of *P. fremontii* and *S. gooddingii* to drought. Some studies predict that *P. fremontii* is more susceptible to drought than *S. gooddingii* [29], [41], whereas others suggest that *S. gooddingii* is less drought-tolerant [42]. Hultine *et al.* [43] found *P. fremontii* to be more sensitive to short-term drought, whereas *S. exigua* had greater sensitivity to long-term drought events.

While it is unclear which native species will be the most susceptible to drought and temperature increases, many studies have found *Tamarix* to be particularly resilient to drought and increased temperature [29], [42], [44] relative to their native counterparts. Together, these studies indicate that climate-induced effects on species' physiological tolerances have the potential to impact their geographic distributions.

In this study we examine effects of CC and exotic species on riparian ecosystems, as quantified by shifts in the geographic distribution of suitable habitat, and test the hypothesis that there will be predictable geographic responses of riparian species to CC. Based on previous studies on the physiological tolerances of native and exotic riparian species to temperature, we predict: 1) *Tamarix* will experience the greatest habitat increase as a result of climate change, with moderate declines of *S. exigua* and extensive habitat reduction of *S. gooddingii*, and *P. fremontii*. This prediction is based on work by Grady et al. [39], which found the productivity of *P. fremontii* to be most sensitive to temperature transfer distance, followed by *S. gooddingii* and *S. exigua*. 2) A greater proportion of future native riparian habitat, relative to current, will be affected by encroachment of *Tamarix* under CC. 3)

Cumulative effects of *Tamarix* invasion and habitat loss due to climate change will be most prevalent at the trailing edges of native species distributions where they are already suffering the greatest abiotic stress. By assessing the impact of CC on the distribution of both native and invasive foundation species, we can begin to address the potential combined effects of CC and invasive species on the viability of threatened riparian species distributions.

Methods

Species Data

We obtained *P. fremontii, S. exigua, S. gooddingii,* and *Tamarix* location points from the Global Biodiversity and Information Facility (GBIF, http://www.gbif.org). Due to the prevalence of hybrids between *T. chinensis* and *T. ramosissima* which are unique to the western United States, we combined occurrences of both species in the U.S. into one *Tamarix* layer [45]. Although there have been questions regarding the accuracy of GBIF data [46], it represents the most comprehensive online database of species location points available. We used an established process to verify locations in the database [47]. Specifically, we removed location points that did not have associated coordinates or were not based on land, and localities where the recorded county did not match with the actual county of collection. A total of 5,758 points were removed in the above process resulting in 739 points for *P. fremontii*, 893 for *S. gooddingii*, 2,092 for *S. exigua* and 1,309 for *Tamarix* that cover their geographic ranges in western North America.

Environmental Variables

Although the species modeled here occupy riparian habitat and are dependent upon environmental variables besides climate, such as flooding, soil salinity, and distance to water [48–50], [37], the focus of this study was to identify areas of climate expansion and contraction across a broad geographic range which could be used to prioritize conservation and restoration efforts. Species within the *Populus* and *Salix* genera are largely dependent upon dynamic river flows and occasional flooding for natural recruitment and survival [51]. However, climate variables typically explain more of the distribution across a large spatial extent than other environmental variables such as soil type and biotic interactions [52]. As such, by including bioclimatic variables to characterize the distribution of these riparian foundation species, we quantify the climatic niche of where *P. fremontii*, *S. exigua* and *S. gooddingii* could occur if flow regimes were suitable.

We used bioclimatic data from the WorldClim database [53] to project the current and future distributions of suitable habitat of all species. To reduce the number of variables included in the final model, we did a preliminary model run for each species and used Maxent's jackknife estimate to examine the permutation importance of the 19 bioclimatic variables (Table 1). Variables with contribution scores <5% were removed from the final model [54]. Variables were averaged for the time periods 1961–1990 and were at a 30 arc-second (~1 km) resolution.

We used a conservative, moderate growth carbon emissions scenario utilized by many scientists [55], A1B [56], and two future time frames, 2050s and 2080s (averaged from 2040 to 2069 and 2070 to 2099, respectively). Because uncertainty in forecasting future climates is mainly attributed to differences in GCMs [57], we used five GCMs that best reflect the current climate of the southwestern US [58]: National Center for Atmospheric Research (NCAR) Community Climate System Model, version 3.0 (CCSM3), Max Planck Institute for Meteorology, Germany (ECHAM5/MPI), Commonwealth Scientific and Industrial Re-

Table 1. The percent contribution and permutation importance (in bold) each bioclimatic variable made to the model building process.

Variable	Species			
	P. fremontii	*S. exigua*	*S. gooddingii*	*Tamarix*
BIO1 (Annual Mean Temperature)	**28.98**	**29.36**	-	-
	18.91	17.66		
BIO3 (Isothermality (BIO2/BIO7))	-	-	**18.87**	**10.04**
			17.68	19.98
BIO4 (Temperature Seasonality (Standard Deviation))	**23.80**	**46.39**	**15.83**	**28.91**
	36.71	61.12	5.96	23.22
BIO5 (Max Temperature of Warmest Month)	**16.17**	-	**4.52**	-
	13.47		9.67	
BIO6 (Min Temperature of Coldest Month)	**18.13**	-	-	**6.64**
	12.12			1.34
BIO7 (Temperature Annual Range (BIO5-BIO6))	-	-	**12.86**	-
			6.22	
BIO10 (Mean Temperature of Warmest Quarter)	-	-	-	**13.62**
				18.05
BIO11 (Mean Temperature of Coldest Quarter)	-	-	-	**17.97**
				2.58
BIO12 (Annual Precipitation)	**5.64**	**3.75**	**9.29**	**17.67**
	7.83	2.67	17.67	31.43
BIO13 (Precipitation of Wettest Month)	-	-	**9.19**	-
			4.32	
BIO14 (Precipitation of Driest Month)	-	-	**13.17**	-
			11.86	
BIO15 (Precipitation Seasonality (Coefficient of Variation))	**7.28**	**5.43**	**16.26**	**5.15**
	10.96	2.32	26.62	3.39
BIO17 (Precipitation of Driest Quarter)	-	**15.07**	-	-
		16.23		
Total Temperature	**87.08**	**75.75**	**52.09**	**77.18**
	81.21	78.78	39.54	65.18
Total Precipitation	**12.92**	**24.25**	**47.91**	**22.82**
	18.79	21.22	60.46	34.82

search Organisation CSIRO, Australia (CSIRO-MK3), Centre National de Recherches Meteorologiques, Meteo France, France (CNRM-CM3), and Hadley Centre for Climate Prediction and Research, Met Office United Kingdom, and Hadley Centre Global Environmental Model, version 1 (HadGEM1). These models were statistically downscaled to a 30 arc-second resolution using the Delta Method [59] and downloaded from Global Climate Model data portal (http://www.ccafs-climate.org/data/).

Model Calibration & Evaluation

We used Maxent version 3.3.3e (http://www.cs.princeton.edu/~schapire/maxent) to forecast potential future shifts in suitable habitat of the four species in response to CC. Maximum entropy (Maxent) is a machine learning algorithm which estimates a species ecological niche by finding the distribution with the maximum entropy subject to constraints by environmental variables chosen *a priori* [36], [60]. Studies have shown Maxent to perform well as compared with other presence-background algorithms [49], [61], [62]. Since the selection of background points in Maxent can influence model performance [63], [64], we restricted the selection of background points to the sampling extent using a Minimum Convex Polygon [65]. Additionally, since the species occurrence data did not come from a random sample, there is the potential for sampling bias in the datasets. To remove potential biases, we followed the methods of Elith et al. [66] to create a Kernel Density Estimator (KDE) surface, rescaled from one to 20, from which to draw the 10,000 random background points. To avoid projecting into environments outside those which the models were trained upon, we specified the 'fade-by-clamping' option in Maxent, which removed heavily clamped pixels from the final model predictions [36].

Output format was specified to a logistic form for easy interpretation. We used a 10-fold cross-validation method, which holds 10% of the data for testing and trains the model on the remaining 90% for 10 iterations, in order to test the predictive performance on withheld data [67]. Model performance was evaluated using partial receiver operator characteristics (pROC) following the methods described in Peterson et al. [68]. We utilized the partial ROC software developed by N. Barve [69] in which we specified 1000 iterations with the omission threshold set at ten percent [70]. T-tests were performed to test for statistical significance of pROC values. Linear, quadratic, product, hinge and threshold functions of predictor variables were employed and variable importance was assessed using a jackknife analysis [36]. A binomial probability test was used to assess the accuracy of each predicted distribution compared to that expected by chance [36]. We evaluated variable importance by examining the permutation importance each bioclimatic variable had on the model building process.

To compare suitable and unsuitable habitats, we specified a 10th percentile training presence threshold (T10), which is a fixed threshold where only the lowest 10 percent of predicted values are rejected [70].

Post-hoc Analyses and Calculation of Species Distribution Overlaps

We performed a number of transformations and analyses on the projected habitat models. First, we used "Raster Calculator" in ArcGIS Desktop 9.3 © [71] to identify areas of habitat loss and expansion by subtracting future habitat layers from the previous time period once a threshold had been applied. Second, to identify areas where *Tamarix* would likely interact with *P. fremontii*, *S. exigua* and *S. gooddingii*, we calculated the projected future overlap between *Tamarix* and the three native species once a threshold had been applied, using "Raster Calculator". Regions where native habitat loss due to CC and future *Tamarix* overlapped were classified as areas where *Tamarix* would have the highest potential to displace *P. fremontii*, *S. exigua* and *S. gooddingii*. To minimize uncertainty pertaining to differences in GCM projections, all calculations using future projected distributions were based on a liberal threshold (T10) agreed upon by four or more GCMs [72].

Results

Model Results

All four species (*P. fremontii*, *S. gooddingii*, *S. exigua*, and *Tamarix*) produced partial ROC values statistically greater than 1.0 (t-test, p<0.0001), with values ranging from 1.169 to 1.484 (Table S1 for supporting information). Binomial probabilities were p<0.0001, suggesting that all models predicting the distribution of suitable habitat of each species provided significant discriminatory ability as compared with random models (Table S1 for supporting information).

Environmental Variable Importance and Habitat Association

For all three native species and *Tamarix* temperature variables had the greatest contribution to the models, as quantified by permutation importance, with the largest effect for *P. fremontii*, *Tamarix* and *S. exigua* (Table 1). The distribution of *S. gooddingii* was influenced by both temperature and precipitation (Table 1).

The response curves indicated that suitable habitat generally decreased with excess precipitation, and increased with increasing temperature, up to a threshold (Fig. 1). Overall, these results indicate that for the riparian species examined here, temperature variables have a larger influence in determining suitable habitat than precipitation.

Impact of Climate Change: Range Contraction and Expansion

Under a liberal threshold (T10), four or more GCMs predict that the distribution of suitable habitat for *Tamarix* is expected to shift a total of 71 percent by the end of the century (Fig. 2, Table 2), with a gain of 62 percent and a loss of 9 percent. Of the native species, suitable habitat for *S. exigua* is projected to be the most stable species through time, with only 39 percent of suitable habitat changing (gain of 9 percent and a loss of 30 percent) (Fig. 2, Table 2). While both *P. fremontii* and *S. gooddingii* are expected to gain suitable habitat, over 62 and 38 percent, respectively, up to 40 percent of the suitable habitat for *S. gooddingii* is projected to be lost between now and 2080s (Fig. 2, Table 2). Together, the three native species are expected to have a net gain of nearly 36 percent of current projected suitable habitat by the 2080s (Fig. 2, Table 2).

Impact of Climate Change in the presence of Tamarix

We found that nearly 140 percent of future projected native riparian habitat (defined as containing a minimum of one of the three native species modeled here) in western North America, relative to current projected habitat, is expected to have *Tamarix* present by 2080 (Fig. 3, Table 3). *Tamarix* invasion is expected to occur across 200 percent of *P. fremontii* and 174 percent of *S. gooddingii* by 2080 (Table 3). By 2080, over 102 percent of current *S. exigua* habitat is expected to have *Tamarix*, with the effect spread out relatively evenly between 2050 and 2080 (Table 3). When combined with the effect of CC, nearly 156 percent of native habitat is predicted to be affected by either factor by 2080 (Fig. 3, Table 3). *P. fremontii* is projected to experience the greatest impact of climate change and *Tamarix* invasion, with 205 percent of habitat being affected by 2080, followed by *S. gooddingii* (194 percent) and *S. exigua* (116 percent) (Fig. 3, Table 3).

Climate change and *Tamarix* will also affect native habitat concurrently. We found that 12 percent of current native riparian habitat may be impacted by both CC and the presence of *Tamarix* by 2080 (Fig. 4, Table 3). Nearly 15 percent of *P. fremontii* and *S. exigua* habitat, and 20 percent of *S. gooddingii* habitat is projected to be affected by both factors by 2080 (Fig. 4, Table 3). Contrary to our original hypothesis, concurrent effects were not constrained to the trailing edges of the distribution of native species, although they were more prevalent at the trailing edge for some species, such as *S. exigua* (Fig. 4).

Discussion

This study demonstrates that CC will likely alter the distribution of three native riparian species in western North America. However, there are two important caveats to address in our study. First, despite statistically significant pROC scores (Table S1 for supporting information), the models developed for *P. fremontii* and *S. exigua* showed a relatively high false negative rate (Fig. 1). This is likely a result of over-fitting in the Maxent algorithm [73], and thus projected changes in both *P. fremontii* and *S. exigua* distributions may be especially conservative. Additionally, all future projections for the species modeled here are based upon only five GCMs and one moderate climate change scenario (SRES A1B). Although our rationale for using the five GCMs chosen is based upon previous research of climatic conditions in the southwestern U.S. [58], we did use a moderate emissions scenario, which is conservative in light of current CO_2 emission rates [74]. Thus, model results presented

P. fremontii

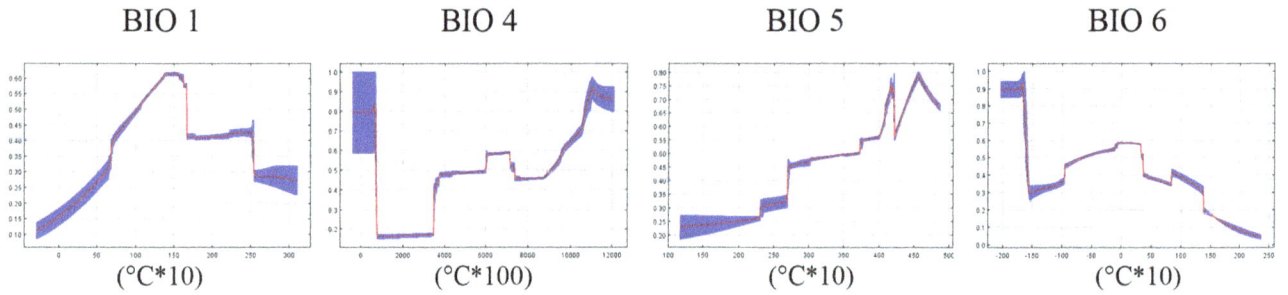

BIO 1	BIO 4	BIO 5	BIO 6
(°C*10)	(°C*100)	(°C*10)	(°C*10)

S. gooddingii

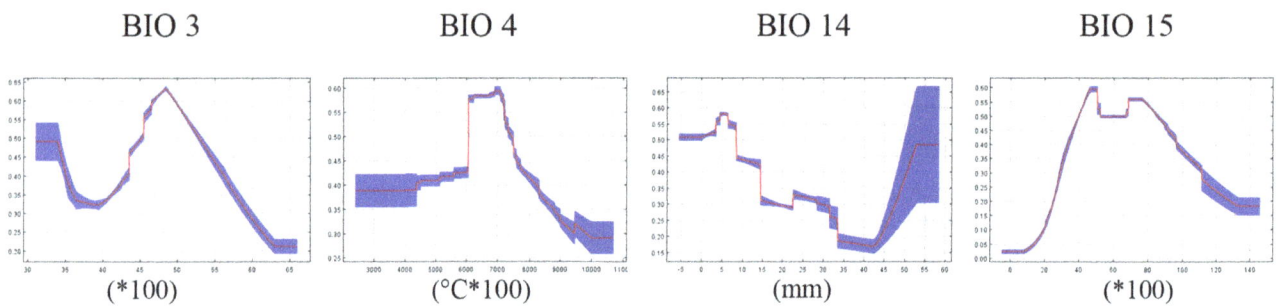

BIO 3	BIO 4	BIO 14	BIO 15
(*100)	(°C*100)	(mm)	(*100)

S. exigua

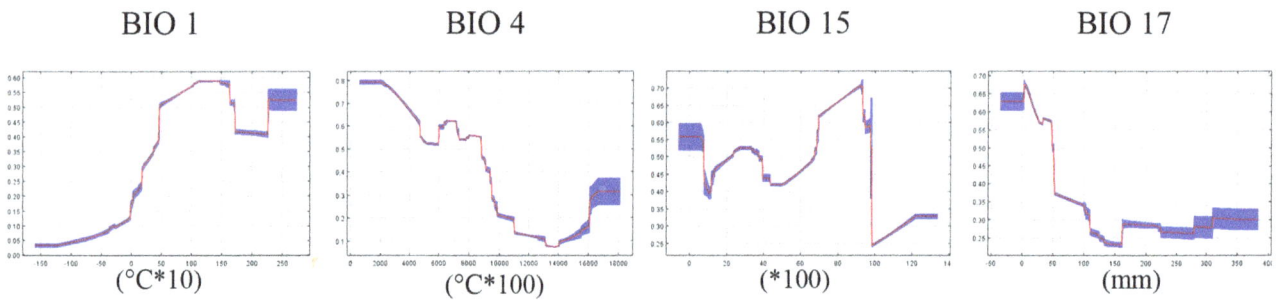

BIO 1	BIO 4	BIO 15	BIO 17
(°C*10)	(°C*100)	(*100)	(mm)

Tamarix

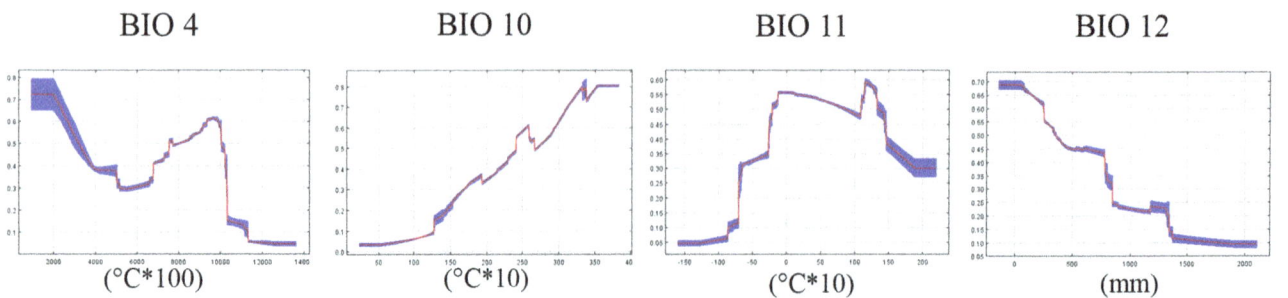

BIO 4	BIO 10	BIO 11	BIO 12
(°C*100)	(°C*10)	(°C*10)	(mm)

Figure 1. Marginal response curves for each climatic variable. Response curves depicting the probability of suitable habitat related to each climatic variable for *Populus fremontii*, *Salix gooddingii*, *Salix exigua* and *Tamarix*. Units of precipitation variables are measured in millimeters (mm), and temperature variables are degrees Celsius times 10 (°C*10). See Table 1 for full variable names.

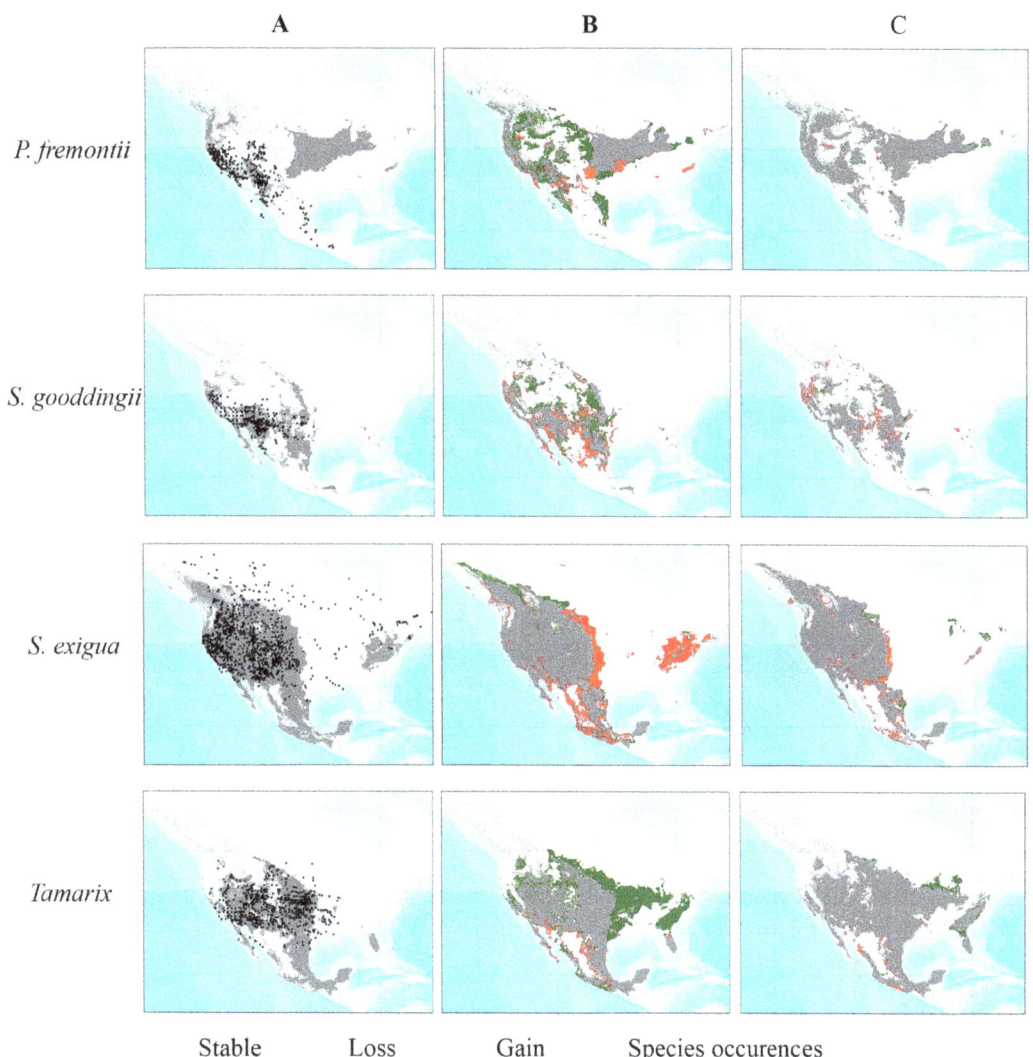

Figure 2. Maxent model outputs showing range expansion and contraction of suitable habitat predicted through time as a result of CC. A) Suitable habitat in current time is shown in grey using a liberal threshold (T10), with the occurrence data overlaid. All occurrence data were downloaded from the Global Biodiversity Information Facility. B-C) Suitable habitat in 2050s and 2080s, predicted by four or more GCMs using a liberal threshold (T10) is also depicted. Areas of loss (red), gain (green), and stable (grey) were calculated by subtracting suitable habitat from the previous time frame.

here are likely conservative estimates of habitat turnover, and actual future climatic conditions may deviate from those used here.

Vulnerability of Native Species to Climate Change

In contrast to our original hypothesis that climate change would have the greatest impact on *P. fremontii*, followed by *S. gooddingii*

Table 2. Predicted percent change (gain and loss) of projected suitable habitat through time using agreement among four or more GCMs and a liberal threshold (T10).

Species	Total % Change Current-2080s	Total % Loss Current-2080s	Total % Gain Current-2080s
P. fremontii	80%	17%	62%
S. exigua	39%	30%	9%
S. gooddingii	77%	40%	38%
Tamarix	70%	9%	62%
Native species	60%	24%	36%

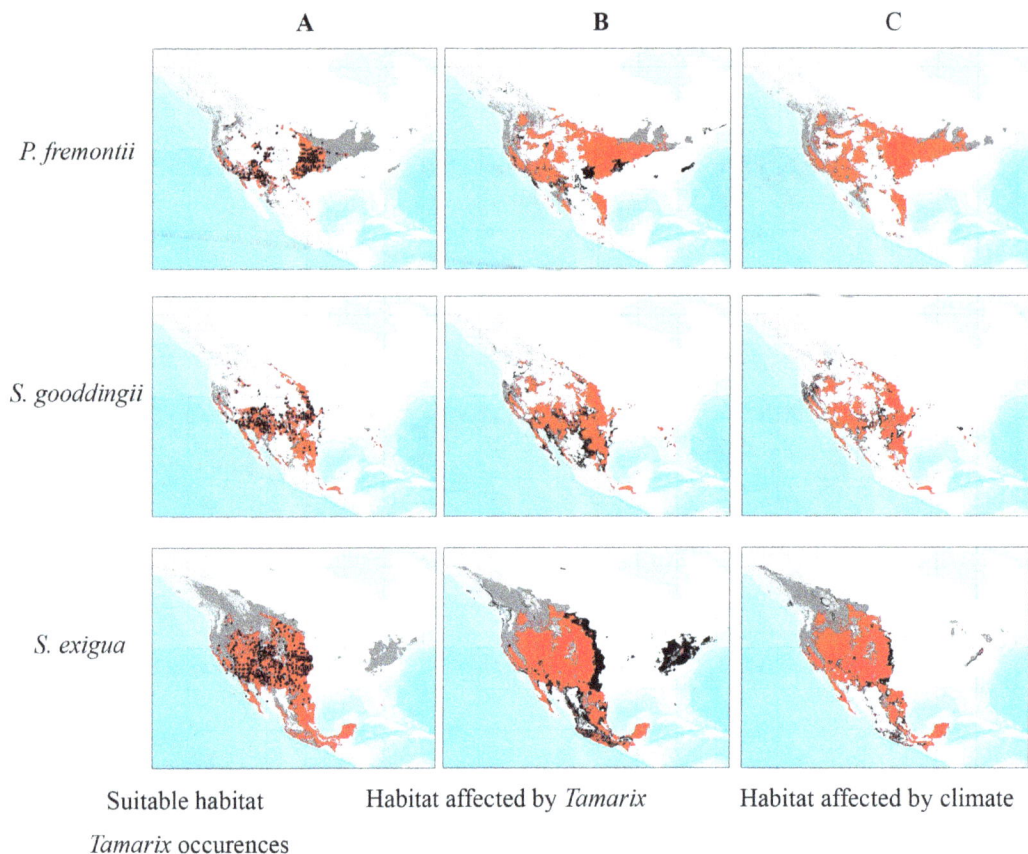

| A | B | C |

P. fremontii

S. gooddingii

S. exigua

Suitable habitat   Habitat affected by *Tamarix*   Habitat affected by climate

Tamarix occurences

Figure 3. Native species distributions affected by CC or *Tamarix*. Projected distributions of the three native species (grey) showing the range predicted to be affected by either *Tamarix* invasion (red) between current and 2080 (A-C) or CC (black) between 2050 and 2080 (B-C).

and *S. exigua*, we found that *S. gooddingii* is projected to experience the greatest loss of suitable habitat due to CC, followed by *S. exigua* and *P. fremontii* (Table 2). These results are in partial agreement to those reported by Grady *et al.* [39], who experimentally demonstrated greater sensitivity in transfer climate distance of *P. fremontii* and *S. gooddingii* relative to *S. exigua*. In our study, the greatest decline in habitat was predicted in the next thirty years, with moderate declines between 2050 and 2080 (Fig. 2). Although it is unclear why the next thirty years would show greater changes than the following 30 years, it may be that a threshold is reached in the first 30 years that triggers major changes. Following a similar pattern with many other species [1], [75], [76], suitable habitat of the three native species is projected to decline along the trailing edges of the species range with little to moderate increases in habitat at the leading edges (Fig. 2).

Two patterns became apparent when examining the bioclimatic variables which correlated with the distribution of these native riparian species. First, variables related to temperature had the greatest permutation importance on the model building process for *P. fremontii*, and *S. exigua* (Table 1), and for both native species increases in temperature correlated with increased habitat suitability (Fig. 1), with some thresholds present. Second, both precipitation and temperature were important for *S. gooddingii*, with suitable habitat generally increasing with temperature up to a threshold (Table 1). However, increases in precipitation led to a decrease in suitable habitat (Fig. 1). Together, these patterns suggest that while variable contribution differs between the native species, drier, warm climates provide more suitable habitat. While this may be surprising, a number of other studies have

documented that water use of riparian species is largely influenced by the availability of groundwater rather than rainfall [77–79].

Tamarix and Climate Change

Current estimates indicate that *Tamarix* occupies an area of 356,241 hectares in the western United States [80]. Our results indicate that the climate niche of *Tamarix* (Fig. 1) is approximately 300 times greater than the realized niche [81], suggesting that there is extensive climatically suitable habitat that this species may yet invade. Additionally, there is substantial overlap in projections of suitable habitat between results generated here and those reported by other studies. Jarnevich *et al.* [37] used 29 climatic, topographic, and geographic variables, including distance to water, to characterize the distribution of *Tamarix*, and found present potential habitat throughout the western US, with distinct patches occurring in Utah and Arizona, along the Pacific crest, regions of Montana and Idaho, and along river systems east of the Rocky Mountain Range. Morisette *et al.* [38] used remote sensing data to examine the current realized niche, and similar to our results, found significant concentration of suitable habitat in the southwestern US, but predicted greater occurrence especially in Texas. The *Tamarix* suitability model developed here predicts suitable habitat in almost all of these regions, despite our use of only five bioclimatic variables, indicating that bioclimatic variables alone can accurately characterize a riparian species niche over broad geographic scales.

The bioclimatic variables describing the niche of *Tamarix* occurrences (Table 1) were consistent with those reported by other

Table 3. The number of cells (km^2) calculated for *P. fremontii*, *S. gooddingii*, *S. exigua*, and all native species together.

	Available habitat current	Available habitat 2050s	Habitat loss 2050s	Habitat Overlap with Tamarix 2050s	Habitat Loss and Tamarix 2050s	Available habitat 2080s	Habitat loss 2080s	Habitat Overlap with Tamarix 2080s	Habitat Loss and Tamarix 2080s	Total % impacted by loss (current-2080s)	Total % impacted by Tamarix (current-2080s)	Total % impacted by either factor (current-2080s)	Total % impacted by both factors (current-2080s)
	a	b	c	d	e	f	g	h	i	((c+g)/a)*100	((e+h)/a)*100	((d+e-f)+(g+h+i))/a)*100	((e+i)/a)*100
P. fremontii	3688174	5E+06	535414	3572125	380305	5E+06	108988	3809569	83554	17%	200%	205%	13%
S. exigua	8806324	7E+06	2072957	4662669	1026756	7E+06	578385	4299096	363179	30%	102%	116%	16%
S. gooddingii	2918525	3E+06	736097	2613738	291126	3E+06	424756	2477806	286337	40%	174%	194%	20%
Native habitat	15413023	2E+07	3344468	10848532	1698187	2E+07	1E+06	10586471	148266	29%	139%	156%	12%

Also shown are the percentages of total habitat impacted by; loss of suitable climatic habitat, *Tamarix*, either loss of suitable climatic habitat or *Tamarix*, and both *Tamarix* and loss concurrently.

studies. We found probability of suitable habitat to increase with mean temperature of the warmest quarter, while the remaining bioclimatic variables concerning precipitation had little influence on habitat suitability (Fig. 1). Jarnevich *et al.* [37] also found mean temperature of the warmest quarter to be a significant predictor of suitable habitat while precipitation variables contributed less than 10 percent.

In support of our original hypothesis, we found that *Tamarix* invasion of riparian habitat is expected to increase with CC (Fig. 2, Table 2). Suitable habitat of *Tamarix* is projected to gain approximately 62 percent of the current suitable habitat by 2080 (Table 2). This finding is contrary to that reported by Bradley *et al.* [24], who predicted *Tamarix* distribution to remain relatively stable through time. The discrepancy is likely due to the fact that Bradley *et al.* [24] used a different modeling algorithm, Mahalanobis distance [82], [83], versus the Maxent algorithm used here. Although no studies directly compare the two modeling approaches, there is evidence to suggest that Maxent performs better than Mahalanobis distance [84]. Some of the habitat gain of *Tamarix* predicted here is projected to occur at latitudes higher than where the species is currently found (Fig. 2). Gaskin and Kazmer [85] found *T. chinensis* and *T. ramosissima* introgression to follow strong latitudinal clines, with a higher occurrence of *T. chinensis* alleles at the southern edge and *T. ramosissima* alleles at the northern edge. This pattern is likely driven by greater frost tolerance in *T. ramosissima* than *T. chinensis* [86]. While projected increases in latitude of *Tamarix* (Fig. 2) should increase the prevalence of *T. ramosissima* alleles, warmer climates could decrease the selection pressure for frost tolerance, resulting in *T. chinensis* alleles becoming more widespread.

Implications for Management and Conservation Efforts

In the context of restoration and conservation efforts, our results highlight three main conclusions. First, mitigation efforts should focus on areas where native habitat is projected to be affected by CC and the presence of *Tamarix* (Fig. 4). We found that 20 percent of the distribution of *S. gooddingii* is projected to be affected by both habitat loss and *Tamarix*, followed by 16 percent of *S. exigua* habitat, and 13 percent of *P. fremontii* habitat (Table 3). Given the projected future anthropogenic impacts on riparian habitat (reviewed in [23]), even relatively small projected effects can have disproportionate impacts. This is especially true when the habitat in question is rare- largely due to anthropogenic factors (e.g., cattle grazing, water diversions, farming, dams) [16]. With such rare habitat types, even relatively low mortality may result in bottlenecks and local extinction. For example, in a survey by Gitlin *et al.* [10], less than two percent of the riparian habitat surveyed had *P. fremontii*, but *P. fremontii* experienced greater than 15 percent mortality. Because riparian areas are predicted to be most susceptible to extirpation of native species and endemic species [87], *Tamarix* removal and restored stream flows, which may alleviate loss from predicted CC and increase native species regeneration [88], [51], should be a priority.

Second, rehabilitation of *Tamarix* altered habitat, involving replanting of native riparian species, should begin in areas where *Tamarix* is expected to decline as a result of CC as reflected in Figure 2. Based on the *Tamarix* response function generated here (Fig. 1), we can expect that *Tamarix* will decrease in regions projected to become wetter. Thus, re-vegetation efforts with *Populus* and *Salix* species may be particularly successful [89], [90], especially when planted in proportions which maximize associated community diversity [91].

Lastly, genetics-based approaches offer a method to restore habitat in suboptimal areas (e.g., in regions with a high likelihood

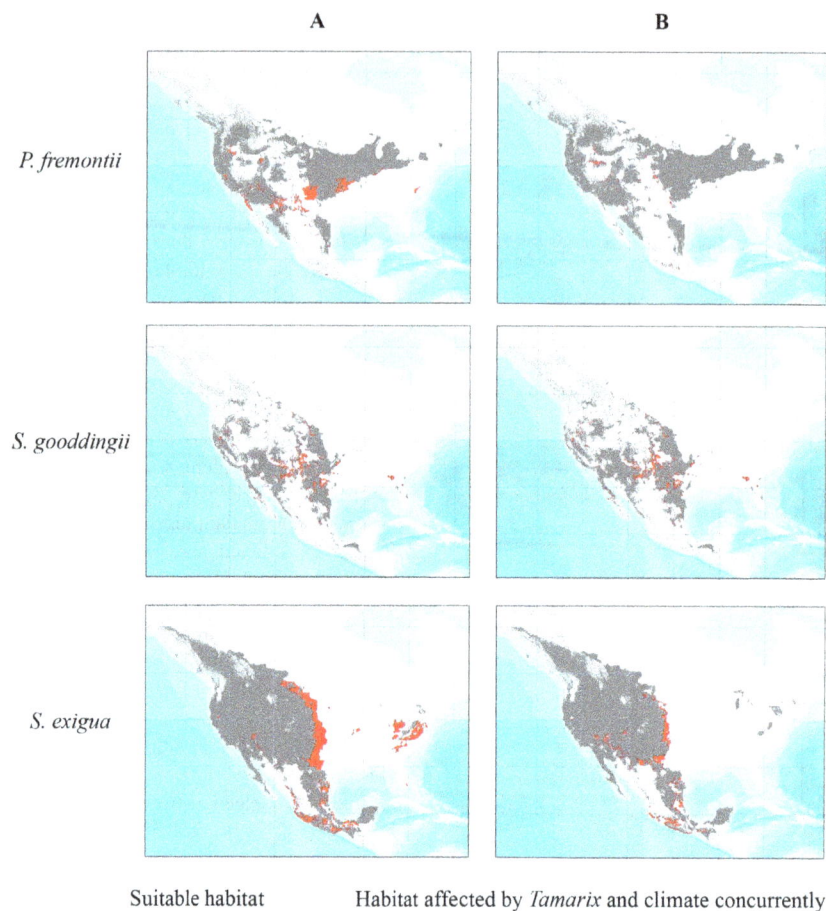

Suitable habitat   Habitat affected by *Tamarix* and climate concurrently

Figure 4. Locations projected to be affected by CC and *Tamarix* concurrently. The projected distributions of three native species depicting the areas with the potential for exotic-by-climate interactions (i.e. the effect of *Tamarix* and habitat loss due to CC) in A) 2050s and B) 2080s.

of excess temperatures, or where *Tamarix* has been present). Planting with source populations known to exhibit increased tolerance to drought and/or higher temperatures, as established through field trials, can increase survival rate and performance [39]. Similarly, managed relocation, based on genetic pre-adaptation, can also be utilized to mitigate the effects of riparian exotics by identifying populations that can better compete with *Tamarix*, as has been shown with other native species and their competing invasives. For example, Goergen *et al.* [92] showed that when native grass species (*Poa secunda* and *Elymus multisetus*) from different source populations were grown with cheatgrass (*Bromus tectorum*), native grasses from source populations of invaded sites better tolerated competition with cheatgrass than native grasses from uninvaded sites. We can take genetics approaches one step further by identifying source populations and genotypes of native species that can simultaneously cope with both CC and invasive species. Such plants may be identified in regions where *Tamarix* occurrences overlap with native species distributions and CC has already resulted in near extirpation of native species [areas with both *Tamarix* occurrences (Fig. 3A) and depicted in black (Fig. 3B)], leaving only genotypes that have persisted in the presence of both *Tamarix* and climate change.

Future Directions: Incorporating Indirect Effects of *Tamarix* and Climate Change on Native Habitat

Although others have reviewed indirect effects of anthropogenic CC on riparian systems in the context of stream flow and human activity (see [23]), there are at least three reasons why *Tamarix* may exert even greater influence on riparian habitats than we predict due to the effects of multiple stressors, community composition and intraspecific variation on the productivity and diversity of native riparian habitat. Thus, our findings may be conservative in light of possible indirect ecological effects between native riparian species, climate and *Tamarix* in the context of CC as follows.

First, *Tamarix* can negatively impact riparian trees already experiencing other stressors. The presence of *Tamarix* seedlings has been shown to be unrelated to *Populus* and *Salix* growth or survival in benign environments [88]; however, *Tamarix* is most abundant along altered streams and rivers where *Populus* recruitment is the lowest [19]. Additionally, Gitlin *et al.* [10] examined the relationship between *Tamarix* density and *P. fremontii* mortality during a severe drought. As *Tamarix* density rose, there was a proportional increase in the mortality of *P. fremontii* ($r^2 = 0.69$) that reached 97 percent when the density of *Tamarix* was approximately 65 percent. Since this mortality relationship was manifested during a record drought, it argues that when *Tamarix* is present, the addition of additional stressor can detrimentally impact a native riparian species. Thus, under CC, we would expect an increase in *Tamarix* due to both an expansion of suitable climatic habitat (as reported here) and projections of escalating human impacts to river systems (reviewed in [23]), and a corresponding rise in mortality of native riparian species due to the presence of *Tamarix* and climate related stress. The combination of these circumstances could act synergistically on *Tamarix* abundance.

Second, *Tamarix* can indirectly impact native riparian habitat by altering the associated community, which can negatively affect both the diversity and productivity of these ecosystems. In *Tamarix* dominated stands, community composition of invertebrates and overall diversity of both invertebrates [93], [94] and birds [95] have been recorded as less than in habitat dominated by native riparian species. Negative effects on biodiversity were also discovered in a stream where aquatic macroinvertebrates were less on *Tamarix* than cottonwood leaf litter [96]. The presence of the tamarisk beetle (*Diorhabda* spp.), which was first released in 2001 as a biocontrol agent for *Tamarix* can also lead to different community assemblages [97]. Furthermore, impacts on the community can occur belowground. Meinhart & Gehring [35] showed that the presence of *Tamarix* altered the arbuscular (AMF) and ectomycorrhizal fungal (EMF) communities of *P. fremontii*, decreasing the colonization rate by half. Since mycorrhizal fungal associations can positively impact the growth of native species [98], and biodiversity promotes ecosystem stability [99], accounting for their presence and structure can further inform how these riparian species and their associated communities might respond to CC.

Lastly, many studies have demonstrated that populations of foundation trees are locally adapted, and intraspecific responses to climate change may be highly variable [77], [100], [101]. Accounting for these population level differences can impact predictions of how species respond to CC. Oney *et al.* [102] compared model results from an ENM (Maxent) and a population-based model (Universal Transfer Functions) for *Pinus contorta* and found that intraspecific genetic variation buffered the species from CC, indicating that ENMs may be under-predicting future suitable habitat. Furthermore, intraspecific genetic variation in productivity will impact associated community members, leading to predictable changes in biodiversity as a result of CC [103]. Although complex, merging the interactions between the indirect effects discussed here and the direct effects reported through our modeling efforts should provide better management strategies to mitigate the effects of both climate change and invasive species on native species and the rich communities they support.

Conclusions

This study is a first step toward integrating multiple environmental challenges (CC and invasive species) to inform conservation and management decisions. Results presented here can help prioritize areas for conservation and restoration by identifying regions where 1) invasive *Tamarix* and CC will jointly impact riparian species in the future, 2) *Tamarix* is expected to increase as a result of climate change, and 3) where the unique evolutionary history of native species is likely to have selected for genotypes that possess high drought tolerance and/or are highly competitive with *Tamarix*. Identifying these superior genotypes may be especially important to utilize in future mitigation efforts. Through provenance trials, we can experimentally identify source populations for use in restoration that are best adapted to future climate, exotics, or climate-by-exotic interactions (e.g., [39], [40]). Together, this information should increase the success of conservation and restoration efforts in the face of multiple stressors.

Acknowledgments

We thank the Cottonwood Ecology group of Northern Arizona University for their valuable feedback in addition to HM Bothwell, MK Lau, and RR Randall. KR Hultine provided insightful comments on an earlier version of this manuscript. We also thank Dr. Bazartseren Boldgiv and four anonymous reviewers for their valuable comments and suggestions.

Author Contributions

Conceived and designed the experiments: DHI KCG. Performed the experiments: DHI. Analyzed the data: DHI. Contributed reagents/materials/analysis tools: DHI KCG SMS TGW. Wrote the paper: DHI KCG TGW SMS.

References

1. Root TL, Price JT, Hall KR, Schneider SH (2003) Fingerprints of global warming on wild animals and plants. Nature 421: 57–60.
2. Walther G-R (2010) Community and ecosystem responses to recent climate change. Philos Trans R Soc Lond B Biol Sci 365: 2019–2024.
3. Parmesan C, Yohe G (2003) A globally coherent fingerprint of climate change impacts across natural systems. Nature 421: 37–42.
4. Walther G-R, Berger S, Sykes MT (2005) An ecological "footprint" of climate change. Proc Biol Sci 272: 1427–1432.
5. Allen CD, Macalady AK, Chenchouni H, Bachelet D, McDowell N, et al. (2010) A global overview of drought and heat-induced tree mortality reveals emerging climate change risks for forests. For Ecol Manage 259: 660–684.
6. Felde VA, Kapfer J, Grytnes J-A (2012) Upward shift in elevational plant species ranges in Sikkilsdalen, central Norway. Ecography 35: 922–932.
7. Bell DM, Bradford JB, Lauenroth WK (2013) Early indicators of change: divergent climate envelopes between tree life stages imply range shifts in the western United States. Glob Ecol Biogeogr 23: 168–180.
8. Williams AP, Allen CD, Macalady AK, Griffin D, Woodhouse CA, et al. (2012) Temperature as a potent driver of regional forest drought stress and tree mortality. Nat Clim Chang 3: 292–297.
9. Breshears DD, Cobb NS, Rich PM, Price KP, Allen CD, et al. (2005) Regional vegetation die-off in response to global-change-type drought. Proc Natl Acad Sci USA 102: 15144–15148.
10. Gitlin AR, Sthultz CM, Bowker MA, Stumpf S, Paxton KL, et al. (2006) Mortality gradients within and among dominant plant populations as barometers of ecosystem change during extreme drought. Conserv Biol 20: 1477–1486.
11. Sánchez-Salguero R, Navarro-Cerrillo RM, Swetnam TW, Zavala MA (2012) Is drought the main decline factor at the rear edge of Europe? The case of southern Iberian pine plantations. For Ecol Manage 271: 158–169.
12. Garfin G, Jardine A, Merideth R, Black M, LeRoy S, editor (2013) Assessment of Climate Change in the Southwest United States: A Report Prepared for the National Climate Assessment. Washington, DC: Island Press.
13. Brusca RC, Wiens JF, Meyer WM, Eble J, Franklin K, et al. (2013) Dramatic response to climate change in the Southwest: Robert Whittaker's 1963 Arizona Mountain plant transect revisited. Ecol Evol 3: 3307–3319.
14. Seager R, Ting M, Held I, Kushnir Y, Lu J, et al. (2007) Model projections of an imminent transition to a more arid climate in southwestern North America. Science 316: 1181–1184.
15. Dominguez F, Rivera ER, Lettenmaier DP, Castro CL (2012) Changes in winter precipitation extremes for the Western United States under a warmer climate as simulated by regional climate models. Geophys Res Lett 39: 1–7.
16. Noss RF, LaRoe III ET, Scott JM (1995) Endangered ecosystems of the United States: a preliminary assessment of loss and degradation. Biological Report 28. U.S. Department of the Interior, National Biological Service, Washington, D.C., USA.
17. Tockner K, Stanford JA (2002) Riverine flood plains: present state and future trends. Environ Conserv 29: 308–330.

18. Davies PM (2010) Climate change implications for river restoration in global biodiversity hotspots. Restor Ecol 18: 261–268.

19. Merritt DM, Poff NL (2010) Shifting dominance of riparian *Populus* and *Tamarix* along gradients of flow alteration in western North American rivers. Ecol Appl 20: 135–152.

20. Durst SL, Theimer TC, Paxton EH, Sogge MK (2008) Temporal variation in the arthropod community of desert riparian habitats with varying amounts of saltcedar (*Tamarix ramosissima*). J Arid Environ 72: 1644–1653.

21. Ellison AM, Bank MS, Clinton BD, Colburn EA, Elliott K, et al. (2005) Loss of foundation species: consequences for the structure and dynamics of forested ecosystems. Front Ecol Environ 3: 479–486.

22. Whitham TG, Bailey JK, Schweitzer JA, Shuster SM, Bangert RK, et al. (2006) A framework for community and ecosystem genetics: from genes to ecosystems. Nat Rev Genet 7: 510–523.

23. Perry LG, Andersen DC, Reynolds LV, Nelson SM, Shafroth PB (2012) Vulnerability of riparian ecosystems to elevated CO_2 and climate change in arid and semiarid western North America. Glob Chang Biol 18: 821–842.

24. Bradley BA, Oppenheimer M, Wilcove DS (2009) Climate change and plant invasions: restoration opportunities ahead? Glob Chang Biol 15: 1511–1521.

25. Walther G-R, Roques A, Hulme PE, Sykes MT, Pysek P, et al. (2009) Alien species in a warmer world: risks and opportunities. Trends Ecol Evol 24: 686–693.

26. Gaskin JF, Schaal BA (2002) Hybrid *Tamarix* widespread in U.S. invasion and undetected in native Asian range. Proc Natl Acad Sci 99: 11256–11259.

27. Friedman JM, Auble GT, Shafroth PB, Scott ML, Merigliano MF, et al. (2005) Dominance of non-native riparian trees in western USA. Biol Invasions 7: 747–751.

28. Howe WH, Knopf FL (1991) On the imminent decline of Rio Grande cottonwoods in central New Mexico. Southwest Nat 36: 218–224.

29. Busch DE, Smith SD (1995) Mechanisms associated with decline of woody species in riparian ecosystems of the southwestern US. Ecol Monogr 65: 347–370.

30. Taylor JP, Wester DB, Smith LM (1999) Soil disturbance, flood management, and riparian woody plant establishment in the Rio Grande floodplain. Wetlands 19: 372–382.

31. Glenn EP, Morino K, Nagler PL, Murray RS, Pearlstein S, et al. (2012) Roles of saltcedar (*Tamarix* spp.) and capillary rise in salinizing a non-flooding terrace on a flow-regulated desert river. J Arid Environ 79: 56–65.

32. Graf WL (1978) Fluvial adjustments to the spread of tamarisk in the Colorado Plateau region. Geol Soc Am Bull 89: 1491–1501.

33. Hultine KR, Bush SE (2011) Ecohydrological consequences of non-native riparian vegetation in the southwestern United States: A review from an ecophysiological perspective. Water Resour Res W07542, doi:10.1029/2010WR010317.

34. Ellis LM, Crawford CS, Molles MC (1997) Rodent communities in native and exotic riparian vegetation in the middle Rio Grande Valley of central New Mexico. Southwest Nat 42: 13–19.

35. Meinhardt KA, Gehring CA (2012) Disrupting mycorrhizal mutualisms: a potential mechanism by which exotic tamarisk outcompetes native cottonwoods. Ecol Appl 22: 532–549.

36. Phillips S, Anderson R, Schapire R (2006) Maximum entropy modeling of species geographic distributions. Ecol Modell 190: 231–259.

37. Jarnevich CS, Evangelista PH, Stohlgren TJ, Morisette JT (2011) Improving national-scale invasion maps: Tamarisk in the western United States. West North Am Nat 71: 164–175.

38. Morisette JT, Jarnevich CS, Ullah A, Cai W, Pedelty, Jeffrey A, et al. (2006) A tamarisk habitat suitability map for the continental United States. Front Ecol Environ 4: 11–17.

39. Grady KC, Ferrier SM, Kolb TE, Hart SC, Allan GJ, et al. (2011) Genetic variation in productivity of foundation riparian species at the edge of their distribution: implications for restoration and assisted migration in a warming climate. Glob Chang Biol: 1–36.

40. Grady KC, Laughlin DC, Ferrier SM, Kolb TE, Hart SC, et al. (2013) Conservative leaf economic traits correlate with fast growth of genotypes of a foundation riparian species near the thermal maximum extent of its geographic range. Funct Ecol 27: 428–438.

41. Stella JC, Battles JJ (2010) How do riparian woody seedlings survive seasonal drought? Oecologia 164: 579–590.

42. Horton JL, Kolb TE, Hart SC (2001) Physiological response to groundwater depth varies among species and with river flow regulation. Ecol Appl 11: 1046–1059.

43. Hultine KR, Bush SE, Ehleringer JR (2010) Ecophysiology of riparian cottonwood and willow before, during, and after two years of soil water removal. Ecol Appl 20: 347–361.

44. Pockman WT, Sperry JS (2000) Vulnerability to xylem cavitation and the distribution of Sonoran Desert vegetation. Am J Bot 87: 1287–1299.

45. Gaskin JF, Shafroth PB (2005) Hybridization of *Tamarix ramosissima* and *T. chinensis* (Saltcedars) with *T. aphylla* (Athel) (Tamaricaceae) in the southwestern USA determined from DNA sequence data. Madroño 52: 1–10.

46. Yesson C, Brewer PW, Sutton T, Caithness N, Pahwa JS, et al. (2007) How global is the global biodiversity information facility? PLoS One 2: e1124.

47. Ramirez-Villegas J, Jarvis A, Touval J (2012) Analysis of threats to South American flora and its implications for conservation. J Nat Conserv 20: 337–348.

48. Evangelista PH, Kumar S, Stohlgren TJ, Jarnevich CS, Crall AW, et al. (2008) Modelling invasion for a habitat generalist and a specialist plant species. Divers Distrib 14: 808–817.

49. Evangelista PH, Stohlgren TJ, Morisette JT, Kumar S (2009) Mapping invasive Tamarisk (*Tamarix*): a comparison of single-scene and time-series analyses of remotely sensed data. Remote Sens 1: 519–533.

50. Kerns BK, Naylor BJ, Buonopane M, Parks CG, Rogers B (2009) Modeling Tamarisk (*Tamarix* spp.) habitat and climate change effects in the northwestern United States. Invasive Plant Sci Manag 2: 200–215.

51. Rood SB, Samuelson GM, Braatne JH, Gourley CR, Hughes FM, et al. (2005) Managing river flows to restore floodplain forests. Front Ecol Environ 3: 193–201.

52. Pearson RG, Dawson TP (2003) Predicting the impacts of climate change on the distribution of species: are bioclimate envelope models useful? Glob Ecol Biogeogr 12: 361–371.

53. Hijmans RJ, Cameron SE, Parra JL, Jones PG, Jarvis A (2005) Very high resolution interpolated climate surfaces for global land areas. Int J Climatol 25: 1965–1978.

54. Sahlean TC, Gherghel I, Papeş M, Strugariu A, Zamfirescu ŞR (2014) Refining climate change projections for organisms with low dispersal abilities: a case study of the Caspian whip snake. PLoS One 9: e91994.

55. Kim HY, Ko J, Kang S, Tenhunen J (2013) Impacts of climate change on paddy rice yield in a temperate climate. Glob Chang Biol 19: 548–562.

56. IPCC (2007) Climate Change 2007: The physical science basis. Cambridge University Press, Cambridge, UK, p. 996.

57. Buisson L, Thuiller W, Casajus N, Lek S, Grenouillet G (2010) Uncertainty in ensemble forecasting of species distribution. Glob Chang Biol 16: 1145–1157.

58. Garfin GM, Eischeid JK, Lenart MT, Cole KL, Ironside K, et al. (2010) Downscaling climate projections in topographically diverse landscapes of the Colorado Plateau in the arid southwestern United States. The Colorado Plateau IV; Shaping Conservation Through Science and Management. pp. 22–43.

59. Ramirez-Villegas J, Jarvis A (2010) Downscaling global circulation model outputs: the Delta method decision and policy analysis working paper No. 1. Cent Trop Agric Colomb.

60. Phillips SJ, Dudik M (2008) Modeling of species distributions with Maxent: new extensions and a comprehensive evaluation. Ecography 31: 161–175.

61. Elith J, Graham CH, Anderson RP, Dudík M, Ferrier S, et al. (2006) Novel methods improve prediction of species' distributions from occurrence data. Ecography 29: 129–151.

62. Kumar S, Spaulding SA, Stohlgren TJ, Hermann KA, Schmidt TS, et al. (2009) Potential habitat distribution for the freshwater diatom *Didymosphenia geminata* in the continental US. Front Ecol Environ 7: 415–420.

63. Phillips SJ, Dudík M, Elith J, Graham CH, Lehmann A, et al. (2009) Sample selection bias and presence-only distribution models: implications for background and pseudo-absence data. Ecol Appl 19: 181–197.

64. VanDerWal J, Shoo LP, Williams SE (2009) New approaches to understanding late Quaternary climate fluctuations and refugial dynamics in Australian wet tropical rain forests. J Biogeogr 36: 291–301.

65. Flory AR, Kumar S, Stohlgren TJ, Cryan PM (2012) Environmental conditions associated with bat white-nose syndrome mortality in the north-eastern United States. J Appl Ecol 49: 680–689.

66. Elith J, Kearney M, Phillips S (2010) The art of modelling range-shifting species. Methods Ecol Evol 1: 330–342.

67. Elith J, Phillips SJ, Hastie T, Dudík M, Chee YE, et al. (2011) A statistical explanation of MaxEnt for ecologists. Divers Distrib 17: 43–57.

68. Peterson AT, Papeş M, Soberón J (2007) Rethinking receiver operating characteristic analysis applications in ecological niche modeling. Ecol Modell 213: 63–72.

69. Barve N (2008) Tool for Partial-ROC (Biodiversity Institute, Lawrence, KS), ver 1.0.

70. Pearson RG, Thuiller W, Araújo MB, Martinez-Meyer E, Brotons L, et al. (2006) Model-based uncertainty in species range prediction. J Biogeogr 33: 1704–1711.

71. ESRI 2009. ArcGIS Desktop: Release 9.3. Redlands, CA: Environmental Systems Research Institute.

72. Araújo MB, New M (2007) Ensemble forecasting of species distributions. Trends Ecol Evol 22: 42–47.

73. Ochoa-Ochoa L, Urbina-Cardona JN, Vázquez L-B, Flores-Villela O, Bezaury-Creel J (2009) The effects of governmental protected areas and social initiatives for land protection on the conservation of Mexican amphibians. PLoS One 4: e6878.

74. Raupach MR, Marland G, Ciais P, Quere C Le, Canadell JG, et al. (2007) Global and regional drivers of accelerating CO2 emissions. Proc Natl Acad Sci 104: 10288–10293.

75. Thomas CD, Franco AMA, Hill JK (2006) Range retractions and extinction in the face of climate warming. Trends Ecol Evol 21: 415–416.

76. Aitken SN, Yeaman S, Holliday JA, Wang T, Curtis-McLane S (2008) Adaptation, migration or extirpation: climate change outcomes for tree populations. Evol Appl 1: 95–111.

77. Kolb TE, Hart SC, Amundson R (1997) Boxelder water sources and physiology at perennial and ephemeral stream sites in Arizona. Tree Physiol 17: 151–160.

78. Snyder KA, Williams DG (2000) Water sources used by riparian trees varies among stream types on the San Pedro River, Arizona. Agric For Meterology 105: 227–240.

79. Cox G, Fisher DG, Hart SC, Whitham TG (2005) Nonresponse of native cottonwood trees to water additions during summer drought. West North Am Nat 65: 175–185.

80. Sher A, Quigley MF (2013) *Tamarix*: A case study of ecological change in the American west. New York: Oxford University Press. 31 p.

81. Hutchinson GE (1959) Homage to Santa Rosalia or why are there so many kinds of animals? Am Nat 93: 145–159.

82. Farber O, Kadmon R (2003) Assessment of alternative approaches for bioclimatic modeling with special emphasis on the Mahalanobis distance. Ecol Modell 160: 115–130.

83. Tsoar A, Allouche O, Steinitz O, Rotem D, Kadmon R (2007) A comparative evaluation of presence-only methods for modelling species distribution. Divers Distrib 13: 397–405.

84. Jeschke JM, Strayer DL (2008) Usefulness of bioclimatic models for studying climate change and invasive species. New York Acad Sci 24: 1–24.

85. Gaskin JF, Kazmer DJ (2008) Introgression between invasive saltcedars (*Tamarix chinensis* and *T. ramosissima*) in the USA. Biol Invasions 11: 1121–1130.

86. Friedman JM, Roelle JE, Gaskin JF, Pepper AE, Manhart JR (2008) Latitudinal variation in cold hardiness in introduced *Tamarix* and native *Populus*. Evol Appl 1: 598–607.

87. Richardson DM, Holmes PM, Esler KJ, Galatowitsch SM, Stromberg JC, et al. (2007) Riparian vegetation: degradation, alien plant invasions, and restoration prospects. Divers Distrib 13: 126–139.

88. Sher AA, Marshall DL, Taylor JP (2002) Establishment patterns of native *Populus* and *Salix* in the presence of invasive nonnative *Tamarix*. Ecol Appl 12: 760–772.

89. Harms RS, Hiebert RD (2006) Vegetation response following invasive Tamarisk (*Tamarix* spp.) removal and implications for riparian restoration. Restor Ecol 14: 461–472.

90. Bay RF, Sher A (2008) Success of active revegetation after *Tamarix* removal in riparian ecosystems of the southwestern United States: a quantitative assessment of past restoration projects. Restor Ecol 16: 113–128.

91. Bangert R, Ferrier SM, Evans L, Kennedy K, Grady KC, et al. (2013) The proportion of three foundation plant species and their genotypes influence an arthropod community: restoration implications for the endangered Southwestern Willow Flycatcher. Restor Ecol 21: 447–456.

92. Goergen EM, Leger EA, Espeland EK (2011) Native perennial grasses show evolutionary response to *Bromus tectorum* (cheatgrass) invasion. PLoS One 6: e18145.

93. Pendleton RL, Pendleton BK, Finch D (2011) Displacement of native riparian shrubs by woody exotics: effects on arthropod and pollinator community composition. Nat Resour Env Iss 1: 1–12.

94. Anderson B, Russell PE, Ohmart R.D (2004) Riparian revegetation: an account of two decades of experience in the arid southwest. Blythe, California: Avvar Books

95. Brand AL, White GC, Noon BR (2008) Factors influencing species richness and community composition of breeding birds in a desert riparian corridor. Condor 110.

96. Bailey JK, Schweitzer JA, Whitham TG (2001) Salt cedar negatively affects biodiversity of aquatic macroinvertebrates. Wetlands 21: 442–447.

97. Strudley S (2009) Impacts of tamarisk biocontrol (*Diorhabda elongata*) on the trophic dynamics of terrestrial insects in monotypic tamarisk stands. Masters of Science thesis. University of Denver, CO.

98. Smith SE, Read DJ (2008) Mycorrhizal symbiosis. Third edition. Academic Press, New York, New York, USA.

99. Tilman D, Downing JA (1994) Biodiversity and stability in grasslands. Nature 367: 363–365.

100. O'Neill GA, Hamann A, Wang T (2008) Accounting for population variation improves estimates of the impact of climate change on species' growth and distribution. J Appl Ecol 45: 1040–1049.

101. Wang T, O'Neill GA, Aitken SN (2010) Integrating environmental and genetic effects to predict responses of tree populations to climate. Ecol Appl 20: 153–163.

102. Oney B, Reineking B, O'Neill G, Kreyling J (2013) Intraspecific variation buffers projected climate change impacts on *Pinus contorta*. Ecol Evol 3: 437–449.

103. Ikeda DH, Bothwell HM, Lau MK, O'Neill GA, Grady KC, et al. (2013) A genetics-based Universal Community Transfer Function for predicting the impacts of climate change on future communities. Funct Ecol 28: 65–74.

7

Molecular and Morphological Evidence Reveals a New Species in the *Phyllomedusa hypochondrialis* Group (Hylidae, Phyllomedusinae) from the Atlantic Forest of the Highlands of Southern Brazil

Daniel P. Bruschi[1]*, **Elaine M. Lucas**[2], **Paulo C. A. Garcia**[3], **Shirlei M. Recco-Pimentel**[1]

1 Departamento de Biologia Estrutural e Funcional, Instituto de Biologia, Universidade Estadual de Campinas - UNICAMP, Campinas, São Paulo, Brazil, **2** Área de Ciências Exatas e Ambientais/Mestrado em Ciências Ambientais, Universidade Comunitária da Região de Chapecó - UNOCHAPECÓ, Chapecó, Santa Catarina, Brazil, **3** Departamento de Zoologia, Instituto de Ciências Biológicas, Universidade Federal de Minas Gerais - UFMG, Belo Horizonte, Minas Gerais, Brazil

Abstract

The taxonomic status of a disjunctive population of *Phyllomedusa* from southern Brazil was diagnosed using molecular, chromosomal, and morphological approaches, which resulted in the recognition of a new species of the *P. hypochondrialis* group. Here, we describe *P. rustica* sp. n. from the Atlantic Forest biome, found in natural highland grassland formations on a plateau in the south of Brazil. Phylogenetic inferences placed *P. rustica* sp. n. in a subclade that includes *P. rhodei* + all the highland species of the clade. Chromosomal morphology is conservative, supporting the inference of homologies among the karyotypes of the species of this genus. *Phyllomedusa rustica* is apparently restricted to its type-locality, and we discuss the potential impact on the strategies applied to the conservation of the natural grassland formations found within the Brazilian Atlantic Forest biome in southern Brazil. We suggest that conservation strategies should be modified to guarantee the preservation of this species.

Editor: Matthias Stöck, Leibniz-Institute of Freshwater Ecology and Inland Fisheries, Germany

Funding: This research was funded by Fundação de Amparo à Pesquisa do Estado de São Paulo (FAPESP; grants 2010/11300-7 and 2010/17464-1), Conselho Nacional de Desenvolvimento Científico e Tecnológico (CNPq) and Coordenadoria de Aperfeiçoamento de Pessoal de Nível Superior (CAPES) for DPB scholarships. The funders had no role in study design, data collection and analysis, decision to publish, or preparation of the manuscript.

Competing Interests: The authors have declared that no competing interests exist.

* Email: daniel_bruschi@hotmail.com

Introduction

The genus *Phyllomedusa* Wagler, 1930 (Anura, Hylidae, Phyllomedusinae) is endemic to the Neotropical region and is currently composed of 30 recognized species [1]. Panama represents the northernmost extreme of the geographic range of this genus, while Argentina and Uruguay constitute its southern limit; species is also being found throughout Colombia, east of the Andes, and in Trinidad [1].

The most recent hypothesis on the phylogenetic relationships of the Phyllomedusinae was presented by Faivovich et al. [2], who recognized the four phenetic groups assigned in the genus as monophyletic groups: the *P. hypochondrialis* group [3], *P. tarsius* group [4], *P. burmeisteri* group [5] and *P. perinesos* group [6]. Nevertheless, some members of this genus have yet to be assigned to a species group [1,2].

A complex taxonomic scenario has been noted in most species of the *P. hypochondrialis* group [2,7,8]. Molecular inferences have revealed two well-supported clades within this group. One subclade includes *P. palliata*, *P. azurea*, *P. hypochondrialis* and *P. nordestina*, while the second comprise *P. rhodei*, *P. ayeaye*, *P. centralis*, *P. megacephala* and *P. oreades* [2].

The use of both morphological and molecular methods is an established and effective approach for the identification of cryptic

biodiversity and the clarification of taxonomic uncertainties (e.g. see references [9–10]). Speciation may not always be accompanied by morphological changes [11–12], in which case the recognition of species might be hampered by the absence of discrete phenotypic traits. The recognition of hidden biodiversity is fundamental for conservation efforts [13], especially the identification of new taxonomic units with small geographic ranges that potentially represent evolutionarily vulnerable lineages [13].

Lucas et al. [14] recorded the presence of a population of *Phyllomedusa* morphologically similar to *P. azurea* in natural grassland formations on a plateau in the highlands of southern Brazil (municipality of Água Doce in Santa Catarina state), but its taxonomic status was unclear. Despite the similar morphology, the authors detected some differences in coloration in comparison with the formal diagnosis of *P. azurea* [3]. Furthermore, this population was found in an unusual location distinct from the savanna formations in which *P. azurea* is known to occur. Bruschi et al. [8] evaluated the taxonomic status of the populations assigned to *P. hypochondrialis* and *P. azurea* from a number of Brazilian regions and included tissue of two specimens from Água Doce. Interestingly, these authors found that the Água Doce population was paraphyletic in relation to the other haplotypes of the *P. azurea* clade, indicating that a more robust analysis, based on a larger number of characters (morphological and genetic) would be needed to identify the true taxonomic status of this population.

In this study, we used morphological, chromosomal and molecular phylogenetic approaches to investigate the taxonomic status of a distinct population of *Phyllomedusa* from southern Brazil. We describe a new species of the genus and infer its phylogenetic relationship as well as present chromosomal data.

Materials and Methods

Population sampling

The individuals examined were collected under authorization number 14468-1/14468-4 issued by SISBIO/Instituto Chico Mendes de Conservação da Biodiversidade. All tissue samples were extracted from euthanized specimens using anesthetic application to the skin (5% Lidocaine) to minimize animal suffering, according to recommendations of the Herpetological Animal Care and Use Committee (HACC) of the American Society of Ichthyologists and Herpetologists (available at:http//www.asih.org/publications), and approved by SISBIO/Institute Chico Mendes de Conservação da Biodiversidade as a condition for the concession license. Specimens were collected in the municipality of Água Doce, Santa Catarina state, southern Brazil (26°35′59.9″S; 51°34′39.4″W; 1330 m above sea level) (Figure 1). Three adult males were collected on 8 January 2009 and 10 adult males on 5 January 2012. These frogs were found in two ponds subjected to anthropogenic impacts; these were separated by an unpaved road (the SC-452 highway). The area is within the Atlantic Forest biome and contains a mosaic of *Araucaria* forests and patches of natural grassland, with distinct swampy areas and patches of mixed rainforest [15]. The natural grassland formations of the highland plateaus of southern Brazil plateau are relics of the drier and colder climates of the Pleistocene [15–17], which were mostly replaced by the subsequent expansion of *Araucaria* forest in the Holocene [18]. Água Doce municipality is located in the transitional zone of the high elevation grasslands known as the Campos de Palmas.

The regional climate is classified as Cfb (temperate) according to the Köppen-Geiger system. Mean annual temperature in this region is 10°C, with mean annual precipitation of 1500–200 mm [19]. The vegetation is predominantly herbaceous at the ponds where the specimens were collected, and composed primarily of plants of the Asteraceae, Cyperaceae and Poaceae families.

Isolation, amplification, and sequencing of DNA

Genomic DNA was extracted from liver or muscle tissue and stored at −70°C in the tissue bank of the Departamento de Biologia Estrutural e Funcional of the Universidade Estadual de Campinas (UNICAMP), in São Paulo state, Brazil, using the TNES method, as applied by Bruschi et al. (2012). The mitochondrial 12S rDNA, tRNA-Val, and 16S ribosomal genes were amplified using the primers MVZ 59(L), MVZ 50(H), 12L13, Titus I (H), Hedges16L2a, Hedges16H10, 16Sar-L, and 16Sbr-H (for primer sequences, see reference [20]). The amplified PCR products were purified using Exonuclease I (10 units) and SAP (1 unit), with a 45-min incubation at 37°C and a 10-min denaturation at 85°C, then used directly as templates for sequencing in an automatic ABI/Prism DNA sequencer (Applied Biosystems, Foster City, CA, USA) with the BigDye Terminator kit (Applied Biosystems, Foster City, CA, USA), as recommended by the manufacturer. The DNA samples were sequenced bi-directionally and edited in Bioedit version 7.0.1 (http://www.mbio.ncsu.edu/BioEdit/bioedit.html) [21].

Analysis of the molecular data

The initial sequence alignment was conducted for each gene separately, using CLUSTALW [22] in Bioedit, version 7.0.1 (http://www.mbio.ncsu.edu/BioEdit/bioedit.html). For each gene, the initial alignment was evaluated using four different gap penalties (5, 10, 15 and 20), and gap length was maintained constant (0.60) to identify regions of ambiguous homology [23]. The regions presenting ambiguous homologies were excluded for our phylogenetic inferences.

The phylogenetic relationships among the species were inferred from the concatenated matrix of the mitochondrial DNA 12S, tRNAval, and 16S rDNA sequences. We selected *Phyllomedusa* sequences available in the GenBank database, encompassing 90% of the species currently recognized for this genus [1]. The outgroup was *Agalychnis granulosa*, which was chosen based on the topology reported by Faivovich et al. [2]. A complete description of the species, sequences and GenBank accession numbers is provided in the Supporting Information (Table S1). Phylogenetic trees were constructed using Bayesian inference and the Maximum Parsimony method. Bayesian inference was based on a Markov chain Monte Carlo (MCMC) analysis conducted in MrBayes 3.1.2 [24] with two independent runs, each with four chains and sampling every 1000 generations for 6 million generations. An adequate burn-in (the first 25% trees were excluded) was determined by examining a plot of the likelihood scores of the heated chain for convergence and stationarity. The evolutionary model most appropriate for each gene was selected by MrMODELTEST [25] using the Akaike Information Criterion (AIC). The trees were sampled every 100 generations, excluding the first 25% of the trees as burn-in, determined by examining a plot of the likelihood scores the heated chain for convergence and stationarity. Tracer software version 1.5 [26] was used to confirm the quality of the parameters of the Bayesian inferences.

The Maximum Parsimony criterion was implemented in TNT v1.1 software [27] using a heuristic search method with tree bisection-reconnection (TBR) swapping and 100 random additional replicates. The bootstrap values of the branches inferred in this analysis were calculated with 1000 non-parametric pseudoreplicates.

Finally, the number of base substitutions per site among the sequences of the species of the *P. hypochondrialis* group was calculated using the maximum composite likelihood model [28] implemented in MEGA5 [29]. Gaps and missing data were eliminated in this analysis, and all parameters were at the default settings.

Cytogenetic analysis

Ten male individuals were studied by cytogenetic methods (UFMG 13353–13362). Metaphase cells were obtained from the intestines and testes of animals that were previously treated with 2% colchicine, following procedures modified from King and Rofe [30] and Schmid [31]. Prior to the removal of the intestine and testes, the animals were deeply anesthetized following the recommendations of the Herpetological Animal Care and Use Committee (HACC) of the American Society of Ichthyologists and Herpetologists.

Cell suspensions were dripped onto clean slides and stored at −20°C. The chromosomes were stained with 10% Giemsa and silver stained using the Ag–NOR method according to Howel and Black [32], in addition to being C-banded as in Sumner [33], with some modifications. To better visualize the heterochromatin, the chromosomes were stained with the fluorochromes AT-specific DAPI (4′, 6-diamidino-2-phenylindole) and GC-specific Mytramicim (MM) sequentially C-banding. To confirm their number

Figure 1. Map of Brazil showing the Atlantic Forest domain (green area). Note the type locality of *Phyllomedusa rustica* sp. n. (square symbol) in the Água Doce municipality, Santa Catarina state.

and positions, the rDNA sites was detected using fluorescence *in situ* hybridization (FISH) as in Viegas-Péquignot [34] using the HM123 probe [35]. The metaphases were photographed under an Olympus microscope and analyzed using Image Pro-Plus software, version 4 (Media Cybernetics, Bethesda, MD, USA). The chromosomes were measured, and the centromere index (CI), relative length (RL), and centromere ratio (CR) were estimated. The chromosomes were ranked and classified according to the scheme of Green and Sessions [36].

Morphological Analyses

The specimens examined for the description of the new species, including those of other taxa used for comparison are deposited in the following Brazilian institutions: CFBH (Célio F. B. Haddad amphibian collection, Universidade Estadual Paulista, Rio Claro, São Paulo state); CAUC (Coleção de Anfíbios da Universidade Comunitária da Região de Chapecó, Chapecó, Santa Catarina state); UFMG (Coleção Herpetológica da Universidade Federal de Minas Gerais, Belo Horizonte, Minas Gerais state), and MNRJ (Museu Nacional, Rio de Janeiro, Rio de Janeiro state) (see list of examined specimens in Appendix S1). The webbing formula

notation followed Savage and Heyer [37], as modified by Myers and Duellman [38].

Measurements of the adult specimens were conducted using a Mytutoyo digital caliper to the nearest 0.01 mm: snout-vent length (SVL), head length (HL), head width (HW), eye diameter (ED), tympanum diameter (TD), eye-to-nostril distance (END), nostril-to-tip of snout distance (NSD), internarial distance (IND), upper eyelid width (UEW), interorbital distance (IOD), tibia length (TL) and foot length (FL) [39]; the distance between the anterior margins of the eyes (AMD) [40]; forearm length (FAL) [41]; hand length (HAL), thigh length (THL) and tarsal length (TAL), following Heyer et al [42]; as well as the disk diameter of the third finger (3FD) and the fourth toe (4TD) [43].

Nomenclatural Acts

The electronic edition of this article adheres to the requirements of the amended International Code of Zoological Nomenclature, and the new names contained herein are therefore available under that Code in the electronic edition of this article. The published work and the nomenclatural descriptions it contains have been registered in ZooBank, the online registration system for the ICZN. The ZooBank LSIDs (Life Science Identifiers) can be

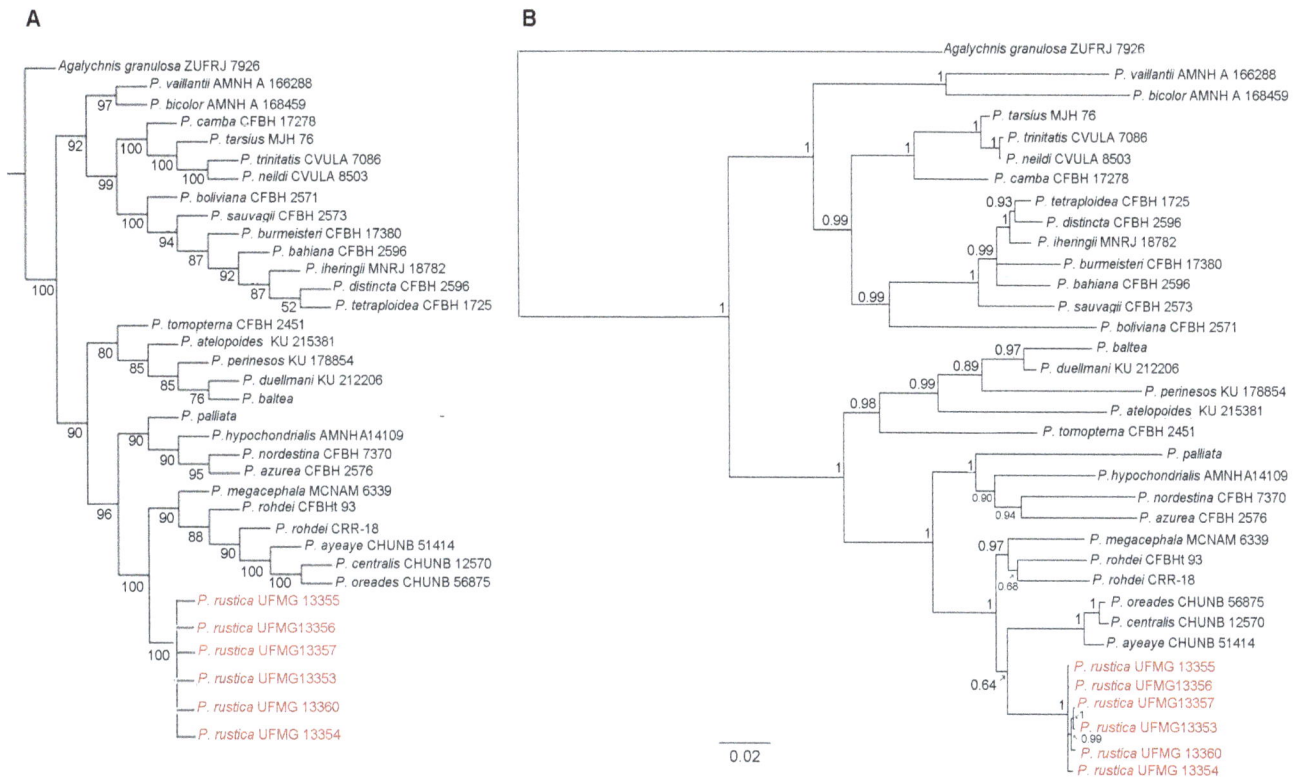

Figure 2. Phylogenetic relationships of the genus *Phyllomedusa* based on 2141 bps of the mitochondrial 12S rDNA, tRNA-val and 16S rDNA mitochondrial genes. (**A**) Strict consensus cladogram inferred using the maximum parsimony (MP) criterion, implemented in the TNT program. Numbers adjacent to the nodes indicate bootstrap values. (**B**) Topology inferred from the Bayesian analysis based on the GTR+R+I model. Bayesian posterior probabilities are shown at each node. Scale bar represents the number of substitutions per site.

obtained and the associated information viewed through any standard web browser by appending the LSID to the prefix "http://zoobank.org/". The LSID for this publication is urn:lsid:zoobank.org:pub:9A74B7B3-32C8-4BAA-86BC-9247128B782E. The electronic edition of this study was published in a journal with an ISSN, has been archived, and is available from the following digital repositories: PubMed Central, LOCKSS.

Results

Phylogenetic Inferences

The combined matrix consisted of 2141 bps. In the Bayesian inference, the GTR+G+I model was identified as being the most appropriate evolutionary model for the data set. The topology indicated similar relationships among the species of the genus to those described by Faivovich et al [2] and Bruschi et al. [8]. Two clades were observed within the current arrangement of the *P. hypochondrialis* group: the first clade included the species *P. palliata*, *P. hypochondrialis*, *P. nordestina* and *P. azurea*, while *P. megacephala*, *P. rohdei*, "*P. rhodei*", *P. centralis*, *P. oreades* and haplotypes of *Phyllomedusa* sp. n. from Água Doce, Santa Catarina State, constituted the second clade. Similar relationships were observed in the MP analyses, though different support values were obtained.

The parsimony and Bayesian trees detected a minor conflict in the internal relationships among the species within the second subclade (Figure 2A–B). In the Parsimony topology, *Phyllomedusa* sp. n. is monophyletic and is a sister group to all of the species of the second clade, with strong bootstrap support (Figure 2A),

whereas in the Bayesian inference (Figure 2B), *Phyllomedusa* sp. n. is grouped with *P. ayeaye*, *P. centralis* and *P. oreades* and this subclade is the sister group of remaining species. However, this arrangement is weakly supported by the posterior probability values.

Considering the genetic divergence in the 1335 positions that make up the final dataset of 12S, t-RNA-val and 16S mitochondrial fragments, the level of sequence divergence between *Phyllomedusa* sp. n. and other species of the *P. hypochondrialis* group ranged from 6% to 12% (Table 1). Within the second clade of the *P. hypochondrialis* group, divergence levels varied from 6% to 11%. The reduced divergence between the sequences of *P. centralis*, *P. oreades*, and *P. ayeaye* (1–2%) has been observed previously in cytochrome b sequences [2].

Cytogenetic Analysis

The analyzed specimens all presented a diploid number of 2n = 26 chromosomes. The karyotype consisted of four metacentric (1, 4, 8, 11), six submetacentric (2, 3, 5, 6, 12, and 13) and three subtelocentric pairs (7, 9 and 10) (Figure 3). Secondary constrictions were observed in the short arms of pair 9 and in one of the homologs in pairs 3 and 4, both in the subterminal short arms (Figure 3A). In some metaphases, secondary constrictions were observed through conventional Giemsa staining of the long arms of pair 7, in the pericentromeric region. The heterochromatin pattern revealed a C-positive block at the centromeric position in all pairs (Figure 3B). C-bands were evident in the pericentromeric region of the long arm of pair 6. When DAPI staining was analyzed (Figure 3C), DAPI-positive bands were observed in the centromeric regions of all chromosomes,

Table 1. Uncorrected pairwise distances between the 12S, tRNA-val and 16S mitochondrial sequences of the species of the *Phyllomedusa hypochondrialis* group.

Species	1	2	3	4	5	6	7	8	9	10
1- *P. palliata*	-									
2- *P. hypochondrialis*	0.11	-								
3- *P. nordestina*	0.11	0.10	-							
4- *P. azurea*	0.12	0.11	0.10	-						
5- *P. megacephala*	0.11	0.10	0.10	0.11	-					
6- *P. rohdei* CFBHt 93	0.12	0.11	0.11	0.12	0.05	-				
7- *P. rohdei* CRR-18	0.11	0.11	0.11	0.12	0.06	0.06	-			
8- *P. centralis*	0.13	0.11	0.11	0.12	0.07	0.07	0.08	-		
9- *P. oreades*	0.13	0.11	0.11	0.12	0.07	0.07	0.07	0.01	-	
10- *P. ayeaye*	0.12	0.11	0.10	0.12	0.07	0.06	0.07	0.02	0.01	-
11- *Phyllomedusa rustica* sp. n.	0.11	0.10	0.11	0.11	0.05	0.06	0.06	0.06	0.06	0.06

presenting a similar pattern to that found by C-banding. MM staining revealed brilliant fluorescence in the same regions as the secondary constrictions that were detected by conventional staining (Figure 3D).

The Ag-NOR and FISH techniques conducted using an rDNA probe revealed that the new species has a single NOR pair located in the subterminal region on chromosome pair 9 (Figure 3E and F). In this chromosome pair, the NOR was detected in the secondary constriction observed through Giemsa and MM-positive banding. The NOR sites presented size heteromorphism in the three specimens analyzed.

Phyllomedusa rustica sp. n. urn:lsid:zoobank.org:act:DD2C2754-AD90-4E63-96E4-C6D61972A4B1.

Phyllomedusa azurea – Lucas, Fortes and Garcia 2010.

Holotype (Figure 4 and 5). UFMG 13360, an adult male from Água Doce municipality (26°35′59.9″ S, 51°34′39.4″ W; 1330 m above sea level), Santa Catarina state, Brazil, collected by Elaine M. Lucas, Daniel Bruschi, and Veluma Debastiani, on 5 January 2012.

Paratopotypes. Nine adult males, UFMG 13353–13359 and 13361–13362, collected together with the holotype; three adult males, UFMG 1585 (Ex UFMG 3222) and CAUC 0864–0865, collected by Elaine M. Lucas, on 8 January 2009, at the type locality.

Type-locality. Água Doce municipality, Santa Catarina state, southern Brazil.

Diagnosis. A small species of *Phyllomedusa* belonging to the *P. hypochondrialis* species group, diagnosed based on the following combination of characters: (1) small to medium size (SVL 33.9–37.1 mm in males); (2) body moderately slender; (3) snout nearly rounded to truncate in dorsal view, vertical to obtuse in profile; (4) flanks and hidden areas of the arms and legs cream to reddish orange with black reticulations; (5) distinct pattern of reddish orange cells encircled by black or dark blue coloring on the concealed surfaces of the limbs; (6) upper lip lacking a reticulate pattern; (7) discreet reticulated pattern on the edge of the lower eyelids; (8) throat and belly whitish to creamy orange, slightly reticulated; (9) dorsal surfaces uniformly green, without spots; (10) dorsal surfaces smooth without granules; (11) translucent palpebral membrane slightly reticulated.

Phyllomedusa rustica sp. n. differs from *P. azurea, P. centralis, P. hypochondrialis* and *P. nordestina* by the distinct pattern of reddish orange small blotches encircled by black or dark blue coloring on the concealed surfaces of the limbs and by the slightly reticulate pattern on the border of the eyelids, and the ventral surfaces of the body (*P. centralis, P. hypochondrialis* and *P. nordestina* are characterized by a pattern of black bars or stripes on the red-orange bottom and an absence of reticulation on the ventral surfaces of the body); from *P. oreades* and *P. ayeaye* based on the absence of a reticulated pattern on the margins of the upper lip (present in these species) and less conspicuous reticulation on the ventral regions of the limbs, jaw, and body (conspicuous in these species); from *P. megacephala* by the presence of a discreet reticulate pattern on the border of the eyelids and the ventral surfaces of the body (absent in this species); from *P. rohdei* and *P. palliata* based on the absence of a whitish stripe along the lateral part of head (present in these species; see [41] regarding *P. palliata*).

Description of the holotype. General aspect slender; head slightly wider than long, SVL approximately three times the head width; snout nearly round to truncate in dorsal view (Figure 4) and vertical-to-slightly obtuse in lateral view (Figure 4 and 5); loreal region concave; nostrils small, circular and anterolaterally

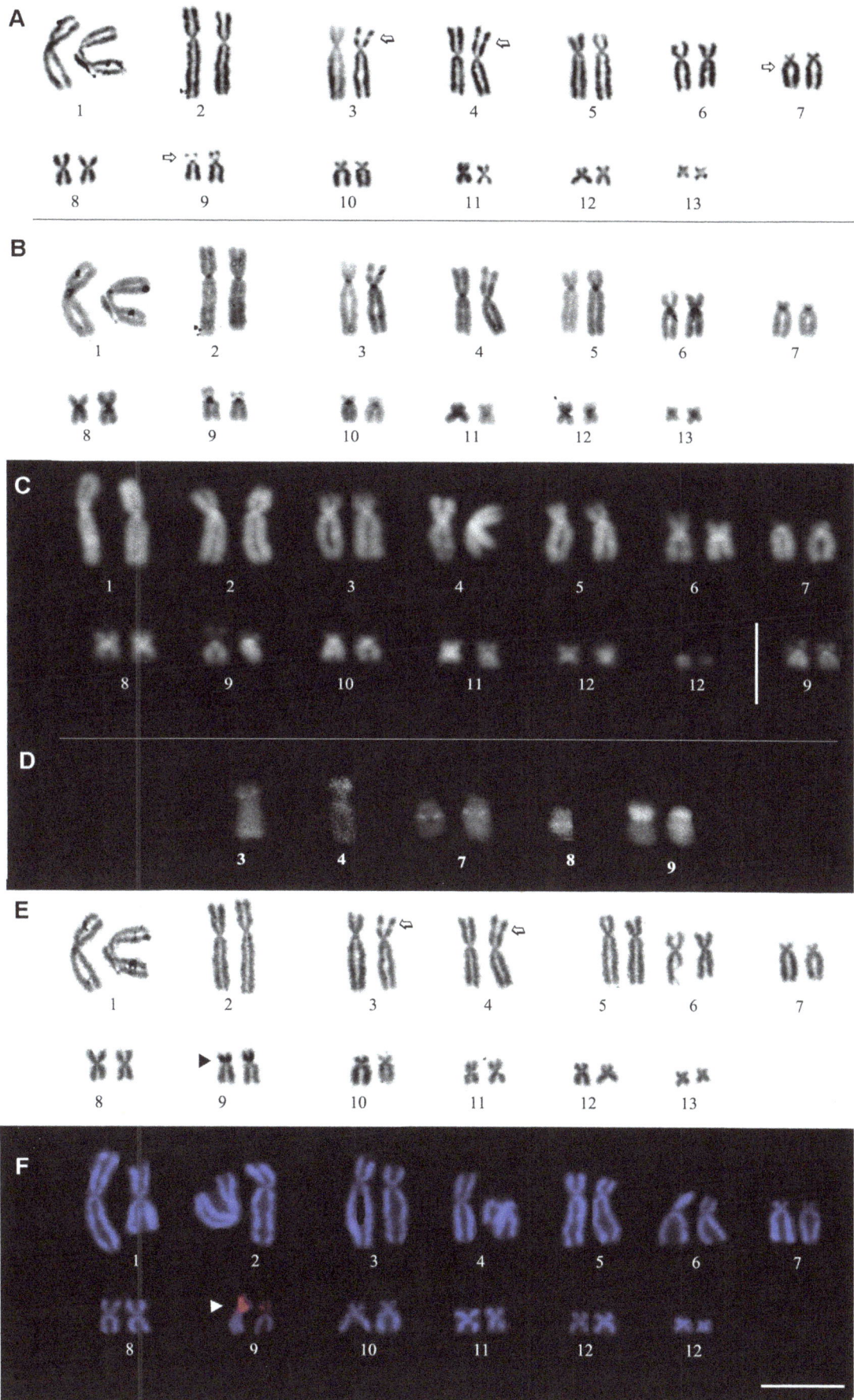

Figure 3. Karyotype of *Phyllomedusa rustica* **sp. n. based on Giemsa staining (A), C-banding (B), DAPI staining after C-banding (C), and chromosomes submitted to Mytramicim (MM) staining after C-banding, silver impregnation via the Ag-NOR method (D) and FISH experiments with a nucleolar 28S rDNA probe.** Arrows in (**A**) indicate secondary constrictions. The inset in (**C**) shows the homomorphic condition of homologue pair 9. In **D**, the chromosome pairs submitted to MM staining exhibit brilliant fluorescence, coinciding with the secondary constrictions detected by Giemsa-staining. The arrowheads indicate NOR sites in (**E**) and (**F**).

directed, near the tip of snout; eyes large, circular and lateral-frontally positioned; vertically elliptical pupil; translucent eyelid membrane slightly reticulated; tympanum nearly circular, with a diameter equal to approximately half of the eye diameter, annulus undefined at the superior border; faint supratympanic fold present; parotoids undistinguished; no vocal slits or external vocal sac; tongue large, piriform, free posteriorly; choanae small, nearly circular; vomerine teeth not visible externally; choanae small, largely separated; arm slender; forearm robust; no finger webbing; comparative finger lengths I<II<IV<III; finger tips not expanded, smaller than the tympanum diameter; palmar callosities large, circular (Figure 4); inner and outer metacarpal tubercles undifferentiated; subarticular tubercles well developed, globose; finger I enlarged at the base, opposed to the other fingers; nuptial pad of horny asperities evident in males; legs short, moderately robust; thigh length slightly longer than tibia length and both shorter than the tarsus-foot length; sum of the thigh and tibia lengths, slightly greater than 80% of SVL; calcar appendage and tarsal fold absent; foot (Figures 4 and 5) with slender toes, not webbed or fringed, with apical disks poorly developed, globose, smaller than the tympanum diameter; comparative toe length II<I = III<V<IV; toes I and II opposed to the others; subarticular and supernumerary tubercles single, large, round (Figure 4), inner and outer metatarsal tubercles undifferentiated; no toe webbing; dorsal skin smooth; ventral skin on the belly and throat granular; ventral surface of thigh granular anteriorly and smooth posteriorly; cloacal region moderately glandulose; cloacal opening not modified. The measurements and proportions of the body parts of the holotype are presented in Tables 2 and 3, respectively.

Coloration in life. Dorsum and dorsal surfaces of the forearm, shank, tarsus, and foot light green, olive or brownish green, varying within an individual (Figure 6); flanks, hidden areas of the arms and legs, and dorsal surface of the hands and feet cream-orange with black or dark blue reticulations; the dorsal surface of the thigh presents only a narrow green strip that tapers toward the groin without contacting it; axial and inguinal areas with black or dark blue reticulation, encircling large vivid reddish orange blotches; tympanic membrane green, similar to the dorsum; posterior border of the maxillae and margin of the lower eyelid with a fine light stripe and discreet dark reticulations on the lower eyelid; throat, belly and ventral surface of thighs whitish, lightly reticulated; iris silver with fine dark reticulation; pupil black; translucent palpebral membrane discretely reticulated.

Color in preservative. Dorsal surfaces of the body, forearms, tarsus and shank bluish gray; blackish purple reticulations on the flanks and hidden parts of the thigh and ventral surface of the shank; reticulated pattern of flanks increasing in size from the angle of the mouth to the groin; belly and chest discretely reticulated; lateral groin reticulations encircling large whitish mottling. A narrow, light stripe with fine dark reticulations on the posterior is found on the margin of the upper jaw and the margin of the lower eyelid.

Variation. Variation was found among individuals in body size (Table 2) and proportions (Table 3). In the paratypes UFMG 13358, 13360 and 13362, most of the ventral region is reticulated.

Natural history. Male *Phyllomedusa rustica* sp. n. were observed perched near the ground during calling activity,

frequently in association with herbaceous vegetation at the margins of ponds (Figure 6). When we handled the specimens, the animals contracted their bodies [44], with the belly, arms, and legs upturned, remaining motionless (Figure 5). We observed the eight other anuran species in the same pond: *Dendropsophus minutus*, *Hypsiboas leptolineatus*, *Hypsiboas pulchellus*, *Sphaenorhynchus surdus*, *Leptodactylus latrans*, *Leptodactylus plaumanni*, *Physalaemus cuvieri* and *Physalaemus* aff. *gracilis*.

Distribution. Known only from the type locality, in the grasslands of southern Brazil, in the municipality of Água Doce in the state of Santa Catarina.

Etymology. The epithet *rustica* originates from the Latin *rusticus* and is used to indicate the characteristics of the fields where this species is found (open fields). The epithet is as an adjective.

Discussion

Specimens of a disjunct population of *Phyllomedusa* from southern Brazil were analyzed through phylogenetic inference, as well as morphological and comparative chromosomal analyses, which allowed us to recognize a new species of the *P. hypochondrialis* group, which we have described here and named *Phyllomedusa rustica* sp. n.

This population was first reported by Lucas et al. [14], who concluded that this was the first population of *P. azurea* to be found outside the savanna biome of South America. Despite this identification, the authors observed the variable morphological characteristics in this population and suggested the existence of a *P. azurea* species complex [14]. Coloration patterns have been used universally for the delimitation of amphibian species, including those of the *Phyllomedusa hypochondrialis* group [45–46]. However, these characteristics do not appear to be consistently useful for the diagnosis of the taxa of this group, and are unable to discriminate reliably groups of species such as *P. hypochondrialis*, *P. azurea*, and *P. nordestina* [2], despite the fact that coloration patterns are the main diagnostic criteria used to differentiate these species [3]. The discrimination of *P. araguari* and *P. oreades* [47], for example, and *P. itacolomi* and *P. ayeaye* [48], has proven similarly problematic. In amphibians, DNA markers have been helpful for the recognition of cryptic lineages [9,49–50] and are especially important in groups which lack clear morphological differences [49,51]. In the present study, the topology obtained from the molecular analyses was fundamental for the recognition of the new species. A similar topology was reported previously [8], based on the comparison of populations from Brazil assigned to *P. hypochondrialis* and *P. azurea*, highlighting the paraphyletic position of the Água Doce population in relation to *P. azurea* from the Pantanal and Chaco formations. The topology obtained in the present study provides strong evidence to justify our taxonomical decision to recognize *Phyllomedusa rustica* sp. n.

The karyotype of *Phyllomedusa rustica* sp. n. had the same diploid number (2n = 26) and conserved chromosome morphology as all other representatives of the genus [8,52–57]. The NOR-bearing chromosomes in *Phyllomedusa rustica* sp. n. are the small submetacentric chromosomes of pair 9 (subtelomeric region), as observed in *P. nordestina* [55], *P. rhodei* [55] and *P. ayeaye*

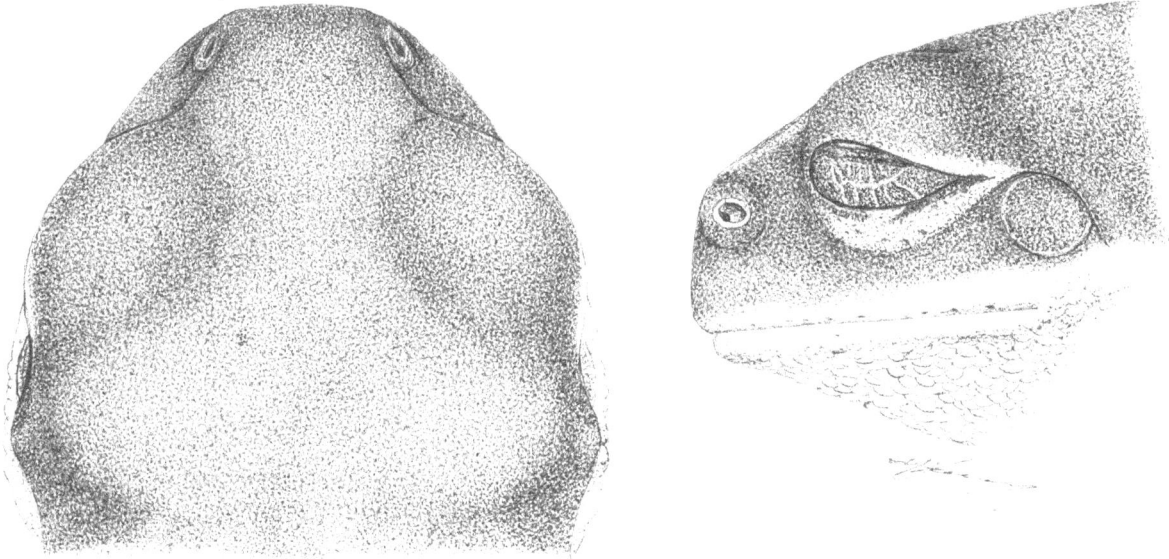

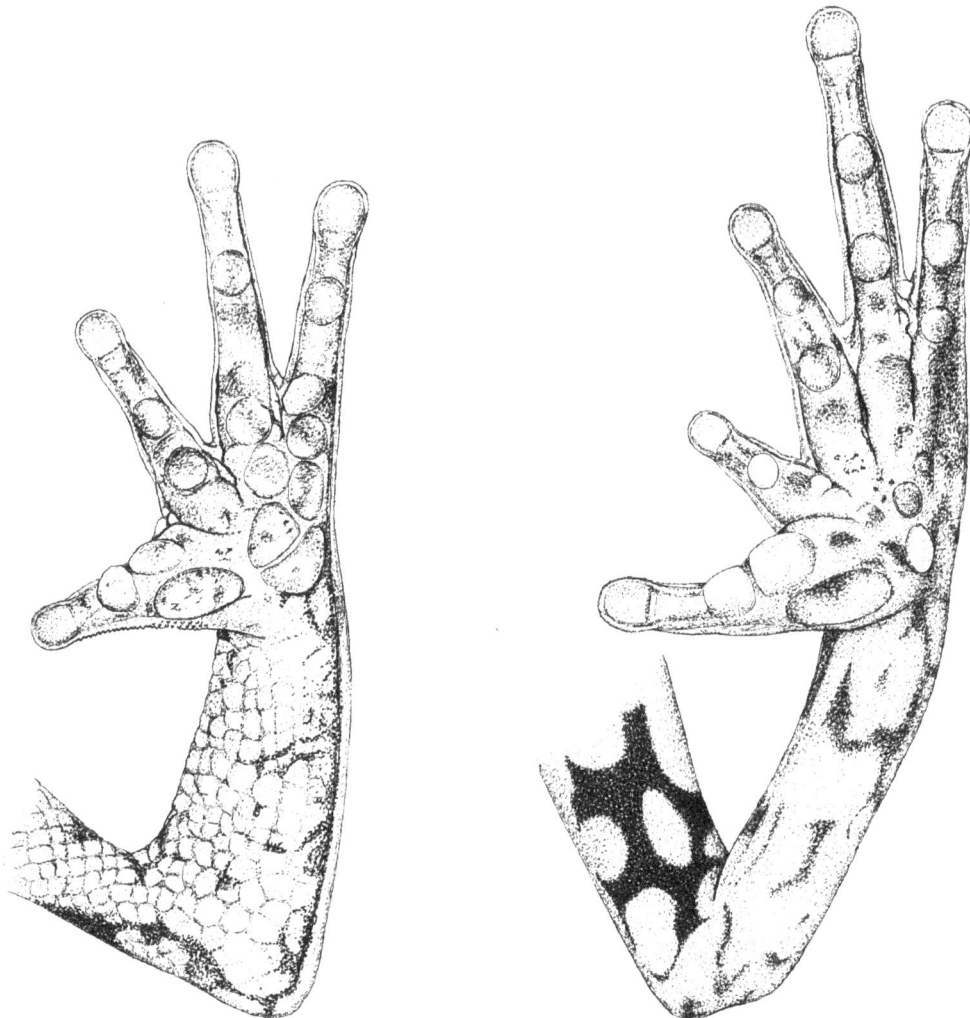

Figure 4. Holotype of *Phyllomedusa rustica* **sp. n. (UFMG 13360), adult specimen: (A) dorsal and lateral views of head and (B) palmar (left) and plantar (right) views.** Scale Bar = 5 mm.

(Bruschi, unpublished data). In the karyotype of the new species, a bright MM fluorescence pattern was observed in the region of the secondary constrictions using conventional staining. Despite this characteristic being consistently associated with NOR sites, with the exception of the NOR on pair 9, no hybridization signals were detected with the 28S rDNA probe in any of the other secondary constrictions (chromosomes 3, 4, 7 and 8). In these chromosomes, the MM-positive marks may be associated with the composition of this class of chromatin, with repetitive sequences rich in GC bases occurring in the chromosomal constrictions. The observed C-patterns were mainly centromeric, which is a common characteristic of the karyotypes of the *P. hypochondrialis* group [8,55].

Phyllomedusa rustica sp. n. was collected at altitude of 1330 m above sea level, within the Atlantic Forest formation of southern Brazil. *Phyllodemusa rohdei* is another species of the same subclade that is also found in the Atlantic Forest. All other species of this clade inhabit plateaus and mountainous regions in the Cerrado savannas of central Brazil [2]. In southern Brazil, the Atlantic Forest biome includes grasslands on the high plateaus that form a forest mosaic in the northern half of Rio Grande do Sul and in the states of Santa Catarina and Paraná. The Água Doce region is included in the Atlantic Forest *sensu lato* [58], which is characterized by seasonal deciduous forest, forming a natural mosaic with grasslands and *Araucaria* forests in western Santa Catarina and Paraná states, along the upper Uruguay River, extending to the Ibicuí and Jacuí basins in the center of the state of Rio Grande do Sul.

Special efforts are needed to define the taxonomic arrangement of the second subclade of *P. hypochondrialis* group, to which *Phyllomedusa rustica* sp. n. belongs, due primarily to difficulties in taxonomic delimitation based only in morphological traits. One example of a taxonomic problem is provided by the different populations of *P. rhodei*, in which strong evidence of mitochondrial sequence divergence was detected [2], indicating the existence of cryptic diversity – at least two distinct species – among the frogs currently assigned to *P. rhodei*. Faivovich et al. [59] corroborated this suggestion based on anatomical differences between populations from the type locality (Rio de Janeiro, Brazil) and Espírito Santo, Brazil. However, this specific question was not addressed within the scope of this work, and we can only note that the resolution of taxonomic questions in this subclade is important and that the identification of new lineages will contribute to a better understanding of their evolutionary history.

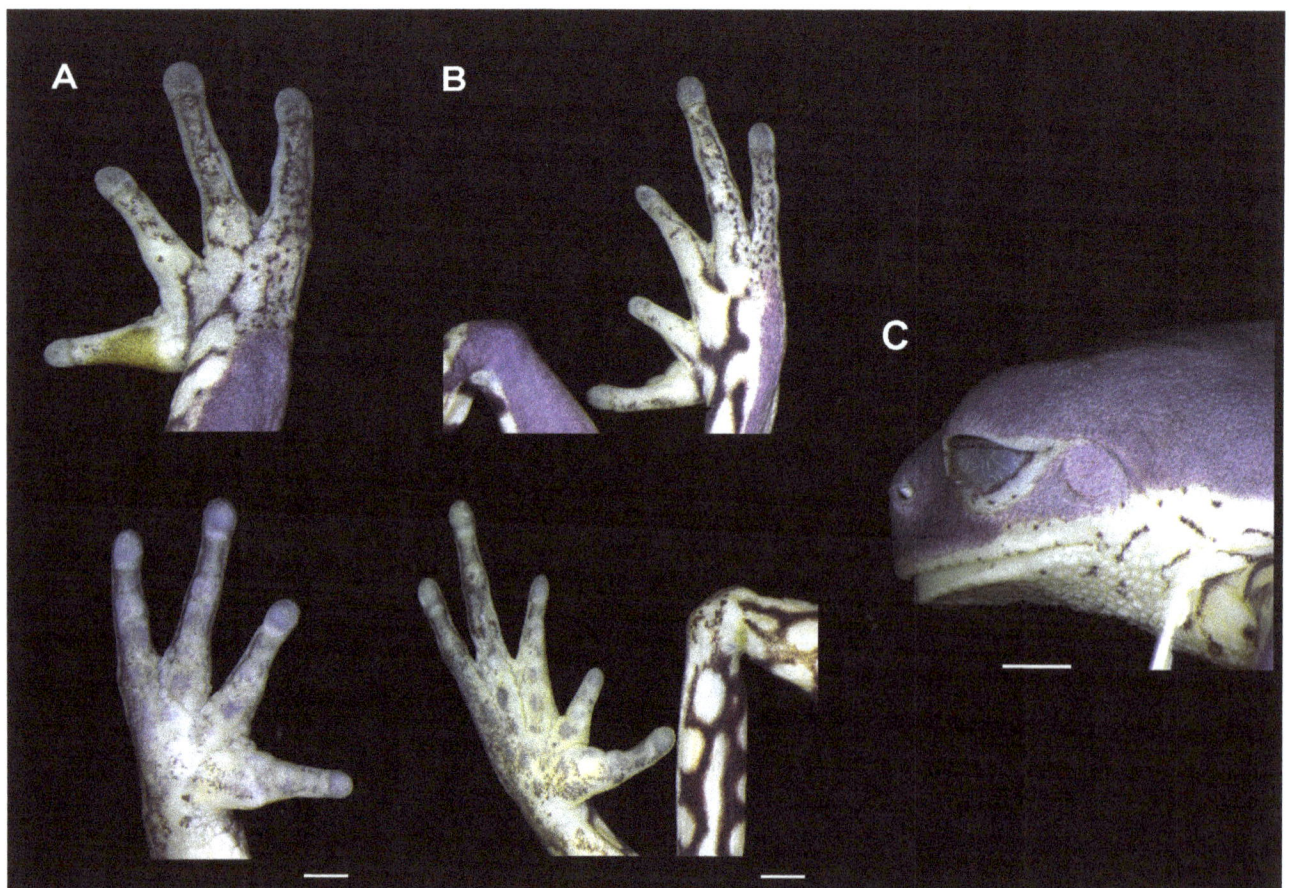

Figure 5. Holotype of *Phyllomedusa rustica* **sp. n. (UFMG 13360), adult specimen: (A) dorsal (top) and ventral (bottom) views of hand, (B) dorsal (top) and ventral (bottom) views of foot (C) lateral view of the head.** Scale bar = 2 mm.

Figure 6. Paratypes of *Phyllomedusa rustica* **sp. n. (A) Coloration pattern in the flank regions; (B) individual during call activity perched near the ground in herbaceous vegetation at the margin of a pond.**

Underestimation of species diversity and conservation of Atlantic Forest frogs

The recognition of the new species has important implications for the development of conservation strategies in southern Brazil, as well as for the understanding of ecological and evolutionary patterns, in particular the biological diversification of the Atlantic Forest. The description of this new species further reinforces the conclusion that the true diversity of the amphibian fauna of the Atlantic Forest biome has been underestimated substantially [60–61]. In recent years, the application of molecular approaches has contributed to the fine-scale taxonomic re-evaluation of many amphibian groups, revealing many new cryptic lineages in the

Table 2. Mean ± standard-deviation (SD), and range of the measurements (in mm) of the males of type series of *Phyllomedusa rustica* sp. n. and the holotype (see Material and methods for abbreviations).

Characters	Type series of *Phyllomedusa rustica* sp. n. (n = 11)	Holotype UFMG 13360
SVL	35.46±1.16 (33.93–37.09)	35.53
HL	10.66±0.47 (9.86–11.4)	11.4
HW	11.85±0.41 (11.17–12.61)	12.16
ED	3.75±0.36 (3.17–4.21)	3.54
TD	2.17±0.18 (1.86–2.36)	2.32
END	2.71±0.22 (2.40–3.10)	2.4
NSD	1.84±0.29 (1.32–2.19)	1.85
IND	3.62±0.34 (2.96–4.14)	3.38
UEW	3.23±0.27 (2.80–3.66)	3.18
AMD	6.96±0.35 (6.44–7.82)	6.94
IOD	4.58±0.38 (4.16–5.45)	4.31
FAL	8.61±0.91 (6.54–9.52)	6.54
HAL	9.88±0.52 (8.93–10.59)	9.9
3FD	0.98±0.12 (0.81–1.25)	0.83
THL	15.33±0.78 (14.21–16.54)	15.31
TL	14.10±0.60 (13.02–14.94)	13.64
TAL	12.73±0.59 (11.96–13.61)	12.29
FL	22.44±1.08 (20.72–23.75)	22.14
4TD	1.07±0.17 (0.81–1.27)	1.07

Table 3. Mean ± standard deviation (SD), and range in parentheses of the ratios of the body parts in the type series of *Phyllomedusa rustica* sp. n. and the holotype (see Material and methods for abbreviations).

Rations	Type series of *Phyllomedusa rustica* sp. n. (n = 11)	Holotype UFMG 13360
HL/SVL	0,30±0,01 (0.28–0.32)	0.32
HW/SVL	0,33+0,01 (0.32–0.35)	0.34
FAL/SVL	0,24±0,02 (0.18–0.27)	0.18
HAL/SVL	0,28+0,01 (0.26–0.30)	0.28
THL/SVL	0,43±0,02 (0.40–0.47)	0.43
TL/SVL	0,40±0,01 (0.38–0.43)	0.38
TAL/SVL	0,36±0,01 (0.35–0.38)	0.35
FL/SVL	0,63±0,02 (0.59–0.68)	0.62
HW/HL	1,11±0,04 (1.03–1.16)	1.07
ED/HL	0,36±0,02 (0.31–0.39)	0.35
TD/HL	0,20±0,02 (0.16–0.23)	0.18
END/HL	0,26±0,02 (0.21–0.29)	0.21
NSD/HL	0,05±0,01 (0.04–0.06)	0.05
IND/HL	0,10±0,01 (0.09–0.12)	0.10
UEW/HL	0,30±0,02 (0.26–0.34)	0.28
AMD/HL	0,65±0,02 (0.61–0.69)	0.61
TD/ED	0,57±0,07 (0.44–0.74)	0.52
3FD/TD	0,10±0,01 (0.08–0.12)	0.08
4TD/3FD	1,10±0,14 (0.90–1.29)	1.29

Atlantic Forest (e.g. see references [9,62–65]). Indeed, this biome has proven to be a biodiversity hotspot with a high level of species endemism in many vertebrate groups [66–68], and the understanding of the mechanisms underlying the biological diversification of this biome is of enormous interest to herpetologists [68–70].

Despite its accentuated species richness and endemism, the Atlantic Forest has suffered high rates of habitat loss throughout its recent history [71]. This biome originally covered an area of over 1,300,000 km^2 of eastern and southern Brazil, extending as far west as Paraguay and Argentina. This area has now been reduced to less than 12% of its original extent, most of which is distributed in small and isolated fragments [71]. Protected areas account for only 1.6% of this domain and do not cover the different vegetation types adequately [71–72]. The type locality of *Phyllomedusa rustica* sp. n., is located within a region classified as "extremely high priority" for conservation actions in the National Plan for the Conservation of the Herpetofauna of Southern Brazil - PAN Herpetofauna do Sul [73]. In spite of this, the natural local grasslands are still being converted into *Pinus* plantations or farmland [74], leading to profound modifications of the structure and composition of local ecological communities, and impacts on the biodiversity of the region. The description of a new Atlantic Forest species contributes to the understanding of the evolutionary history of the biota of this formation [75] and the phylogenetic diversity of its different ecosystems [76–77]. A better understanding of a region's biodiversity [78], the life-history traits of its species

[79–81] and their evolutionary history [76–77,79,82–83] may be especially important for the identification of priority areas in conservation planning.

Acknowledgments

We are also grateful to Veluma Debastiane for helping to collect the specimens and Jaime Somera for drawing the holotype. We also thank Luciana Bolsoni Lourenço and Ana Cristina Prado Veiga-Menoncello for discussions and/or information provided. PCAG thanks CNPq for a research productivity fellowship. We are indebted to Matthias Stöck and two anonymous reviewers for their valuable contributions to the improvement of this manuscript. We thank the Willi Hennig Society for subsidizing the TNT program and making it freely available.

Author Contributions

Conceived and designed the experiments: DPB EML PCAG SMRP. Performed the experiments: DPB EML PCAG SMRP. Analyzed the data: DPB EML PCAG SMRP. Contributed reagents/materials/analysis tools: SMRP. Wrote the paper: DPB EML PCAG SMRP.

References

1. Frost DR (2014) Amphibian Species of the World: an online reference. Version 5.6 Am Museum Nat Hist. Available: http://research.amnh.org/vz/herpetology/amphibia/. Accessed 2014 January 26.

2. Faivovich J, Haddad CFB, Baêta D, Jungfer KH, Álvares GFRA, et al. (2010) The phylogenetic relationships of the charismatic poster frogs, Phyllomedusinae (Anura, Hylidae). Cladistics 25: 1–35.

3. Caramaschi U (2006) Redefinição do grupo de *Phyllomedusa hypochondrialis*, com redescrição de *P. megacephala* (Miranda-Ribeiro, 1926), revalidação de *P.*

azurea Cope, 1826 e descrição de uma nova espécie (Amphibia, Anura, Hylidae). Arq Mus Nac 64: 159–179.

4. Barrio-Amorós CL (2006) A new species of *Phyllomedusa* (Anura: Hylidae: Phyllomedusinae) from northwestern Venezuela. Zootaxa 1309: 55–68.

5. Pombal JP, Haddad CFB (1992) Espécies de *Phyllomedusa* do grupo bumeisteri do Brasil oriental, com descrição de uma espécie nova (Amphibia, Hylidae). Rev Bras Biol 52: 217–229.

6. Cannatella DC (1982) Leaf-frogs of the *Phyllomedusa perinesos* group (Anura: Hylidae). Copeia 1982: 501–513.

7. Faivovich J, Haddad CFB, Garcia PCA, Frost DR, Campbell JA, et al. (2005) Systematic review of the frog family Hylidae, with special reference to Hylinae: phylogenetic analysis and taxonomic revision. Bull Amer Mus Nat Hist 294: 1–240.

8. Bruschi DP, Busin CS, Toledo LF, Vasconcellos GA, Strussmann C, et al. (2013) Evaluation of the taxonomic status of populations assigned to *Phyllomedusa hypochondrialis* (Anura, Hylidae, Phyllomedusinae) based on molecular, chromosomal, and morphological approach. BMC Gen 14: 70.

9. Fouquet A, Gilles A, Vences M, Marty C, Blanc M, et al. (2007) Underestimation of species richness in Neotropical frogs revealed by mtDNA analyses. PloS ONE 2: e1109.

10. Funk WC, Caminer M, Ron SS (2012) High levels of cryptic species diversity uncovered in Amazonian frog. Proc R Soc B 279: 1806–1814.

11. Bickford D, Lohman DJ, Sodhi NS, Ng PKL, Meier R, et al. (2007) Cryptic species as a window on diversity and conservation. Trends Ecol Evol 22: 148–155.

12. McLeod DS (2010) Of least concern? Systematic of a cryptic species complex: *Limnonectes kuhlii* (Amphibia: Anura: Dicroglossidae). Mol Phylogenet Evol 56: 991–1000.

13. Lips KR, Reeve JD, Witters LR (2003) Ecological traits predicting Amphibian populations declines in Central America. Conserv Biolo 17: 1078–1088.

14. Lucas EM, Fortes VB, Garcia PCA (2010) Amphibia, Anura, Hylidae, *Phyllomedusa azurea* Cope, 1862: Distribution extension to southern Brazil. Check List 6: 164–166.

15. Maack V (1981) Geografia física do estado do Paraná. Livraria José Olympio, Rio de Janeiro.

16. Behling H (1997) Late Quaternary vegetation, climate and fire history in the Araucaria forest and campos region from Serra Campos Gerais (Parana), S Brazil. Rev Palaeobot Palynol 97: 109–121.

17. Behling H, Lichte M (1997) Evidence of dry and cold climatic conditions at glacial times in tropical Southeastern Brazil. Quat Res 48: 348–358.

18. Behling H, Pillar VD, Orlóci L, Bauermann SG (2004) Late Quaternary Araucaria forest, grassland (Campos), fire and climate dynamics, studied by high-resolution pollen, charcoal and multivariate analysis of the Cambara do Sul core in southern Brazil. Palaeogeogr Palaeocl 203: 277–297.

19. Overbeck SCM, Fidelis A, Pfadenhauer J, Pillar VP, Blanco CC, et al. (2009) Os Campos Sulinos: um bioma negligenciado. In: Pillar VP, Müller SC, Castilhos ZMS, Jacques AV, editors. Campos Sulinos, conservação e uso sustentável da biodiversidade. Brasília: MMA. 403 p.

20. Goebel AM, Donnelly JM, Atz ME (1999) PCR primers and amplification methods for 12S ribosomal DNA, the control region, cytochrome oxidase I, and cytochrome b in bufonids and other frogs, and an overview of PCR primers which have amplified DNA in amphibians successfully. Mol Phylogenet Evol 11: 163–199.

21. Hall T (1999) Bioedit. Available: http://www.mbio.ncsu.edu/BioEdit/bioedit.html.

22. Thompson JD, Higgins DG, Gibson TJ (1994) CLUSTAL W: improving the sensitivity of progressive multiple sequence alignments through sequence weighting, position specific gap penalties and weight matrix choice. Nucleic Acids Res 22: 4673–4680.

23. Gatesy J, Desalle R, Wheeler W (1993) Alignment-ambiguous nucleotide sites and the exclusion of systematic data. Mol Phylogenet Evol 2: 152–157.

24. Ronquist F, Huelsenbeck JP (2003) MrBayes 3: Bayesian phylogenetic inference under mixed models. Bioinformatics 19: 1572–1574.

25. Nylander JAA (2004) MrModeltest v2. Program distributed by the author. Evolutionary Biology Centre, Uppsala University.

26. Rambaut A, Drummond A (2007) TRACER. MCMC trace analysis tool, Version 1.4. University of Oxford. Available: http://tree.bio.ed.ac.uk/software/tracer.

27. Goloboff PA, Farris JS, Nixon KC (2003) T.N.T.: Tree analysis using new technology. Program and documentation. Available: www.zmuc.dk/public/phylogeny.

28. Tamura K, Nei M, Kumar S (2004) Prospects for inferring very large phylogenies by using the neighbor-joining method. PNAS 101: 11030–11035.

29. Tamura K, Peterson D, Peterson N, Stecher G, Nei M, et al. (2011) MEGA5: Molecular evolutionary genetics analysis using Maximum Likelihood, evolutionary distance, and Maximum Parsimony methods. Mol Biol Evol 28: 2731–2739.

30. King M, Rofe R (1976) Karyotypic variation in the Australian gecko *Phyllodactylus marmoratus* (Gray) (Gekkonidae: Reptilia). Chromosoma 54: 75–87.

31. Schmid M (1978) Chromosome banding in Amphibia. I. Constitutive heterochromatin and nucleolus organizer regions in *Bufo* and *Hyla*. Chromosoma 66: 361–368.

32. Howell WM, Black DA (1980) Controlled silver staining of nucleolar organizer regions with a protective colloidal developer: a −1 step method. Experientia 36: 1014–1015.

33. Sumner AT (1972) A simple technique for demonstrating centromeric heterochromatin. Exp Cell Res 83: 438–442.

34. Viegas-Péquignot E (1992) In situ hybridization to chromosomes with biotinylated probes. In: In situ hybridization: a practical approach. Edited by Willernson D. Oxford: Oxford University Press 137–158.

35. Meunier-Rotival M, Cortadas J, Macaya G, Bernardi G (1979) Isolation and organization of calf ribosomal DNA. Nucleic Acids Res 6: 2109–2123.

36. Green DM, Sessions SK (1991) Nomenclature for chromosomes. In: Amphibian cytogenetics and evolution. San Diego: Academic Press 431–432.

37. Savage JM, Heyer WR (1967) Variation and distribution in the tree-frog genus *Phyllomedusa* in Costa Rica, Central America. Beitrage zur Neotropischen Fauna 5: 111–131.

38. Myers CW, Duellman WE (1982) A new species of *Hyla* from Cerro Colorado, and other tree frog records and geographical notes from western Panama. Am. Mus. Novitates 2752: 1–32.

39. Duellman WE (1970) The hylid frogs of Middle America. Monographs of the Museum of Natural History University of Kansas 1: 1–753.

40. Garcia PCA, Vinciprova G, Haddad CFB (2003) The taxonomic status of *Hyla pulchella joaquini* B. Lutz, 1968 (Anura: Hylidae), with description of tadpole, vocalization, and comments on its relationships. Herpetologica 59: 350–363.

41. Duellman WE (2005) Cusco Amazonico. The lives of amphibians and reptiles in an Amazonian Rainforest. Cornell University Press, Ithaca.

42. Heyer WR, Rand AS, Cruz CAG, Peixoto OL, Nelson CE (1990) Frogs of Boracéia. Arq Zool 31: 231–410

43. Napoli MF (2005) A new species allied to *Hyla circumdata* (Anura: Hylidae) from Serra da Mantiqueira, Southeastern Brazil. Herpetologica 61: 63–69.

44. Toledo LF, Sazima I, Haddad CFB (2010) Is it all death feigning? Case in anurans. J Nat Hist 44: 1979–1988.

45. Caramaschi U, Cruz CAG, Feio RN (2007) A new species of *Phyllomedusa* Wagler, 1830 from the state of Minas Gerais, Brazil (Amphibia Anura, Hylidae). Bol Mus Nac 524: 1–8.

46. Giaretta AA, Filho JCO, Kokubum MN (2007) A new *Phyllomedusa* Wagler (Anura, Hylidae) with reticulated pattern on flanks from Southeastern Brazil. Zootaxa 1614: 31–41.

47. Brandão RA, Álvares GFR (2009) Remarks on "a new *Phyllomedusa* Wagler (Anura: Hylidae) with reticulated pattern on flanks from Southeastern Brazil". Zootaxa 2044: 61–64.

48. Baêta D, Caramaschi U, Cruz CAG, Pombal JP (2009) *Phyllomedusa itacolomi* Caramaschi, Cruz & Feio, 2006, a junior synonym of *Phyllomedusa ayeaye* (B. Lutz, 1966) (Hylidae, Phyllomedusinae). Zootaxa 2226: 58–65.

49. Jansen M, Bloch R, Schulze A, Pfenninger M (2012) Integrative inventory of Bolivia's lowland anurans reveals hidden diversity. Zool Scr 40: 567–583.

50. Jungfer KH, Faivovich J, Padial JM, Castroviejo-Fisher S, Lyra MM, et al. (2013) Systematics of spiny-backed treefrogs (Hylidae: *Osteocephalus*): an Amazonian puzzle. Zool Scr 42: 351–380.

51. Stuart SN, Chanson JS, Cox NA, Young BE, Rodrigues ASL, et al. (2004) Status and trends of amphibian declines and extinctions worldwide. Science 306: 1783–1786.

52. Morando M, Hernando A (1997) Localización cromossômica de genes ribosomales activos em *Phyllomedusa hypochondrialis* y *P. sauvagii* (Anura, Hylidae). Cuad Herpetol 11: 31–36.

53. Barth A, Solé M, Costa MA (2009) Chromosome polymorphism in *Phyllomedusa rohdei* populations (Anura, Hylidae). J Herpetol 43: 676–679.

54. Paiva CR, Nascimento J, Silva APZ, Bernarde OS, Ananias F (2010) Karyotypes and Ag-NORs in *Phyllomedusa camba* De La Riva, 1999 and *P. rohdei* Mertens, 1926 (Anura, Hylidae, Phyllomedusinae): cytotaxonomic considerations. Ital J - Zool 77: 116–121.

55. Bruschi DP, Busin CS, Siqueira S, Recco-Pimentel SM (2012) Cytogenetic analysis of two species in the *Phyllomedusa hypochondrialis* group (Anura, Hylidae). Hereditas 149: 34–40.

56. Barth A, Souza VA, Solé M, Costa MA (2013) Molecular cytogenetics of nucleolar organizer regions in *Phyllomedusa* and *Phasmahyla* species (Hylidae, Phyllomedusinae): a cytotaxonomic contribution. Gen Mol Res 122400–2408.

57. Gruber SL, Zampieri AP, Haddad CFB, Kasahara S (2013) Cytogenetic analysis of *Phyllomedusa distincta* Lutz, 1950 (2n = 2x = 26), *P. tetraploidea* Pombal and Haddad, 1992 (2n = 4x = 52), and their natural triploid hybrids (2n = 3x = 39) (Anura, Hylidae, Phyllomedusinae). BMC Genetics 14: 75.

58. Oliveira-Filho AT, Fontes MAL (2000) Patterns of floristic differentiation among Atlantic forests in southeastern Brazil and the influence climate. Biotropica 32: 793–810.

59. Faivovich J, Baêta D, Vera Candioti F, Haddad CFB, Tylers MJ (2011) The submandibular musculature of Phyllomedusinae (Anura: Hylidae): a reappraisal. J Morphol 272: 354–362.

60. Myers N, Mittermeier RA, Mittermeier CG, da Fonseca GAB, Kent J (2000) Biodiversity hotspots for conservation priorities. Nature 403: 853–858.

61. Giam X, Scheffers BR, Sodhi NS, Wilcove DS, Ceballos G, et al. (2012) Reservoirs of richness: least disturbed tropical forests are centres of undescribed species diversity. Proc Roy Soc B 279: 67–76.

62. Fouquet A, Loebmann D, Castroviejo-Fisher S, Padial JM, Orrico VGD, et al. (2012) From Amazonia to the Atlantic forest: Molecular phylogeny of Phyzelaphryninae frogs reveals unexpected diversity and a striking biogeo-

graphic pattern emphasizing conservation challenges. Mol Phylogenet Evol 65: 547–561.

63. Fusinatto LA, Alexandrino J, Haddad CFB, Brunes TO, Rocha CFD, et al. (2013) Cryptic genetic diversity is paramount in small-bodied Amphibians of the genus *Euparkerella* (Anura: Craugastoridae) endemic to the Brazilian Atlantic Forest. PloS ONE 8(11): e79504.

64. Gehara M, Canedo C, Haddad CFB, Vences M (2013) From widespread to microendemic: molecular and acoustic analyses show that *Ischnocnema guentheri* (Amphibia: Brachycephalidae) is endemic to Rio de Janeiro, Brazil. Conserv Genet 14: 973–982.

65. Teixeira M, Recoder RS, Amaro RC, Damasceno RP, Cassimiro J, et al. (2013) A new *Crossodactylodes* Cochran, 1938 (Anura: Leptodactylidae: Paratelmatobiinae) from the highlands of the Atlantic Forests of southern Bahia, Brazil. Zootaxa 3702: 459–472.

66. Costa LP, Leite YLR (2000) Biogeography of South American Forest mammals: endemism and diversity in the Atlantic Forest. Biotropica 32: 872–881.

67. Silva JMC, Sousa MC, Castelletti CHM (2004) Areas of endemism for passerine birds in the Atlantic forest, South America. Global Ecol Biogeogr 13: 85–92.

68. Carnaval AC, Hickerson MJ, Haddad CFB, Rodrigues MT, Moritz C (2009) Stability predicts genetic diversity in the Brazilian Atlantic forest hotspot. Science 323: 785–789.

69. Thomé MTC, Zamudio KR, Giovanelli JGR, Haddad CFB, Baldiserra FA Jr, et al. (2010) Phylogeography of endemic toads and post-Pleistocene persistence of the Brazilian Atlantic Forest. Mol Phylogenet Evol 55: 1018–1031.

70. Tonini JFR, Costa LP, Carnaval AC (2013) Phylogeographic structure is strong in the Atlantic Forest; predictive power of correlative paleodistribution models, not always. J Zool Syst Evol Res 51: 114–121.

71. Ribeiro MC, Metzger JP, Martensen AC, Ponzoni FJ, Hirota MM (2009) The Brazilian Atlantic Forest: How much is left, and how is the remaining forest distribuited? implications for conservation. Biol Conserv 142: 1141–1153.

72. MMA (2007) Áreas prioritárias para conservação, uso sustentável e repartição de benefícios da biodiversidade Brasileira: Atualização - Portaria MMA n°9, de 23 de janeiro de 2007. Ministério do Meio Ambiente, Secretaria de Biodiversidade e Florestas, Brasília.

73. MMA (2012) Plano de ação nacional para conservação de Répteis e Anfíbios ameaçados da região Sul do Brasil - PAN Herpetofauna do Sul – Portaria ICMBIO N° 25, DE 17 DE FEVEREIRO DE 2012. Instituto Chico Mendes da Conservação da Biodiversidade, Brasília.

74. Behling H, Pillar VP (2007) Late Quaternary vegetation, biodiversity and fire dynamics on the southern Brazilian highland and their implication for conservation and management of modern Araucaria forest and grassland ecosystems. Philos T Roy Soc B 362: 243–251.

75. Fouquet A, Blotto BL, Maronna MM, Verdade VK, Juncá FA, et al. (2013) Unexpected phylogenetic positions of the genera *Rupirana* and *Crossodactylodes* reveal insights into the biogeography and reproductive evolution of leptodactylid frogs. Mol Phylogenet Evol 67: 445–457.

76. Forest F, Grenyer R, Rouget M, Davies TJ, Cowling RM, et al. (2007) Preserving the evolutionary potential of floras in biodiversity hotspots. Nature 445: 757–760.

77. Loyola R, Oliveira G, Diniz-Filho JAF, Lewinsohn TM (2008) Conservation of Neotropical carnivores under different prioritization scenarios: mapping species traits to minimize conservation conflicts. Diversity distrib 14: 949–960.

78. Faith DP (1992) Conservation evaluation and phylogenetic diversity. Biol Conserv 61: 1–10.

79. Loyola RD, Becker CG, Kubota U, Haddad CFB, Fonseca CR, et al. (2008) Hung out to dry: choice of priority ecoregions for conserving threatened neotropical anurans depends on life-history traits. PloS ONE 5: e2120.

80. Becker CG, Loyola RD, Haddad CFB, Zamudio KR (2010) Integrating species life-history traits and patterns of deforestation in amphibian conservation planning. Divers Distrib 16: 10–19.

81. Becker CG, Fonseca CR, Haddad CFB, Batista RF, Prado PI (2007) Habitat split and the global decline of Amphibians. Science 318: 1775–1777.

82. Sechrest W, Brooks TM, da Fonseca GAB, Konstant WR, Mittermeier RA, et al. (2002) Hotspots and the conservation of evolutionary history. PNAS 99: 2067–2071.

83. Rolland J, Cadotte MW, Davies J, Devictor V, Lavergnes S, et al. (2011) Using phylogenies in conservation: new perspectives. Biol Lett 8: 692–694.

8

Patterns of Tree Species Diversity in Relation to Climatic Factors on the Sierra Madre Occidental, Mexico

Ramón Silva-Flores[1], Gustavo Pérez-Verdín[2], Christian Wehenkel[3]*

1 Universidad Juárez del Estado de Durango, Ciudad Universitaria, Durango, México, **2** Instituto Politécnico Nacional, CIIDIR Durango, Durango, México, **3** Instituto de Silvicultura e Industria de la Madera, Universidad Juárez del Estado de Durango, Ciudad Universitaria, Durango, México

Abstract

Biological diversity can be defined as variability among living organisms from all sources, including terrestrial organisms, marine and other aquatic ecosystems, and the ecological complexes which they are part of. This includes diversity within species, between species, and of ecosystems. Numerous diversity indices combine richness and evenness in a single expression, and several climate-based explanations have been proposed to explain broad-scale diversity patterns. However, climate-based water-energy dynamics appears to be an essential factor that determines patterns of diversity. The Mexican Sierra Madre Occidental occupies an area of about 29 million hectares and is located between the Neotropical and Holarctic ecozones. It shelters a high diversity of flora, including 24 different species of *Pinus* (ca. 22% on the whole), 54 species of *Quercus* (ca. 9–14%), 7 species of *Arbutus* (ca. 50%) and many other trees species. The objectives of this study were to model how tree species diversity is related to climatic and geographic factors and stand density and to test the Metabolic Theory, Productivity-Diversity Hypothesis, Physiological Tolerance Hypothesis, Mid-Domain Effect, and the Water-Energy Dynamic Theory on the Sierra Madre Occidental, Durango. The results supported the Productivity-Diversity Hypothesis, Physiological Tolerance Hypothesis and Water-Energy Dynamic Theory, but not the Mid-Domain Effect or Metabolic Theory. The annual aridity index was the variable most closely related to the diversity indices analyzed. Contemporary climate was found to have moderate to strong effects on the minimum, median and maximum tree species diversity. Because water-energy dynamics provided a satisfactory explanation for the patterns of minimum, median and maximum diversity, an understanding of this factor is critical to future biodiversity research. Quantile regression of the data showed that the three diversity parameters of tree species are generally higher in cold, humid temperate climates than in dry, hot climates.

Editor: Keping Ma, Institute of Botany, Chinese Academy of Sciences, China

Funding: These authors have no support or funding to report.

Competing Interests: The authors have declared that no competing interests exist.

* Email: wehenkel@ujed.mx

Introduction

Biological diversity can be defined as variability among living organisms from all sources, including terrestrial organisms, marine and other aquatic ecosystems, and the ecological complexes which they are part of; this includes diversity within species, between species, and of ecosystems" [1]. McNeely [2] considered biodiversity as an umbrella term for the degree of variety in nature, including the number and frequency of ecosystems, species or genes in a given assemblage. Biodiversity is usually considered at three different levels: "genetic diversity", "species diversity" and "ecosystem diversity" [3], [4]. Biodiversity is not simply the number of different genes, species, ecosystems, or any other group of things in a defined area. The composition, structure and function determine and also constitute the biodiversity of an area [5].

Various diversity indices have been established, but very few are commonly applied in ecological studies, e.g. richness [6], the Shannon index [7], Simpson index [8]. However, many of these measures can be converted into members of a family of explicit diversity indices, also known as Hill family [9], [10] or Rényi-diversity [11], [12], [13], [14].

Climate is a key factor that determines the distribution of plant species [15]. Global climate change is an enormous challenge to those responsible for developing conservation strategies for forest species [16] because it can modify the distribution of genes and species as well as the composition of vegetation and also create new biogeoclimatic zones with individual species [17], [18]. The amount of predicted decoupling between biomes and their suitable climatic habitat will vary greatly between geographic areas and will depend on the level of greenhouse gas emissions in this century [19]. Wright et al., [20] and Hawkins et al., [21] have suggested that one of the most important patterns in ecology is the variation in broad-scale variation in taxonomic richness with climate and geography.

Several climate-based explanations have been proposed to explain broad-scale diversity patterns, e.g., the Mid-Domain Effect [22], Productivity-Diversity hypothesis (the more individuals hypothesis), the Physiological Tolerance Hypothesis, the Speciation Rates Hypothesis and Species-Temperature Hypotheses such as the Metabolic Theory [23], [24]. However, climate-based water-energy dynamics also appears to be an essential factor that determines patterns of diversity [25]. According to this theory, the influence of water decreases and the influence of energy increases with absolute latitude [26].

Recent studies have demonstrated a strong relationship between total species richness and temperature, precipitation, and net

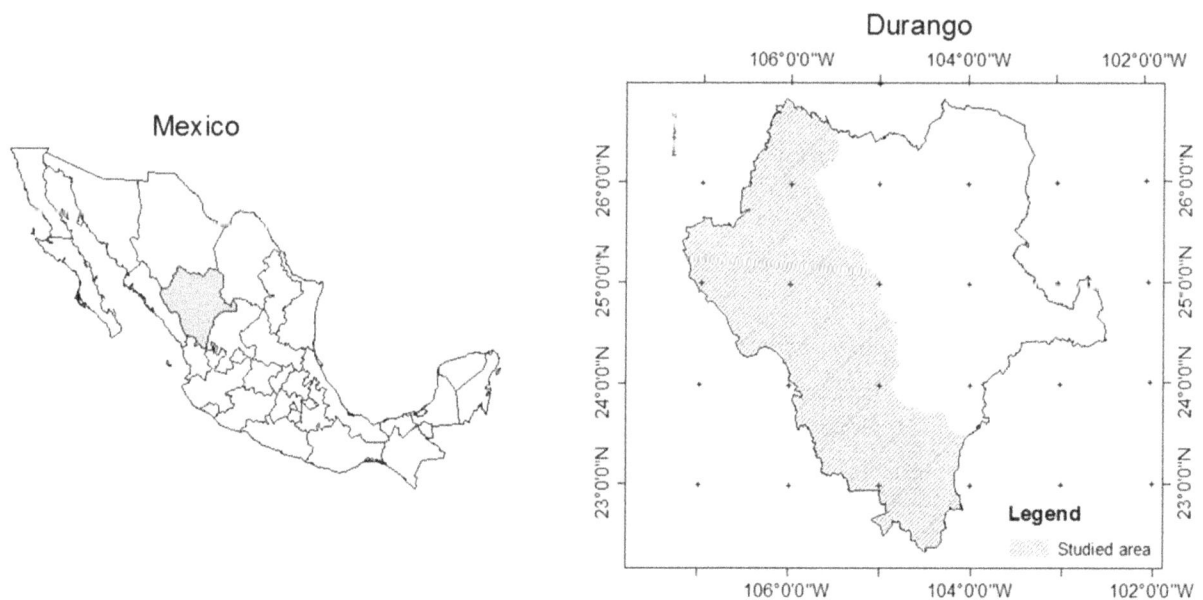

Figure 1. Location of study area.

primary productivity [21], [23], e.g., in South Africa [24], China [27], [28], [29], Ecuador, Costa Rica, Mexico and Tanzania [30], USA and Canada [29], India [31], [32], and Europe [33]. In a meta-analysis of 46 broad-scale data sets of species richness for a wide range of terrestrial plant, invertebrate, and ectothermic vertebrate groups throughout the world, the authors found that the relationship between richness and temperature is both taxonomically and geographically conditional [34]. However, there is no evidence of a universal response of diversity to temperature. For strictly tropical taxa such as palms [26], it has been confirmed that the influence of water and energy on species richness varies across large climatic gradients spanning tropical to temperate and arctic zones and also within megathermal climates. In a study carried out in the Amazon rainforest, Steege et al., found that dry season length, while only weakly correlated with average tree α-diversity, is a strong predictor of tree density and of maximum tree α-diversity [35]. Denser forests are more diverse than sparser forests, even when a diversity measure is used to correct for sample size [36].

The Mexican Sierra Madre Occidental occupies an area of about 29 million hectares and is located between the Neotropical and Holarctic ecozones. It shelters a high diversity of flora and fauna [37], [38] including 24 different species of *Pinus* (ca. 22% on the whole), 54 species of *Quercus* (ca. 9–14%), 7 species of *Arbutus* (ca. 50%) and many other trees species [39], [40]. Although this ecosystem is the largest forest biomass reserve in the country, little is known about the diversity of tree species [41].

Therefore, the objective of this study was to model how tree species diversity is related to climatic factors and stand density in the Sierra Madre Occidental and to test the Metabolic Theory, Productivity-Diversity Hypothesis, Physiological Tolerance Hypothesis [23], [24], Mid-Domain Effect [22], and the Water-Energy Dynamic Theory [25], [26]. The results may serve as tool for evaluating tree species diversity on the basis of climatic variables.

Material and Methods

The CONAFOR (National Forestry Commission), Mexico (http://www.conafor.gob.mx/portal/) provides the data set. No specific permissions were required for these locations/activities. We confirm that the field studies did not involve endangered or protected species and provide the specific location of your study (e.g. GPS coordinates). There no were vertebrate studies.

Study area

The study was conducted in the State of Durango (22°20′49″ N - 26°46′33″ N; 103°46′38″ W -107°11′36″ W), which occupies about 23% of the Sierra Madre Occidental ecosystem (Figure 1). The area covers a surface of approximately 6.33 million ha. The elevation above sea level varies between 363 and 3,190 mm (average 2,264 m). The climate ranges from temperate to tropical, with a total annual rainfall varying from 443 to 1,452 mm and an annual average of 917 mm. The mean annual temperature varies from 8.2 to 26.2°C, and the annual average is 13.3°C [42], [43]. The predominant forest types are uneven-aged pine-oak, often mixed with *Pseudotsuga menziesii*, *Arbutus* spp., *Juniperus* spp. and other tree species. The relative frequency of 67 tree species out of the 327 existing in the 1,632 sample plots is shown in Figure 2. These 67 trees species represent 97% of the cumulative relative frequency.

About 2 million ha of the forest land is mainly managed by selective removal, with only a small amount of other harvesting methods (less than 5% of the productive forest area). Clear felling is almost unknown, except in some parts of central and northern Durango [39]. The forest structure and density are due to the particular climatic and soil conditions and also to the specific management practices, which include use of the forest as pasture. Cattle grazing generate a large amount of the forest owners' income, which requires fairly open forest conditions [44].

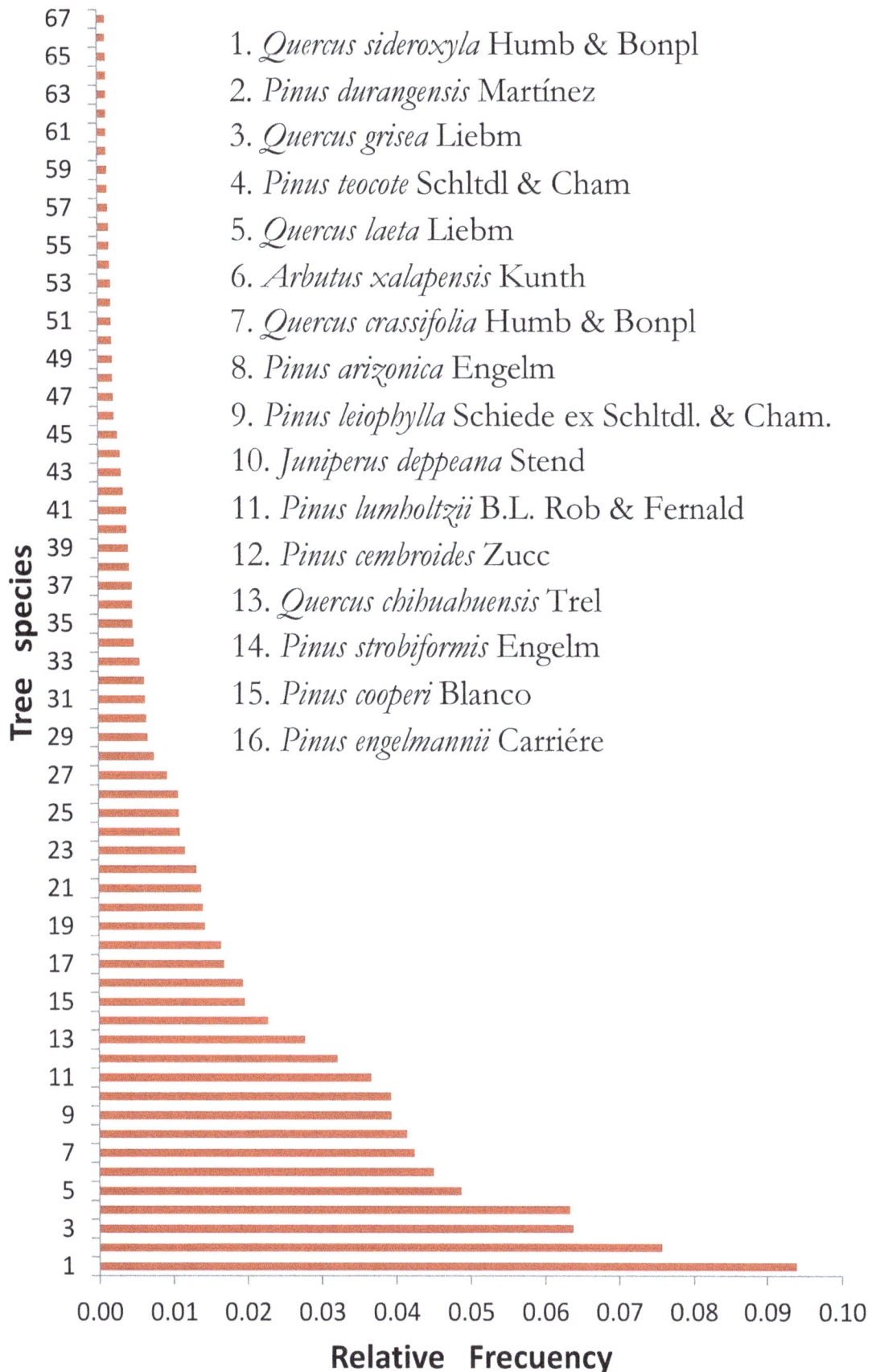

Figure 2. Relative frequency of trees species in the Sierra Madre Occidental. The names of the 16 most frequent species found in the study area are shown.

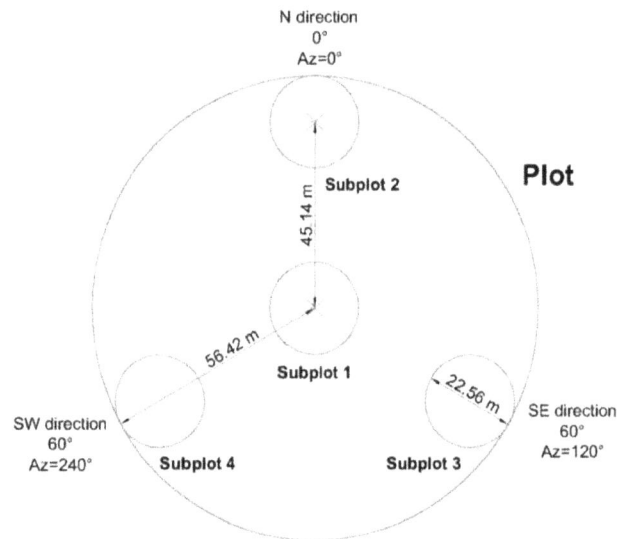

Figure 3. Design of a single plot included in the National Forest and Soil Inventory (modified from CONAFOR, 2004). The plot is composed of four 400 m² subplots distributed as an inverse 'Y'. Abbreviations: Az = Azimuth, m = meters, N = North; SE = Southeast; SW = Southwest

Sampling sites

The vegetation data were obtained from the National Forestry Commission, which is responsible for implementing the National Forest and Soil Inventory in Mexico. The sampling design included approximately 25,000 plots [43], distributed throughout the forest area in 5 by 5 km grids (Figure 3). Of the 1,737 plots located in the State of Durango, 1,632 were used in this study. A plot (1,600 m²) consists of four 400 m²-circular subplots, as shown

Table 1. Basic statistics of the geographical, climatic and species diversity variables studied.

Variable	Acronym	Units	Means	Std. Dev.	Min	Max
Longitude	LONG	Deg	−105.6	0.690	−107.2	−104.1
Latitude	LAT	Deg	24.5	1.066	22.4	26.8
Elevation above sea level	ELEV	m	2,239.0	506.1	390.1	3,156.0
Mean annual temperature	MAT	°C	13.5	3.8	8.3	26.2
Mean annual precipitation	MAP	mm	916.8	218.7	444.6	1,450.0
Growing season precipitation, April-September	GSP	mm	694.3	143.3	378.0	1,033.0
Mean temperature in the coldest month	MTCM	°C	8.3	4.0	2.9	21.4
Minimum temperature in the coldest month	MMIN	°C	−0.867	4.6	−6.8	12.0
Mean temperature in the warmest month	MTWM	°C	18.1	3.8	12.6	30.5
Maximum temperature in the warmest month	MMAX	°C	26.5	3.4	20.6	40.2
Julian date of the last freezing date of spring	SDAY	Days	112	49	1	182
Julian date of the first freezing date of autumn	FDAY	Days	303	29	251	365
Length of the frost-free period	FFP	Days	192	76	79	365
Degree days >5°C	DD5	Days	3,200	1,317	2,725	7,640
Degree days >5°C accumulating within the frost-free period	GSDD5	Days	2,373	1,537	1,780	7,551
Julian date when the sum of degree days >5°C reaches 100	D100	Days	36	19	36	84
Degree days <0°C	DD0	Days	8	11	1	65
Minimum degree days <0°C	MMINDD0	Days	503	367	484	1,323
Annual aridity index (the ratio of square root of DD5 to MAP)	AAI	Days$^{0.5}$*mm^{-1}	0.064	0.020	0.060	0.154
Number of individuals per plot	NIP	Trees	70	43	62	348
Species richness index per plot	v_0		6.5	2.9	6	17
Effective species number per plot	v_2		3.5	1.6	3.3	9.7
Amount of prevalent tree species	v_∞		2.42	0.98	2.30	6.92

Table 2. Matrix of Spearman correlation (r_S) between climate variables.

VARIABLES	LONG	LAT	ELEV	MAT	MAP	GSP	MTCM	MMIN	MTWM	MMAX	SDAY	FDAY	FFP	DD5	GSDD5	D100	DD0	MMINDD0	AAI
LAT	−0.786																		
ELEV	0.022ns	0.206																	
MAT	0.183	−0.296	−0.902																
MAP	−0.271	−0.331	−0.153	−0.017ns															
GSP	−0.137	−0.458	−0.194	0.066	0.982														
MTCM	0.230	−0.405	−0.886	0.981	0.101	0.188													
MMIN	0.212	−0.432	−0.831	0.938	0.193	0.275	0.980												
MTWM	0.117	−0.182	−0.905	0.983	−0.096	−0.026	0.935	0.873											
MMAX	0.054ns	−0.035ns	−0.823	0.895	−0.237	−0.179	0.813	0.718	0.950										
SDAY	−0.239	0.372	0.835	−0.967	−0.041ns	−0.132	−0.973	−0.965	−0.924	−0.789									
FDAY	0.200	−0.411	−0.894	0.965	0.160	0.241	0.985	0.977	0.918	0.777	−0.972								
FFP	0.214	−0.398	−0.883	0.971	0.119	0.203	0.984	0.975	0.926	0.786	−0.985	0.996							
DD5	0.178	−0.289	−0.904	0.999	−0.022ns	0.059ns	0.978	0.934	0.986	0.901	−0.965	0.962	0.968						
GSDD5	0.188	−0.319	−0.895	0.995	0.020ns	0.103	0.984	0.955	0.971	0.864	−0.982	0.977	0.984	0.994					
D100	−0.225	0.385	0.877	−0.984	−0.069	−0.158	−0.997	−0.976	−0.939	−0.818	0.979	−0.981	−0.984	−0.982	−0.986				
DD0	−0.296	0.364	0.807	−0.931	0.031ns	−0.057	−0.948	−0.926	−0.884	−0.753	0.926	−0.935	−0.937	−0.928	−0.933	0.948			
MMINDD0	−0.224	0.402	0.837	−0.960	−0.117	−0.205	−0.984	−0.991	−0.903	−0.757	0.987	−0.981	−0.986	−0.957	−0.975	0.987	0.935		
AAI	0.271	0.087	−0.451	0.647	−0.741	−0.674	0.538	0.446	0.708	0.778	−0.584	0.479	0.517	0.652	0.617	−0.566	−0.568	−0.520	
NIP	−0.251	0.238	0.399*	−0.467*	0.108	0.046ns	−0.447*	−0.407*	−0.463*	−0.426*	0.458*	−0.434*	−0.446*	−0.467*	−0.458*	0.453*	0.422*	0.437*	−0.381*

Notes: ns = no significant correlations (p>0.01), * = Association of variables with significant correlation (Spearman coefficient >|0.3| and p<0.01).

Table 3. Matrix of Spearman correlation (r_s) calculated from the climate and diversity variables.

VARIABLES	LONG	LAT	ELEV	MAT	MAP	GSP	MTCM	MMIN	MTWM	MMAX	SDAY	FDAY	FFP	DD5	GSDD5	D100	DD0	MMINDD0	AAI	NIP
v_0	-0.127	0.021ns	0.272	-0.317*	0.248	0.213	-0.276	-0.228	-0.338*	-0.358*	0.295	-0.258	-0.271	-0.318*	-0.302*	0.288	0.265	0.265	-0.412*	0.472*
v_2	-0.082	-0.003ns	0.237	-0.266	0.206	0.178	-0.232	-0.188	-0.287	-0.310*	0.244	-0.216	-0.225	-0.267	-0.253	0.241	0.220	0.219	-0.345*	0.267
v_∞	-0.069	-0.007ns	0.230	-0.256	0.187	0.161	-0.225	-0.186	-0.275	-0.296	0.237	-0.211	-0.218	-0.257	-0.245	0.234	0.213	0.214	-0.323*	0.220

Notes: ns = no significant correlations ($p > 0.01$),* = Association of variables with significant correlation (Spearman coefficient $>|0.3|$ and $p < 0.01$).

in Figure 3 [45]. In each plot, the total number of trees, genera and species were recorded along with altitude, latitude and longitude. Tree data were taken from individuals with at least 3 m tall and diameter at breast height (1.30 m) of 7.5 cm.

Climate model

The climate model of Rehfeldt [45], [46], [47], which is based on the thin plate splines of Hutchinson [42], [47] and [48] was used to estimate climate variables in each plot. The model produces climate surfaces from normalized monthly values of total precipitation and mean, maximum and minimum temperatures collected between 1961 and 1990 from approximately 6,000 weather stations (183 stations in Durango State). Hutchinson's software was used to predict the following: (i) the climate at specific points, identified by latitude, longitude and elevation, and (ii) the climate along gridded surfaces. Point estimates were obtained using a national database run by the University of Idaho (http://forest.moscowfsl.wsu.edu/climate/) which requires point coordinates (latitude, longitude, and elevation) as the main inputs (see [45] and [46] for technical procedures). Sixteen variables were derived from the original data, which also included an annual aridity index (AAI), defined as the ratio of square root of degree days $>5°C$ (DD5) to mean annual precipitation (MAP). Higher values of the aridity index indicate more arid climate, whereas smaller values indicate either very warm and very humid or cold and humid climates. The AAI index is a powerful climatic variable for describing and predicting distributions of pine species [47]. Other variables are described in Table 1.

Calculation of diversity

The species diversities (Table 1) were calculated by the so-called Hill numbers, Hills family, or diversity profile v_a [9], [10], where a is a real number ranging from zero to infinity. The general concept underlying classification with v_a is that with increasing a, the most frequent types of species increasingly determine the diversity of a collection to a greater degree than the less frequent types and that the extent to which this is true increases with increasing values of parameter a. Among the most desirable characteristics of a measure of diversity is that v_a satisfies the following requirements, irrespective of the value of a: (i) for a given number of variants it assumes its largest value exactly when all these variants are equally frequent, and this value equals the number of tree species, (ii) it increases as two variants approach equal frequencies, and (iii) it increases when one variant is subdivided into several varieties.

Considered as a function of a, v_a describes a diversity profile for each frequency distribution. The following are the most illustrative values of the subscript a in such diversity profiles: (i) $a = 0$, where the diversity is equivalent to the total number of variants; (ii) $a = 2$ as the effective number used in most genetic studies, and (iii) $a = \infty$, where only the relative frequency of the most frequent variant determines the diversity. In the present study, the diversity profiles are represented by all three diversities for each sample plot. Thus, each population was characterized by the *total* number, the *effective* number [49], inherent in Simpson diversity ($D = \sum p_i^2$) [8], and the amount of *prevalent* variants [10]. All variants have the same abundance when v_0, v_2 and v_∞ have the same value.

Table 4. The parameters and probability of error (p) of polynomial nonlinear quantile regression models of diversity indices with climatic variables at the 0.9, 0.5 and 0.1 quantile levels (τ).

Variables	τ	r^2	Parameters	a	b	c	d	e
v_0 – MAT	0.9	0.68	value	-0.0009	0.0628	-1.6188	17.2090	-53.471
			p	0.1676	0.1142	0.0807	0.0647	0.1159
			error	0.0006	0.0397	0.9261	9.3097	33.9910
	0.5	0.36	value	-0.0009	0.0618	-1.5001	15.0798	-45.9928
			p	0.0070	0.0048	0.0037	0.0040	0.0163
			error	0.0003	0.0219	0.5162	5.2265	19.1328
	0.1	0.76	value	-0.0005	0.0308	-0.7001	6.3320	-15.0739
			p	0.2092	0.2280	0.2680	0.3494	0.5672
			error	0.0004	0.0255	0.6318	6.7644	26.3377
v_0 – MTWM	0.9	0.70	value	-0.0007	0.0678	-2.3462	34.2400	-168.4842
			p	0.2653	0.1872	0.1324	0.0978	0.0957
			error	0.0006	0.0514	1.5583	20.6660	101.0643
	0.5	0.46	value	**-0.0013**	**0.1176**	**-3.7609**	**51.5186**	**-247.7348**
			p	0.0006	0.0003	0.0002	0.0001	0.0001
			error	0.0004	0.0324	0.9900	13.1527	64.0229
	0.1	0.79	value	-0.0007	0.0587	-1.8170	23.9195	-109.1866
			p	0.0412	0.0412	0.0455	0.0572	0.0895
			error	0.0003	0.0287	0.9077	12.5663	64.2657
v_0 – MMAX	0.9	0.80	value	**-0.0014**	**0.1764**	**-8.2776**	**168.7058**	**-1252.0169**
			p	0.0069	0.0025	0.0009	0.0003	0.0001
			error	0.0005	0.0582	2.4830	46.6776	326.5024
	0.5	0.43	value		0.0057	-0.4649	12.0452	-92.7230
			p		0.0003	0.0006	0.0020	0.0121
			error		0.0501	2.1334	40.0340	279.2817
	0.1	0.81	value	-0.0006	0.0753	-3.2486	60.8045	-411.7345
			p	0.0986	0.0907	0.0869	0.0886	0.0998
			error	0.0004	0.0445	1.8964	35.6846	250.0008
v_0 – AAI	0.9	0.66	value	**-571021.2704**	**195722.2198**	**-23593.7069**	**1132.4081**	**-7.6657**
			p	0.0000	0.0000	0.0000	0.0000	0.0134
			error	187.8099	8876.5940	2096.1069	150.2528	3.0954
	0.5	0.48	value	**-453491.3979**	**158284.4650**	**-19007.7266**	**860.1545**	**-4.8716**
			p	0.0000	0.0000	0.0000	0.0000	0.0008
			error	572.1734	3589.5689	863.7811	63.9525	1.4532
	0.1	0.85	value	**-426868.9341**	**131061.4899**	**-13434.8946**	**476.2212**	**0.2906**

Table 4. Cont.

Variables	t	r²	Parameters	a	b	c	d	e
v_0 – NIP*	0.9	0.90	value	**−1.7209E-08**	**1.1462E-05**	**−2.5805E-03**	**0.24392**	**2.67529**
			error	1530.9391	7867.5688	1840.7669	134.5190	3.0364
			p	0.0000	0.0000	0.0000	0.0004	0.5238
	0.5	0.61	value	−4.7616E-09	3.8079E-06	−1.1208E-03	0.14230	0.00000
			error	0.00000	0.00000	0.00047	0.00000	1.56533
			p	0.00622	0.00016	0.00000	0.02793	0.41146
	0.1	0.87	value	−2.4871E-09	1.5035E-06	0.15491	0.06309	0.60922
			error	0.00000	0.00000	−0.00037	0.01590	0.21781
			p	0.20723	0.03372	0.00028	0.00000	0.00000
v_2 – MMAX	0.9	0.67	value	**−0.0010**	**0.1180**	**−5.3449**	**105.9564**	**−769.6654**
			error	0.00000	0.00000		0.01239	0.15939
			p	0.52797	0.41469	0.00026	0.00000	0.00014
	0.5	0.37	value	**−0.0005**	**0.0627**	**−2.8916**	**58.0009**	**−423.3663**
			error	0.0002	0.0242	1.0382	10.6058	137.2664
			p	0.0000	0.0000	0.0000	0.0000	0.0000
	0.1	0.75	value	−0.0003	0.0355	−1.5887	30.9829	−220.1202
			error	0.0002	0.0176	0.7624	14.5742	103.5761
			p	0.0009	0.0004	0.0002	0.0001	0.0001
v_2 – AAI	0.9	0.64	value	**−413660.0818**	**143466.7997**	**−17458.6761**	**855.7508**	**−8.3143**
			error	0.0001	0.0169	0.7364	14.1736	101.6441
			p	0.0428	0.0357	0.0311	0.0290	0.0305
	0.5	0.36	value	−149515.0681	53535.0264	−6662.4690	310.2461	−0.7709
			error	1451.8510	4644.0345	1119.7915	84.8163	1.9990
			p	0.0000	0.0000	0.0000	0.0000	0.2000
	0.1	0.81	value	−172936.4467	52569.6039	−5359.5093	191.1053	0.3302
			error	851.3523	2033.2694	500.2416	40.0305	1.0217
			p	0.0000	0.0000	0.0000	0.0000	0.4506
v_∞ – AAI	0.9	0.65	value	**−292007.5163**	**101441.4217**	**−12274.8255**	**594.2765**	**−5.7624**
			error	1565.8892	4205.3104	978.0459	70.0480	1.5484
			p	0.0000	0.0000	0.0000	0.0000	0.8312
	0.5	0.31	value	**−50472.4434**	**17865.9822**	**−2180.6198**	**91.9919**	**1.4579**
			error	0.0000	2923.4180	651.1020	46.4309	
			p	0.0000	0.0000	0.0000	0.0000	0.0000
			error	486.8971	942.9610	234.4576	18.5449	0.4577
			p	0.0000	0.0000	0.0000	0.0000	0.0015

Table 4. Cont.

Variables	r	r^2	Parameters	a	b	c	d	e
	0.1	0.79	value	−115122.9857	36829.1202	−4014.9091	163.7877	−0.5152
			p	0.0000	0.0000	0.0000	0.0000	0.3975
			error	247.6810	1317.2320	316.3406	24.4349	0.6087

Notes: * The parameters obtained in the models are valid for a range in the number of trees per plot from 1 to 120. The statistically significant parameters in the models ($p<0.05$) are indicated by bold numbers.

Formally (p_i = relative frequency of a tree species i):

$$v_a = v_a(p) = \left(\sum_i p_i^a \right)^{\frac{1}{1-a}} \qquad (1)$$

Data analysis

Plots without trees were excluded. IBM SPSS Statistics and SAS/STAT software [50] were used to calculate descriptive statistics, and to carry a Principal Component Analysis (PCA) and Spearman's Correlation Analysis.

Principal Component Analysis (PCA) with varimax rotation [51], [52] was used to reduce the number of variables into underlying factors (components) of climatic and geographic variables. Two factors were clearly identified and accounted for 85% of the total variance. Bartlett's test of sphericity, which evaluates the hypothesis that the correlations in the correlation matrix are zero, was equal to 0.864 (p<0.001), i.e. it was highly significant. The factor loadings indicated that the main factor (factor 1) included MAT, MTCM, FDAY, DD5, FFP, GSDD5, SDAY, MMIN, D100, MTWM, MMINDD0, ELEV, MMAX and DD0. This factor was identified as the temperature group. The second factor only included GSP, MAP, LAT, and AAI and was designated as the precipitation group. The 14 variables included in the temperature group represented 69% of the total variance of the diversity of tree species in the Sierra Madre Occidental, while the precipitation group comprised four variables representing 16% of the total variance.

SAS/STAT software [50] was also used to calculate the well-known Spearman's coefficient (r_s) to determine how diversity is correlated with climatic and geographic variables and the number of individuals per plot (NIP).

In addition, R software, version 2.13.1 [53] and the 'quantreg' module were used to construct quantile models for tree species diversity with climatic and geographic variables that were included in the PCA main factor (factor 1), and for which the absolute values of r_s were larger than 0.3 (Table 2). Additionally, quantile models for tree species diversity with NIP and AAI were generated because of a strong correlation (r_s) between these variables and tree species diversity (Table 2). This method was also used for nonlinear regression because the variables were analyzed without prior transformation.

Quantile regression is useful for analysis of non-normally distributed data sets. This nonparametric method is used to examine specific segments of the conditional distribution and upper or lower quantile function of several covariates of interest. The technique is based on minimizing the absolute error and estimated functions for the median (as a robust version of the mean) and other quantiles conditional for a random variable with distribution function $F(y) = \text{Prob }(Y \leq y)$. The quantile ($\tau$) is defined by the inverse $Q(\tau) = \inf\{y: F(y) \geq \tau\}$ where $0 < \tau < 1$. The median equals $Q(0.5)$ [54], [55]. The quantile regression procedure computes the quantile function $Q(\tau \mid X = x)$.

In the nonlinear quantile regressions, parameters a, b, c, d and e were used with the form $y = dx+e$ to $y = ax^4+bx^3+cx^2+dx+e$ to analyze the polynomial family of univariate models from 2 to 5. After preselection of various linear and nonlinear multivariate models, the linear function $v_a = aMAP+bMMAX+c$ was used to test the water energy dynamics theory (WED). The mean annual precipitation (MAP) and the maximum temperature in the warmest month (MMAX) were selected for the following reasons: i) MAP is most closely correlated with v_a, in the set of precipitation

Table 5. Points of local minimum (Min) and maximum values (Max) of the tree species richness (v_0), the effective tree species number (v_2) and the number of prevalent tree species (v_∞) with (τ) 0.5, and 0.9 as the quantile levels.

Diversity index	Climatic variable	Unit	$\tau = 0.5$		$\tau = 0.9$	
			Min	**Max**	**Min**	**Max**
v_0	MAT	°C	(18.48; 4.39)	(9.96; 7.51)	(18.57; 7.57)	(10.48; 10.96)
	MTWM	°C	(23.30; 3.65)	(14.90; 7.94)	(23.90; 6.46)	(15.31; 10.52)
	MMAX	°C	(32.92; 3.73)	(21.35; 8.12)	(32.20; 3.17)	(23.80; 9.58)
	AAI	Days$^{0.5}$/mm	(0.10; 3.99)	(0.04; 8.11)	(0.10; 8.26)	(0.04; 11.01)
v_2	MMAX	°C	(32.75; $\approx$1.00)	(23.50; 3.60)	(29.50; 5.91)	(24.50; 6.83)
	AAI	Days$^{0.5}$/mm	–	(0.04; 4.02)	(0.09; 4.72)	(0.05; 6.22)
v_∞	AAI	Days$^{0.5}$/mm	–	(0.03; 2.70)	(0.09; 3.09)	(0.04; 4.17)

The climatic variables are as follows: mean annual temperature (MAT), mean temperature in the warmest month (MTWM), maximum temperature in the warmest month (MMAX), and the annual aridity index (AAI).

Table 6. The parameters a, b and c, probability of error (p) and the error of linear multivariate models of diversity indices (error) with climatic variables MMAX and MAP at the 0.1, 0.5 and 0.9 quantile levels (τ).

Variables	τ	r^2	Parameters	a*	b*	c*
v_0 – MMAX - MAP	0.90	0.504	value	0.00223	−0.14530	11.77380
			p	0.00038	0.00807	0.00000
			error	0.00063	0.05478	1.72616
	0.50	0.431	value	0.00286	−0.32059	12.25270
			p	0.00000	0.00000	0.00000
			error	0.00034	0.02233	0.69262
	0.10	0.834	value	0.00270	−0.27780	8.18577
			p	0.00000	0.00000	0.00000
			error	0.00047	0.03881	1.24521
v_2 – MMAX - MAP	0.90	0.578	value	0.00039	−0.10047	6.87903
			p	0.00010	0.00097	0.00000
			error	3.89923	0.03040	0.92411
	0.50	0.353	value	0.00139	−0.15818	6.20199
			p	0.00000	0.00000	0.00000
			error	0.00024	0.02065	0.62554
	0.10	0.754	value	0.00078	−0.09694	3.52116
			p	0.00000	0.00000	0.00000
			error	0.00013	0.00795	0.27403
v_∞ – MMAX-MAP	0.90	0.551	value	0.00093	−0.06663	4.64010
			p	0.00135	0.00001	0.00000
			error	0.00029	0.01504	0.46712
	0.50	0.297	value	0.00075	−0.08048	3.70658
			p	0.00000	0.00000	0.00000
			error	0.00016	0.01216	0.36610
	0.10	0.712	value	0.00035	−0.04265	2.13045
			p	0.00000	0.00000	0.00000
			error	0.00007	0.00466	0.14981

Notes: * The parameters (a, b, c) are highly statistically significant in the models ($p<0.01$).

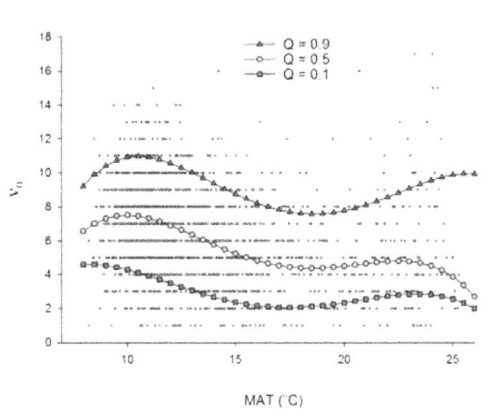

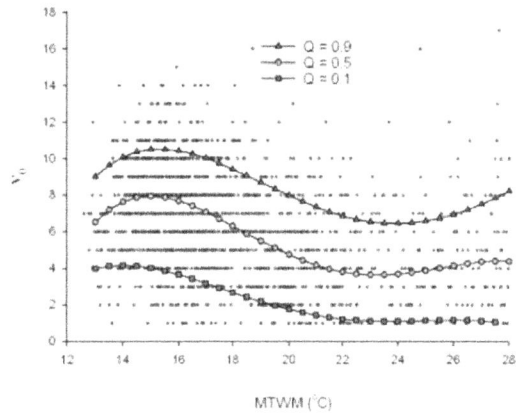

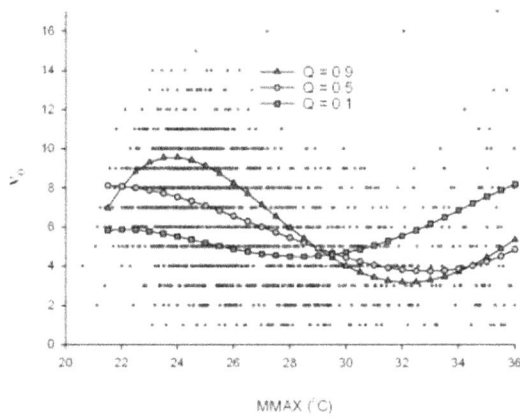

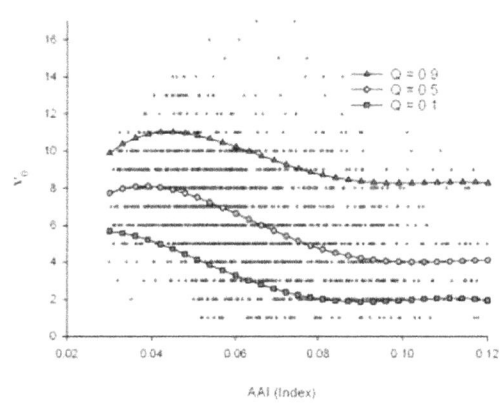

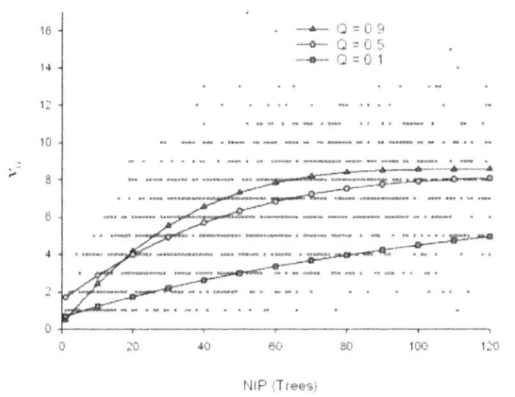

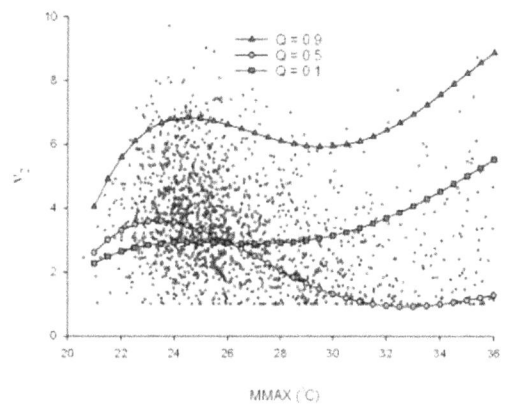

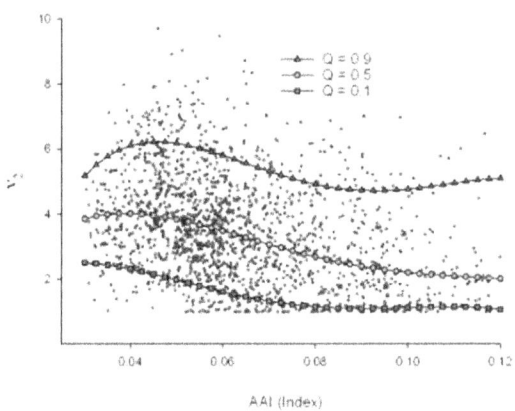

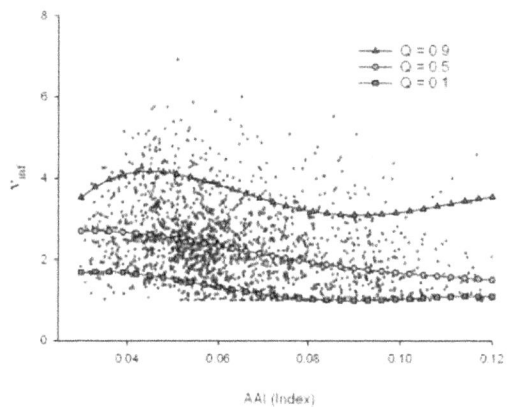

Figure 4. Series of curves of the quantile regression for tree species richness (v_0), the effective tree species number (v_2) and the amount of prevalent tree species (v_∞) for quantile levels of (τ) 0.1, 0.5, and 0.9. The climatic variables are as follows: mean annual temperature (MAT), mean temperature in the warmest month (MTWM), maximum temperature in the warmest month (MMAX), annual aridity index (AAI), and finally, the number of individuals per plot (NIP).

variables, and MMAX, in the group of temperature variables with v_a (Table 3) calculated by stepwise multiple regression, and ii) these two variables display a very low degree of collinearity (Table 2).

The values of the quantiles (τ) were set at 0.9 for the upper limit, 0.5 for the median, and 0.1 for the lower limit. The quantile 0.9 was used to estimate the trend of maximum diversity with climate variables and the quantile 0.1 for minimum diversity with climate variables and NIP. The maximum and minimum diversity should serve as additional diversity measures. The linear and nonlinear model functions (*object<- lm* and *object<-nlrq*) in R [53] and the correlation (r) between the predicted and observed v_a were used to compute the coefficient of determination (r^2) of the bi- and multivariate quantile and non-quantile models (R-code for (pseudo) r^2 for quantile regression: *res<-residuals(object, type = "rho") R2<- (cor(predict.nlrq(object), res + predict.nlrq(object)))^2*).

Results

Tree species richness (v_0) was almost twice as high as effective tree diversity (v_2) and v_2 was almost one and a half times greater than the number of prevalent tree species (v_∞), i.e. the abundance of the tree species was different (Table 1, Figure 2). The matrix of Spearman coefficients (r_s) revealed significantly weak to moderate correlations between climatic variables and also between climatic variables and diversity indices (Table 3). Climate had a greater effect on v_0 than on v_2 and v_∞, i.e., the climate had a greater influence in determining the number of rare tree species (which is the main component of the species richness index) than in determining the number of more frequent species.

The most significant relationships were the annual aridity index (AAI), maximum temperature in the warmest month (MMAX), the number of individuals per plot (NIP) and the tree species diversity (v_a), with r_s values ranging from -0.32 to -0.41, from -0.31 to -0.36, and 0.22 to 0.47, respectively. The temperature variables mean annual temperature (MAT), mean temperature in the warmest month (MTWM), degree days $>5°C$ (DD5) and degree days $>5°C$ accumulated within the frost-free period (GSDD5) were also highly significantly and negatively correlated with v_a (Table 3). The three linear multivariate non-quantile models of v_a = function (MAP, MMAX) showed weak coefficients of determination ($r^2 = 0.08$ (v_∞), 0.09 (v_2) and 0.13 (v_0)). However, the nine non-linear multivariate quantile models of v_a with MAP and MMAX exhibited moderate to high coefficient values ($r^2 = 0.55$–0.71 (v_∞), 0.58–0.75 (v_2) and 0.50–0.83 (v_0)) (Table 6). In addition, the non-linear bivariate quantile models of v_a with climatic factors as well as stand density presented moderate to high coefficient values ($r^2 = 0.31$–0.79 (v_∞), 0.36–0.81 (v_2) and 0.36–0.90 (v_0)) (Table 4, Figure 4). Plots with minimal tree diversity were often located between 1,800 and 2,200 m above sea level (at AAI≈0.120–0.154 index value) at the border of the Chihuahuan desert, facing inland.

Analysis of ten non-linear quantile (τ) regressions of the three ranges of tree species diversity (v_0, v_2 and v_∞) and six climate variables as well as the number of individuals per plot (NIP) were carried out for absolute r_s values >0.30 between the three tree species diversities and climate variables and NIP (Tables 2 and 3).

The results showed five significant fits for regression of v_0 and v_2 with MMAX for the maximum, median, and minimum diversity of tree species (quantile levels (τ) = 0.1, 0.5 and 0.9). The regression of tree species diversity with AAI and NIP resulted in five good-fit models (Table 4). In contrast, DD5 and GSDD5 were not significantly correlated with the three diversity indices according to the polynomial models. Quantile regressions of species richness (v_0) with statistically significant parameters such as a, b, c, d, and e were mainly found for MAT, MTWM, MMAX, AAI and NIP. Quantile regressions of the effective species number (v_2) with statistically significant parameters were only observed for MMAX and AAI. The number of prevalent tree species (v_∞) only yielded a significant quantile regression with AAI (Table 4, Figure 4).

All temperature-diversity-curves were hollow-shaped (with a local minimum) for the median ($\tau = 0.5$) and maximum diversity ($\tau = 0.9$). Tree species diversity is generally higher in colder temperate climates and in hot climates, but tends to fall to a local minimum in milder climates. The local maxima of the tree species richness and effective diversity (v_0 and v_2 at $\tau = 0.9$) were found in temperate climate areas. In some cases, the curves were sinuous, with a local maximum and local minimum of tree species diversity. However, the relationship between AAI and the diversity variables was essentially linear or constrained within a linear limit. Finally, the NIP-tree diversity curves formed saturation curves (Figure 4, Table 5).

The resulting curves in Figure 5 showed that the minimum NIP occurred at the point at which the values of the most diversity-influential variables MAT, MMAX, and AAI (18.5 °C, 32.9 °C, and 0.10 days$^{0.5}$*mm^{-1}, respectively) produced almost exactly the local minimum of the tree species richness (v_0) at quantile (τ) 0.5 (Figure 4, Table 5). In the final graph in Figure 5, the local maximum of AAI value clearly coincided with MMAX (32.9 °C), generating a local minimum of the median tree species richness (v_0) at the quantile level 0.5 (Table 5).

The multivariate quantile regression of tree species diversity (v_0, v_2 and v_∞ at $\tau = 0.1$, 0.5 and 0.9) with MAP and MMAX showed that tree species diversity is generally higher in colder, humid temperate than in dry, hot climates in all nine models. MAP and MMAX together are good predictors of the minimum, median and maximum tree species diversity. These models show that the tree species diversity increases with MAP but decreases with MMAX (Table 6, Figure 6).

Discussion and Conclusion

The results of this study showed that nearly all climate variables, particularly those identified in the temperature group, were weakly to moderately related to the tree species diversity in the study region, indicated using non-quantile calculations (Table 3). Similar results have been found in other studies [23], [56]. The findings also demonstrated that there is a strong relationship between the minimum, median and maximum tree species diversity and climate variables calculated by quantile regression models (Tables 4 and 6). Therefore, quantile regression could provide a more complete view of possible causal relationships between species diversity and climate variables [57].

Results support the Productivity-Diversity Hypothesis, Physiological Tolerance Hypothesis and Water-Energy Dynamic Theo-

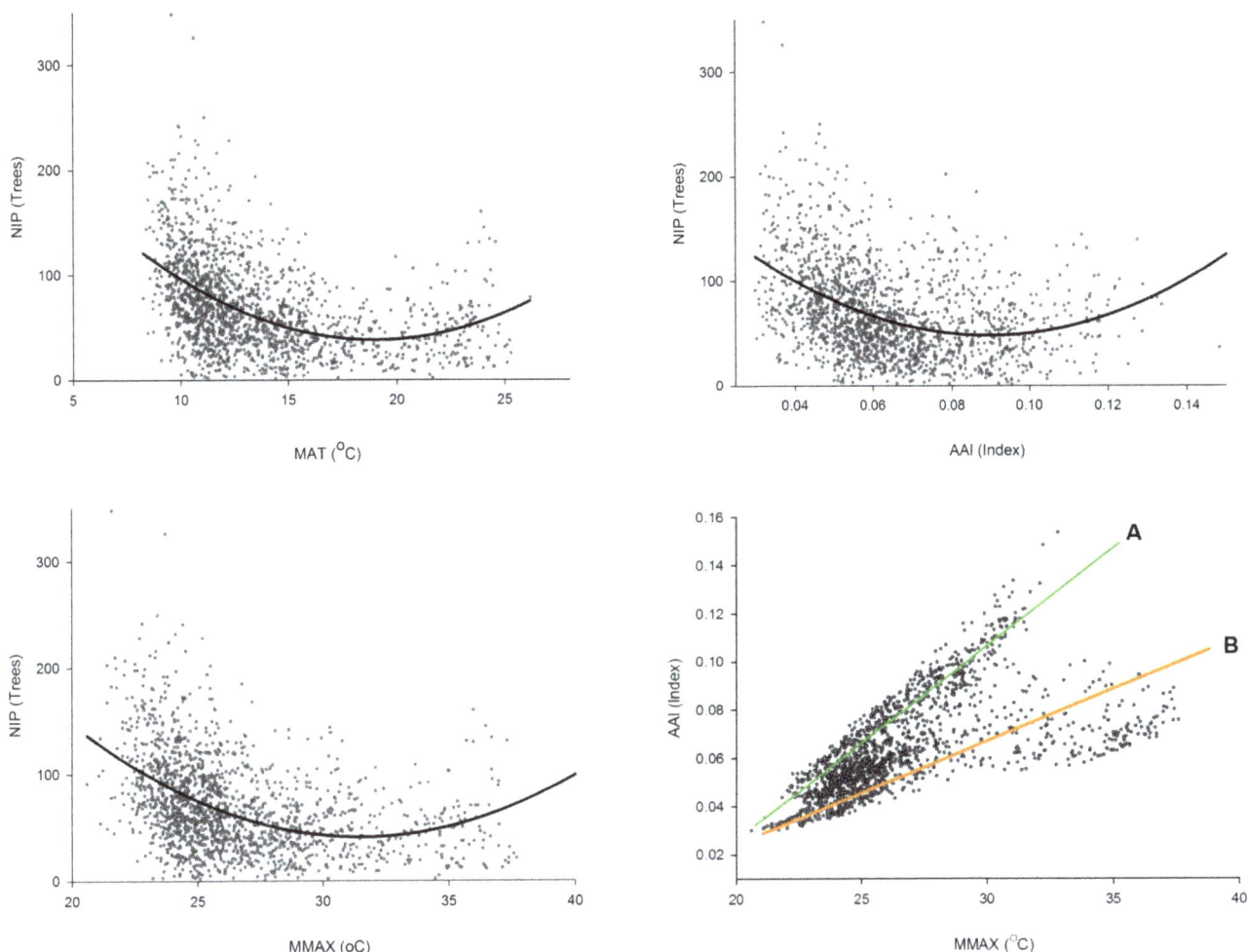

Figure 5. Curves of quantile regression with combinations and relationship between MMAX and AAI, Sierra Madre Occidental, Durango. Those shown are number of individuals per plot (NIP) with mean annual temperature (MAT, $r^2 = 0.52$); annual aridity index (AAI, $r^2 = 0.47$) and maximum temperature in the warmest month (MMAX, $r^2 = 0.46$). The correlation was calculated with the quadratic function by quantile regression ($\tau = 0.5$). Relationship between MMAX and AAI on the Sierra Madre Occidental, Durango, A = part facing inland and B = part facing Pacific Ocean.

ry, but not the Mid-Domain Effect or Metabolic Theory. The annual aridity index (AAI) was the variable most closely related to the diversity indices analyzed, although the temperature variables MAT, MTWM and MMAX were also closely associated with these variables (Table 3), as observed in other studies [21], [58], [59]. The explanation is not consistent with the Hypothesis of Evolutionary Rates and Biotic Interactions [23], or even the Metabolic Theory of Ecology [24] because diversity should be positively correlated with temperature.

However, in the present study, the relationship between diversity and temperatures in degrees Celsius was almost negative and nonlinear (Table 3, Figure 4). Hawkins et al., [34] observed significant heterogeneity in slopes among data sets, and concluded that the combined slopes across studies were significantly lower than the range of slopes predicted by metabolic theory. However, the moderate positive diversity-density relationship observed in the study (Table 3, Figure 4) confirmed that the energy-diversity relationship limits the number of individuals since climate moderately affects the net primary productivity (NPP) and thus diversity [23], [24]. The results also supported the Tolerance-Diversity Hypothesis as all diversity indices were negatively

affected by the degree of aridity (Table 3, Figure 4), i.e., the harsher the climate conditions, the fewer the number of species that can maintain local conditions [58], [23]. Because of the elevation gradient, which was strongly correlated with temperature, the results also provided evidence against the Mid-Domain Effect [22] as an explanation for species diversity patterns. Furthermore, the water-energy dynamics (WED) theory [25] was also confirmed since all tree species diversities were weakly and simultaneously affected by precipitation (MAP) and temperature (MMAX) (Table 6, Figure 6). Not surprisingly, the AAI, which combined water and energy, proved to be the strongest correlate of tree diversity (Table 3). The model parameters obtained (Table 6) were highly significant for the three levels of diversity. The results obtained for the WED – tree diversity relationship calculated by linear non-quantile models was lower than the values reported by Hawkins et al., [21]. They found on average over (60%) of the variation in the richness of a wide range of plant and animal groups computed also by non-quantile statistic methods, but similarly analyzed by non-linear quantile models (Table 6). In our study, the overall weak to moderate relationship between WED and tree species richness was probably due to the small

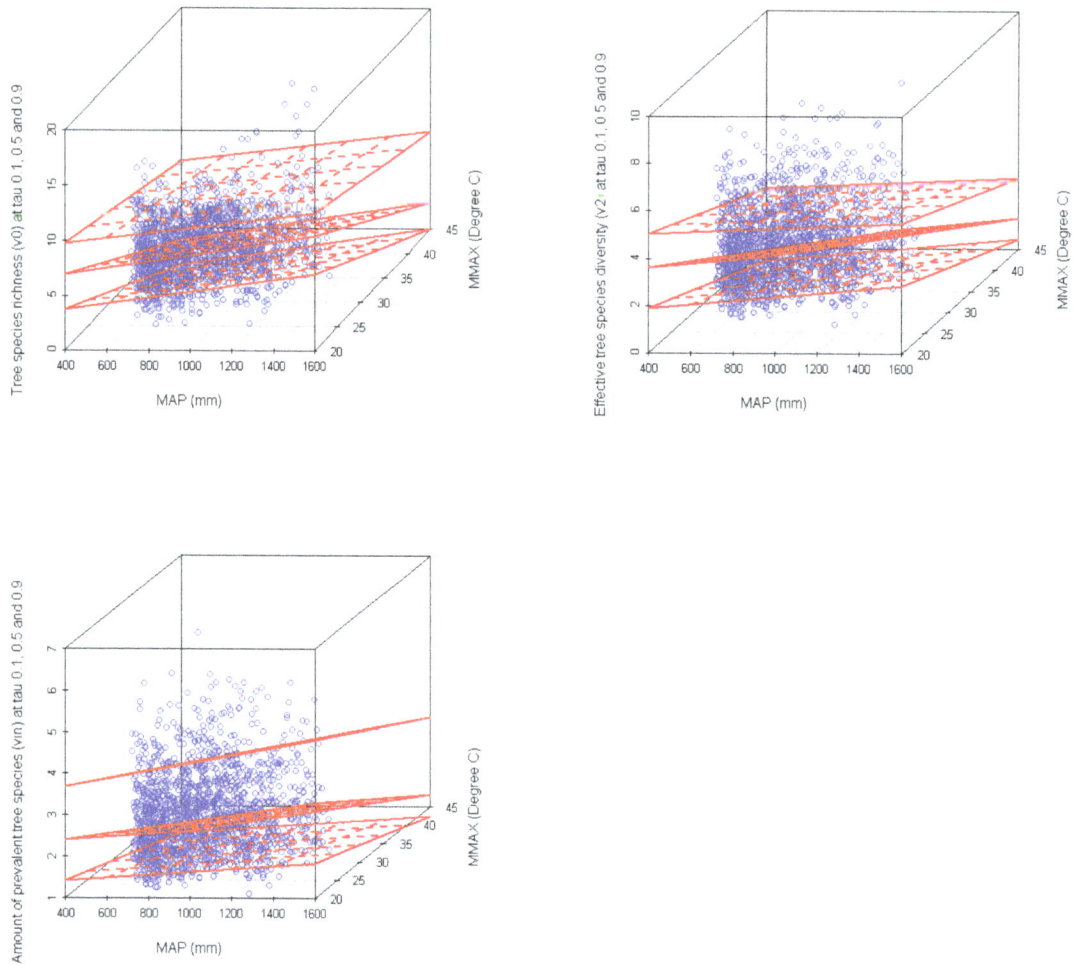

Figure 6. Linear multivariate model of tree species diversity with temperature and precipitation. Included tree species richness (v_0), effective tree species number (v_2) and the number of prevalent tree species (v_∞) with mean annual precipitation (MAP) and maximum temperature in the warmest month (MMAX) for quantile levels of (τ) 0.1, 0.5, and 0.9.

geographic scale, the species studied, and the methodology applied.

The saturation shape of the tree number per plot (NIP)-tree diversity relationship (Figure 4) may be caused by a combination of the accumulation effect [36] and increasing competition in denser plots [60], [61]. While the accumulation effect resulted in higher diversity, the self-thinning processes that took place as a result of competition in dense conditions [62] led to saturation in tree species diversity.

Interestingly, the quantile regression analysis showed that various temperature-diversity curves form a statistically significantly hollow-shaped pattern (Tables 4 and 5, Figure 4), i.e. there was a local minimum in the center of the curve. As already mentioned, this is well explained by the WED theory, which addresses the delicate balance between the temperature available to plants for growth and the available moisture (precipitation). It also provides a good explanation for the unusual regression curves because the local minimum diversity occurred almost exact when the aridity and temperature were highest and the precipitation was lowest (Figures 5 and 6). This was because lower water availability led to reduced species diversity by decreasing productivity and increasing drought stress [29], [35], [58], [63], [65].

However, the trend in the relationship between MAT and diversity was consistent with that shown in the scatter plots constructed by Hawkins et al., [34]. After transformation of the special temperature index to degrees Celsius, these scatter plots showed a local minimum diversity at a temperature of about 20°C for Californian plants, Southern African woody plants, New world ants, and of about 21–23°C for Californian and Australian butterflies and Australian amphibians.

Using non-quantile statistic methods, climatic factors had a greater effect on tree species richness (v_0), in which rare tree species have greater weighting, than on the effective species number (v_2) and the amount of prevalent tree species (v_∞) (Table 3). The number of rare tree species, which generated the difference between values of v_0 and v_2, was more dependent on climate than the number of more frequent species was. In contrast, the authors of a study of Scottish grassland species concluded that the richness of rare species may be intrinsically less well explained by environmental variables than the richness of common species [65]. Small populations are more exposed to genetic erosion, demographic and environmental stochasticity and natural catastrophes [66], and they are therefore at a high risk of falling below the minimum viable population size [67] and will be the first to disappear. However, in non-linear quantile regression models climatic variables had a similar effect on v_0, v_2 and v_∞, but affected the minimum, median and maximum tree species diversity differently (Tables 4 and 6).

The minimum ($\tau = 0.1$), and maximum ($\tau = 0.9$) diversity can be well predicted using climate variables (Tables 4 and 6). Therefore, the minimum and maximum diversity could be very sensitive indicators that can detect large-scale environmental changes (such as human-derived fragmentation and climate change) as target diversity, in contrast with the actual (observed) diversity.

Forecasts of future climate scenarios for Mexico predict an average annual temperature increase of 3.7–3.8°C, a decrease in annual precipitation of 18.2%, and an increase in aridity (AAI) of about 26% by the end of the century (in 2090) [47]. In this scenario, the diversity of tree species in the state of Durango will be drastically reduced. Assuming that these predictions are accurate, the median diversity-WED-models calculate reductions of 26% in species richness (tree species number), 25% in effective species number, and 19% in prevalent tree species in a mean plot by the year 2090. However, the model error was moderate ($r^2 = 0.30$–0.43; Table 6).

In conclusion, contemporary climate affected the minimum, median and maximum tree species diversity moderately to strongly. Water-energy dynamics provided a satisfactory explanation for the pattern of minimum, median and maximum diversity, and an understanding of this factor is therefore critical to future biodiversity research [21]. The quantile regression could be a useful tool to accurately describe the curve shape of minimum, median and maximum species diversity.

Acknowledgments

We are grateful to the Comisión Nacional Forestal (CONAFOR) for sharing the National Forest and Soil Inventory data and to the Consejo Nacional de Ciencia y Tecnología (CONACYT) for the support awarded for a doctoral research project. We also thank María del Socorro González-Elizondo and Daniel Moya-Navarro, for their helpful suggestions in an early manuscript.

Author Contributions

Conceived and designed the experiments: CW RSF. Performed the experiments: RSF CW. Analyzed the data: CW RSF GPV. Contributed reagents/materials/analysis tools: RSF CW. Wrote the paper: RSF CW GPV.

References

1. CBD (Convention on Biological Diversity) (2011) Handbook of the Convention on biological diversity. 3a edition. 28 p. Available: http://www.cbd.int/doc/handbook/cbd-hb-a-en.pdf. Accesed 2011 Nov 6.
2. McNeely JA (1988) Economics and Biological Diversity: Developing and Using Economic Incentives to Conserve Biological Resources. IUCN, Gland, Switzerland. XIV, 232 pp.
3. Norse EA, Rosenbaum KL, Wicove DS, Wilcox BA, Rome WH, et al. (1988) Conserving BioScience Diversity in our National Parks. The Wilderness Society. Washington, D.C.
4. Gregorius HR, Gillet EM (2008) Generalized Simpson-diversity. Ecol. Model. 2:90–96.
5. Noss R (1990) Indicators for monitoring Biodiversity: A hierarchical approach. Conserv. Biol. 4:355–364.
6. Whittaker RH (1972) Evolution and Measurement of Species Diversity. Taxon, 21:213–251.
7. Shannon CE (1948) A mathematical theory of communication. The Bell System Technical Journal, 27:379–423 and 623–656.
8. Simpson EH (1949) Measurement of diversity. Nature 163: 688.
9. Hill MO (1973) Diversity and evenness: a unifying notation and its consequences. Ecology 54:427–432.
10. Gregorius HR (1978) The concept of genetic diversity and its formal relationship to heterozygosity and genetic distance. Math. Biosci. 41:253–271.
11. Zyczkowski K (2003) Rényi Extrapolation of Shannon Entropy. Open Sys & Information Dyn. 10:297–310. Kluwer Academic Publishers. Netherlands.
12. Jost L (2006) Entropy and diversity. OIKOS. 113(23):363–375.
13. Jost L (2007) Partitioning diversity into independent alpha and beta components. Ecology. 88:2427–2439.
14. Gregorius HR (2010) Linking diversity and differentiation. Diversity 2:370–394. doi:10.3390/d2030370.
15. Woodward FI (1987) Climate and Plant Distribution. Cambridge University Press, Cambridge, UK.
16. Aitken SN, Yeaman S, Holliday JA, Wang T, Curtis-McLane S (2008) Adaptation, migration or extirpation: climate change outcomes for tree populations. Evol. Appl. 1:95–111.
17. Hebda RJ (1994) Future of British Columbia's flora. In Biodiversity in British Columbia: Our changing environment. L.E. . Harding and E. . McCullum (editors). Canadian Wildlife Service Ottawa ON., pp 343–352.
18. Richardson BA, Rehfeldt GE, Kim MS (2009) Congruent climat-related genecological responses from molecular markers and quantitative traits for western white pine (Pinus monticola). Int. J. Plant Sci. 170(9):1120–1134.
19. Rehfeldt GE, Crookston N, Sáenz-Romero C, Campbell E (2012) North American vegetation model for land-use planning in a changing climate: a solution to large classification problems. Ecol. Appl. 22: 119–141.
20. Wright DH, Currie DJ, Maurer BA (1993) Energy supply and patterns of species richness on local and regional scales. Species Diversity in Ecological Communities: Historical and Geographical Perspectives (eds. Ricklefs, R.E. & Schluter, D.). University of Chicago Press, Chicago II. pp. 66–74.
21. Hawkins BA, Field R, Cornell HV, Currie DJ, Guégan J-F, Kaufman DM, et al. (2003) Energy, water, and broad-scale geographic patterns of species richness. Ecology 84:3105–3117.
22. Colwell RK, Hurtt GC (1994) Nonbiological gradients in species richness and a spurious Rapoport effect. Amer. Nat. 144:570–595.
23. Currie DJ, Mittelbach GG, Cornell HV, Field R, Guégan J-F, et al. (2004) Predictions and tests of climate-based hypotheses of broad-scale variation in taxonomic richness. Ecol. Lett. 7:1121–1134.
24. Sanders NJ, Lessard JP, Fitzpatrick MC, Dunn RR (2007) Temperature, but not productivity or geometry, predicts elevational diversity gradients in ants across spatial grains. Global Ecol. Biogeogr. 16:640–649.
25. O'Brien EM (1998) Water-Energy Dynamics, Climate, and Prediction of Woody Plant Species Richness: An Interim General Model. J. Biogeog. 25:379–398.
26. Eiserhardt WL, Bjorholm S, Svenning J-C, Rangel TF, Balslev H (2011) Testing the water-energy theory on American palms (Arecaceae) using geographically weighted regression. PLoS ONE 6(11), e27027.
27. Chen X, Li BL (2003) Effect of global climatic change and human disturbances on tree diversity of the forest regenerating from clear-cuts of mixed broadleaved Korean pine forest in Northeast China. Chemosphere 51:215–226.
28. Sang W (2009) Plant diversity patterns and their relationships with soil and climatic factors along an altitudinal gradient in the middle Tianshan Mountain area, Xinjiang, China. Ecol. Res. 24:303–314.
29. Wang Z, Brown JH, Tang Z, Fang J (2009) Temperature dependence, spatial scale, and tree species diversity in eastern Asia and North America. Proceedings of the National Academy of Sciences 106:13388–13392.
30. Karger N, Kluge J, Krömer H, Lehnert M, Kessler M (2011) The effect of area on local and regional elevational patterns of species richness. J. Biogeogr. 38:1177–1185.
31. Sharma CM, Suyal S, Gairola S, Ghildiyal SK (2009) Species richness and diversity along an altitudinal gradient in moist temperate forest of Garhwal Himalaya. J. American Science 5:119–128.
32. Bharali S, Ashish P, Mohamed LK, Lal BS (2011) Species diversity and community structure of a temperate mixed Rhododendron forest along an altitudinal gradient in West Siang District of Arunachal Pradesh, India. Nature and Science 9:125–140.
33. Alkemade R, Bakkenes M, Eickhont B (2011) Towards a general relationship between climate change and biodiversity: an example for plant species in Europe. Reg Environ Change 11:143–150.
34. Hawkins BA, Albuquerque FS, Araujo MB, Beck J, Bini LM, et al. (2007) A global evaluation of metabolic theory as an explanation for terrestrial species richness gradients. Ecology 88:1877–1888.
35. Steege HT, Pitman N, Sabatier D, Castellanos H, Hout PVD, et al. (2003) A spatial model of tree α diversity and tree density for the Amazon. Biodivers. Conserv. 12:2255–2277.
36. Gotelli NJ, Colwell RK (2001) Quantifying biodiversity: procedures and pitfalls in the measurement and comparison of species richness. Ecol. Lett. 4:379–391.
37. WWF (World Wildlife Fund) (2001) Sierra Madre Occidental pine-oak forests (NA0302) Prepared by Valero A, Schipper J, Allnutt T and Burdette Ch. Available: http://www.worldwildlife.org/wildworld/profiles/terrestrial/na/na0302_full.html Accesed 2011 Dec 14.
38. Rzedowski J (2006) Vegetación de México. 1a. edición digital. Comisión Nacional para el Conocimiento y Uso de la Biodiversidad. México. 504 p.
39. Wehenkel C, Corral-Rivas JJ, Hernández-Díaz JC, Gadow KV (2011) Estimating balanced structure areas in multi-species forests on the Sierra Madre Occidental. Mexico. Ann. Forest Sci. 68:385–394.

40. González-Elizondo MS, González-Elizondo M, Tena-Flores JA, Ruacho-González L, López-Enríquez L (2012) Vegetación de la Sierra Madre Occidental, México: Una Síntesis. Acta Bot. Mex. 100:351–403.

41. SRNyMA (Secretaría de Recursos Naturales y Medio Ambiente) (2006) Programa Estratégico Forestal 2030 para Durango. Gobierno del Estado de Durango, México. 209 p.

42. Hutchinson ME (1991) Continent-wide data assimilation using thin plate smoothing splines. Pages 104–113 in JD Jasper, ed. Data assimilation systems. Meteorology, Melbourne.

43. CONAFOR (Comisión Nacional Forestal) (2004) Inventario Nacional Forestal y de Suelos.

44. Wehenkel C, Corral-Rivas JJ, Gadow KV (2014) Quantifying differences between ecosystems with particular reference to selection forests in Durango/Mexico. Forest Ecology and Management 316:117–124.

45. Rehfeldt GE (2006) A spline model of climate for the Western United States. Gen Tech Rep. RMRS-GTR-165. U.S. Department of Agriculture, Forest Service, Rocky Mountain Research Station, Fort Collins, Colorado, USA.

46. Rehfeldt GE, Crookston NL, Warwell MV, Evans JS (2006) Empirical analyses of plant-climate relationships for the Western United States. Int. J. Plant. Sci. 167:1123–1150.

47. Sáenz-Romero C, Rehfeldt GE, Crookston NL, Duval P, St-Amant R, Beaulieu J, et al. (2010) Spline models of contemporary, 2030, 2060 and 2090 climates for Mexico and their use in understanding climate-change impacts on the vegetation. Climatic Change 102:595–623.

48. Hutchinson ME (2004) Anusplin Version 4.3. Centre for Resource and Environmental Studies. The Australian National University, Canberra, Australia.

49. Crow JF, Kimura M (1970) An introduction to population genetics theory. Burges Pub. Co. Science 591 p.

50. Statistical Analysis System SAS Institute Inc. (2004) SAS/ETS 9.1 User's Guide. Cary, NC. SAS Institute Inc.

51. Wolfgang KH, Leopold S (2012) Applied Multivariate Statistical Analysis. Third Edition. Springer Heidelberg Dordrecht London New York. 514 p.

52. Tabachnick BG, Fidell LS (2001) Using multivariate statistics. Needham Heights, MA: Allyn & Bacon. 966 p.

53. R version 2.13.1 (2011) The R Foundation for Statistical Computing. ISBN 3-900051-07-0. Available: http://www.r-project.org/ Accesed 2011 Nov 6.

54. Koenker R, Bassett G Jr (1978) Regression Quantiles. Econometrica 46:33–50.

55. Koenker R (2011) Quantile regression in R: a Vignette. Available: http://cran.r-project.org/web/packages/quantreg/vignettes/rq.pdf Accesed 11th March 2011.

56. Badii MH, Landeros J, Cerna E (2008) Patrones de asociación de especies y sustentabilidad (Species association patterns and sustainibility). Daena: International Journal of Good conscience 3:632–660.

57. Cade BS, Noon BR (2003) A gentle introduction to quantile regression for ecologists. Frontiers in Ecology and the Environment 1:412–420.

58. Kalin-Arroyo MT, Squeo FA, Armesto JJ, Villagran C (1988) Effects of Aridity on Plant Diversity in the Northern Chilean Andes: Results of a Natural Experiment. Ann. Mo. Bot. Gard. 75:55–78.

59. Allen AP, Brown JH, Gillooly JF (2002) Global biodiversity, biochemical kinetics, and the energetic-equivalence rule. Science 297:1545–1548.

60. Begon M, Harper JL, Townsend CR (1996) Ecology: Individuals, populations and communities. Blackwell Science. Third edition. USA.

61. Brooker RW (2006) Plant-plant interactions and environmental change. New Phytol. 171:271–284.

62. Zeide B (1991) Self-thinning and Stand-Density. Forest. Sci. 37:517–523.

63. Currie DJ, Paquin V (1987) Large-scale biogeographical patterns of species richness of trees. Nature 329:326–327.

64. O'Brien EM (2006) Biological relativity to water–energy dynamics. J. Biogeogr. 33:1868–1888.

65. Lennon JJ, Beale CM, Reid CL, Kent M, Pakeman RJ (2010) Are richness patterns of common and rare species equally well explained by environmental variables? Ecography 34:529–539.

66. Wehenkel C, Sáenz-Romero C (2012) Estimating genetic erosion using the example of *Picea chihuahuana* Martínez. Tree Genet Genomes 8(5):1085–1094.

67. Traill LC, Bradshaw LJA, Brook BW (2007) Minimum viable population size: A meta-analysis of 30 years of published estimates. Biol. Conserv. 139:159–166.

9

Unmanned Aerial Survey of Fallen Trees in a Deciduous Broadleaved Forest in Eastern Japan

Tomoharu Inoue[1]*, Shin Nagai[1], Satoshi Yamashita[2], Hadi Fadaei[1], Reiichiro Ishii[1], Kimiko Okabe[2], Hisatomo Taki[2], Yoshiaki Honda[3], Koji Kajiwara[3], Rikie Suzuki[1]

1 Department of Environmental Geochemical Cycle Research, Japan Agency for Marine-Earth Science and Technology (JAMSTEC), Yokohama, Japan, 2 Department of Forest Entomology, Forestry and Forest Products Research Institute (FFPRI), Tsukuba, Ibaraki, Japan, 3 Center of Environmental Remote Sensing, Chiba University, Chiba, Japan

Abstract

Since fallen trees are a key factor in biodiversity and biogeochemical cycling, information about their spatial distribution is of use in determining species distribution and nutrient and carbon cycling in forest ecosystems. Ground-based surveys are both time consuming and labour intensive. Remote-sensing technology can reduce these costs. Here, we used high-spatial-resolution aerial photographs (0.5–1.0 cm per pixel) taken from an unmanned aerial vehicle (UAV) to survey fallen trees in a deciduous broadleaved forest in eastern Japan. In nine sub-plots we found a total of 44 fallen trees by ground survey. From the aerial photographs, we identified 80% to 90% of fallen trees that were >30 cm in diameter or >10 m in length, but missed many that were narrower or shorter. This failure may be due to the similarity of fallen trees to trunks and branches of standing trees or masking by standing trees. Views of the same point from different angles may improve the detection rate because they would provide more opportunity to detect fallen trees hidden by standing trees. Our results suggest that UAV surveys will make it possible to monitor the spatial and temporal variations in forest structure and function at lower cost.

Editor: Dale A. Quattrochi, NASA Marshall Space Flight Center, United States of America

Funding: This study was funded by the Environment Research and Technology Development Fund (S-9) of the Ministry of the Environment of Japan. The funders had no role in study design, data collection and analysis, decision to publish, or preparation of the manuscript.

Competing Interests: The authors have declared that no competing interests exist.

* Email: tomoharu@jamstec.go.jp

Introduction

Fallen trees are an ecologically relevant indicator of forest biodiversity [1], since they provide habitat for many species, such as small animals (e.g., birds, mammals, and insects) and fungi [2–4]. In addition, the decomposition of fallen trees is an important mechanism driving biogeochemical cycles such as nutrient and carbon cycling [5–6]. Thus, information about their spatial distribution is of use in indicating species distribution, biodiversity, and biogeochemical cycles in forest ecosystems [7–8]. Fallen trees are typically assessed by field surveys [9]. However, because field surveys are time consuming and labour intensive [10], it is expensive to assess the spatial distribution of fallen trees over a wide area. One way to reduce costs is to use remote-sensing technologies such as airborne and satellite imagery (e.g., [10–14]). For example, some studies have utilized the remote-sensing technologies for assessment of post-hurricane forest damage (e.g., [15–16]).

Recent advances in the technology of unmanned aerial vehicles (UAVs) have made UAVs ideal for remote sensing [17–20]. Equipped with sensors such as digital cameras, UAVs can gather aerial photographs with fine spatial and temporal resolution [20].

They also have greater flexibility in flying height and schedule and lower operating costs than manned aircraft [18,19,21]. For these reasons, aerial surveys using UAVs have been suggested for low-cost ecological research (e.g., vegetation monitoring and wildlife surveys) [19,20,22–26].

In this study, we photographed a deciduous broadleaved forest from a UAV, compared the positions of fallen trees identified in the images by eye with those recorded on the ground, and compared the number of the fallen trees detected between the original images and an orthorectified mosaic. The purpose of this study was to show the applicability and limits of this technique.

Materials and Methods

Study site

The study was conducted in a 300-m × 200-m plot in the Ogawa Forest Reserve (OFR) in Kitaibaraki, Ibaraki, Japan (36°56′10″N, 140°35′18″E, 650–700 m above sea level; Figs. 1, 2). The forest was dominated by deciduous broadleaved trees such as oak (*Quercus serrata*) and beech (*Fagus japonica* and *F. crenata*) [27], with a patchy distribution of dwarf bamboos on the forest floor [28]. The OFR is described in detail by Nakashizuka and

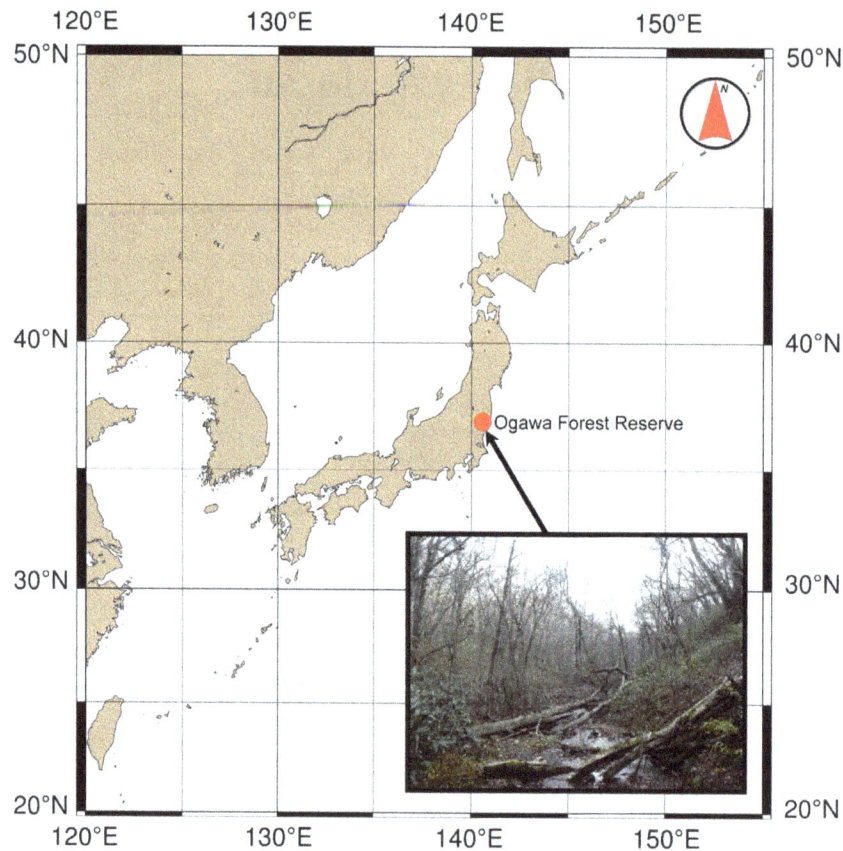

Figure 1. Location of the Ogawa Forest Reserve (OFR).

Matsumoto [29]. This study was conducted including a national forest under the permission of the Ibaraki district forest office of the Forestry Agency in Japan.

Collection of low-altitude aerial photographs from a UAV

On 29 November 2011, when the trees were bare, we flew a UAV (RMAX-G1 helicopter, Yamaha-Motor Co. Ltd., Shizuoka, Japan; Fig. 3; a model commonly used in Japan (e.g., [30–32])) over the OFR plot at 30 to 70 m above the ground in a north–south orientation at 3 m s^{-1}. A consumer-grade digital camera with a 35-mm lens (EOS Kiss X5, Canon, Tokyo, Japan; image sensor 14.9 mm × 22.3 mm) mounted beneath the UAV pointing straight down took images (5184 × 3456 pixels) every 5 s. The

UAV was also equipped with a global positioning system (GPS) detector that recorded altitude, latitude, and longitude every 0.1 s. The GPS position with the timestamp closest to that of each photograph was used as the position of the UAV.

Generation of a digital elevation model (DEM)

The UAV was also equipped with a laser range finder (LRF; SkEyesBOX MP-1, SkEyes Unlimited Corp., Washington, PA, USA), which assembled a 3D point cloud. We divided the plot (300 m × 200 m) into 1-m × 1-m cells, and the lowest elevation in each cell was used in the creation of a digital elevation model (DEM) of the site.

Table 1. Relationship of maximum diameter between ground-surveyed and visually identified fallen trees.

Maximum diameter* of fallen tree (m)	Number of ground-surveyed fallen trees	Number of visually identified fallen trees	Identification rate (%)
0.30≤ x	8	7	88
0.20≤ x <0.30	7	2	29
0.10≤ x <0.20	28	2	7
0.05≤ x <0.10	1	0	0

*Maximum of diameters at each end and middle (see Table S1).

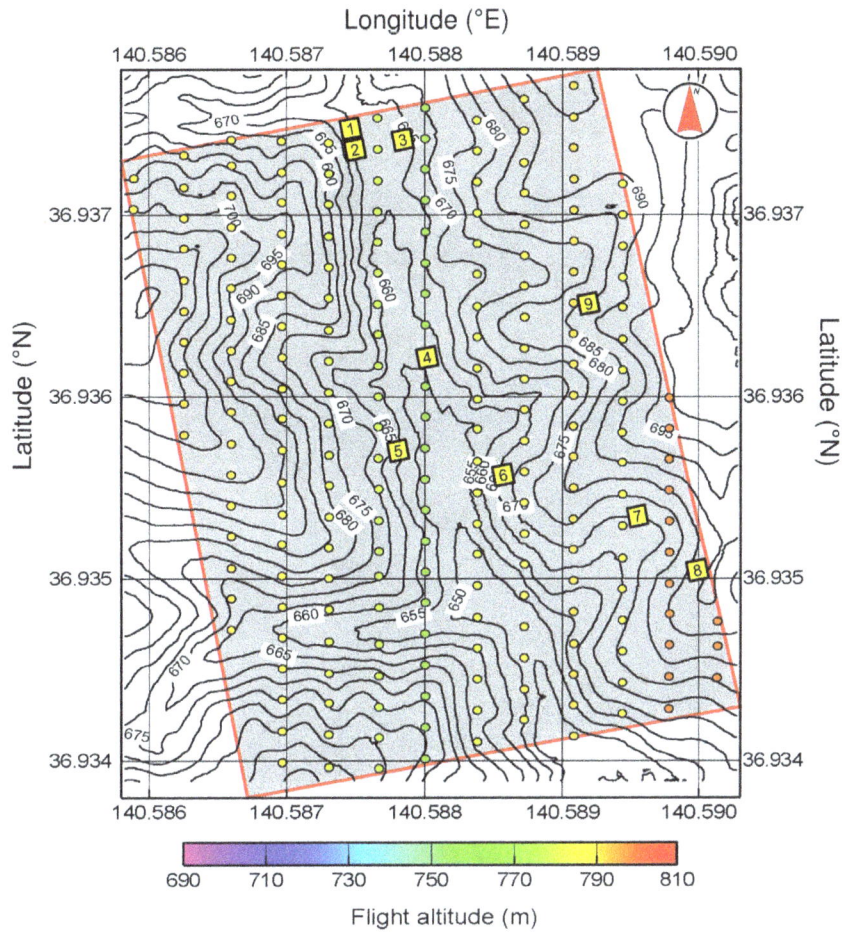

Figure 2. OFR plot (grey area). Yellow squares mark nine sub-plots for ground survey. Circles indicate UAV photograph points. Circle colour indicates flight altitude.

Orthorectification and mosaic of aerial photographs

The aerial photographs were orthorectified according to the DEM and the position (longitude and latitude) and attitude (pitch, roll, and heading) of the UAV, and assembled into one mosaic image with the help of "tie points" in overlapping photographs.

Ground survey

On 31 May 2012 in nine 10-m × 10-m sub-plots (Fig. 2), ground observers recorded the position, stem diameters (each end and midpoint) and length of all fallen trees with a diameter of >

5 cm. The ground survey was conducted in leafy season because it was easier to check the tree species in the study site when there were leaves on the trees, as opposed to a leaf-off season.

Detection of fallen trees in aerial photographs

Fallen trees were identified by eye in the original aerial photographs on a computer monitor. Positions were compared with those of fallen trees mapped in the ground survey and those identified by eye in the orthorectified mosaic.

Table 2. Relationship between lengths of ground-surveyed and visually identified fallen trees.

Length of fallen tree (m)	Number of ground-surveyed fallen trees	Number of visually identified fallen trees	Identification rate (%)
$10 \leq x$	9	7	78
$5 \leq x < 10$	15	3	20
$0 \leq x < 5$	20	1	5

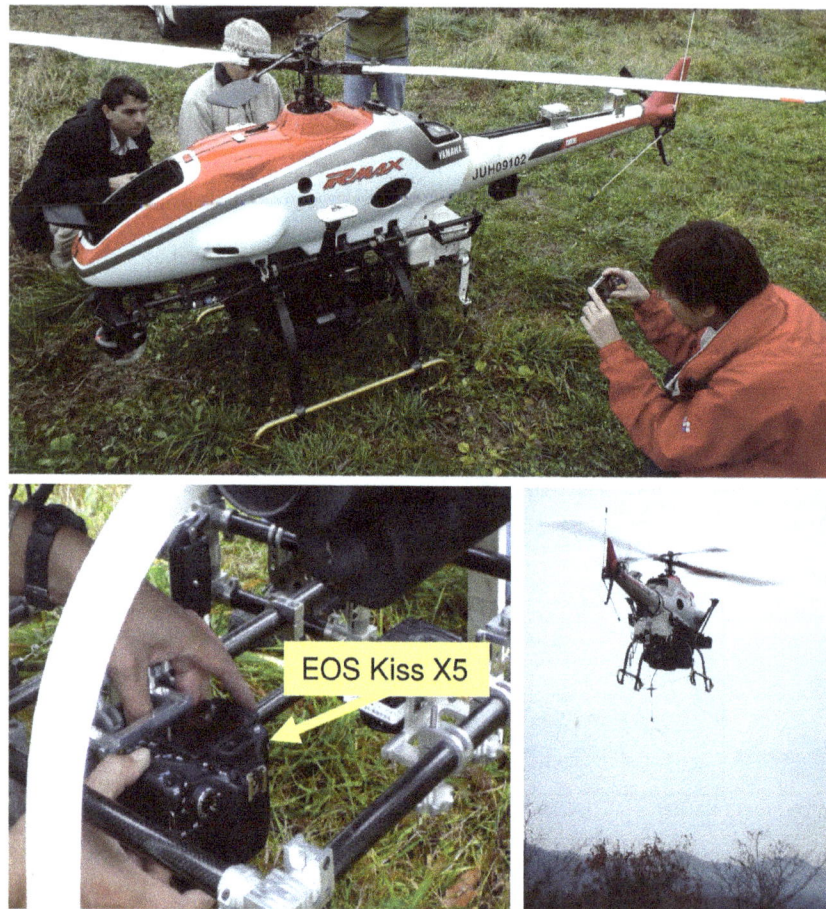

Figure 3. UAV (RMAX-G1) equipped with a digital camera (EOS KISS X5).

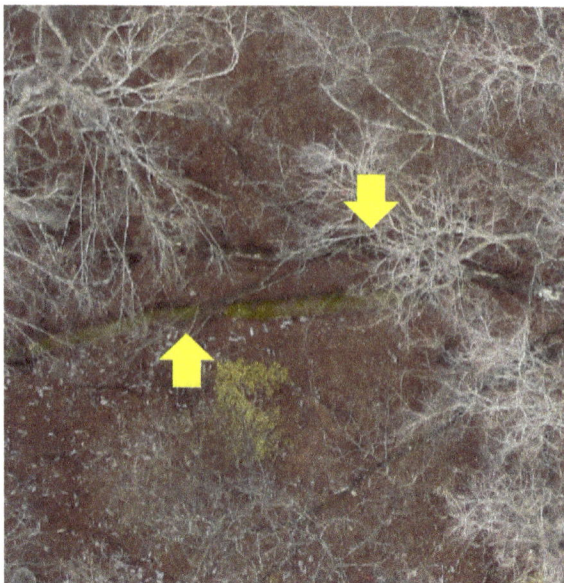

Figure 4. Close-up image of part of the forest floor. Two fallen trees are detectable.

Results

The OFR plot was covered by 211 aerial photographs (Fig. 2) with a spatial resolution of 0.5 to 1.0 cm per pixel. Since the trees were bare in late November, fallen trees were visible (Fig. 4). A DEM of the plot was generated from a cloud of 5445612 points (Fig. 5). The DEM showed a valley with an elevation range of 640 m (in the south) to 720 m (in the north). From the DEM and the position of the UAV, an orthorectified mosaic image was generated (Fig. 6).

In the ground survey, we found a total of 44 fallen trees in the sub-plots (Table S1). By eye, however, we identified only 11 fallen trees in the sub-plots on the original images (Table S1). We identified 80% to 90% of fallen trees which were >30 cm in diameter or >10 m in length, but few that were thinner or shorter (Tables 1, 2). Over the whole plot, we detected 244 fallen trees on the original photographs and 209 on the orthorectified mosaic (Fig. 6).

Discussion

Because fallen trees are generally defined as being >2.5 cm in diameter [2], we assumed that the high spatial resolution of our aerial photographs (0.5–1.0 cm per pixel) would allow us to detect

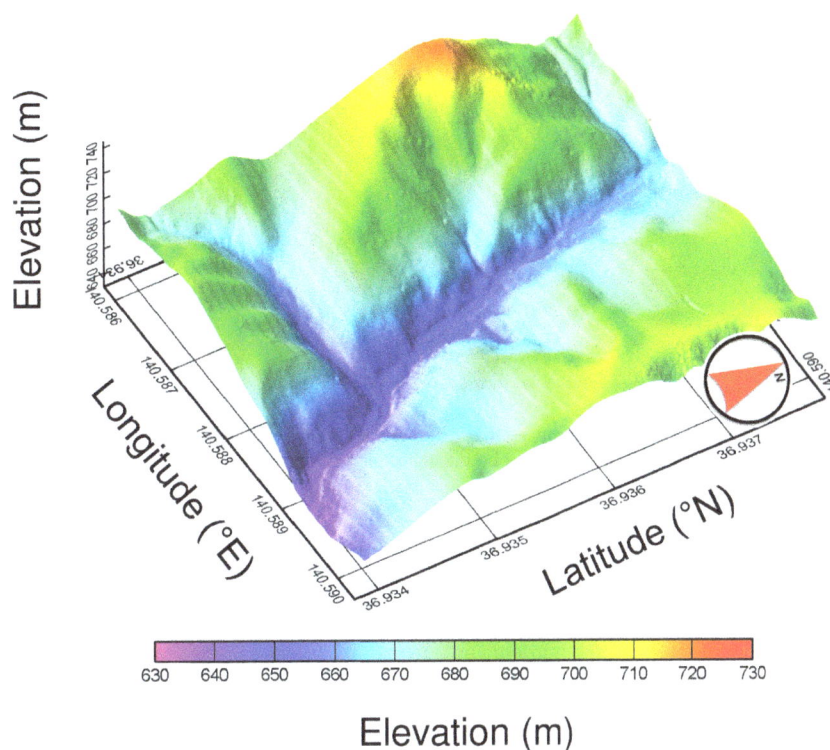

Figure 5. Digital elevation model (DEM) of the OFR plot. This DEM was used for orthorectification of the aerial photographs.

Figure 6. Fallen trees (yellow circles) detected by eye in the orthorectified mosaic.

them. However, our results suggest difficulty in identifying narrow or short fallen trees (Tables 1, 2). This failure may be due to the similarity of fallen trees to trunks and branches of standing trees, and masking of fallen trees by the branches of standing trees and forest floor vegetation. Hodgson et al., who surveyed marine mammals by UAV in Australia, reported the usefulness of overlaps between photographs for detecting animals that are masked by sun glitter [24]. For a similar reason, views of the same point from different angles may provide more opportunity to detect fallen trees hidden by standing trees (Fig. 7). The memory capacity of the camera [24] will determine the balance between coverage and overlap. The optimal degree of overlap will also depend on flying speed, flight altitude, and camera specifications (e.g., frames per second).

Another factor contributing to the poor rate of visual identification of fallen trees might be ambiguity in colour. One of the clues we used in detecting fallen trees was the colour of mosses growing on them. Trees with mosses are easy to detect visually, but freshly fallen trees with no mosses might be confused with tree branches or fallen leaves.

The orthorectification and mosaicking of aerial photographs according to the DEM require much labour and time, even for experts, and can also require specialized and expensive software. However, we identified fewer fallen trees from the orthorectified mosaic than from the original photographs. Thus, non-orthorectified photographs of the same point from different angles would allow better identification of fallen trees, at no additional cost, than orthorectified mosaics.

Conclusions

We showed the applicability of aerial photographs captured from a UAV for the detection of large fallen trees in a deciduous

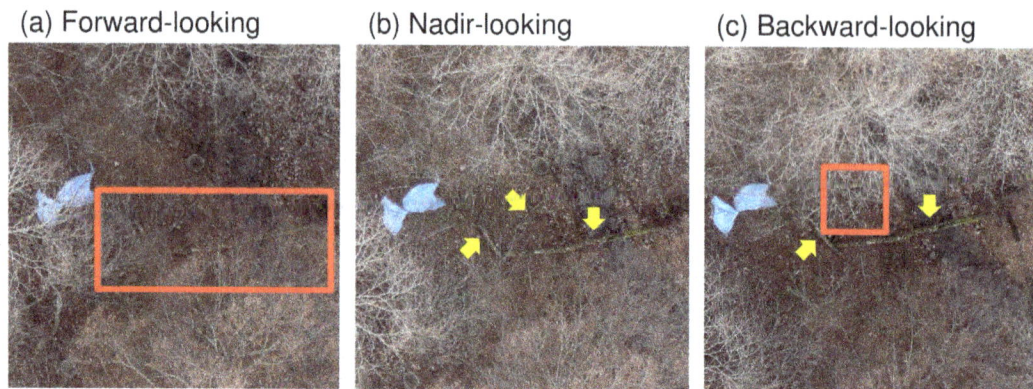

Figure 7. Example of overlapping images of the same point taken from different angles. In this example, the visual detection of three fallen trees from the forward- and backward-looking images may be difficult owing to masking by tree branches (within red boxes). The fallen trees are clearly visible in the nadir-looking image (yellow arrows).

broadleaved forest in eastern Japan. Because much tree death is episodic and irregular [33], high-frequency monitoring at multiple points is necessary for the detection of newly fallen trees and for the understanding of species distribution, biodiversity, and nutrient and carbon cycling in forest ecosystems. UAVs now permit high-frequency monitoring at low cost. This approach has great potential for forest ecology, especially for measuring temporal and spatial variations in forest structure and functioning. Furthermore, as UAVs are advancing and the payload of them is increasing, new sensors for forest monitoring on an UAV will become more common in the future [26]. Installation of a multi-angle photographing system, like a PRISM (Panchromatic Remote-sensing Instrument for Stereo Mapping) on-board the Japanese satellite ALOS (Advanced Land Observing Satellite), on an UAV would potentially help to find more opportunity for detection of fallen trees hidden by standing trees and it may increase the accuracy of fallen tree identification.

Acknowledgments

The authors are thankful to the administrators of the OFR site of the Forestry and Forest Products Research Institute (FFPRI) for their cooperation during the field data collection. We also thank the anonymous reviewers and editor for their helpful and constructive comments.

Author Contributions

Conceived and designed the experiments: RS SY KK HT SN RI HF KO YH. Performed the experiments: RS SY KK HT SN RI HF KO YH. Analyzed the data: KK RS TI SN HF. Contributed to the writing of the manuscript: TI SN RS.

References

1. Biała K, Condé S, Delbaere B, Jones-Walters L, Torre-Marín A (2012) Streamlining European biodiversity indicators 2020: Building a future on lessons learnt from the SEBI 2010 process. EEA Tech Rep No.11/2012. doi: 10.2800/55751.

2. Harmon ME, Franklin JF, Swanson FJ, Sollins P, Gregory SV, et al. (1986) Ecology of coarse woody debris in temperate ecosystems. Adv Ecol Res 15: 133–302. doi: 10.1016/S0065-2504(08)60121-X.

3. Bunnell FL, Houde I (2010) Down wood and biodiversity – implications to forest practices. Environ Rev 18: 397–421. doi: 10.1139/A10-019.

4. Stokland JN, Siitonen J, Jonsson BG (2012) Biodiversity in dead wood. Cambridge University Press, Cambridge.

5. Herrmann S, Prescott CE (2008) Mass loss and nutrient dynamics of coarse woody debris in three Rocky Mountain coniferous forests: 21 year results. Can J For Res 38: 125–132. doi: 10.1139/X07-144.

6. Ohtsuka T, Shizu Y, Hirota M, Yashiro Y, Shugang J, et al. (2014) Role of coarse woody debris in the carbon cycle of Takayama forest, central Japan. Ecol Res 29: 91–101. doi: 10.1007/s11284-013-1102-5.

7. Edman M, Jonsson BG (2001) Spatial pattern of downed logs and wood-decaying fungi in an old-growth *Picea abies* forest. J Veg Sci 12(5): 609–620. doi: 10.2307/3236900.

8. Wu JB, Guan DX, Han SJ, Zhang M, Jin C (2005) Ecological functions of coarse woody debris in forest ecosystem. J For Res 16: 247–252. doi: 10.1007/BF02856826.

9. Ståhl G, Ringvall A, Fridman J (2001) Assessment of coarse woody debris: A methodological overview. Ecol Bull 49: 57–70.

10. Bütler R, Schlaepfer R (2004) Spruce snag quantification by coupling colour infrared aerial photos and a GIS. For Ecol Manage 195: 325–339. doi: 10.1016/j.foreco.2004.02.042.

11. Meentemeyer RK, Rank NE, Shoemaker DA, Oneal CB, Wickland AC, et al. (2007) Impact of sudden oak death on tree mortality in the Big Sur ecoregion of California. Biol Invasions 10: 1243–1255. doi: 10.1007/s10530-007-9199-5.

12. Chambers JQ, Fisher JI, Zeng H, Chapman EL, Baker DB, et al. (2007) Hurricane Katrina's carbon footprint on U.S. Gulf Coast forests. Science 318: 1107. doi: 10.1126/science.1148913.

13. Kupfer JA, Myers AT, McLane SE, Melton GN (2008) Patterns of forest damage in a southern Mississippi landscape caused by Hurricane Katrina. Ecosystems 11: 45–60. doi: 10.1007/s10021-007-9106-z.

14. Pasher J, King DJ (2009) Mapping dead wood distribution in a temperate hardwood forest using high resolution airborne imagery. For Ecol Manage 258: 1536–1548. doi: 10.1016/j.foreco.2009.07.009.

15. Wang F, Xu YJ (2010) Comparison of remote sensing change detection techniques for assessing hurricane damage to forests. Environ Monit Assess 162: 311–326. doi: 10.1007/s10661-009-0798-8.

16. Wang W, Qu JJ, Hao X, Liu Y, Stanturf JA (2010) Post-hurricane forest damage assessment using satellite remote sensing. Agric For Meteorol 150: 122–132. doi: 10.1016/j.agrformet.2009.09.009.

17. Valavanis KP (ed) (2007) Advances in unmanned aerial vehicles: State of the art and the road to autonomy. Springer, Dordrecht, Netherlands.

18. Watts AC, Perry JH, Smith SE, Burgess MA, Wilkinson BE, et al. (2010) Small unmanned aircraft systems for low-altitude aerial surveys. J Wildl Manag 7: 1614–1619. doi: 10.1111/j.1937-2817.2010.tb01292.x.

19. Watts AC, Ambrosia VG, Hinkley EA (2012) Unmanned aircraft systems in remote sensing and scientific research: Classification and considerations of use. Remote Sens 4: 1671–1692. doi: 10.3390/rs4061671.

20. Anderson K, Gaston KJ (2013) Lightweight unmanned aerial vehicles will revolutionize spatial ecology. Front Ecol Environ 11: 138–146. doi: 10.1890/120150.

21. Rango A, Laliberte A, Herrick JE, Winters C, Havstad K, et al. (2009) Unmanned aerial vehicle-based remote sensing for rangeland assessment, monitoring, and management. J Appl Remote Sens 3: 033542. doi: 10.1117/1.3216822.

22. Koh LP, Wich SA (2012) Dawn of drone ecology: low-cost autonomous aerial vehicles for conservation. Trop Conserv Sci 5: 121–132.

23. Sardà-Palomera F, Bota G, Viñolo C, Pallarés O, Sazatornil V, et al. (2012) Fine-scale bird monitoring from light unmanned aircraft systems. Ibis 154: 177–183. doi: 10.1111/j.1474-919X.2011.01177.x.

24. Hodgson A, Kelly N, Peel D (2013) Unmanned aerial vehicles (UAVs) for surveying marine fauna: a dugong case study. PLoS One 8(11): e79556. doi: 10.1371/journal.pone.0079556.

25. Vermeulen C, Lejeune P, Lisein J, Sawadogo P, Bouché P (2013) Unmanned aerial survey of elephants. PLoS One 8(2): e54700. doi: 10.1371/journal.pone.0054700.

26. Getzin S, Nuske RS, Wiegand K (2014) Using unmanned aerial vehicles (UAV) to quantify spatial gap patterns in forests. Remote Sens 6: 6988–7004. doi: 10.3390/rs6086988.

27. Masaki T, Suzuki W, Niiyama K, Iida S, Tanaka H, et al. (1992) Community structure of a species-rich temperate forest, Ogawa Forest Reserve, central Japan. Vegetatio 98: 97–111. doi: 10.1007/BF00045549.

28. Nakashizuka T, Iida S, Tanaka H, Shibata M, Abe S, et al. (1992) Community dynamics of Ogawa Forest Reserve, a species rich deciduous forest, central Japan. Vegetatio 103: 105–112. doi: 10.1007/BF00047696.

29. Nakashizuka T, Matsumoto Y (eds.) (2002) Diversity and interaction in a temperate forest community: Ogawa Forest Reserve of Japan. Ecological Studies vol. 158, Springer, Tokyo.

30. Kaneko T, Koyama T, Yasuda A, Takeo M, Yanagisawa T, et al. (2011) Low-altitude remote sensing of volcanoes using an unmanned autonomous helicopter: an example of aeromagnetic observation at Izu-Oshima volcano, Japan. Int J Remote Sens 32: 1491–1504. doi: 10.1080/01431160903559770.

31. Koyama T, Kaneko T, Ohminato T, Yanagisawa T, Watanabe A, et al. (2013) An aeromagnetic survey of Shinmoe-dake volcano, Kirishima, Japan, after the 2011 eruption using an unmanned autonomous helicopter. Earth Planets Sp 65: 657–666. doi: 10.5047/eps.2013.03.005.

32. Sanada Y, Kondo A, Sugita T, Nishizawa Y, Yuuki Y, et al. (2014) Radiation monitoring using an unmanned helicopter in the evacuation zone around the Fukushima Daiichi nuclear power plant. Explor Geophys 45: 3–7. doi: 10.1071/EG13004.

33. Franklin JF, Shugart HH, Harmon ME (1987) Tree death as an ecological process: the causes, consequences, and variability of tree mortality. BioScience 37(8): 550–556. doi: 10.2307/1310665.

10

Large Spatial Scale Variability in Bathyal Macrobenthos Abundance, Biomass, α- and β-Diversity along the Mediterranean Continental Margin

Elisa Baldrighi[1], Marc Lavaleye[2], Stefano Aliani[3], Alessandra Conversi[3,4], Elena Manini[1]*

1 Institute of Marine Sciences, National Research Council (ISMAR-CNR), Ancona, Italy, **2** Department of Marine Ecology, Royal Netherlands Institute for Sea Research (NIOZ), Texel, The Netherlands, **3** Institute of Marine Sciences, National Research Council (ISMAR-CNR), La Spezia, Italy, **4** Marine Institute, Plymouth University, Plymouth, United Kingdom

Abstract

The large-scale deep-sea biodiversity distribution of the benthic fauna was explored in the Mediterranean Sea, which can be seen as a miniature model of the oceans of the world. Within the framework of the BIOFUN project ("Biodiversity and Ecosystem Functioning in Contrasting Southern European Deep-sea Environments: from viruses to megafauna"), we investigated the large spatial scale variability (over >1,000 km) of the bathyal macrofauna communities that inhabit the Mediterranean basin, and their relationships with the environmental variables. The macrofauna abundance, biomass, community structure and functional diversity were analysed and the α-diversity and β-diversity were estimated across six selected slope areas at different longitudes and along three main depths. The macrobenthic standing stock and α-diversity were lower in the deep-sea sediments of the eastern Mediterranean basin, compared to the western and central basins. The macrofaunal standing stock and diversity decreased significantly from the upper bathyal to the lower bathyal slope stations. The major changes in the community composition of the higher taxa and in the trophic (functional) structure occurred at different longitudes, rather than at increasing water depth. For the β-diversity, very high dissimilarities emerged at all levels: (i) between basins; (ii) between slopes within the same basin; and (iii) between stations at different depths; this therefore demonstrates the high macrofaunal diversity of the Mediterranean basins at large spatial scales. Overall, the food sources (i.e., quantity and quality) that characterised the west, central and eastern Mediterranean basins, as well as sediment grain size, appear to influence the macrobenthic standing stock and the biodiversity along the different slope areas.

Editor: Brian R. MacKenzie, Technical University of Denmark, Denmark

Funding: This study is part of ESF-EuroDEEP project BIOFUN (CTM2007-28739-E) and for final version by BALMAS (IPA ADRIATIC project; 1°STR/0005). The funders had no role in study design, data collection and analysis, decision to publish, or preparation of the manuscript.

Competing Interests: The authors have declared that no competing interests exist.

* Email: e.manini@ismar.cnr.it

Introduction

Different studies have been conducted worldwide to define latitudinal and longitudinal diversity patterns of marine biodiversity [1–3], which have often been coupled to the bathymetric trends of organisms [4–7]. Nevertheless, these patterns and the mechanisms involved in their generation are still far from being understood [8–11].

Rex and co-authors [12] presented the first global-scale analysis of the bathymetric patterns of the standing stock (i.e., abundance, biomass) for four major size classes of deep-sea biota: prokaryotes, metazoan meiofauna, macrofauna and megafauna. For the last three of these benthic components, they reported that the community standing-stock decreases with depth, and interpreted this to be a universal phenomenon. This is, however, controversial, and should be related to the taxon considered each time within each benthic size component [13–15]. Similarly, for the bathymetric trends in the standing stock, the well-known 'hump-shape' distribution in species richness with a diversity maximum at mid-slope depths might not always be the rule [16–18].

The spatial heterogeneity of benthic communities is usually related to the different environmental conditions encountered [19], although our understanding of the mechanisms that might act as drivers for the benthic fauna distribution and diversity in the deep sea is still limited [20]. Nevertheless, some factors are usually invoked, including: substrate heterogeneity [21,22]; water circulation [4,23]; oxygen availability [19]; productivity and microbial activity [24]; and food resources [25,26]. Food availability in particular, which is mainly determined by the surface-water primary production [27], and which can decrease sharply with depth, appears to be a major factor that influences the standing stock and the diversity of the deep-benthic communities that depend on this allochthonous organic-matter input [10,28,29]. Degradation processes in the water column that affect the quantity and quality of the organic matter that reaches the bottom have also been suggested to have an influence on benthic communities [30].

In the Mediterranean Sea, an overall decrease in benthic abundance, biomass and species richness has been observed from northwestern to southeastern areas for the meiofauna [29,31], macrofauna [4,32–34] and megafauna [35,36]. According to different studies [37,38], the west-east gradient of decreasing surface-water productivity of the Mediterranean Sea is reflected in an increasing paucity of the food that reaches the sea floor moving eastwards. Such a gradient might thus be responsible for the decrease in deep benthic fauna abundance and biomass from west to east. Danovaro and co-authors [39] have shown that the effects of the food supply, and consequently the derived longitudinal trend in the Mediterranean, might be inconsistent across different components of the benthic diversity.

There have been few recent quantitative studies that have dealt with the large-scale patterns of distribution and diversity of the deep Mediterranean macrofauna and/or have addressed the influence of environmental conditions on these macrofauna communities. After a qualitative review by Fredj and Laubier [40] and some descriptive studies that were conducted on bathyal and abyssal macrofaunal organisms or focused on specific taxonomic groups [34,41–54], the most recent studies have reported that the bathyal macrofauna communities in the eastern basin are characterised by low abundance and low diversity, with respect to the western basin; these decrease sharply with depth and are strongly related to food availability [32,55,56]. The scant information regarding the macrofauna of the western basin comes from a limited number of slope and canyon areas [23,26,57,58] on the northwestern side of this basin, and shows a decrease in both biomass and density with depth.

We hypothesised that the macrofauna standing stock and diversity change with longitude and depth and according to the major influential environmental variables that characterise the systems investigated.

The main aims were thus to study the deep Mediterranean, in order to:

- assess the longitudinal-related (over >1000 km, and from 3° E to 25° E) and depth-related (1200 m to 2800 m water depth) trends in the macrofaunal abundance, biomass and diversity (i.e., structural and functional diversity), and the influence of the environmental variables on the macrobenthic populations;

-investigate and quantify the macrofaunal β-diversity [60] between: (i) the three basin areas; (ii) the slopes within the same basin; and (iii) stations at different depths.

Materials and Methods

Ethics statement

All of the field activities were approved by the local national authorities. The sampling areas were not privately owned or protected in any way, and no endangered or protected species were involved in this study.

Sampling plan

To achieve our aims, sediment samples were collected from the deep Mediterranean Sea, covering a large spatial scale (over >1000 km) of investigation across different depths (from 1200 m to 2800 m) and longitudes (from 3° E to 25° E). Six cruises (see Table S1 for details) were performed in the Mediterranean Sea on board the R/V Urania (2008–2010), R/V Pelagia (2009) and R/V Meteor (2010), within the framework of the BIOFUN project ("Biodiversity and Ecosystem Functioning in Contrasting Southern European Deep-sea Environments: from viruses to megafauna"), to collect biological and environmental samples from different continental slope systems. The relationship(s) between the macrofauna standing stock and diversity with a number of environmental variables that characterise the investigated areas was assessed. According to the sampling strategy of the BIOFUN project, a total of six selected slopes were chosen along a gradient of increased oligotrophy in the three main Mediterranean basins; i.e., the western (WM; Algero-Provençal basin), central (CM;

Ionian Sea) and eastern (EM; northern Levantine basin) Mediterranean basins (see Fig. 1). All of the selected open-slope systems were from topographically regular settings, with well-oxygenated bottom waters. Three of the slopes were in the WM basin (WM-1, Balearic slope 1; WM-2, Balearic slope 2; WM-3, Sardinia slope), two in the CM basin (CM-1, Maltese slope; CM-2, Ionian slope) and one in the EM basin (EM, Cretan slope) (Fig. 1). For each slope, three stations at three different depths were sampled. The three different station depths always fell into three depth ranges: upper bathyal (1200 m), mid-bathyal (from 1800 m to 1900 m), and lower bathyal (from 2400 m to 2700 m). CM-1 was not sampled at the lower bathyal, and was substituted by a station at a depth of 2120 m. At each station and with the employment of cylindrical box-corers (see details below), independent replicate samples were taken for the analyses of the macrobenthos (n = 3 replicates), microbial (n = 3 replicates) and environmental (n = 3 replicates) variables. We selected the heterogeneity of the substrate (grain size), the organic matter content of the sediments, and the prokaryotic abundance and biomass as recognized drivers that influence the benthic fauna distribution and diversity [10,17]. The details of the sampling locations and the environmental features of all of the stations are given in Table 1 and Table S1.

Environmental variable sampling

The near-bottom temperature and salinity were recorded, using a conductivity–temperature–depth (CTD) SBE 911 plus probe mounted on a CTD rosette system (Table 1). To analyse grain size and biochemical composition of the organic matter content, subsamples of sediments from each box-corer were collected using plexiglass cores of 3.6-cm internal diameter. As analysis of the top 1-cm layer has been shown to represent a feasible proxy for the whole trophic status of a sediment [61,62], only the top 1-cm of subsamples was collected and frozen at −20°C, for the analysis of chlorophyll-a, phaeopigment and organic matter content.

Grain size

The subsamples for the grain size analysis (the top 20 cm) were preserved at +4°C. Aliquots of fresh sediment were sieved over a 63-μm mesh. The two fractions (>63 μm, sand; <63 μm, silt and

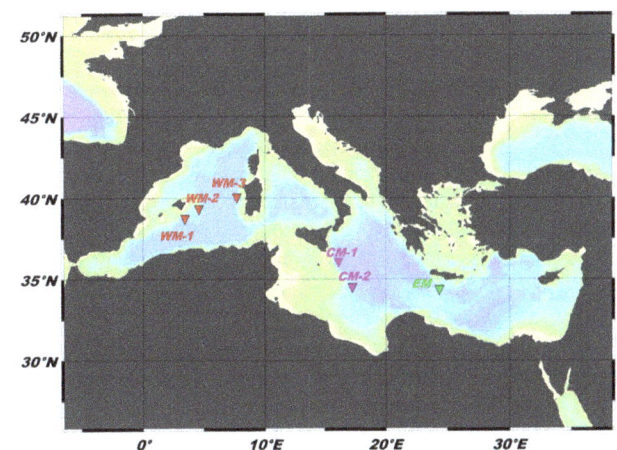

Figure 1. Map of the study area and sampling sites. Red triangles, western Mediterranean basin (WM-1, -2, -3); purple triangles, central Mediterranean basin (CM-1, -2); green triangle, eastern Mediterranean basin (EM). WM-1, Balearic slope 1; WM-2, Balearic slope 2; WM-3, Sardinia slope; CM-1, Maltese slope; CM-2, Ionian slope; EM, Cretan slope.

Table 1. Main characteristics of the sampling slopes studied along the six Mediterranean continental slope areas.

Slope	Water depth (m)	Bottom temperature (°C)	Salinity	POC flux (mgC m^{-2} d^{-1})	Sand (%)	Silt (%)
WM-1	1224	13.1	38.5	18.2	9.0	91.0
WM-1	1803	13.2	38.5	11.3	13.8	86.2
WM-1	2362	13.3	38.5	9.0	12.3	87.7
WM-2	1179	13.1	38.5	12.7	6.3	93.7
WM-2	1862	13.2	38.5	8.8	8.9	91.1
WM-2	2758	13.3	38.5	7.4	6.3	93.7
WM-3	1258	13.3	38.5	12.2	8.0	92.0
WM-3	1890	13.2	38.5	8.2	14.1	85.9
WM-3	2448	13.3	38.5	6.1	18.7	81.3
CM-1	1236	13.7	38.7	22.1	3.3	90.5
CM-1	1798	13.7	38.7	14.7	1.3	98.7
CM-1	2120	13.8	38.7	10.9	2.6	97.4
CM-2	1219	13.7	38.6	13.4	14.3	85.7
CM-2	1924	13.7	38.6	6.6	39.9	60.2
CM-2	2693	13.8	38.6	4.6	33.7	66.3
EM	1237	14.7	38.8	11.0	3.6	96.4
EM	1907	14.7	38.8	7.0	10.1	89.9
EM	2766	14.7	38.7	5.1	3.0	97.0

WM, CM, EM, western, central, eastern Mediterranean.
POC, particle organic carbon.

clay) were dried in an oven at 60°C and weighed. Data were expressed as percentages of the total sediment dry weight.

Phytopigment contents and seafloor particulate organic carbon flux

Chlorophyll-a and phaepigments were determined according to standard tecniques [63]. The sum of the chlorophyll-a and phaeopigment concentrations were defined here as chloroplastic pigment equivalents (CPE). The concentrations of these total phytopigments were converted into carbon (C) equivalents using the conversion factor of 40 [64], and expressed as mgC g^{-1}.

Johnson et al. [65] showed that the estimated particulate organic C (POC) flux from the surface represents a good predictor of the benthic standing stock, even at large spatial scales. We extracted the surface primary production data (as mgC m^{-2} d^{-1}) from the ocean productivity database (http://www.science. oregonstate.edu/ocean.productivity/index.php). These data were used to estimate the flux of C to the seafloor, using Equation (1), as reported in Lutz et al. [66], and introduced by Suess [67]:

$$Cflux(z) = Cprod/(0.0238z + 0.212),\qquad(1)$$

where the C flux to depth $C_{flux}(z)$ is described as a function of the primary production of organic carbon in the surface waters C_{prod}, scaled to the depth below the sea surface, z.

Quantity and biochemical composition of the organic matter

The contents of carbohydrate, protein and lipid were determined according to standard techniques [63]. These concentrations were then converted into C equivalents using the conversion factors of 0.40, 0.49 and 0.75 µgC µg^{-1}, respectively [63], and normalised to the sediment dry weight after desiccation (60°C,

24 h). Biopolymeric organic C (BPC) was calculated as the sum of the C equivalents of carbohydrate, protein and lipid [68]. The contributions of phytopigment C (CCPE) and protein C (CPRT) to the BPC concentrations (CCPE/BPC and CPRT/BPC ratios, respectively) and the protein/carbohydrate (PRT/CHO) ratio were then calculated and used as descriptors of the aging, origin and nutritional quality of the sediment organic matter [62]. PRT/CHO ratios >1.0 indicate relatively high quality and high food availability for the organisms [62].

Prokaryotic abundance and biomass

For the analyses of the prokaryotic abundance and biomass, subsamples of sediments from each box-corer were collected using plexiglass cores of 3.6-cm internal diameter. Circa 1 ml of the wet surface sediment layer (0–1 cm) was fixed using buffered formaldehyde (2% final concentration, in sterile, filtered seawater [v/v]), and stored at 4°C until processed [69].

The total prokaryotic number (TPN) was determined using a staining technique with acridine orange [70], and analysed using epifluorescence microscopy (magnification, 1000×). The total prokaryotic biomass (TPB) was estimated using an ocular micrometer, assigning the prokaryotic cells into different size classes based on their maximum length and width [71]. These were converted to biovolumes on the assumption of an average C content of 310 fgC µm^{-3} [71]. The TPN and TPB were normalised to the sediment dry weight after desiccation (24 h, 60°C).

Macrofaunal sampling

At each station, three independent replicates of undisturbed sediment samples were collected using cylindrical box-corers (Ø 50 cm, WM-2 and CM-2; Ø 32 cm, all other stations). From each box-corer sample, the top 20 cm of the sediment, along with their overlying water, were gently sieved over a 300-μm mesh sieve to retain all of the macrobenthic organisms [72]. The residual left behind on the sieve was immediately fixed in 10% buffered formalin solution, and stained with Rose Bengal.

Macrofauna abundance, biomass and biodiversity estimation

All multicellular organisms (including Nematoda, Copepoda and Ostracoda; macrofauna *sensu lato*, [73,74]) and Foraminifera that were retained on a 300-μm mesh sieve were sorted under a stereomicroscope, and identified to the lowest possible taxonomic level according to the main literature [59,75–78]. The taxon names of the organisms were cross-checked with the World Register of Marine Species (WoRMS, www.marinespecies.org). For each species the total number was calculated and the wet-weight biomass measured; the number of individuals and weight were expressed as abundance and biomass per square meter. The wet biomass (g wet weight m^{-2}) was converted to ash-free dry weight and organic carbon content using standard conversion factors [79]. In accordance with the literature [59,75,76–78], four major macrofaunal trophic (functional) groups were identified: surface deposit feeders (SDFs), subsurface deposit feeders (SSDFs), carnivores/scavangers, and filter feeders/suspension feeders.

Biodiversity was measured as α-diversity (or 'sample' diversity) by calculating several indices: species richness (SR), or total number of species collected in each boxcorer sample; Shannon-Weaner index (H': $\log_2$) [80]; and Pielou's [81] index of equitability (J'). Moreover, the species-abundance data were converted into rarefaction diversity indices ([82], as modified by Hurlbert [83]), and the expected number of species ES(n) for theoretical samples of $n = 30$ and $n = 50$ individuals were calculated for each station. This method of rarefaction provides a good tool for comparisons of species richness among samples that have different total abundances [84]. The number of higher taxonomic groups identified (e.g., Polychaeta, Isopoda, Tanaidacea, Bivalvia, and others) in each of the samples was also considered. To characterise the macrobenthic community structure, the percentage contribution of each of the higher taxonomic groups to the total abundance and biomass was calculated.

The degree of change in the species composition between habitats or along an environmental gradient is usually defined as the turnover (β)-diversity. The macrofaunal turnover diversity between the different depths and longitudes was measured by the dissimilarity coefficients, based on a Bray-Curtis similarities matrix. The statistical differences in the macrofauna composition among all sampling sites was tested by the analysis of similarities (ANOSIM) [85].

Statistical analyses

To test for differences in the patterns of environmental (i.e., temperature, salinity, grain size, quantity and quality of organic matter) and biological (i.e., macrofauna, microbial components) variables between different longitudes and stations at different depths, distance-based permutational multivariate analysis of variance was used (PERMANOVA; [86,87]). The design included two factors: slope location (six levels, fixed, from the west to the east basin) and depth (three levels, fixed). The analysis was based on Euclidean distances of previously normalised data, using 999 random permutations of the appropriate units [88]. The tests were carried out using the permutation of residuals under a reduced model. As there was a restricted number of unique permutations in the pair-wise tests, the p values were obtained from Monte Carlo tests [89]. When significant differences were observed between

stations at different longitudes and/or depths, pair-wise comparisons were also performed.

The β-diversity in the macrobenthic organism composition and trophic structure was estimated: (i) between basins; (ii) between stations at the same depth; (iii) between slopes within the same basin; and (iv) between stations at different depths. The turnover diversity was estimated through Bray-Curtis dissimilarity coefficients. The SIMPER analysis was used to identify the organisms that contributed the most to the dissimiarity between longitudes and depths.

To test for the presence of statistical differences in the macrobenthic organism compositions and functional structures, analysis of similarities (ANOSIM) was performed, as above: (i) between basins; (ii) between stations along the same depth; (iii) between slopes within the same basin; and (iv) between stations at different depths. All of the macrofaunal abundance data were presence/absence transformed prior to the analysis. When significant differences were observed, a non-metric multidimensional scaling ordination was carried out to visualise similarities between basins, slopes and depths along the same slope area. PERMANOVA, ANOSIM, SIMPER and nMDS analyses were performed using the PRIMER version 6 software package [90].

To determine whether the investigated environmental variables influence changes in the macrofaunal standing stock, trophic composition and diversity between basins and between slopes in the same basin, non-parametric multivariate multiple regression analysis was used, with the DISTLM forward routine [87]. The regression analysis was based on Euclidean distances when abundance, biomass and percentage of different trophic groups were considered, and on Bray-Curtis distances when diversity indices were tested. The forward selection of the predictor variables was carried out with tests by permutation. P- values were obtained using 4999 permutations of raw data for the marginal tests (tests of individual variables), while for all of the conditional tests, the routine used 4999 permutations of residuals under a reduced model. Bottom temperature, salinity and grain size were used as environmental parameters. BPC content, phytopigments content, microbial standing stock (i.e., abundance and biomass) and estimates of POC fluxes to the bottom were selected as indicators of food quantity, and the CCPE/BPC, CPRT/BPC, and PRT/CHO ratios as proxies for the quality of the sedimentary organic matter.

Results

Environmental features and trophic state of the sampling sites

The water mass features (temperature, salinity) and sediment grain sizes (sand, silt) are reported in Table 1. The bottom water temperature and salinity increased significantly moving eastwards (pair-wise tests; p<0.01), with values that ranged from 13.1°C and 38.5 for the WM basin to 14.7°C and 38.8 in the EM basin. The dissolved oxygen content ranged between 3.7 ml l^{-1} and 4.8 ml l^{-1}, with the lowest concentration registered in the EM basin at 1200 m in depth, and the highest in the WM basin (WM-1) at 2400 m in depth. Most of the sediment was silt (Table 1, range, 60%–97.4%) at all depths and for all sites. A significantly higher percentage of silt fraction was seen for the CM-1 slope in the central basin, compared to those of the WM basin (pair-wise tests, p<0.01). The variability in the water mass and grain size occurred mainly at different longitudes, while there were no significant changes with depth (Table S2, PERMANOVA results). Significant changes in the quantity of the organic matter (Tables 1, 2, CPE, POC flux, BPC) were detectable at both different longitudes and different depths (Table S2, PERMANOVA tests). However, the organic matter content did not show any clear increasing or decreasing trends with depth or moving eastwards. The only exception was the POC flux, which clearly declined with increasing depth (pair-wise tests, p<0.01). Relatively high organic matter quantities characterised both the WM (WM-2) and CM (CM-1) basins, compared to the other slope systems (pair-wise tests, p<0.05). Significant changes in the organic matter quality (Table 2, CCPE, CPRT, PRT/CHO ratio) were detected at different longitudes, but not at different depths (Table S2, PERMANOVA tests). In particular, a high organic matter quality was seen for the sediments along the CM-1 slope in the CM basin (pair-wise tests, p<0.01). The prokaryotic standing stock (Table 2, abundance, biomass) varied significantly with longitude and depth (Table S2, PERMANOVA tests), although most of the variability was explained by the effect of longitude (see pseudo-F values, Table S2). The highest values for both the prokaryotic abundance and biomass were seen for the CM basin (i.e., CM-1 slope; pair-wise tests, p<0.01).

Macrofauna abundance, biomass and community structure

The total macrofaunal abundance and biomass are shown in Figure 2 and reported in Table 3. Significantly higher values for the macrobenthic standing stock were seen for two of the slope areas in the WM basin: WM-2 and WM-3 (pair-wise tests, p< 0.01, vs. all of the other slopes) Differences in the standing stock with depth were generally seen for all of the slope areas between the shallower stations (1200 m) and the deeper stations (pair-wise tests, p<0.05), except for the WM-1 slope area, where no significant differences between the depths were detected. The PERMANOVA tests carried out on the macrofaunal biomass and abundance showed significant differences according to both longitude and depth (Table S1). Longitude explained most of the variability in the macrofaunal abundance (65%), while both longitude and depth explained the variability in the macrofaunal biomass (32% and 30%, respectively) (Fig. 3).

A total of 22 higher taxa were identified (i.e., Foraminifera, Porifera, Hydrozoa, Scyphozoa, Nematoda, Nemertea, Oligochaeta, Polychaeta, Priapulida, Sipuncula, Echiura, Ostracoda, Copepoda, Cumacea, Tanaidacea, Isopoda, Amphipoda, Aplacophora, Scaphopoda, Gastropoda, Bivalvia, Bryozoa), with the highest mean number present of 16 seen for the WM-2 slope (Table 3). The different contributions in terms of the abundance and biomass of the most represented groups are reported in Figure 4 (grouped as Polychaeta, Oligochaeta, Crustacea, Mollusca, Nematoda, Sipuncula, Foraminifera, and others), for all of the slopes investigated. There were clear changes in the community compositions between the slopes at different longitudes, in terms of both abundance (Fig. 4a) and biomass (Fig. 4b). While the WM basin slopes were dominated by a high number of Foraminifera (range, 23%–67%), these were almost completely absent in the other Mediterranean basins. Polychaeta were relatively important in all of the stations, and in EM and CM-1, they were the dominant group (range, 31%–67%). Mollusca (i.e., mostly bivalves) were always at relatively low levels, and these peaked for the Maltese slope (CM-1), with a range of 9% to 21%. Sipuncula (range, 6%–14%), and particularly the macrobenthic Nematoda (range, 16%–40%), showed relatively high abundance along the CM-2 slope; the first group also had a relatively high abundance at the shallowest station, of EM (23%). Crustacea had the highest relative abundance at the CM-1 slope (range, 18%–30%; mostly Isopoda and Tanaidacea) and at the deepest station, of EM (Amphipoda, Ostracoda; 25%). Following these above-

mentioned groups, Hydrozoa was the only other group of importance, but only for the EM slope, with a range of 12%–22%.

Polychaeta contributed the highest biomass for almost all of the stations (Fig. 4b; range, 13%–91%). The four exceptions (Fig. 4b) were: the WM-2 1200-m-deep station, with 61% of the biomass formed by branched Foraminifera; the WM-2 2700-m-deep station, with the bivalve *Nucula* sp.1 forming 31% of the biomass; the CM-2 1200-m-deep station, with the bivalve *Cuspidaria* sp.1 forming 47% of the biomass; and the CM-2 1900-m-deep station, with 90% of the biomass formed by Sipuncula (*Golfingia spp.* and *Phascolosoma spp.*) (Fig. 4b).

There was a significant difference in the community structure among the Mediterranean basins (ANOSIM, p = 0.001), but not between the depths. The Bray-Curtis coefficient of dissimilarity detected major changes in the community composition between the WM and CM basins (32%) and between the WM and EM basins (42%). In terms of the biomass contributions, again, significant differences were detected between the WM and CM basins (ANOSIM, p<0.05; dissimilarity coefficient 29%) and between the WM and EM basins (ANOSIM, p<0.01; dissimilarity coefficient 39%), without any differences across the depths.

Macrofaunal α-diversity and trophic composition

The macrofaunal diversity indices are reported in Table 3. A total of 274 macrobenthic organisms were identified (Table S3). Significantly lower macrofaunal α-diversity was reported from the EM basin compared to the WM and CM basins (pair-wise tests, p<0.01). Along all of the slopes, with the sole exception of WM-1, the diversity decreased with depth, and significant differences were detected mostly between the 1200-m-deep stations and the deeper stations (pair-wise tests, p<0.05; Table 3). The macrofaunal α-diversity varied mostly with longitude (67%), and to a lesser degree with depth (8%) (see Fig. 3; Table S2). The variability in the equitability index J′ (Fig. 3; Table S2) was along the west-east axis, with significantly higher values for the CM and EM basin slopes compared to the WM basin slopes (pair-wise tests, p<0.01), which is converse to the result for the α-diversity.

The trophic composition of the macrofauna is shown in Figure 5, i.e., for the four major functional groups which we considered. The surface deposit feeders (SDFs), which were mainly represented by Crustacea and Polychaeta, were dominant at all depths and for all of the areas (always >40%). The contribution of the subsurface deposit feeders (SSDFs), such as the Polychaeta of the families Capitellidae, Fauveliopsidae and Cossuridae, to total abundance decreased moving eastwards (range, 24%–5%), except for the 2700-m-deep station in the EM basin, which did not fit this trend. Carnivores (range, 5%–40%) had a peak in CM-2, which was mainly caused by Polychaeta of the families Eunicidae, Syllidae and Glyceridae, and Nematoda of the genera *Pareurystomina*, *Oncholaimellus*, *Trissonchulus* and *Pheronus*. Filter feeders were more abundant for CM-2 and EM (range, 8%–30%), which was due to small Hydrozoa (Fig. 5). Significant differences in the trophic structure compositions were detected along the longitudinal axis (Table S4), but not between different depths. Indeed, high similarities between the stations at increasing water depths were detected by the SIMPER analysis, which characterised each of the slope areas (range, 77%–96%, in the EM basin and along WM-2). There were major changes in the trophic compositions between the WM basin and the CM and EM basins. Pair-wise comparisons of the WM and CM basins and the WM and EM basins were highly significant, whereas those for the CM and EM basins were not (p = 0.16).

Table 2. Organic matter quantity, quality and prokaryotic standing stock.

Slope	Depth (m)	CPE (µg g⁻¹)	CCPE/BPC (%)	BPC (mgC g⁻¹)	CPRT/BPC (%)	PRT/CHO	TPN (ncell g⁻¹)	TPB (µgC g⁻¹)
WM-1	1224	0.42±0.18	2.26±0.96	0.73±0.03	25.15±7.38	0.40±0.19	4.06E+07±2.56E+06	0.81±0.05
	1803	0.62±0.01	3.84±0.48	0.66±0.06	23.24±0.71	0.31±0.04	2.53E+07±4.08E+06	0.65±0.13
	2362	0.60±0.09	3.70±0.67	0.65±0.02	17.04±3.32	0.19±0.04	2.85E+07±4.35E+06	0.57±0.09
WM-2	1179	6.19±0.18	11.70±0.77	2.13±0.11	35.30±1.31	0.67±0.07	1.30E+08±8.85E+06	3.46±0.57
	1862	1.98±0.22	6.51±0.51	1.21±0.04	36.68±3.55	0.78±0.15	6.94E+07±1.69E+06	2.82±0.60
	2758	2.71±0.50	8.82±2.07	1.25±0.08	37.54±4.94	0.96±0.21	1.08E+08±7.80E+06	3.94±0.81
WM-3	1258	0.71±0.04	3.72±0.11	0.67±0.10	42.67±8.08	1.11±0.40	8.71E+07±9.41E+06	1.74±0.19
	1890	1.12±0.08	6.86±1.18	0.68±0.07	34.13±11.40	0.51±0.18	8.67E+07±1.60E+07	1.73±0.32
	2448	0.96±0.17	5.50±1.31	0.80±0.29	40.79±5.70	1.02±0.33	5.94E+07±8.23E+06	1.19±0.16
CM-1	1236	4.16±0.34	10.50±1.70	1.64±0.19	50.10±1.41	3.68±0.08	1.64E+08±1.43E+07	5.33±0.26
	1798	8.35±0.27	23.75±1.01	1.41±0.05	54.86±2.05	2.92±0.21	1.82E+08±1.20E+07	7.61±0.13
	2120	1.63±0.15	5.95±0.72	1.11±0.07	56.32±1.86	2.20±0.32	3.01E+08±4.00E+07	10.58±0.85
CM-2	1219	1.83±0.19	8.59±0.97	0.85±0.01	28.11±2.68	0.46±0.05	8.45E+07±5.40E+06	1.69±0.11
	1924	2.63±0.69	14.89±4.26	0.72±0.03	39.89±3.37	0.80±0.15	1.60E+08±1.12E+07	3.20±0.22
	2693	1.73±0.80	8.64±4.23	0.82±0.03	21.03±2.58	0.29±0.04	1.82E+08±3.31E+07	3.65±0.66
EM	1237	1.96±0.17	10.34±0.85	0.76±0.03	22.74±3.57	0.26±0.05	5.49E+07±9.43E+06	1.15±0.21
	1907	0.42±0.00	1.93±0.05	0.87±0.03	24.84±1.54	0.40±0.03	8.05E+07±4.07E+06	2.15±0.13
	2766	0.55±0.08	2.20±0.32	1.01±0.04	22.32±0.42	0.31±0.01	1.42E+07±7.76E+05	0.39±0.07

Data are means ± standard deviation.
CPE, total phytopigments; CCPE/BPC, contribution of phytopigment C to biopolymeric C; BPC, biopolymeric C; CPRT/BPC, contribution of protein C to biopolymeric C; PRT/CHO, protein to carbohydrate ratio; TPN, total prokaryotic abundance; TPB, total prokaryotic biomass.

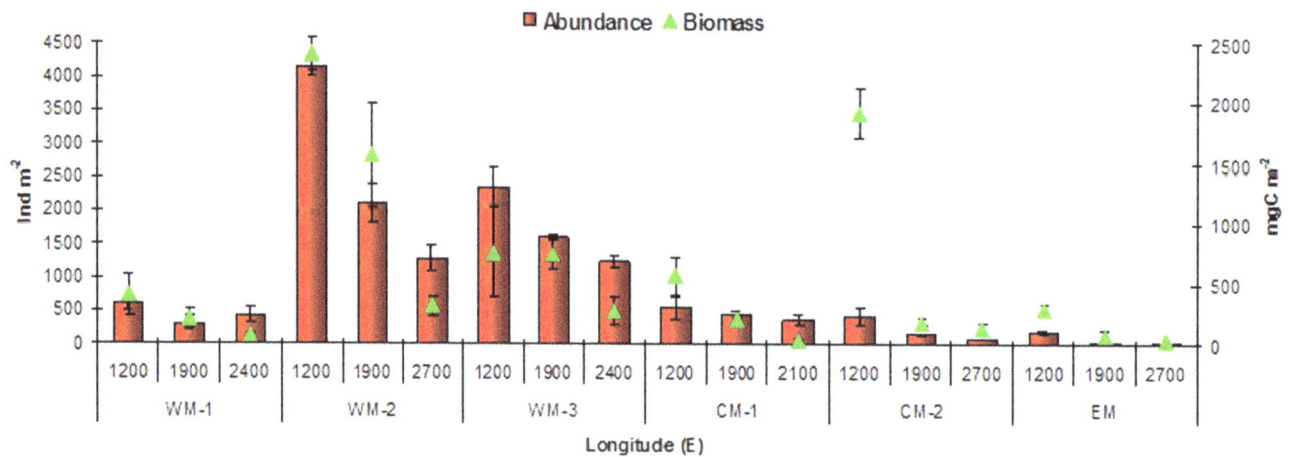

Figure 2. Macrofaunal standing stock. Mean total macrofauna abundance (bars; ind/m^2) and biomass (triangles; mgC/m^2) for each station in the WM, CM and EM basins. Data are means $\pm$ standard deviation (n = 3).

β-diversity in the macrofaunal composition: longitudinal and bathymetric trends

Significant differences in β-diversity were found between: (i) the basins (Table S5A); (ii) the slopes of the different basins (Table S5B); and (iii) the slopes within each basin (Table S5C).

The macrobenthic community composition changed significantly also considering stations at different water depths within the same slope system. The main differences were observed between the communities inhabiting the upper bathyal stations and the mid- and deep bathyal stations (Table S6). A high β-diversity emerged at all levels, expressed by the coefficient of dissimilarity values: between basins (from 65% to 82%); between slopes (from 61% to 85%); between slopes within each basin (from 43% to 71%) (Table S5); and between stations at different depths (from 39% to 98%) (Table S6). The largest dissimilarities were seen when the EM basin communities were compared to those of the WM and CM basins. The overall dissimilarity between depths and basins was driven by small contributions of many species (Tables S7A, S7B). In the case of dissimilarities between basins, these species belonged mostly to Polychaete families, such as Maldanidae sp1 (1.85%; WM-CM basins), Cirratulidae sp1, and Spionidae sp1 (both 1.95%; WM and EM basins) and Paraonidae sp1 (3.49%; CM and EM basins). The dissimilarities between stations at different depths along the slopes were driven by organisms that belonged to different higher taxa, such as Bivalvia, Sipuncula, Foraminifera, Polychaeta and Tanaidacea, depending on the slope system considerered. The highest contribution came from *Golfingia* sp1, which drove the dissimilarity along the EM slope between both the 1200 m to 1900 m stations (11.09%) and the 1200 m to 2400 m stations (10.69%), which was absent in the two deeper stations. The high rates of turnover diversity were also evident in the nMDS (Fig. 6), which grouped slopes according to longitude and, to a lesser extent, according to depth, at a similarity of only 40% and 20%, respectively. In a multi-dimensional scaling representation, it emerged that slopes at different longitudes and stations at different depths differed in macrofaunal composition. Only a few organisms, 19 of a total of 274 organisms identified (i.e., 7%), were reported for all of the basins (Table S3). Most of these were Polychaeta (such as *Glycera* sp1; *Capitellidae* sp1; *Syllidae* sp1 or *Heterospionidae* sp1), Crustacea (Copepoda; the amphipods *Eusiridae* sp1 and Ostracoda), Nematoda (*Linhystera* sp1 and *Bathyeurystomina* sp1) and Sipuncula (such as *Golfingia*

sp1). The density of these organisms changed from one basin to another, with them being widely represented in one or two basins and less represented in the others. Usually, they reached very low abundances in the EM basin. However, one explanation for this may be that the EM basin was under-sampled compared to the other basins.

Macrofauna relationships with environmental variables

To determine whether and how the environmental features and the trophic state of the system might influence the variability in the macrobenthic communities, multivariate multiple regression analysis (DISTLM forward) was carried out. These data are reported in Figure 7a and Table S8, for all of the slopes investigated, and in Figure 7b–d and Table S9 for each basin separately. Overall the most important factors that influenced the variability of the macrofauna abundance and biomass according to the above-mentioned analyses appeared to be the quantity of the food sources (BPC, 27%; TPN, 6%) and the heterogeneity of the substrate (grain size, 13%; Fig. 7a). The macrofaunal diversity appeared to correlate well not only with the quantity, but also with the quality (i.e., CPRT) of the organic matter in the sediment, and secondly with the grain size. The trophic compositions of the macrobenthic communities were only weakly influenced (range, 15%–26%) by the trophic sources, expressed as the microbial stock and the quality of food (Table S8). However, the high percentages of the variability in abundance, biomass and diversity did not correlate with our environmental variables (Fig. 7, grey).

As the trophic state of the system changed between the basins along the west-east axis, DISTLM forward analyses were also carried out considering each basin separately. Indeed, different drivers might be involved in the variance of the macrofauna distribution and diversity within each basin, compared to the data obtained for the whole of the Mediterranean Sea. Along the WM basin slopes, the macrofauna standing stock and diversity correlated with the quality of the organic matter (i.e., CPRT, PRT/CHO ratio), and to a lesser extent with the microbial abundance (range, 8%–11%) and the grain size (11%) (Fig. 7b; Table S9). Considering the available food, the quantity (i.e., TPN, BPC) and quality (i.e., CCPE) contributed significantly to the variance in the macrofaunal trophic groups, although overall they accounted for a limited amount of the variance (see Fig. 7b; Table S9).

Table 3. Macrofaunal variables.

Slope	Depth (m)	Abundance (Ind m⁻²)	Biomass (mg m⁻² WW)	Structural diversity					
				N° taxa	SR	ES(30)	ES(50)	Shannon H'	Pielou J'
WM-1	1224	600±102	405.9±170.6	8±1	59±5	11±2	19±2	4.17±0.27	0.71±0.01
	1803	314±104	209.8±77.3	6±1	60±3	13±1	24±1	4.90±0.20	0.83±0.01
	2362	434±103	72.5±8.5	5±0	46±1	8±4	16±2	3.22±0.66	0.58±0.16
WM-2	1179	4153±128	2412.8±141.4	16±1	148±3	14±1	25±1	5.25±0.14	0.73±0.02
	1862	2104±287	1567.1±432.8	13±1	96±5	13±2	25±1	5.04±0.17	0.77±0.02
	2758	1281±204	321.9±81.5	14±1	74±3	12±1	21±0	4.48±0.12	0.72±0.02
WM-3	1258	2349±300	765.2±371.1	12±0	97±4	11±1	21±1	4.54±0.14	0.69±0.01
	1890	1615±45	752.1±127.2	10±0	77±2	10±1	18±1	4.13±0.13	0.66±0.02
	2448	1248±88	274.7±106.5	8±1	63±1	8±1	15±1	3.27±0.16	0.55±0.03
CM-1	1236	556±180	565.6±163.9	10±1	51±7	14±4	26±4	5.07±0.44	0.89±0.01
	1798	452±41	207.0±13.2	9±1	46±2	14±1	24±1	4.85±0.14	0.88±0.00
	2120	365±83	40.0±11.9	8±1	37±3	12±2	21±2	4.37±0.27	0.84±0.03
CM-2	1219	438±133	1926.0±215.9	8±1	51±3	14±1	24±2	4.88±0.19	0.86±0.01
	1924	145±13	186.7±27.3	9±0	24±1	10±1	17±1	3.80±0.16	0.83±0.01
	2693	84±4	137.5±47.9	5±0	17±1	10±2	15±1	3.62±0.19	0.89±0.01
EM	1237	199±25	284.7±59.2	4±1	20±0	10±1	16±0	3.71±0.14	0.86±0.03
	1907	37±19	72.5±53.0	2±0	4±0	4±1	-	1.66±0.33	0.83±0.00
	2766	33±8	37.9±16.1	2±0	7±1	7±0	-	2.75±0.32	0.98±0.03

Data are means ± standard deviation.
WW, wet weight; SR, species richness; ES(30), ES(50), Hulbert indices.

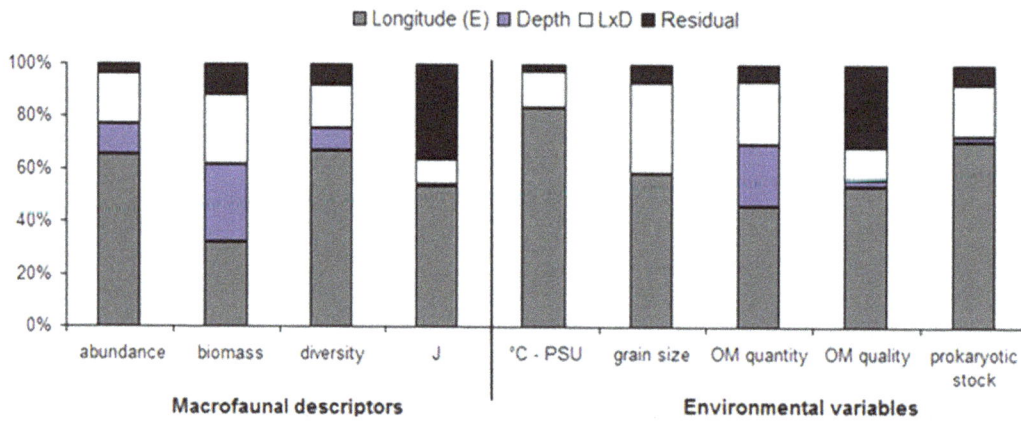

Figure 3. PERMANOVA results. Contributions of the components of variance (longitude, depth, longitude × depth [L×D]) according to the macrofaunal descriptors and the main environmental features. Diversity, macrofaunal diversity (SR, H':log2, Hulbert index); Pielou evenness (J'); bottom temperature (°C); bottom salinity; organic matter (OM); and prokaryotic stock.

a)

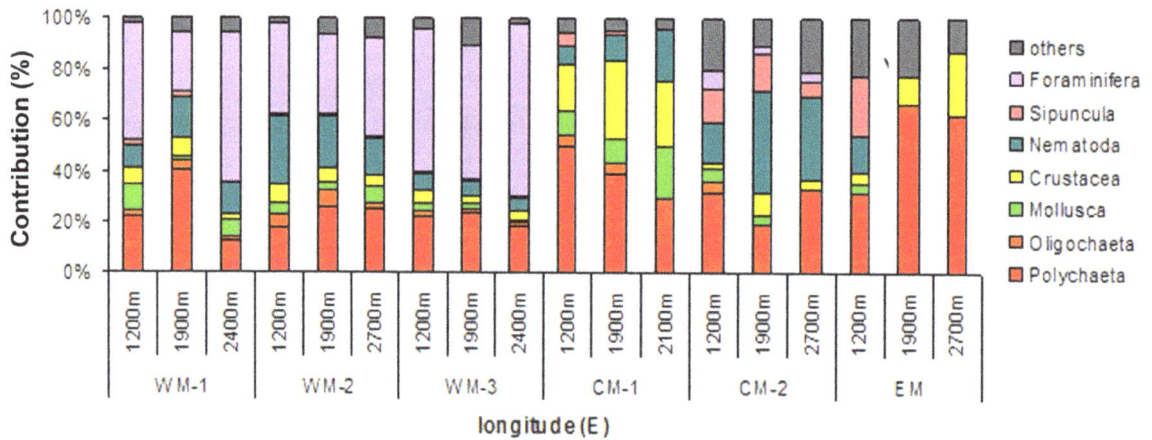

b)

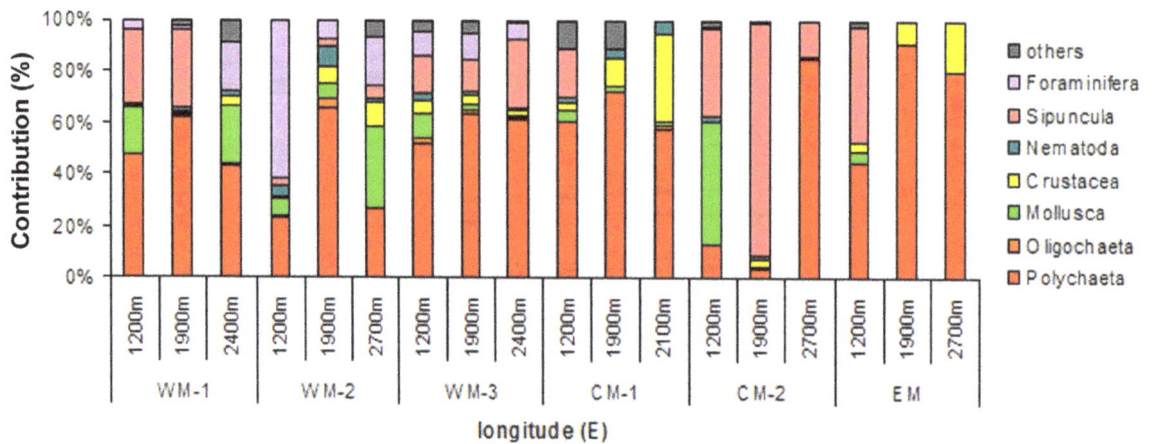

Figure 4. Macrobenthic community structure. Macrofaunal community structure in terms of (a) abundance contribution (%), and (b) biomass contribution (%) of the major taxonomic groups represented in the graphs as Polychaeta, Oligochaeta, Mollusca, Crustacea, Nematoda, Sipuncula, Foraminifera and others, at all of the investigated stations. The group of 'others' includes: Nemertea, Caudofoveata, Porifera, Bryozoa, Hydrozoa, Scyphozoa, Priapulida and Echiura.

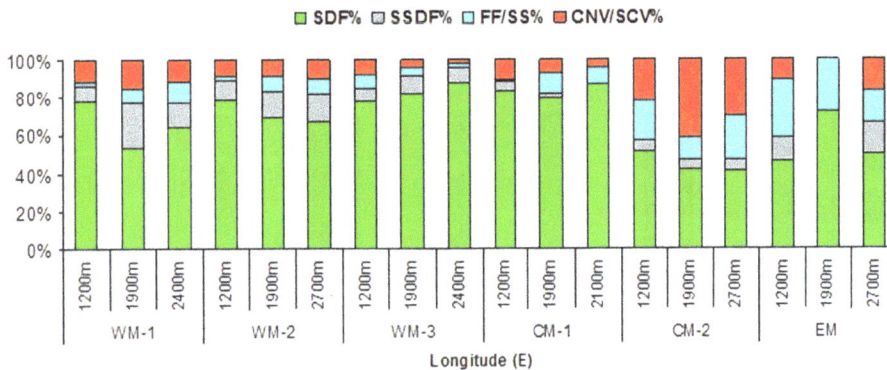

Figure 5. Macrofaunal functional composition. Trophic structure composition as the percentage contribution of each trophic group, from the WM basin to the EM basin and along all of the investigated slope areas. SDF, surface deposit feeder; SSDF, subsurface deposit feeder; FF/SF, filter feeder/suspension feeder; CNV/SCV, carnivore/scavanger; WM, western Mediterranean; CM, central Mediterranean; EM, eastern Mediterranean.

In the CM basin, both the quantity (BPC) and quality (CCPRT) of the putative food sources might be drivers for the macrofauna standing stock and diversity variability (Fig. 7c; Table S9). The prokaryotic abundance in the sediment and the POC flux to the sea bottom were highly significantly correlated with the trophic group variability.

In the EM basin, the variance of the macrofaunal stock appeared to be influenced by the quantity of organic matter in the sediment (i.e., BPC). The diversity was correlated to the BPC content and to the heterogeneity of the substrate (Fig. 7d; Table S9). Once again, with the only exception being the SDFs, the variability of the different functional groups was related to the quantity of the available food sources (i.e., BPC, POC flux), and the grain size.

Discussion

Longitudinal and bathymetric trends in the macrofauna abundance, biomass, community and trophic structure

For the deep Mediterranean macrobenthos, there are no comparable datasets in terms of the spatial scale, because almost all of the available information is scattered and often restricted to specific areas (e.g., canyons or a single slope system). Since the investigations conducted in the EM basin from 1989 to 1998 across the continental shelf and at various bathyal depths [4,32,55,91], and the limited set of data for the WM basin collected between 1988 and 1996 and in 2007 [23,26,58], there have been no other more recent studies on the deep Mediterranean macrobenthic infauna. Another point is that sediment has been sieved through mesh sizes ranging from 250 μm to 500 μm for the deep macrofauna. In effect, the discussion among

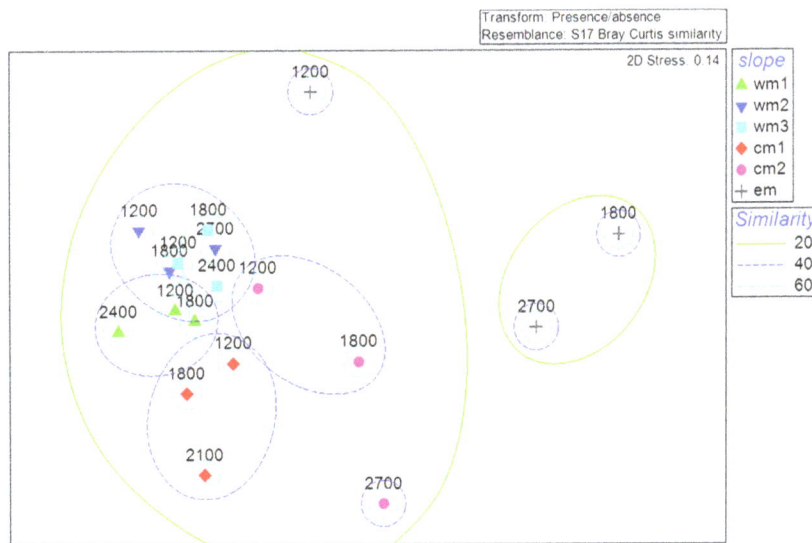

Figure 6. Non-metric multidimensional scaling. Non-metric multidimensional scaling ordination plot based on macrobenthic organisms composition, showing the similarity among the slopes at different longitudes and stations at different depths. wm1, wm2, wm3, WM basin; cm1, cm2, CM basin; em, EM basin. Numbers above symbols indicated station depths.

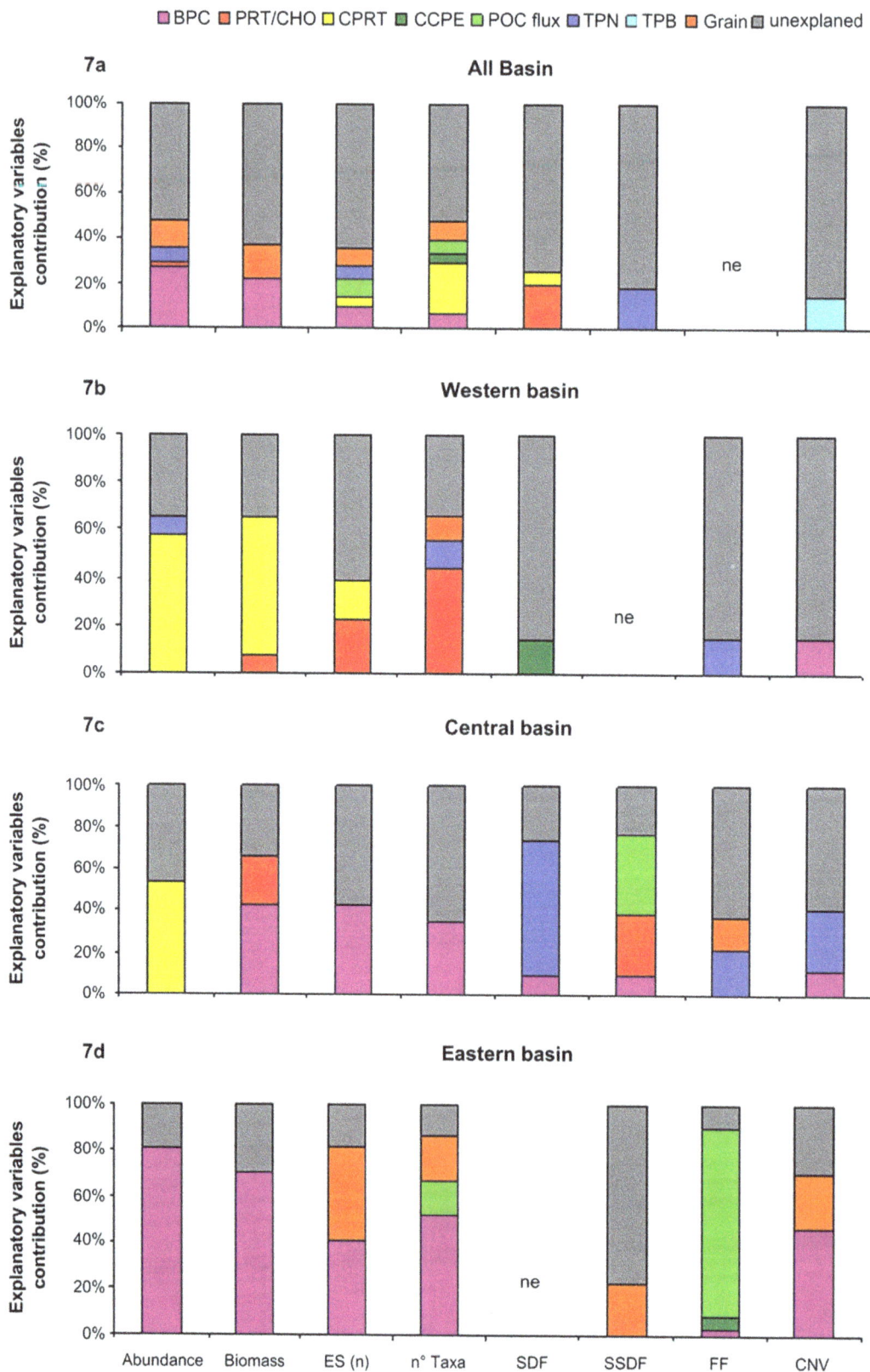

Figure 7. DISTLM forward results. Results of multivariate multiple regression analysis of the macrofaunal abundance, biomass, expected species number (ES(30, 50)), number of taxa, trophic community structure (%SDFs, %SSDFs, %FF/SF, %CNV/SCV), for (a) all of the investigated areas, and for the (b) WM, (c) CM, and (d) EM basins. The contributions are shown for the significant environmental variables (i.e., the explanatory variables) according to the variability of each of the macrofauna descriptors. BPC, biopolymeric organic C; PRT/CHO, protein/carbohydrate ratio; CPRT, protein C; CCPE, phytopigment C; POC, particulate organic C; TPN, total prokaryotic number; TPB, total prokaryotic biomass; Grain, grain size; ne, not explained.

European scientists in terms of what mesh size to use is still open, as studies in different countries use different mesh sizes [6]. This makes it difficult to directly compare quantitative data on deep-sea macrobenthic fauna from the Mediterranean Sea, and also worldwide.

Previous research within the Mediterranean Sea that included open slope systems has reported a general decline in the abundance and biomass of the benthic fauna (i.e. meiofauna, macrofauna, megafauna) with increasing depth and longitude [4,36,92–94]. One of the main causes of this decreasing trend in macrobenthic standing stock is the increasing oligotrophy of the water masses from the west to the east Mediterranean basin [92,98]. The present study also recorded a decreasing trend in the macrofauna standing stock from the WM to the CM and EM basins, the only exception being the evident drop along the Balearic slope area (WM-1). Similarly, Tselepides et al. [95] reported lower values for the bathyal meiofaunal abundance from the south Balearic Islands area. They suggested that the paucity in the meiobenthic population can be ascribed to the oligotrophy of that area, which is influenced by the food-depleted Liguro-Provençal current. This observation is consistent with the general lower values for the food sources in the present study, which includes the microbial component in WM-1 relatively to WM-2 and WM-3. We can infer that the quantity and quality of the food (expressed as the BPC content) influence the macrobenthic population, as already shown for the Mediterranean Sea and for oceans worldwide [93,99,100], as well as for other benthic components (i.e., megafauna, meiofauna, prokaryotes) [28,95,101].

In the present study, an increase in the water depth was associated with a decrease in the macrobenthic stock, especially in the biomass. Similarly, other studies have reported sharper decreases with depth in the biomass rather than in the number of macrofauna organisms [12,32], which appears to be mainly due to the rapid depletion of the food sources [96,97].

The macrofaunal community composition changed with longitude rather than with depth, such that every slope system is a naturally heterogeneous system, including its fauna population [102]. The group of Polychaeta was not always the most abundant, in contrast to what has been reported in other studies [4,32,57]. In the WM basin, Foraminifera such as *Hoeoglundina elegans*, *Uvigerina mediterraean*, *Ammolagena clavata* and *Truncorotalia sp.* were present in higher levels (range, 23%–67%). Rosso and co-authors [103] found that some of these Foraminifera species were associated with deep-water corals from Santa Maria di Leuca (Italy); however, most other studies conducted in the Mediterranean Sea have excluded this group in macrobenthic studies [58,103]. However, the Foraminifera can be an abundant and widespread component of deep-sea benthic populations, and they cover specific functional roles [104–107].

The other abundant groups found, such as Crustacea, Bivalvia, macrobenthic Nematoda and Sipuncula, have often been reported as important and diversified [34,49,50,53] components of the Mediterranean bathyal fauna [4,58], and of other seas [108,109]. The identified Amphipoda families (e.g., Eusiridae, Phoxocephalidae, Lyssianassidae) and the species *Harpinia truncata* and *Paracentromedon crenulatum* usually inhabit the WM basin and the EM basin [51,53], as well as for the Cumacea genera and the species recognized (e.g., *Cyclaspis longicaudata*, *Diastyloides bacescoi*, *Diastyloides serratus*) [49,50,52,54]. A large contribution of Sipuncula, particularly in terms of biomass, was also documented by Cosson et al. [25] in their comparison of stations at increasing oligotrophy in the Atlantic Ocean. The Sipuncula can catch and bury food deeper in the sediment, in this way they

can cope with conditions of low food sources [110]. Nematoda occurred widely from the WM basin to the EM basin slope areas. This is another group that has rarely been included in macrofauna studies [4,26,111], because it is considered an exclusively meiofaunal taxon [18]. They cover different functional roles and represent an important and distinct assemblage in the macrobenthos [111] and in comparison with nematodes from meiofauna. Indeed, with the sole exception of the highly represented genus *Halalaimus* also in our samples, the most abundant meiobenthic nematode genera reported by Pape et al. [29] (e.g. *Acantholaimus*, *Amphimonistrella*, *Monhystrella*, *Neochromadora*) were rarely or never found in our macrobenthic samples.

The trophic structure of a population can provide information on the trophic status of a system, and on the structural complexity of a community [32]. The dominance of SDF, followed by SSDF, for all of our slopes and depths confirms that the deposit feeding mode is one of the best feeding strategies in environments that generally have low food sources, such as the deep sea [105,112]. Major changes in the trophic structure composition occurred moving eastwards, where the contributions of carnivores/scavengers and filter feeders gained importance, as reported previously [4,32]. The carnivore/scavenger feeding mode is considered advantageous in a nutrient-limited environment, as their mobility is necessary to locate the more scarce food sources [4,114]. The number of carnivores in our study substantially increased by including the many predatory/omnivore genera of large-sized Nematoda, as has been documented in other areas where food is relatively scarce [111,115]. The relatively high percentage of filter feeders/suspension feeders in the EM basin, which was also reported by Tselepides et al. [32], was explained by Kröncke et al. [4] in terms of the large-scale hydrodynamic features of the open basin (e.g., lateral transport of organic material from coastal regions). The link between the filter feeders/suspension feeders group and the organic carbon fluxes was confirmed by the strong correlation observed between these filter feeders/suspension feeders and the POC flux in the EM slope area.

Overall, the data in the present study show that the trophic diversity of the Mediterranean macrobenthic populations might be influenced only partially by food availability and the heterogeneity of the substrate (range, 15% to 26%). However, it has been noted [22,105] that correlations between food availability and feeding strategies of benthic organisms might be more the result of a combination of factors (e.g. hydrodynamic conditions, small-scale physical events), which can influence the availability of food sources for the benthic populations. Even though the communities in the WM basin appeared to be less affected by food availability, this appears to have an important role in the CM and EM basins for the determination of the functional structure of the macrofauna [55,58]. In the CM basin, the highest microbial and organic matter qualities were reported, which may point to the influence of the available food on the macrofaunal trophic structure, and in particular, of the grazing activity of the macrofauna on the microbial organisms [113].

Macrofaunal α-diversity versus β-diversity

For the Mediterranean Sea, the species diversity from meiofauna to megafauna has been reported to show an overall decline both with increasing water depth and with longitude, even though some exceptions have been observed [98]. Patterns in the faunal biodiversity are usually ascribed to a decrease in food availability with increasing water depth, and in the case of the Mediterranean Sea, also with an eastwards trend in food sources [20].

Our data confirm the longitudinal decreasing trend in macrofaunal species richness from west to east and with depth for the Mediterranean Sea, along all of the slopes investigated. The food availability and heterogeneity of the substrate appear to influence the diversity of the macrofauna (i.e., ES(30, 50), number of taxa), especially when the effects of these variables are tested within each basin of the Mediterranean Sea. The potential drivers that are usually mentioned to explain patterns in faunal biodiversity (i.e., food availability, sediment grain size) have important roles also in the present study, particularly at a within-basin spatial scale. Indeed, some drivers act differently, but simultaneously, on smaller or larger spatial scales, where they might often be hidden by the effects of depth, longitude and/or latitude [8,84,116,117]. The equitability index (J) showed an opposite trend to that of the diversity indices. This means that moving eastwards, the dominance of some Foraminifera (e.g., *Uvigerina mediterranea*) and Polychaeta (e.g., Cirratulidae, Fauveliopsidae) disappeared.

It has been demonstrated that a simple analysis of local α-diversity is not enough to evaluate the biogeographic differences in deep-sea species compositions, and therefore does not provide a real picture of the biodiversity that characterises different systems, as well as the factors that control them [85,118]. Here we have quantified for the first time the deep Mediterranean basin β-diversity of macrofauna across different depths and longitudes. While the α-diversity showed significant differences between the WM, CM and EM basins and some variability along the bathymetric gradient, the β-diversity revealed large changes in the macrobenthic organism compositions at the different levels: (i) between the basins and slopes; (ii) between the slopes within the same basin; and (iii) at different depths. No clear spatial overlap emerged between slope systems or depths. In contrast to what was reported by Vanreusel et al. [119], but similar to the findings of Serpetti et al. [109], the organisms that generated the high rates of turnover diversity were not necessarily organisms that were dominant along one slope or at one particular depth. The low overlap in the species compositions can be ascribed to the high habitat heterogeneity that is typically reported for the continental margins [19]. The reasons for this heterogeneity might include differences in food supply, substrate heterogeneity, hydrological features and/or geographic position [8].

The high macrofauna β-diversity is comparable to that reported for nematodes [31,120] and for megafauna [36], which demonstrates a highly variable macrofauna composition at different longitudes and depths. For this reason, we can hypothesise that macrofaunal biodiversity is determined locally (i.e., on smaller spatial scales), and even more, regionally (i.e., on larger spatial scales). This is in agreement with findings reported for other benthic compartments, and it indicates that each region in the Mediterranean Sea can be distinguished according to the presence of a specific assemblage and species composition.

Conclusions

From our large spatial scale investigation of the macrofauna that inhabit the deep Mediterranean Sea, it has emerged that:

- The macrobenthic abundance and biomass show a general longitudinal decreasing trend from the WM basin to the EM basin. Biomass, rather than abundance, is negatively affected by increasing water depth;
- The macrobenthic community and trophic structure change significantly with longitude; there were no significant changes here between depths;

- The macrofaunal standing stock, diversity and trophic structure are differently influenced by the quantity and quality of the food sources and the habitat features (e.g., grain size), which depend on the basin or slope system investigated. From our analysis, we can infer that the influence of the food source or substrate heterogeneity on the benthic fauna might be modulated or partially masked by the multiplicity of interactions between 'local' ecological characteristics and environmental factors, as opposed to those considered here, for each specific basin and/or slope environment;
- The high β-diversity through the Mediterranean basins and for different depths suggests notable large (i.e., between basins) and smaller (i.e., across depths along the same slope system) spatial scale diversity in the macrofauna composition that is not detectable by estimating the α-diversity alone.

The present study also highlighted the following gaps in the study of the deep Mediterranean Sea macrofauna:

- A lack of recent comparable datasets (e.g., for standing stock, α-diversity, β-diversity) on large spatial scales;
- A lack of a unified sampling technique for the macrobenthos;
- The *a-priori* exclusion of some organisms, such as macrobenthic Foraminifera and Nematoda, even though they often constitute an important and distinct component of the macrobenthos, in terms of their abundance, biomass, and structural and functional diversity.

Supporting Information

Table S1   Sampling details.

Table S2   PERMANOVA results carried out to ascertain multivariate differences in environmental features, prokaryotes and macrofauna at different longitudes and depths.

Table S3   List of the identified macrobenthic organisms.

Table S4   Dissimilarity in trophic groups composition between basins and variables responsible for the estimated differences.

Table S5   Dissimilarities in macrobenthic organisms composition between all investigated A) basins; B) slopes; C) slopes within each basin.

Table S6   Dissimilarities in the macrofaunal organisms composition between depths at all investigated slopes, from west to east basin.

Table S7   A) Contribution of macrobenthic organisms responsible for the dissimilarity between depths. B) Contribution of macrobenthic organisms responsible for the dissimilarity between basin.

Table S8   Results of the multivariate multiple regression analysis carried out on the macrofaunal descriptors.

Table S9   Results of the multivariate multiple regression analysis carried out separately in the western, central and eastern basins on the macrofaunal descriptors.

Acknowledgments

The authors are indebted to the crews of the ships R/V Pelagia (The Netherlands), R/V Urania (Italy) and R/V Meteor (Germany) for their valuable help during the sampling activities. We acknowledge constructive criticism of the reviewers that helped us to significantly improve an early version of our manuscript.

Author Contributions

Conceived and designed the experiments: EB EM ML SA. Performed the experiments: EB EM ML. Analyzed the data: EB EM. Contributed reagents/materials/analysis tools: EB EM ML. Wrote the paper: EB EM ML SA AC.

References

1. Arvanitidis C, Bellan G, Drakopoulos P, Valavanis V, Dounas C, et al. (2002) Seascape biodiversity patterns along the Mediterranean and the Black Sea: lessons from the biogeography of benthic polychaetes. Mar Ecol Prog Ser 244: 139–152.

2. Tittensor DP, Mora C, Jetz W, Lotze HK, Ricard D, et al. (2010) Global patterns and predictors of marine biodiversity across taxa. Nature 466: 1098–1101.

3. Beaugrand G, Rombouts I, Kirby RR (2013) Towards an understanding of the pattern of biodiversity in the oceans. Glob Ecol Biogeogr 22: 440–449.

4. Kröncke I, Turkay M, Fiege D (2003) Macrofauna Communities in the Eastern Mediterranean Deep Sea. Mar Ecol 24: 193–216.

5. Gambi C, Lampadariou N, Danovaro R (2010) Latitudinal, longitudinal and bathymetric patterns of abundance, biomass of metazoan meiofauna: importance of the rare taxa and anomalies in the deep Mediterranean Sea. Adv Oceanogr Limnol 1: 167–197.

6. Narayanaswamy BE, Renaud PE, Duineveld GCA, Berge J, Lavaleye MSS, et al. (2010) Biodiversity trends along the western European margin. PLoS One 5: e14295.

7. Leduc D, Rowden A, Bowden D, Nodder S, Probert P, et al. (2012) Nematode beta diversity on the continental slope of New Zealand: spatial patterns and environmental drivers. Mar Ecol Prog Ser 454: 37–52.

8. Levin LA, Etter RJ, Rex MA, Gooday J, Smith CR, et al. (2001) Environmental influences on regional deep-sea species diversity. Annu Rev Ecol Syst 32: 51–93.

9. Rex MA, McClain CR, Johnson NA, Etter RJ, Allen JA, et al. (2005) A source-sink hypothesis for abyssal biodiversity. Am Nat 165: 163–178.

10. Ramirez-Llodra E, Brandt A, Danovaro R, De Mol B, Escobar E, et al. (2010) Deep, diverse and definitely different: unique attributes of the world's largest ecosystem. Biogeosciences 7: 2851–2899.

11. Carney RS (2005) Zonation of deep biota on continental margins. Oceanogr Mar Biol An Annu Rev 43: 211–278.

12. Rex M, Etter R, Morris J, Crouse J, McClain C, et al. (2006) Global bathymetric patterns of standing stock and body size in the deep-sea benthos. Mar Ecol Prog Ser 317: 1–8.

13. Galéron J, Sibuet M, Mahaut M, Dinet A (2000) Variation in structure and biomass of the benthic communities at three contrasting sites in the tropical Northeast Atlantic. Mar Ecol Prog Ser 197: 121–137.

14. Hughes DJ, Gage JD (2004) Benthic metazoan biomass, community structure and bioturbation at three contrasting deep-water sites on the northwest European continental margin. Prog Oceanogr 63: 29–55.

15. Pavithran S, Ingole BS, Nanajkar M, Raghukumar C (2009) Composition of macrobenthos from the central Indian Ocean Basin. J Earth Syst Sci 118: 689–700.

16. Rex MA (1981) Community structure in the deep-sea benthos. Annu Rev Ecol Syst 12: 331–353.

17. Levin LA, Gage JD (1998) Relationships between oxygen, organic matter and diversity of bathyal macrofauna. Deep Sea Res Part II 45: 129–163.

18. Flach E, Vanaverbeke J, Heip C (1999) The meiofauna:macrofauna ratio across the continental slope of the Goban Spur (north-east Atlantic). J Mar Biol Assoc UK 79: 233–241.

19. Levin LA, Sibuet M (2012) Understanding Continental Margin Biodiversity: A New Imperative. Ann Rev Mar Sci 4: 79–112.

20. Snelgrove PVR, Smith CR (2002) A riot of species in an environmental calm: the paradox of the species-rich deep-sea floor. Oceanogr Mar Biol 40: 311–342.

21. Etter RJ, Grassle JF (1992) Patterns of species diversity in the deep sea as a function of sediment particle size diversity. Nature 360: 576–578.

22. Leduc D, Rowden AA, Probert PK, Pilditch CA, Nodder SD, et al. (2012) Further evidence for the effect of particle-size diversity on deep-sea benthic biodiversity. Deep Sea Res Part I 63: 164–169.

23. Tahey TM, Duineveld GCA, Berghuis EM, Helder W (1994) Relation between sediment-water fluxes of oxygen and silicate and faunal abundance at continental shelf, slope and deep-water stations in the northwest Mediterranean. Mar Ecol Prog Ser 104: 119–130.

24. Tyler PA (2003) The peripheral deep seas. Ecosystems of the world. Amsterdam: Elsevier. pp. 261–293.

25. Cosson N, Sibuet M, Galeron J (1997) Community structure and spatial heterogeneity of the deep-sea macrofauna at three contrasting stations in the tropical northeast Atlantic. Deep Sea Res Part I 44: 247–269.

26. Mamouridis V, Cartes JE, Parra S, Fanelli E, Saiz Salinas JI (2011) A temporal analysis on the dynamics of deep-sea macrofauna: Influence of environmental variability off Catalonia coasts (western Mediterranean). Deep Sea Res Part I 58: 323–337.

27. Corliss BH, Brown CW, Sun X, Showers WJ (2009) Deep-sea benthic diversity linked to seasonality of pelagic productivity. Deep Sea Res Part I 56: 835–841.

28. Cartes JE, Maynou F, Fanelli E, Romano C, Mamouridis V, et al. (2009) The distribution of megabenthic, invertebrate epifauna in the Balearic Basin (western Mediterranean) between 400 and 2300 m: Environmental gradients influencing assemblages composition and biomass trends. J Sea Res 61: 244–257.

29. Pape E, Jones DOB, Manini E, Bezerra TN, Vanreusel A (2013) Benthic-pelagic coupling: effects on nematode communities along southern European continental margins. PLoS One 8: e59954.

30. Hasemann C, Soltwedel T (2011) Small-scale heterogeneity in deep-sea nematode communities around biogenic structures. PLoS One 6: e29152.

31. Danovaro R, Gambi C, Lampadariou N, Tselepides A (2008) Deep-sea nematode biodiversity in the Mediterranean basin: testing for longitudinal, bathymetric and energetic gradients. Ecography 31: 231–244.

32. Tselepides A, Papadopoulou K-N, Podaras D, Plaiti W, Koutsoubas D (2000) Macrobenthic community structure over the continental margin of Crete (South Aegean Sea, NE Mediterranean). Prog Oceanogr 46: 401–428.

33. Cartes J, Maynou F, Sardà F, Company J, Lloris D, et al. (2004) The Mediterranean deep-sea ecosystem. Part I. An overview of their diversity, structure, functioning and anthropogenic impacts. In: The Mediterranean deep-sea ecosystems: an overview of their diversity, structure, functioning and anthropogenic impacts, with a proposal for conservation. pp. 9–38.

34. Galil BS (2004) The limit of the sea: the bathyal fauna of the Levantine Sea. Sci Mar 68: 63–72.

35. Company JB, Maiorano P, Tselepides A, Politou C, Plaity W, et al. (2004) Deep-sea decapod crustaceans in the western and central Mediterranean Sea: preliminary aspects of species distribution, biomass and population structure. Sci Mar 68: 73–86.

36. Tecchio S, Ramírez-Llodra E, Sardà F, Company J, Palomera I, et al. (2011) Drivers of deep Mediterranean megabenthos communities along longitudinal and bathymetric gradients. Mar Ecol Prog Ser 439: 181–192.

37. D' Ortenzio F, d' Alcal MR (2009) On the trophic regimes of the Mediterranean Sea: a satellite analysis. Biogeosciences 6: 139–148.

38. Santinelli C, Sempéré R, Van Wambeke F, Charriere B, Seritti A (2012) Organic carbon dynamics in the Mediterranean Sea: An integrated study. Global Biogeochem Cycles 26: 1–13.

39. Danovaro R, Company JB, Corinaldesi C, D'Onghia G, Galil B, et al. (2010) Deep-sea biodiversity in the Mediterranean Sea: the known, the unknown, and the unknowable. PLoS One 5: e11832.

40. Fredj G, Laubier L (1985) The deep Mediterranean benthos. Mediterranean marine ecosystems. New York: Plenum Press. pp. 109–146.

41. Carpine C (1970) Écologie de l'étage bathyal dans la Méditerranée occidentale. Mémoires de l'Istitute océanographique, Monaco 2. 146 p.

42. Vamvakas CNE (1970) Peuplements benthiques des substrats meubles du sud de la mer Egée. Tethys 2: 89–130.

43. Laubier L, Picard C, Ramos J (1973) Les heterospionidae (Annélides polychètes sédentaires) de Méditerranée occidentale. Vie Milieu 23: 243–254.

44. Chardy P, Laubier L, Reyss D, Sibuet M (1973) Dragages profonds en mer Egèe - donnèes préliminaires (1). Rapp Comm int Mer Médit 22: 107–108.

45. Chardy P (1974) Deux especes nouvelles d'isopodes asellotes récoltées en méditerranée profonde. Vie Milieu 24: 409–420.

46. Wagele JW (1980) Anthuridea (Crustacea, Isopoda) aus dem Tyrrhenischen Meer. Zool Scr 9: 53–66.

47. Gutu M (2001) Apparence and reality in the knowledge of the tanaidacean crstaceans from the Mediterranean basin. Trav du Muséum Natl d'Histoire Nat 43: 59–63.

48. Gutu M (2002) Contributions to the knowledge of the genus Apseudes Leach, 1814 (Crustacea: Tanaidacea, Apseudomorpha) from the Mediterranean basin and the North African Atlantic. Trav du Muséum Natl d'Histoire Nat 44: 19–39.

49. Reyss D (1972) Résultats scientifiques de la campagne du N.O. "Jean Charcot" en Méditerranée occidentale, Mai-Juin-Juillet 1970. Cumacés. Crustaceana Suppl. III: 362–377.

50. Reyss D (1974) Cumacés. Resultats scientifiques de la campagne "Polymède II" du N.O. "Jean Chracot" en Mer Ionienne et en Mer Egéé (Avril-Mai 1972). Crustaceana 27: 275–282.

51. Cartes JE, Sorbe JC (1993) Les Communautés suprabenthiques bathyales de la mer Catalene (Méditerranée occidenatele): données préliminaires sue la répartition bathymétrique et l'abondance des crustacés péracarides. Crustaceana 64: 155–170.

52. Corbera J, Galil BS (2001) Cumaceans (Crustacea, Peracarida) from the lower slope of the northern Israel coast, with a discussion on the status of *Platysympus typicus*. Isr J Zool 47: 135–146.

53. Sorbe JC, Galil BS (2002) The Bathyal Amphipoda Off the Levantine Coast, Eastern Mediterranean. Crustaceana 75: 957–968.

54. Muhlenhardt-Siegel U (2009) Cumacea (Crustacea, Peracarida) in the deep Mediterranean, with the description of one new species. Zootaxa 2096: 413–432.

55. Tselepides A, Eleftheriou A (1992) South aegean (eastern mediterranean) continental slope benthos: macroinfaunal - environmental relationships. Deep-Sea food chains and the global carbon cycle. Kluwer Academic Publishers. pp. 139–156.

56. Basso D, Thomson J, Corselli C (2004) Indications of low macrobenthic activity in the deep sediments of the eastern Mediterranean Sea. Sci Mar 68: 53–62.

57. Gerino M, Stora G, Poydenot F, Bourcier M (1995) Benthic fauna and bioturbation on the Mediterranean continental slope: Toulon Canyon. Cont Shelf Res 15: 1483–1496.

58. Stora G, Bourcier M, Arnoux A, Gerino M, Campion J Le, et al. (1999) The deep-sea macrobenthos on the continental slope of the northwestern Mediterranean Sea: a quantitative approach. Deep Sea Res Part I 46: 1339–1368.

59. Gage JD, Tyler PA (1991) Deep-Sea biology: A natural history of organisms at the deep sea floor. Cambridge, UK: Cambridge University Press. 493 p.

60. Anderson MJ, Crist TO, Chase JM, Vellend M, Inouye BD, et al. (2011) Navigating the multiple meanings of β diversity: a roadmap for the practicing ecologist. Ecol Lett 14: 19–28.

61. Pusceddu A, Dell'Anno A, Fabiano M, Danovaro R (2009) Quantity and bioavailability of sediment organic matter as signatures of benthic trophic status. Mar Ecol Prog Ser 375: 41–52.

62. Pusceddu A, Bianchelli S, Canals M, Sanchez-Vidal A, Durrieu De Madron X, et al. (2010) Organic matter in sediments of canyons and open slopes of the Portuguese, Catalan, Southern Adriatic and Cretan Sea margins. Deep Sea Res Part I Oceanogr Res Pap 57: 441–457.

63. Danovaro R (2010) Methods for the study of deep-sea sediments, their functioning and biodiversity. Section V. Deep-sea benthic functioning. Taylor and. New york. 162 p.

64. De Jonge VN (1980) Fluctuations in the Organic Carbon to Chlorophyll a Ratios for Estuarine Benthic Diatom Populations. Mar Ecol Prog Ser 2: 345–353.

65. Johnson NA, Campbell JW, Moore TS, Rex MA, Etter RJ, et al. (2007) The relationship between the standing stock of deep-sea macrobenthos and surface production in the western North Atlantic. Deep Sea Res Part I 54: 1350–1360.

66. Lutz M, Dunbar R, Caldeira K (2002) Regional variability in the vertical flux of particulate organic carbon in the ocean interior. Global Biogeochem Cycles 16: 1–18.

67. Suess E (1980) Particulate organic carbon flux in the oceans: surface productivity and oxygen utilization. Nature 288: 260–263.

68. Fabiano M, Danovaro R, Fraschetti S (1995) A three-year time series of elemental and biochemical composition of organic matter in subtidal sandy sediments of the Ligurian Sea (northwestern Mediterranean). Cont Shelf Res 15: 1453–1469.

69. Manini E, Danovaro R (2006) Synoptic determination of living/dead and active/dormant bacterial fractions in marine sediments. FEMS Microbiol Ecol 55: 416–423.

70. Luna GM, Manini E, Danovaro R (2002) Large Fraction of Dead and Inactive Bacteria in Coastal Marine Sediments: Comparison of Protocols for Determination and Ecological Significance. Appl Environ Microbiol 68: 3509–3513.

71. Fry JC (1990) Determination of biomass. Methods in aquatic bacteriology. New york: John Wiley & Sons. pp. 27–72.

72. Pavithran S, Ingole B, Nanajkar M, Goltekar R (2009) Importance of sieve size in deep-sea macrobenthic studies. Mar Biol Res 5: 391–398.

73. Gray JS (1994) is deep-sea species diversity really so high? Species diversity of the Norwegian continental shelf. Mar Ecol Prog Ser 112: 205–209.

74. Dinet A, Desbruyères D, Khripounoff A (1985) Abondance des peuplements macro- et meio-benthiques: repartition et stratégie d'échantillonage. In: Laubier L, Monniot C, editors. Peuplements profonds du golfe du Gascogne. Brest: Institut Francaise de Recherche pour l''Exploitation de la Mer. pp. 121–142.

75. Sars GO (1882) An account of crustacea of Norway with short description and figures of all the species. Bergen: The Bergn Museum. 255 p.

76. Norman AN, Stebbing TRR (1889) On the crustacea isopoda of the lightning. Porcupine and valorous expeditions. Washington: Smithsonian Institution. 141 p.

77. Romero-Wetzel MB (1987) Sipunculans as inhabitants of very deep, narrow burrows in the deep-sea sediments. Mar Biol 96: 87–91.

78. Fauchald K, Jumars PA (1979) The diet of worms: a study of polychaete feeding guilds. Ocean Mar Biol Ann Rev 17: 193–284.

79. Rowe GT (1983) Deep-sea biology. New york: John Wiley & Sons. 565 p.

80. Shannon CE, Weaver W (1949) The mathematical Theory of communication. Urbana, IL: University of Illinois.

81. Pielou EC (1975) Ecological diversity. New york: Wily. 165 p.

82. Sanders HL (1968) Marine benthic diversity: A comparative study. Am Nat 102: 243–282.

83. Hurlbert SH (1971) The Nonconcept of Species Diversity: A Critique and Alternative Parameters. Ecology 52: 577–586.

84. Gage JD, Lamont PA, Kroeger K, Paterson GLJ, Gonzalez Vecino JL (2000) Patterns in deep-sea macrobenthos at the continental margin: standing crop, diversity and faunal change on the continental slope off Scotland. Hydrobiologia 440: 261–271.

85. Gray JS (2000) The measurement of marine species diversity, with an application to the benthic fauna of the Norwegian continental shelf. J Exp Mar Bio Ecol 250: 23–49.

86. Anderson MJ (2001) A new method for non-parametric multivariate analysis of variance. Austral Ecol 26: 32–46.

87. McArdle BH, Anderson MJ (2001) Fitting multivariate models to community fitting multivariate models to community data: a comment on distance-based redundancy analysis. Ecology 82: 290–297.

88. Anderson MJ, ter Braak CJF (2003) Permutation tests for multi-factorial analysis of variance. J Stat Comput Simul 73: 85–113.

89. Anderson MJ, Robinson J (2003) Generalized discriminat analysis based on distances. Aust NZJ Stat 45: 301–318.

90. Clarke KR, Gorley RN (2006) PRIMER v6: User Manual/Tutorial. PRIMER-E. Plymouth.

91. Karakassis I, Eleftheriou A (1997) The continental shelf of Crete:structure of macrobenthic communities. Mar Ecol Prog Ser 160: 185–196.

92. Sardà F, Calafat A, Flexas MMAR, Tselepides A, Canals M, et al. (2004) An introduction to Mediterranean deep-sea biology. Sci Mar 68: 7–38.

93. Pape E, Bezerra TN, Jones DOB, Vanreusel A (2013) Unravelling the environmental drivers of deep-sea nematode biodiversity and its relation with carbon mineralisation along a longitudinal primary productivity gradient. Biogeosciences 10: 3127–3143.

94. Tecchio S, Ramírez-Llodra E, Sardà F, Company JB (2011) Biodiversity of deep-sea demersal megafauna in western and central Mediterranean basins. Sci Mar 75: 341–350.

95. Tselepides A, Lampadariou N (2004) Deep-sea meiofaunal community structure in the Eastern Mediterranean: are trenches benthic hotspots? Deep Sea Res Part I 51: 833–847.

96. De Bovée F, Guidi LD, Soyer J (1990) Quantitative distribution of deep-sea meiobenthos in the northwestern Mediterranean (Gulf o f Lions). Cont Shelf Res 10: 1123–1145.

97. Bouchet P, Taviani M (1992) The Mediterranean deep-sea fauna: pseudopopulations of Atlantic species? Deep Sea Res 39: 169–184.

98. Coll M, Piroddi C, Steenbeek J, Kaschner K, Ben Rais Lasram F, et al. (2010) The biodiversity of the Mediterranean Sea: estimates, patterns, and threats. PLoS One 5: e11842.

99. Wei C, Rowe G, Hubbard G, Scheltema A, Wilson G, et al. (2010) Bathymetric zonation of deep-sea macrofauna in relation to export of surface phytoplankton production. Mar Ecol Prog Ser 399: 1–14.

100. Ellingsen KE, Gray JS (2002) Spatial patterns of benthic diversity: is there a latitudinal gradient along the Norwegian continental shelf? J Anim Ecol 71: 373–389.

101. Giovannelli D, Molari M, d'Errico G, Baldrighi E, Pala C, et al. (2013) Large-scale distribution and activity of prokaryotes in deep-sea surface sediments of the mediterranean sea and the adjacent atlantic ocean. PLoS One 8: e72996.

102. Levin LA, Sibuet M, Gooday AJ, Smith CR, Vanreusel A (2010) The roles of habitat heterogeneity in generating and maintaining biodiversity on continental margins: an introduction. Mar Ecol 31: 1–5.

103. Rosso a., Vertino a., Di Geronimo I, Sanfilippo R, Sciuto F, et al. (2010) Hard-and soft-bottom thanatofacies from the Santa Maria di Leuca deep-water coral province, Mediterranean. Deep Sea Res Part II 57: 360–379.

104. Gooday AJ, Hughes JA, Levin LA (2001) The foraminiferan macrofauna from three North Carolina (USA) slope sites with contrasting carbon flux: a comparison with the metazoan macrofauna. Deep Sea Res Part I 41: 1709–1739.

105. Flach E, Lavaleye M, de Stigter H, Thomsen L (1998) Feeding types of the benthic community and particle transport across the slope of the N.W. European continental margin (Goban Spur). Prog Oceanogr 42: 209–231.

106. Lipps JH (1983) Biotic interactions in benthic foraminifera. Biotic Interactions in recent and fossil benthic communities. New york: Plenum Press. pp. 331–376.

107. Lee JJ (1980) Nutrition and physiology of the foraminifera. Biochemistry and Physiology of Protozoa. New york: Academic Press. pp. 43–66.

108. Witte U (2000) Vertical distribution of metazoan macrofauna within the sediment at four sites with contrasting food supply in the deep Arabian Sea. Deep Sea Res Part II 47: 2979–2997.

109. Serpetti N, Gontikaki E, Narayanaswamy BE, Witte U (2012) Macrofauna community inside and outside of the Darwin Mounds SAC, NE Atlantic. Biogeosciences 9: 16907–16932.

110. Galéron J, Menot L, Renaud N, Crassous P, Khripounoff a., et al. (2009) Spatial and temporal patterns of benthic macrofaunal communities on the deep continental margin in the Gulf of Guinea. Deep Sea Res Part II 56: 2299–2312.

111. Sharma J, Baguley J, Bluhm B a, Rowe G (2011) Do meio- and macrobenthic nematodes differ in community composition and body weight trends with depth? PLoS One 6: e14491.

112. Kröncke I, Vanreusel A, Vincx M, Wollenburg J, Mackensen A, et al. (2000) Different benthic size-compartments and their relationship to sediment chemistry in the deep Eurasian Arctic Ocean. Mar Ecol Prog Ser 199: 31–41.

113. Boetius A, Scheibe S, Tselepidess A, Thiel H (1996) Microbial biomass and activities in deep-sea sediments of the Eastern Mediterranean: trenches are benthic hotspots. Deep Sea Res Part I 43: 1439–1460.

114. Gage JD (2003) Food inputs, utilization, carbon flow and energetics. In: Tyler PA, editor. Ecosystems of the world. Ecosystems of the deep ocean. Amsterdam: Elsevier. pp. 313–426.

115. Sharma J, Bluhm BA (2011) Diversity of larger free-living nematodes from macrobenthos (>250 μm) in the Arctic deep-sea Canada Basin. Mar Biodivers 41: 455–465.

116. Leduc D, Rowden A, Bowden D, Nodder S, Probert P, et al. (2012) Nematode beta diversity on the continental slope of New Zealand: spatial patterns and environmental drivers. Mar Ecol Prog Ser 454: 37–52.

117. Narayanaswamy BE, Coll M, Danovaro R, Davidson K, Ojaveer H, et al. (2013) Synthesis of knowledge on marine biodiversity in European Seas: from census to sustainable management. PLoS One 8: e58909.

118. Zajac RN (2008) Macrobenthic biodiversity and sea floor landscape structure. J Exp Mar Bio Ecol 366: 198–203.

119. Vanreusel A, Fonseca G, Danovaro R, da Silva MC, et al. (2010) The contribution of deep-sea macrohabitat heterogeneity to global nematode diversity. Mar Ecol 31: 6–20.

120. Danovaro R, Bianchelli S, Gambi C, Mea M, Zeppilli D (2009) Alpha-, beta-, gamma-, delta- and epsilon-diversity of the deep-sea nematodes in canyons and open slopes of Northeast Atlantic and Mediterranean margins. Mar Ecol Prog Ser 396: 197–209.

11

The Crowded Sea: Incorporating Multiple Marine Activities in Conservation Plans Can Significantly Alter Spatial Priorities

Tessa Mazor[1]*, **Hugh P. Possingham**[1,2], **Dori Edelist**[3], **Eran Brokovich**[4], **Salit Kark**[1]

1 ARC Centre of Excellence for Environmental Decisions, School of Biological Sciences, The University of Queensland, Brisbane, Queensland, Australia, 2 Grand Challenges in Ecosystems and the Environment, Silwood Park, Imperial College, London, United Kingdom, 3 Leon Recanati Institute for Maritime Studies, Department of Maritime Civilizations, University of Haifa, Mount Carmel, Haifa, Israel, 4 Department of Geography, The Hebrew University of Jerusalem, Mount Scopus, Jerusalem, Israel

Abstract

Successful implementation of marine conservation plans is largely inhibited by inadequate consideration of the broader social and economic context within which conservation operates. Marine waters and their biodiversity are shared by a host of stakeholders, such as commercial fishers, recreational users and offshore developers. Hence, to improve implementation success of conservation plans, we must incorporate other marine activities while explicitly examining trade-offs that may be required. In this study, we test how the inclusion of multiple marine activities can shape conservation plans. We used the entire Mediterranean territorial waters of Israel as a case study to compare four planning scenarios with increasing levels of complexity, where additional zones, threats and activities were added (e.g., commercial fisheries, hydrocarbon exploration interests, aquaculture, and shipping lanes). We applied the marine zoning decision support tool Marxan to each planning scenario and tested a) the ability of each scenario to reach biodiversity targets, b) the change in opportunity cost and c) the alteration of spatial conservation priorities. We found that by including increasing numbers of marine activities and zones in the planning process, greater compromises are required to reach conservation objectives. Complex plans with more activities incurred greater opportunity cost and did not reach biodiversity targets as easily as simplified plans with less marine activities. We discovered that including hydrocarbon data in the planning process significantly alters spatial priorities. For the territorial waters of Israel we found that in order to protect at least 10% of the range of 166 marine biodiversity features there would be a loss of ~15% of annual commercial fishery revenue and ~5% of prospective hydrocarbon revenue. This case study follows an illustrated framework for adopting a transparent systematic process to balance biodiversity goals and economic considerations within a country's territorial waters.

Editor: Elliott Lee Hazen, UC Santa Cruz Department of Ecology and Evolutionary Biology, United States of America

Funding: This research was conducted with the support of funding from the Australian Research Council Centre of Excellence for Environmental Decisions. T. M was a recipient of a University of Queensland Library OA Award, was funded by an Australian Postgraduate Award and gratefully acknowledges the financial support of the Australian-Israel Scientific Exchange Foundation (AISEF). The funders had no role in study design, data collection and analysis, decision to publish, or preparation of the manuscript.

Competing Interests: The authors have declared that no competing interests exist.

* Email: t.mazor@uq.edu.au

Introduction

Implementing marine conservation plans is a major challenge. Plans that determine priority areas for conservation are often based solely on biological and ecological information [1] One of the main factors inhibiting the uptake of marine conservation plans by decision makers is inadequate consideration of the broader social and economic context within which conservation operates [2–4]. Marine waters and their biodiversity are shared by a host of stakeholders and interest groups, such as commercial fishers, recreational users and offshore developers [5]. Inclusion of the activities of these multiple marine users within conservation plans is critical for achieving plans which are realistic and achievable in the real world, thereby moving from paper to action [2].

Conservation planners must try to explicitly consider other marine activities within conservation plans, to ensure no time is wasted over trying to conserve areas essential for other uses [6]. Competition for ocean space is becoming increasing intensified as resource extraction and developments are expanding to include the marine realm [7]. Offshore activities such as commercial fishing, aquaculture facilities, sand mining, desalination plants, offshore wind farms and offshore power plants, provide countries with substantial economic gains [5]. Currently, hydrocarbon operations are one of the largest economic stakeholders in the sea [8], and provide countries with huge potential and realized monetary benefits, and are expected to increase economic and political independence [9,10]. However, incorporation of such economic activities is often absent from marine conservation planning literature. Despite the little willingness for countries to protect marine areas that are deemed economically important

[11], excluding other marine activities in conservation planning means we may not be able to design a marine reserve network that is representative or economically viable [12].

Disregarding other marine activities in marine conservation planning may also mean that anthropogenic threats to biodiversity are being ignored. When planning marine reserves that aim to reap sustainable long-term benefits it is important to examine the threats to biodiversity of the system that could impair this goal. However, reserve planning should not be solely based upon threat data [13]. Examples of threats to biodiversity for consideration in reserve planning include: shipping lanes which pose a collision risk to marine mammals [14], trawlers and demersal longliners which are damaging to benthic environments and responsible for the majority of annual sea turtles deaths via by-catch [15], and marine energy installations which have been linked to habitat loss, noise pollution and invasive species [16]. In some cases marine users have made changes or modifications, such as altering the path of shipping lanes for cetaceans [17]. However, in cases where compromises cannot be met, conservation planners must be able to incorporate the potential threats to biodiversity into the planning process.

A common misconception is that marine zoning itself is a conservation planning tool. Marine zoning is the allocation of particular activities to specified marine areas [5,18]. This practice can help reduce user conflict by separating incompatible activities [7,18,19]. Several countries have stepped up to implement zoning strategies for their waters, the largest and perhaps most successful example of marine zoning is the Great Barrier Reef Marine Park off the coast of Queensland, Australia [20,21]. More recent zoning efforts occurring around the globe include the United Kingdom Irish Sea Pilot [22], the Belgian Exclusive Economic Zone [11], the waters of Norway [18], Australia's entire commonwealth waters [23] and the zoning of China's territorial sea [24]. However, key elements are often missing from some zoning plans to ensure biodiversity goals are met. For marine zoning to be used as an appropriate method or tool for protecting marine biodiversity it must enable an explicit consideration of the trade-off between biodiversity and socio-economic objectives [25]. Furthermore, zoning plans need to ensure that the zoning system provides protection that is representative of as many biodiversity features as possible [12,25].

The concept of including other activities within marine conservation planning is slowly emerging. Unlike marine spatial planning (MSP) which aims to plan water spaces to meet objectives of multiple marine users and stakeholders, [19,26], marine conservation planning (MCP) is centred on one primary goal - achieving biodiversity protection [18]. Recently, several systematic conservation plans in the marine realm have focused on a hybrid approach; reaching conservation objectives while also minimizing the opportunity cost to fishery stakeholders [25,27,28]. However, only some of these plans have been expanded to other social and economic contexts (e.g., [3,29]). Facilitating the inclusion of other activities into marine conservation planning is the emerging development of zoning software that enables multiple objectives to be considered (e.g., Marxan with Zones [27]). Up to now there has been little application of these new tools to address the complexity of marine conservation planning at regional scales or an entire country scale. As many countries around the globe aim to implement conservation measures by zoning their waters [18], it is important to develop an explicit zoning process which integrates the current spatial occupancy of other marine activities and where possible their economic objectives. The inclusion of other marine uses in marine conservation planning means that we need to carefully consider the trade-offs that underpin the resulting

conservation plans and ensure that biodiversity goals are adequately achieved.

In this study we follow a framework (Fig. 1) using a systematic approach for zoning territorial waters to achieve the protection of marine biodiversity in the face of multiple anthropogenic threats and economic activities. Within this context, we aim to test how increased complexity (by the inclusion of zones, multiple activities and economic factors) in marine conservation planning alters: a) the ability to reach biodiversity targets, b) the opportunity cost, and c) the spatial conservation priorities. Furthermore, we aim to examine the explicit incorporation of prospective hydrocarbon extraction into marine conservation planning [8].

Methods

Here our methods follow the steps outlined in Figure 1.

Spatial setting and study area

As a case study, we examined Israel's complete Mediterranean territorial waters. Israel is located in the eastern Mediterranean Sea and has relatively small territorial waters ($\sim$4200 km^2) compared with other coastal countries around the world. Currently, it faces rapid exploitation of its marine resources and aims to expand its protection of marine biodiversity [30]. Israel's Mediterranean Sea territorial waters are defined by the National Planning Authority of Israel and are used by The Israel Nature and Parks Authority (NPA) for marine reserve planning. The territorial waters of Israel's Mediterranean Sea spreads along a coastline $\sim$190 km long, and extends outwards for 12 nautical miles from the coast to a depth of $\sim$1000 m, covering an area of $\sim$4200 km^2 [30]. For our analyses, we divided this study area into 1×1 km planning units, resulting in a total of 4,205 planning units.

Compiling biodiversity features

In order to select marine areas which will fulfil a representative reserve network where all types of biodiversity are protected we compiled available distribution data of Israel's Mediterranean territorial waters of biotic and abiotic features. These included 166 biodiversity features, comprising of vertebrate marine species (153 fishes, 2 turtles, 1 cetacean), and 10 geomorphologic features (Fig. 2a; see Table S1 in File S1. for a list of species and features included in this study).

Marine species distribution data.   We compiled data from currently available published studies on native cartilaginous and bony fishes whose distribution lies within Israel's Mediterranean waters [31–38]. All native (non-alien) fish species (153 species) present in these publications were included in our study. We digitized the documented depth ranges of these native fish species using ArcGIS ([39]; Fig. S1 in File S1.; Table S2 in File S1.) and sea floor bathymetry [40], following methods in Tognelli et al. [41] and Clark & Tittensor [42]. We derived the distributions via a number of sources; locations and depth ranges from the above eight studies, data from the Hebrew University of Jerusalem Fish Collection (accessed 2012), ranges as documented in Golani et al. [43], and by expert opinion (for further details see File S2).

The distribution of sea turtles within Israel's marine waters has not been well documented and their preferred feeding, foraging and mating areas are currently poorly known. Therefore, we used the locations of established nesting sites (within [44] in Israel for both the green (*Chelonia mydas*) and loggerhead (*Caretta caretta*) sea turtle species. The targeted nesting habitats for protection in this study were chosen as planning units adjacent to nesting beaches with over 20 nest counts (from 1993–2011) and a persistence of more than five years of nesting at a particular site, in

a)

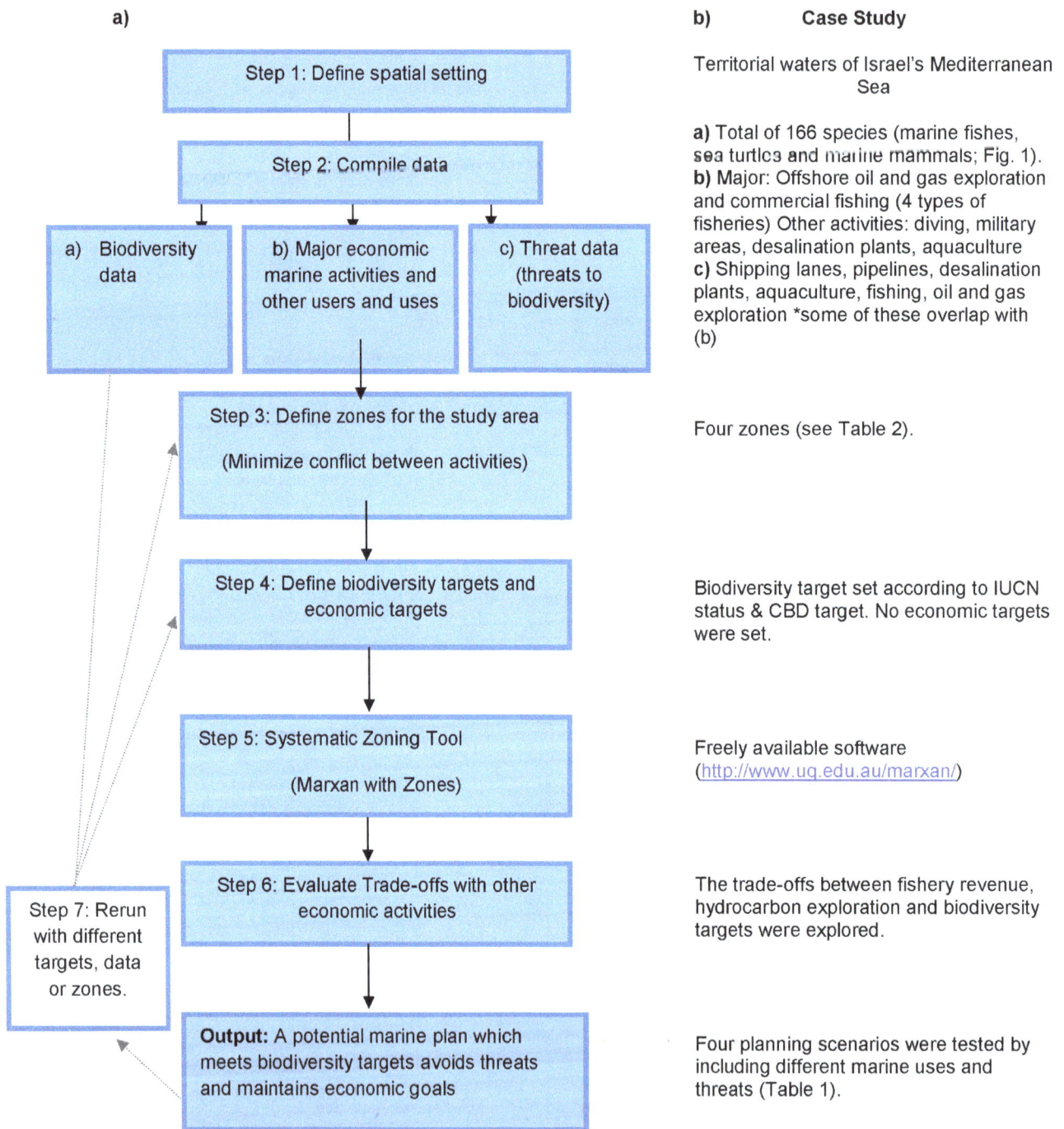

b)   Case Study

Territorial waters of Israel's Mediterranean Sea

a) Total of 166 species (marine fishes, sea turtles and marine mammals; Fig. 1).
b) Major: Offshore oil and gas exploration and commercial fishing (4 types of fisheries) Other activities: diving, military areas, desalination plants, aquaculture
c) Shipping lanes, pipelines, desalination plants, aquaculture, fishing, oil and gas exploration *some of these overlap with (b)

Four zones (see Table 2).

Biodiversity target set according to IUCN status & CBD target. No economic targets were set.

Freely available software (http://www.uq.edu.au/marxan/)

The trade-offs between fishery revenue, hydrocarbon exploration and biodiversity targets were explored.

Four planning scenarios were tested by including different marine uses and threats (Table 1).

Figure 1. Proposed framework for incorporating multiple activities and threats into marine conservation planning. These show the steps followed in the case study presented in this paper that encompasses Israel's entire Mediterranean territorial waters.

accordance with expert opinion from rangers and scientists at Israel's Nature and Parks Authority and Sea Turtle Rescue Centre.

We included the distribution of the common bottlenose dolphin (*Tursiops truncatus*), the most common cetacean species in Israel's territorial waters. Other cetacean species exist in Israel's waters but not enough observational data exists to determine priority habitats for these species. The common bottlenose dolphin has been sighted throughout Israel's territorial waters, therefore to better direct our conservation efforts we have considered

important habitat areas as the species distribution. Scheinin [45], identified three core areas for feeding and foraging, an area at a depth of 40–50 m near Ashkelon, an area at a depth of 30–60 m between Ashdod and Palmachim beaches and another area off the coast of Netanya at a depth of 90–120 m. These three core habitat areas cover in total 213.64 km² (for additional information see File S2.).

Geomorphological features.   In order to represent different types of marine habitats we included geomorphologic features to serve as surrogate "biodiversity features". We used ten geomor-

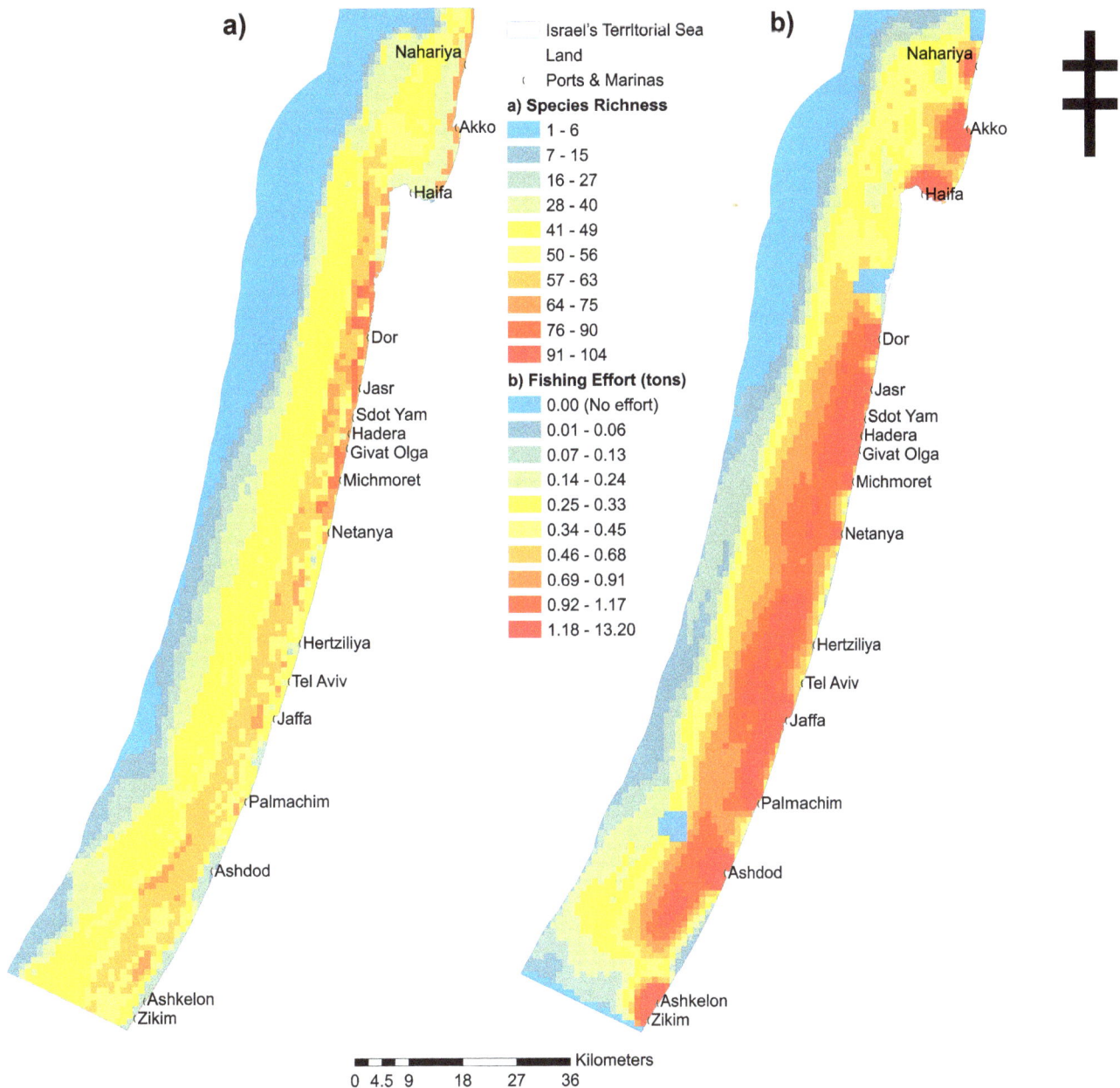

Figure 2. Biodiversity features and fishing effort in Israel's Mediterranean Sea territorial waters; a) species richness of 166 biodiversity features (species and geomorphologic features), b) combined fishing effort (entangling nets, longliners, purse seiners and trawlers), where the blue areas (no effort) are restricted fishing areas; marine reserves, military areas and aquaculture.

phologic features within Israel territorial waters that were mapped (in 2008) and provided by The Israel Nature and Parks Authority. These features include: shallow rocks, kurkar (calcareous aeolianite) ridges, kurkar bustan, deep kurkar ridges, continental shelf silt, continental shelf sand, continental ridges, large canyons, continental slope and canyons, deep sea [46].

Setting biodiversity targets. Biodiversity targets were set to protect a percentage of the species distribution according to its level of global threat based on the IUCN red list criteria ([46], Table S1 in File S1.) and current range size. We set a 10% target for species that were listed "Least Concern" by the IUCN and all other fish species that have not been evaluated by IUCN. This target was increased to 15% for species listed "Vulnerable" by the IUCN [47] and to 20% for species listed "Endangered" by the by

the IUCN. Species listed "Endangered" that had a distribution of less than 1% of the study area were given a target of 50%. For the geomorphological features, we set a target to protect 5% of all features and those that are represented by an area less than 1% of Israel's territorial waters were given a 10% target. We also set a constraint that at least 5% of the distribution of all species and features must be placed within the no-take zone (Conservation Zone), meaning that the rest of the biodiversity target could be fulfilled in other zones. While our target setting approach does not consider whether the target is adequate at conserving the species or maintaining population viability, it aims to address the IUCN criteria that defines the risk of species extinction as applied in Kark et al. [48] and Lieberknecht et al [49]. To test the sensitivity of our results we also used a 10% target for each species and a 5% target for each geomorphologic feature.

Incorporating economic activities in the sea

We included the two major economic activities in the Mediterranean waters of Israel (commercial fishing and hydrocarbon operations) in the conservation planning exercise. While there are other localized marine activities and features (addressed below; Table 1) commercial fishing and hydrocarbon operations are activities that span across Israel's territorial waters and rely on resource extraction. Thus, we focused on these activities which are likely to be the main source of opportunity cost incurred when implementing marine protected areas and zones. We translated these activities into opportunity cost layers for use within Marxan. Opportunity cost in this study was defined as the value of forgone economic activities (commercial fishing and hydrocarbon operations) when a particular area (planning unit) is made into a protected area that excludes these economic activities. As spatial opportunity cost data were unavailable for these activities, we developed surrogates to represent the annual revenue (approximation of annual opportunity cost) of each economic activity within our 1 km^2 planning units. Here we used annual values to reflect the relative opportunity cost differences across the territorial waters of Israel. We used the most current available data for Israel's territorial waters for each of these activities, specifically the year 2009 for commercial fisheries and year 2012 for hydrocarbon operations. The minimal fluctuation of Israel's annual commercial fishing catch and value over the last few years suggests that the available data of these activities is relatively comparable [50].

Opportunity cost of commercial fisheries. We developed surrogate opportunity cost layers of commercial fishing by spatially mapping fishing effort for the four major commercial fishing gears used in Israel; entangling nets, longliners, purse seiners and trawlers (see Fig. S2 in File S1; Fig. S3 in File S1; File S2. for

detailed methods). We derived effort maps by equations which assume effort is proportional to the number of fishing vessels at each port for each gear type and effort decreases exponentially with distance from port (methods described in Mazor et al. [51]). For each gear type we used expert opinion (total of 25 experts) to refine our effort layers. We did this by constraining our effort layer by the maximum depth that each fishing gear is used and incorporating weightings over habitats and areas that are targeted by particular gear types. For entangling nets we constrained our effort layer by a depth of 50 m (maximum depth that entangling nets are used in Israel's as confirmed by 15 entangling net fishers in Israel). For longliners, fishing effort was weighted by both distance from port and rocky habitat (targeted fishing areas) and confined to 50 m depth (confirmed by 6 longline fishers in Israel; Fig. S3 in File S1.). For purse seiners, effort was weighted across two distinct areas in the north and south at a depth between 10–50 m as determined by expert opinion (6 purse seine fishers; Fig. S2 in File S1.). Trawling effort was based on data collected from on-board GPS devices by Edelist [38] between the years 2009–2011 and trawling data from Israel's Department of Fisheries and Aquaculture ([52]; Fig. S2 in File S1.) Using these effort maps we created surrogate opportunity cost layers by overlaying the annual revenue (year 2009) reported by Edelist et al. [50] for each fishing gear type, thereby assigning monetary values to each planning unit for each fishing gear type (Fig. 2b).

Opportunity cost of hydrocarbon operations. Spatial data identifying offshore oil and gas operations and leased and licensed marine extraction areas was provided by Israel's Ministry of Interior from the National Master Plan of Israel (Tama 34b). Areas of Israel's Mediterranean waters are licensed to several oil and gas companies (e.g., Noble Energy, Shemen, Delek) for hydrocarbon exploration for a period of seven years [53]. If economically viable resources are found within these licensed areas they can then be leased by energy companies with a fifty year

Table 1. Four zones for Israel's territorial waters that restrict and permit different activities.

Activities	Zones			
	Conservation Zone "No-Take"	Benthic Protection Zone	Exploration Zone	Economic Zone "General use"
Trawling	X	X	X	✓
Purse Seiners	X	✓	X	✓
Gillnetting	X	✓	X	✓
Long liners	X	✓	X	✓
Oil and Gas Exploration	X	X	✓	✓
Additional threats and marine activities				
Aquaculture	X	X	X	✓
Current protected areas[1]	✓	X	X	X
Desalination plants	X	X	X	✓
Diving	✓	✓	✓	✓
Military areas	✓	✓	✓	✓
Pipelines	X	X	✓	✓
Safety area[2]	X	X	X	X
Shipping lanes	X	✓	X	✓

[1]Rosh HaNikra
[2]Mari B Platform
Additional threats and marine activities (listed below) in Israel's territorial waters have been locked to particular zones as per the four scenarios.
A "✓" in the column means that this activity was permitted in this zone, where an "x" it is prohibited.

production permit. Unexplored "blank" areas will be temporarily left aside as Israel is trying to limit exploration into these new areas. The licensed areas that were not explored will be recycled if there is no exploration in them.

As no reliable data sources were available for a total estimation of the value of Israel's offshore oil and gas reserves we performed calculations using data from Israel's Ministry of Energy and Water Resources [53,54] (Table S3 in File S1.) and converted these estimated reserve quantities into monetary values. We multiplied the annual average international market price of oil (NIS per barrel = 404.52 in 2012; World Bank http://www.worldbank.org/) and natural gas (NIS per thousands of cubic meters = 399.33; International Monetary Fund http://www.imf.org/external/index.htm) with Israel's estimated reserve volumes. These calculations resulted in a static estimate (year 2012 values) of the value of Israel's oil and gas reserves (not including extraction cost), but we realize that prices will fluctuate annually and are expected to reach higher values in the future, thus our calculated values are expected to be an under estimate (unless estimated reservoirs will be smaller than predicted). We have estimated the value of Israel's offshore oil and gas reserves at US$ ~324 billion (~1,250 billion NIS; Table S3 in File S1.), with 15% of this amount retrieved from the territorial waters (US$ 50 billion). Our resulting equation gives a greater weighting to the opportunity cost of leased areas (known sources of oil and gas; $\alpha = 1$) compared to licensed areas (half weighting $\alpha = 0.5$):

$$Cost\ of\ one\ exploration\ unit(EU) =$$

$$\frac{\alpha(Area\ of\ EU)}{\sum Area\ of\ all\ EU} * Value\ of\ oil\ and\ gas(US\$),$$

where, leased areas $\alpha = 1$ and licensed areas $\alpha = 0.5$,

and, the Cost of each planning unit(PU) =

$$\frac{Area\ PU}{Area\ EU} * Value\ of\ EU\ unit(US\$).$$

Considering additional marine activities in conservation planning. There are many features to consider when planning marine conservation within territorial waters. Israel has a relatively small territorial water area with a large number of marine activities (Fig. 3). In addition to the fishing and hydrocarbon operations (included as opportunity cost) we included eight additional marine activities. These include: aquaculture, desalination plants, dive sites, current protected areas, exploration safety zone (500 m buffer around hydrocarbon exploration sites), military areas (fire zones), shipping lanes and pipelines (Table 1; see File S2. for a full description of these activities and their data sources). We included these other activities by assigning their usage to specific zones (see Table 1).

Systematic planning tools and planning scenarios

Marxan with Zones is a conservation decision-support tool that enables the user to prioritize places for different zones to achieve multiple objectives [25,27]. This tool is an extension of Marxan [55], a globally used conservation planning tool for marine and terrestrial realms [27]. Marxan works by minimizing one variable (e.g., the opportunity cost of commercial fishing), creating a system that is separated into areas which are protected or non-protected [25]. In comparison, Marxan with Zones can minimize multiple variables (i.e. incorporating more than two opportunity cost layers) and enables the user to develop a more complex system of zones

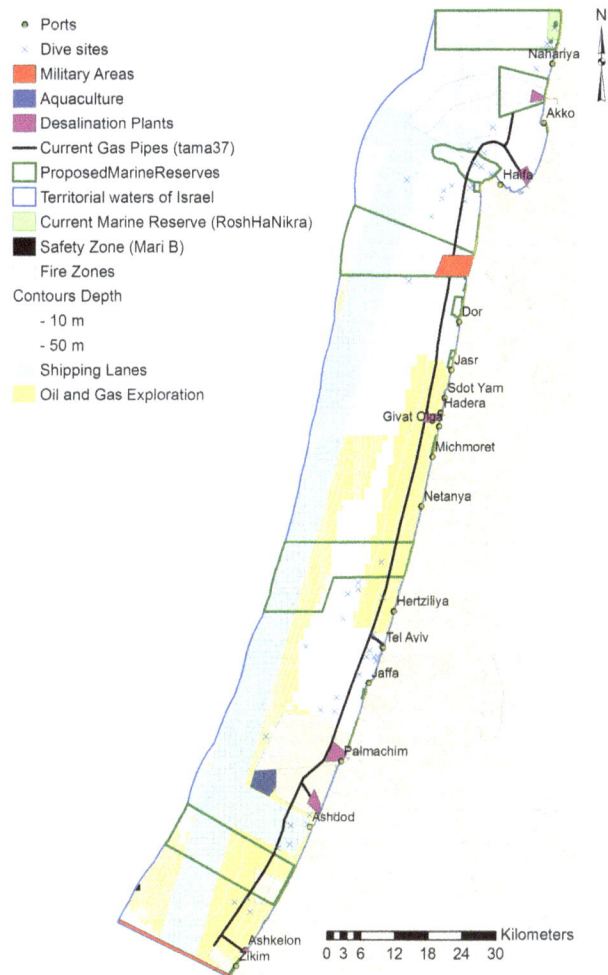

Figure 3. A map of the activities of Israel's Mediterranean territorial waters included in this study.

that provide varying degrees of protection and have zone specific actions, objectives and restrictions [27].

We applied Marxan [55] and Marxan with Zones [27] to compare four planning scenarios for Israel's Mediterranean territorial waters (see Table 2). For each planning scenario we aimed to meet the same biodiversity targets while minimizing the opportunity cost incurred by other marine activities, as described below. The four scenarios increase in complexity with the inclusion of human activities (threats) and economic objectives; Simple Planning, Basic Zoning, Intermediate Zoning and Complex Zoning (Table 2). We define the term "activities" as any other activity within Israel's marine waters that is not biodiversity protection as proposed in this study. For the first scenario, Simple Planning, we used Marxan (without zoning) and tested two sub-scenarios; Simple Planning A with six activities and commercial fishing opportunity cost, and Simple Planning B with seven activities and combined commercial fishing and hydrocarbon extraction opportunity cost. Our second scenario, Basic Zoning, used Marxan with Zones and included three zones and six other activities. The third scenario, Intermediate Zoning, used Marxan with Zones and included four zones and seven activities (three sub-scenarios A, B and C for protection effectiveness of the Exploration Zone; see File S2. for full explanation). In the fourth scenario, Complex Zoning, we used Marxan with Zones with four

Table 2. Four planning scenarios that were examined using Marxan and Marxan with Zones for Israel's territorial Mediterranean waters.

		Planning Scenarios						
		1. Simple Planning		2. Basic Zoning	3. Intermediate Zoning			4. Complex Zoning
		A	B		A	B	C	
Planning Tool:	Marxan	+	+					
	Marxan with Zones			+	+	+	+	+
Zones	Conservation Zone (No-Take)			+	+	+	+	+
	Economic Zone (General Use)			+	+	+	+	+
	Benthic Protection Zone			+	+	+	+	+
	Exploration Zone (% of effectiveness at protecting species[3])				+ (25%)	+ (50%)	+ (75%)	+ (50%)
Marine activities *(Included as opportunity cost)*[4]	Commercial Fisheries: Trawlers, Purse Seiners, Gill Nets, Long liners	+	+	+	+	+	+	+
(Assigned to specific zones)[5]	Hydrocarbon Operations		+	+	+	+	+	+
	Aquaculture	+	+	+	+	+	+	+
	Current Protected Areas	+	+	+	+	+	+	+
	Diving areas	+	+	+	+	+	+	+
	Military areas	+	+	+	+	+	+	+
	Safety platform	+	+	+	+	+	+	+
	Shipping lanes							+
	Desalination plants							+
	Pipelines							+
Number of Zones		0		3	4			4
Total number of marine activities included in the analysis		6	7		7			10

[3]We assigned possible percentages of biodiversity protection that may be achieved by the Exploration Zone. The Conservation Zone assumes 100% protection of .biodiversity but due to the unknown impacts of hydrocarbon exploration we tested different values (25%, 50%, and 75%). See File S2 for further details.

[4]The opportunity cost layers are the variables where are minimized in Marxan software: In Marxan this is treated as one minimized cost layer (summed together), and in Marxan with Zones these opportunity cost layers are separately minimized.

[5]See Table 1 for the zones that each marine activity is permitted or restricted within and File S2 for a detailed explanation of each activity and their data references.

The inclusion of features and data in each scenario is represented by a plus sign (+). Planning scenarios increase (from the Simple Planning to Complex Zoning scenario) in complexity by the planning tool, zones and number of activities included. For more detailed information on each of the zones see Table 1.

zones (for descriptions of each zone see Table 1) and ten other activities. For a detailed description of each scenario see File S2.

Comparing planning scenarios

Four planning scenarios (Table 2) were compared. The Simple Planning scenario (without zoning) was run using Marxan and the other three scenarios (Basic Zoning, Intermediate Zoning and Complex Zoning) used Marxan with Zones, all scenarios with 1,000 runs each. Based on the results of the 1,000 runs we calculated the average opportunity cost and number of 1 km^2 planning units within each zone that were needed to meet our biodiversity targets. We tested the ability of planning scenarios to meet all biodiversity targets. In cases where targets were unable to be reached for a particular species, we eliminated the constraint for 5% of their protection to be met in the Conservation Zone. Thus, we re-ran our results with the same altered targets for all scenarios. We then mapped the selection frequency outputs (number of time a planning unit is selected in Marxan for a particular zone) for each planning scenario and each zone. To compare between zoning configurations and scenarios we also mapped the best solution that Marxan could find. To test the similarity between the selection frequency outputs for each scenario we used the Spearman Rank Correlation (ρ). Higher values indicate a more similar spatial pattern in selection frequencies, meaning that these plans will require similar conservation actions.

Evaluating trade-offs

We evaluated the trade-off between meeting biodiversity targets and maximizing annual fishery revenue for each of the four fisheries in Israel's Mediterranean Sea, following methods described in Klein et al. [25]. These trade-offs can only be evaluated for scenarios using Marxan with Zones that enables multiple variables to be considered. We set fishery targets where we aimed to preserve an equal percentage of the total fishing revenue (from the fishing effort maps) for each of the four fishery gear types. These targets could only be met within zones that did not restrict that type of fishery (Table 1). Expanding this analysis, we tested the trade-off with areas that are leased and licensed for oil and gas (using the hydrocarbon opportunity cost layer described above). We therefore included a hydrocarbon target (preserving hydrocarbon industry revenue) as well as both biodiversity and fishery targets. Fishery targets were extracted from the previous trade-off analysis; the highest target where all biodiversity targets were met.

Results

Comparing planning scenarios for territorial waters

Here we compared our four planning scenarios (Table 2) by: a) the ability to reach biodiversity targets, b) the change in opportunity cost and c) the alteration of spatial conservation priorities.

a) Biodiversity targets. We found that meeting the same biodiversity targets became more difficult as our planning scenarios included more marine activities. The Simple Planning (without zoning and six activities) and Basic Zoning (three zones and six activities) scenarios met all biodiversity targets, the Intermediate Zoning scenario (four zones and seven activities) met 98% of targets and the Complex Zoning scenario (four zones and ten activities) met 96% of targets (Table 3). Our constraint (5% target in the Conservation Zone – no-take area) was unable to be met in the Intermediate and Complex Zoning scenarios for nine species (Table S4 in File S1.) that had restricted distribution

ranges that overlapped with prospective hydrocarbon exploration areas. For each of these nine species we eliminated the constraint; however the overall biodiversity target for these nine species remained and was met within other zones. Targets were then able to be met for all planning scenarios.

b) Opportunity cost. We found that more complex planning scenarios incurred greater opportunity cost (Table 3). When comparing the two Simple Planning scenarios (Simple Planning A with six users and commercial fishing opportunity cost, and Simple Planning B with seven users and combined commercial fishing and potential hydrocarbon opportunity cost) we found that a reserve network that only included the opportunity cost of fishing had a substantially lower cost (Simple Planning A = US$2.05 million) compared to a plan that included the opportunity cost of hydrocarbon operations (Simple Planning B = US$595,132.38 million). Comparing our zoning scenarios (when targets are met 100% in each scenario) we found that the most expensive zoning scenario is the Intermediate Zoning scenario A that assumes the Exploration Zone can provide a zone effectiveness measure of twenty-five percent. This opportunity cost decreased as the Exploration Zone's ability to protect biodiversity (zone effectiveness) was increased to fifty percent (Intermediate Zoning scenario B 10.5% cost decrease) and seventy-five percent (Intermediate Zoning scenario C 14.9% cost decrease); allowing targets to be met more easily within the Exploration Zone. The Intermediate Zoning scenario increased opportunity cost by 27.8% from the Basic Zoning scenario (three zones and six activities) as we introduced the opportunity cost of prospective oil and gas reserves as well as a fourth zone (Exploration Zone). The Complex Zoning scenario also increased the opportunity cost of the Basic Zoning plan by 35.7% and Intermediate Zoning B plan by 6.2%.

c) Conservation priorities. Selection frequency outputs from our analysis indicated that spatial configurations are substantially altered by the inclusion of hydrocarbon opportunity cost (Fig. 4; Fig. 5). The scenarios Simple Planning A (without zoning and six activities) and Basic Zoning (three zones and six users), which did not include hydrocarbon opportunity cost had a high Spearman's rank correlation of $\rho = 0.84$ (p<0.001; Table 4).

We found that priority areas for no-take reserves (Conservation Zone) were mainly concentrated in the north and south of Israel's territorial waters. From the best solution outputs (Fig. 6) no-take areas moved from areas in the south to areas in the north with the inclusion of potential hydrocarbon extraction data. Similarly, the three scenarios that included hydrocarbon opportunity cost (Simple Planning B, Intermediate Zoning and Complex Zoning) had selection frequency outputs that were significantly correlated (Table 4). The most similar spatial outputs were between Simple Planning B and Intermediate Zoning B ($\rho = 0.86$, p<0.001; Table 4). In these three scenarios, we discovered that spatial priorities were much more distinct (higher spectrum of selection frequency; see Fig. 4) than the Simple Planning A and Basic Zoning scenarios. Areas with high selection frequency for placing no-take reserves were off the coast of Jaffa, in coastal waters between Dor and Haifa Bay and along northern border with Lebanon (Fig. 4).

Priority areas for each zone become more pronounced as planning scenarios became more complex and restricted by the inclusion of other marine activities. Conservation priorities for the Benthic Zone were most similar between the Intermediate Zoning B and Complex Zoning ($\rho = 0.82$, p<0.001; Table 4). In all scenarios we find that Benthic Protection Zone has higher selection frequency in the northern part of the Sea. In the best solution outputs we also notice how benthic protection becomes confined to the north with the inclusion of the Exploration Zone

Table 3. Results showing average opportunity cost for 1,000 Marxan runs for each planning scenario.

Planning Scenario	Opportunity cost (US$ million)	Percent of targets met	Percent of conservation zone in entire reserve design (no-take areas)
Simple Planning A	2.05	100	22
Simple Planning B	595,132.38	100	21
Marxan with Zones			
Basic Zoning	4.09	100	22
Intermediate Zoning B 50%	333,004.37	98	17
Complex Zoning	4.20	96	14
Marxan with Zones (minus nine species for the 5% Conservation Zone target)			
Basic Zoning	3.92	100	22
Intermediate Zoning A 25%	5.59	100	18
Intermediate Zoning B 50%	5.01	100	17
Intermediate Zoning C 75%	4.76	100	17
Complex Zoning	5.32	100	14

The constraint/target that 5% of the distribution of all features needs to be within the Conservation Zone (no-take zone) was unable to be reached for nine species. This constraint was removed for these species so targets could all be met. This table shows the opportunity cost of each planning scenario, the percentage of biodiversity targets met in the scenario and the percentage of "no-take area" surface coverage of the entire reserve system. For a description of planning scenarios see Table 2. Targets were set according to IUCN criteria and the size of a species distribution range (as described in the methods section).

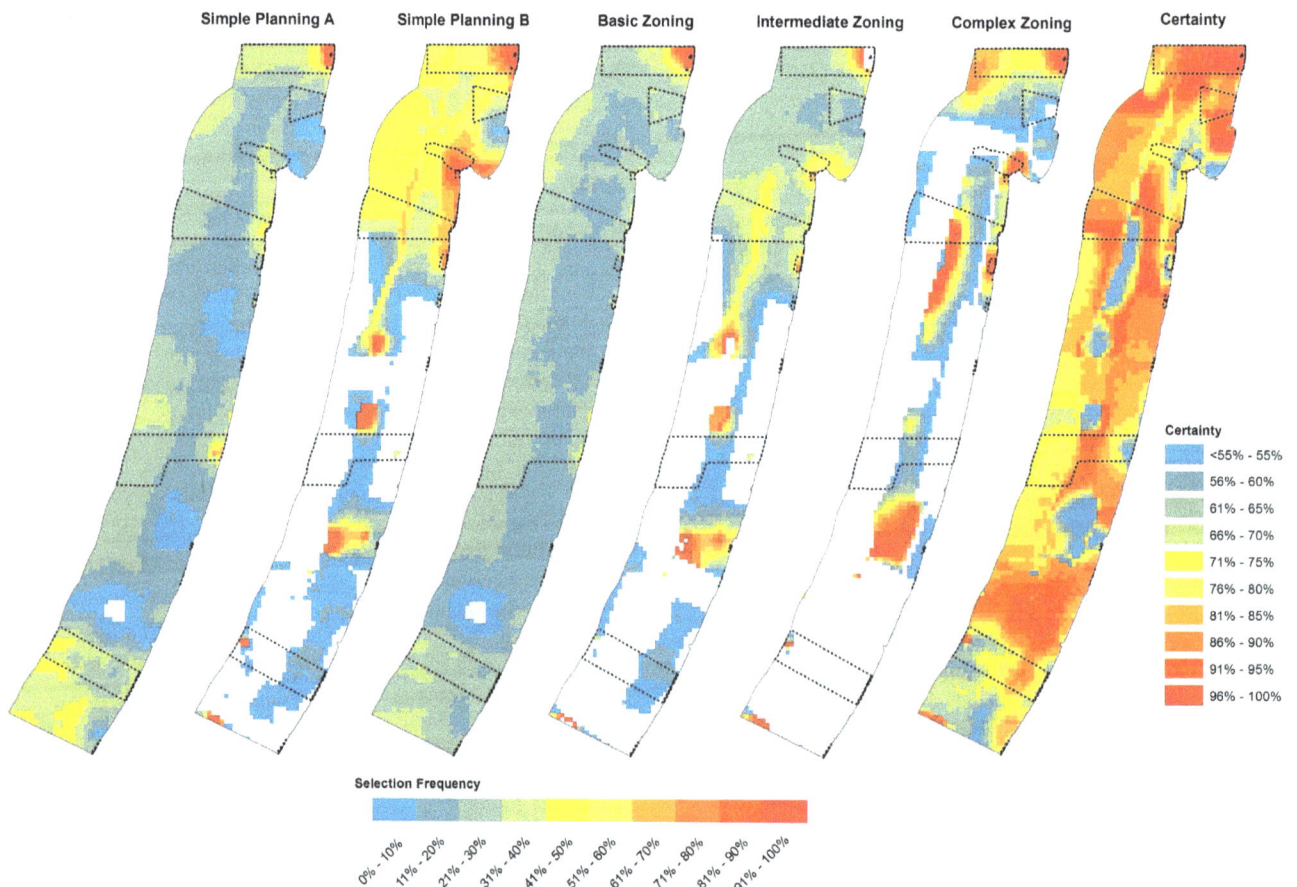

Figure 4. Selection frequency output maps (shows the percentage of times a planning unit was selected when run in Marxan 1000 times) from Marxan with Zones for each Zone and each zoning scenario. All scenarios meet biodiversity targets. The dashed black lines represent the proposed marine reserve system by Israel's Nature and Parks Authority [45]. The certainty map expresses the level of certainty/agreement of planning units selected (either highly selected for no-take areas or low selection) across all planning scenarios. Therefore, the higher the percentage of certainty means there is more agreement between scenarios.

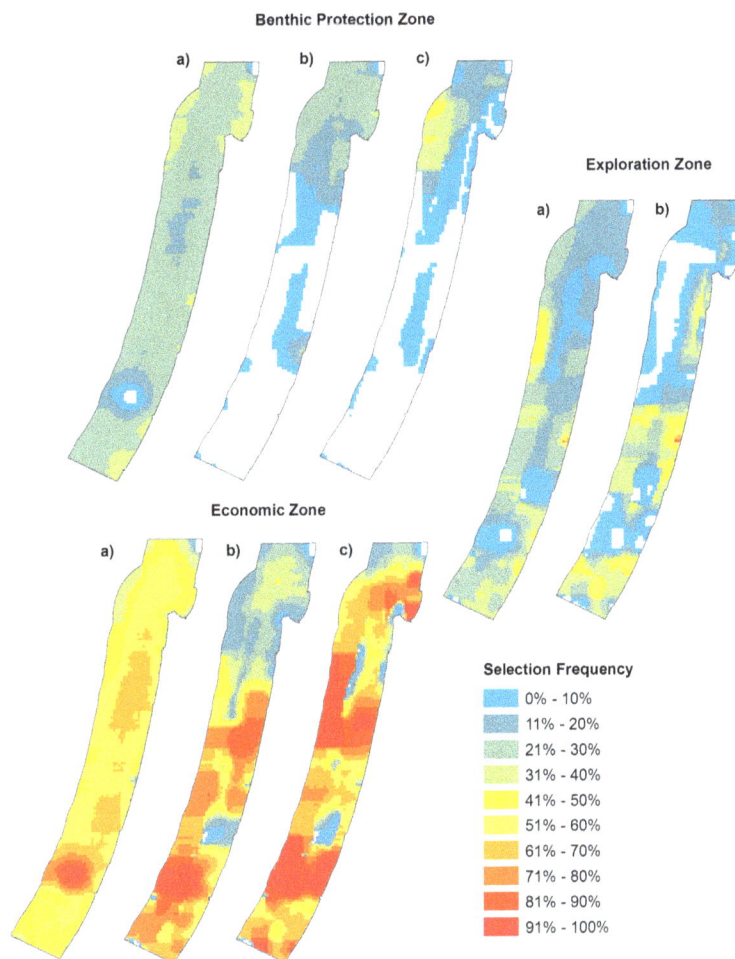

Figure 5. Selection frequency output maps (shows the percentage of times a planning unit was selected when run in Marxan 1000 times) from Marxan with Zones for each Zone and each zoning scenario. For the Benthic Protection Zone and Economic Zone the three scenarios are a) Basic Zoning, b) Intermediate Zoning, c) Complex Zoning. For the Exploration Zone the two scenarios are a) Intermediate Zoning and b) Complex Zoning.

Table 4. Spearman rank correlation (ρ) of the similarity between the selection frequency outputs of each planning scenario.

Zone	Planning Scenario	Simple Planning A	Simple Planning B	Basic Zoning	Intermediate Zoning B	Complex Zoning
Conservation Zone	Simple Planning A	-	0.69	0.84	−0.09	−0.09
	Simple Planning B	0.69	-	0.11	0.86	0.67
	Basic Zoning	0.84	0.11	-	−0.04	−0.06
	Intermediate Zoning B	−0.09	0.86	−0.04	-	0.73
	Complex Zoning	−0.09	0.67	−0.06	0.73	-
Benthic Zone	Basic Zoning	-	-	-	0.45	0.33
	Intermediate Zoning B	-	-	0.45	-	0.82
	Complex Zoning	-	-	0.33	0.82	-
Economic Zone	Basic Zoning	-	-	-	0.41	0.29
	Intermediate Zoning B	-	-	0.41	-	0.58
	Complex Zoning	-	-	0.29	0.58	-
Exploration Zone	Intermediate Zoning B	-	-	-	-	0.42
	Complex Zoning	-	-	-	0.42	-

High values (closer to 1) indicate a more similar spatial pattern in selection frequencies, meaning that these plans will require similar conservation actions. All scenarios show significant correlations (p<0.001).

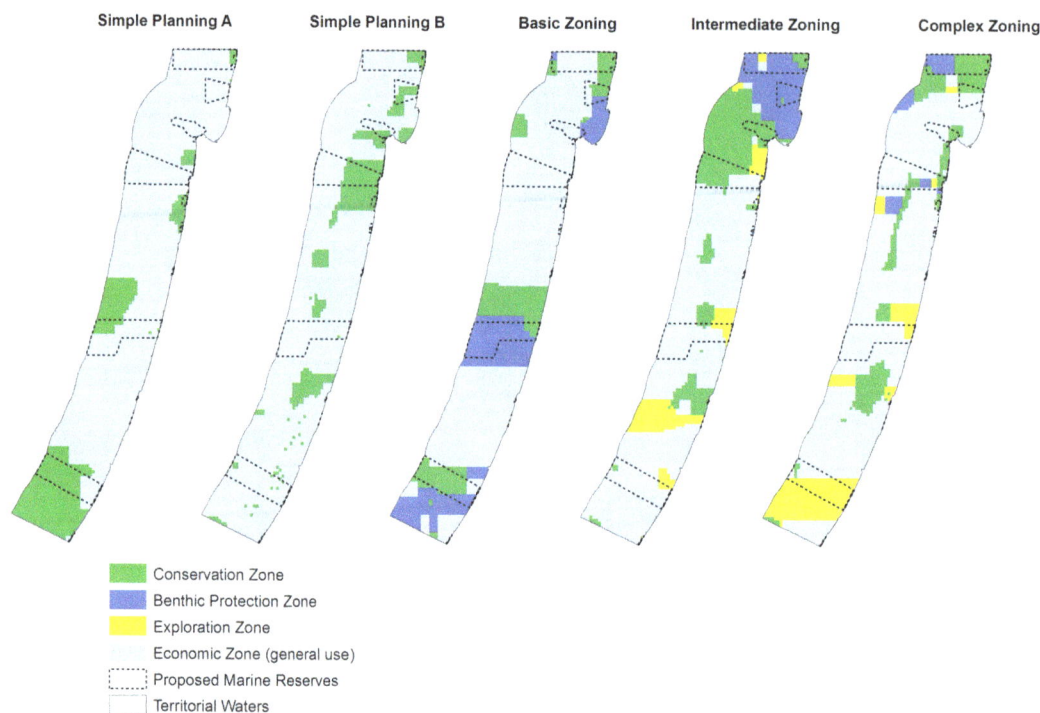

Figure 6. Marxan best solution outputs (the reserve configuration that best reduces opportunity cost and meets biodiversity targets from 1000 Marxan runs) for each planning scenario. The four colours designate the four types of zones (see Table 1).

(Fig. 6). The Economic Zone has highest selection frequency in the south for the Basic Zoning scenario where high fishing pressure is evident. In the Intermediate Zoning B scenario the high selection frequency of this zone extends over the south and central region where hydrocarbon is included. Further expansion of this zone's high selection frequency extends to the north as shipping lanes and pipelines are included in the Complex Zoning scenario. The Exploration Zone's priority areas were dissimilar between the Intermediate and Complex Planning scenarios, ($\rho = 0.42$, p< 0.001; Table 4). The inclusion of other marine activities affected the available area for the Exploration Zone.

Evaluating trade-offs between conservation and economic objectives

In the Basic Zoning scenario (three zones and six users) all biodiversity targets were met with a loss of 7% of commercial fishing revenue (Fig. 7a). By increasing the complexity of our planning scenarios (including more marine activities) we found that our biodiversity targets could only be met by decreasing the area of fishery grounds, consequently decreasing the revenue. Hence, the resulting fishing revenue loss was 12% for the Intermediate Zoning scenario (zoning network that includes four zones) and 15% for the Complex Zoning scenario.

In comparison, by including the economic objectives of hydrocarbon operations while meeting biodiversity and fishery targets (all four fishing gear types targeted 88% (Intermediate Zoning) and 85% (Complex Zoning) of revenue; values obtained from Fig. 7a), a small revenue loss was incurred (Fig. 7b). For the Intermediate Zoning scenario 5% of hydrocarbon revenue was lost. Similarly, the Complex Zoning scenario kept biodiversity and fishery targets with revenue losses of 6%. Therefore, for a loss of ~5% of hydrocarbon revenue, fishery and biodiversity targets could be fully met. Interestingly, we found that the drop-off rate of

not meeting biodiversity and fishery targets was very minimal for the hydrocarbon industry in comparison with the rate at which biodiversity and fishery targets were traded off. Moreover, if hydrocarbon revenue was not traded-off (100% revenue was maintained), biodiversity and fishery targets could reach ~98% (Fig. 7b). However, if fishery revenue was not traded off (100% revenue was maintained), biodiversity targets could only reach between 93–84% (Fig. 7a).

Discussion

This study demonstrates how conservation objectives can be achieved while considering economic objectives where there are multiple marine activities. We found that the inclusion of many activities in marine conservation plans can significantly alter spatial priorities (Table 4; Fig. 4; Fig. 5). Economic goals are more compromised (in this case for the fisheries and hydrocarbon industries; Fig. 7) to achieve biodiversity targets when there are more marine activities in the planning process. Moreover, complex plans with more activities involved greater opportunity cost and did not reach biodiversity targets as easily as more simplified plans with less marine activities. Given that a complex plan is working with a more constrained problem, this result is expected [3,56]. Despite the increased opportunity cost and lack of spatial flexibility to achieve biodiversity goals with more complex conservation plans, planning that incorporates other activities can steer us towards areas which are feasible (greater potential for implementation success), minimize conflict with other users and reduce threats to biodiversity.

Conservation planning and zoning with multiple activities is challenging. Our case study shows that decisions made by conservation planners such as the number of zones or number of marine activities included in the planning process can substantially shape the resulting zone and reserve configuration.

(a)

(b)

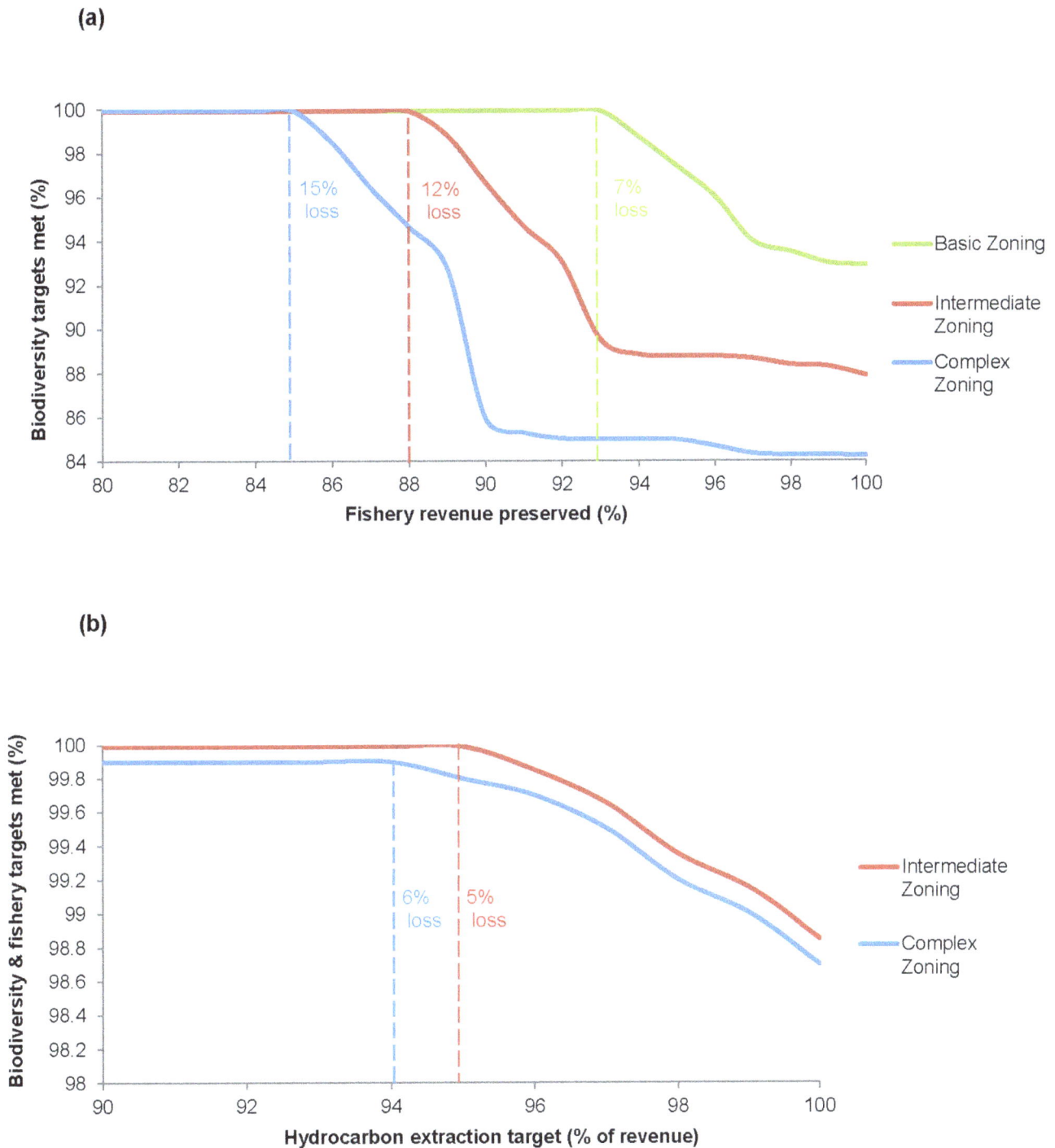

Figure 7. The trade-off between meeting biodiversity targets and maintaining economic objectives for each zoning scenario. (a) biodiversity targets are met when the fishery targets (percentage of annual fishery revenue) are less than 93% (7% revenue loss) in the Basic Zoning scenario (three zones and six activities), less than 88% (12% revenue loss) in the Intermediate Zoning B scenario (four zones and seven activities), and less than 85% (15% revenue loss) is the Complex Zoning scenario (four zones and ten activities), (b) biodiversity targets are met when hydrocarbon operations (leased and licensed expected revenue) are less than ≤95% (5% revenue loss) in the Intermediate Zoning scenario and less than 94% (6% revenue loss) in the Complex Zoning scenario.

Therefore, it is important to first identify the impact that each activity and feature has on marine biodiversity in the study system and the appropriate conservation action to take [57]. Here we follow a framework (Fig. 1) to help conservation planners address offshore activities and their potential threat in the marine realm.

This framework outlines the steps needed to comprehensively zone for biodiversity protection while maintaining economic goals and can be a useful guide for countries currently striving to zone their waters [18]. One of the most important steps is testing the sensitivity of the results to user decisions (e.g., the inclusion of data, number of zones, the targets, see Step 7 Fig. 1; [58]). Other challenges that need to be accounted for when zoning include: the lack of shared information between stakeholders [59], the unknown expansion and objectives of industries [60], the unknown value of economic industries [5], and unforseen threats or disasters [18]. Given that some of these challenges can be overcome, in reality, conservation planning is largely shaped by the willingness to trade-off economic and conservation objectives. Moreover, there is no one correct solution to planning within a complex system [61].

Trade-off analysis is an important step to include in conservation planning [62,63]. It enables us to determine how much of a commercial activity must be forgone in order to achieve biodiversity targets. It also helps to address the implementation gap (the gap between conservation planning and real-world action) inherent in many conservation plans [2]. However, Hirsch et al. [62] cautions that not every problem can be solved by compromise. For example, we assume in our study that a portion of the hydrocarbon leased and licensed areas and commercial fishing grounds are available for trade-off, whereas stakeholders may disagree and reject any compromise. In marine conservation, fishing trade-offs have been the focus of several studies [3,25,28]. In this study we have incorporated economic trade-offs for both the commercial fishing and hydrocarbon industries, indicating the necessary compromises that are needed to meet our biodiversity targets in each planning scenario. Specifically, we have triaged targets for nine species that were unable to be met within no-take zones (of the Intermediate Zoning and Complex Zoning scenarios) and enabled them to be met within other zones. We suggest that future work should expand this type of analysis to examine the trade-offs with other social, economic and cultural activities where appropriate.

Marine features and activities which are confined in space may be difficult, or impossible, to trade-off. In this study we introduced a range of features into marine conservation planning in addition to the more traditionally used fishing such as pipelines, shipping lanes, desalination plants and aquaculture. In comparison to the full coverage of commercial fishing practises and the wide cover of hydrocarbon exploration across the study area, other features are restricted to a specific area (Fig. 3). Such restricted features are difficult to plan around as they often cannot be traded-off. Fishing effort for example can be redispersed to other spatial areas when an area is declared a marine reserve [64,65], whereas aquaculture farms are more difficult to relocate. We also found that linear-shaped features such as pipelines and shipping lanes influence the shape of marine zones, causing thin elongated zones (Fig. 6). Therefore, conservation planners must decide whether such features are planned around, planned with, or ignored. Performing a cost-benefit analysis of altering some of these features (e.g., rerouting shipping lanes, planning reserves over pipelines or moving planned aquaculture cages) within various planning scenarios could be a way to examine their potential flexibility within the reserve system. We suggest that future research explores the way that such features are included in conservation planning as they can have an influence on the selection of conservation priorities.

We found that the incorporation of oil and gas exploration can substantially alter spatial priorities and the opportunity cost of conservation. This is first time that offshore hydrocarbon operations are explicitly incorporated in marine conservation planning. Possible reasons for its absence in previous conservation plans are because a) the economic gains that are at stake are so large that these areas are "off-limits" to all other marine activities (e.g., Australian commonwealth zoning plan; [12]), b) uncertainty

as to how to incorporate hydrocarbon information and c) the uncertain future of the industry that is dependent on new discoveries and may quickly demand large marine space (e.g., new discoveries in the Mediterranean Sea [66]). If we incorporate hydrocarbon information by assuming such areas cannot be protected, we may not be able to achieve a representative reserve network. One of the problems we encountered with including prospective hydrocarbon exploration is that sometimes biodiversity targets could not be achieved because a few species substantially overlap with hydrocarbon interests. Thus, we must carefully assess our targets and understand the compromises or actions that need to be taken in order to ensure conservation-worthy species are maintained in the face of hydrocarbon operations. We suggest that conservation plans endeavour to incorporate mining and fossil fuel data where possible to avoid costly conservation mistakes.

The ability of hydrocarbon exploration areas to provide some level of protection for biodiversity is unknown. In this study we tested different levels of protection from the "Exploration Zone" (see Intermediate Zoning Table 3; "zone effectiveness" see [67]), and found that opportunity cost is reduced if hydrocarbon areas are able to contribute to biodiversity protection. This is a novel conservation planning example that incorporates the notion of hydrocarbon areas providing some conservation benefit. The impacts of oil spills and gas leaks on marine biodiversity are severe and are well documented [68–70]. Likewise, there is some understanding of the impacts of offshore construction and extraction e.g., drilling impacts that are damaging to benthic structures [71]. However we have little understanding of the impacts posed by the ongoing maintenance of a drilling site that is dormant (leased or licensed without current activity). We suggest that future research focuses on better understanding the impacts that hydrocarbon operations pose on marine biodiversity and further develop ways to include hydrocarbon information into marine conservation plans.

Our study has interesting implications for Israel. We found that for Israel's territorial waters we can meet all our biodiversity targets (but not all no-take zone targets) for a loss of ~15% of annual commercial fishery revenue and ~5% of potential hydrocarbon revenue. A reduction of 7% of fishery revenue was needed to meet our biodiversity targets if hydrocarbon exploration is ignored. Our planning scenarios indicate that a surface area of 14–22% (Table 3) of Israel's territorial waters needs to be protected to meet biodiversity targets. The marine area reserved in Israel is currently less than 1% of the territorial waters [72], although none of these are considered no-take areas. Efforts are currently being undertaken to expand Israel's reserve network. The proposed network has been planned using species gradients and the representation of geomorphological features, without the use of Marxan or other similar tools [72]. We found that there is some overlap between the proposed marine reserves and the high priority areas found in our study (Fig. 4). The primary overlapping areas include: the proposed reserve in the north (Rosh Hanikra), the Haifa headland, the proposed reserve near Atlit and the smaller sized reserve near Dor. While different methods have been used for these two plans, some results are overlapping and we recommend that these areas that overlap should be targeted as initial reserve priorities for Israel as they are robust to the kind of process used to define priorities. However, it should be cautioned that, while overlapping priorities could be a good starting point, they will not necessarily provide a representative network that meets biodiversity targets.

Marine conservation planning often lacks good quality spatial data and must therefore rely on surrogate measures [6,59,73]. In this study, the surrogate fishing effort layers were generated with large involvement and input from experts. In comparison, our opportunity cost layer for hydrocarbon operations, although based on available government data, may less accurately reflect unpredictable shifts in future opportunity cost due to the fluctuating price of fossil fuels. Here we also set relatively low biodiversity targets because very minimal marine protection exists in this area, thus, these targets are potentially achievable (20% of Israel's Mediterranean Sea needs to be protected to meet our biodiversity targets (Table 3), corresponding with Israel's proposed target by the Israel Nature and Parks Authority [72]). These targets do not guarantee species persistence, but increasing these target may mean that other targets become unachievable, particularly within the Intermediate Zoning and Complex Zoning scenarios. We have included several novel features in our planning (e.g., aquaculture farms, desalination plants, shipping lanes and pipelines), yet there are other features that could be incorporated in future work, for example sand mining, offshore power plants, tourism and recreational fishing. Similarly, management and monitoring cost can also be included in future studies [6]. The aim of this study was to evaluate the impact of including multiple features and activities into marine conservation, however we do intend for these results to serve as useful baseline plans for the territorial waters of Israel. To improve the selection of conservation priorities in Israel's Mediterranean waters future work should attempt to build upon these scenarios, including additional species data for which data is currently limited, incorporate additional marine activities and create more robust cost layers with the availability of new data.

This case study can serve as an example for many other countries around the world, which are faced with the need to carefully balance economic considerations while protecting marine biodiversity. It is particularly relevant for countries surrounding the Mediterranean Sea that share common challenges and arising threats from developing offshore hydrocarbon exploration to biodiversity and ecosystems [66]. Our results suggest that planning with more complexity (e.g., multiple economic objectives, multiple threats and multiple zones) will be slightly more costly, have higher trade-offs with other marine activities and will require more input data. Despite these inefficiencies, a complex plan considers the objectives of more stakeholders (marine activities) and is more likely to result in successful implementation of conservation

outcomes [2] and better compliance than a plan which ignores other activities. In the Mediterranean region with its many marine users, this is particularly important where compliance is often a major limiting factor in reserve design and implementation success [74]. We propose that countries aiming to protect marine biodiversity in their territorial waters should move from a single objective approach to one that links to the broader socioeconomic context incorporating multiple activities. A way forward may be the incorporation of lessons from marine spatial planning [11,19] into marine conservation planning, while aiming at maintaining biodiversity goals and examining trade-offs. Explicitly quantifying trade-offs can provide an initial starting point for discussion between stakeholders [62] and ultimately enable successful conservation outcomes which other marine users are willing to comply with.

Acknowledgments

We thank the Israel Nature and Parks Authority, particularly R. Yahel for advice and G. Weil for providing GIS data. For advice on the collation of fish species data and distribution we thank D. Golani, M. Gorem, & G. Rilov. For advice and data on sea turtles and cetaceans we thank Y. Levy and A. Scheinin. We thank the Israel Department of Fisheries and Aquaculture, especially O. Sonin and N. Mozes for providing data, advice and expert opinions and N. Mazor for Hebrew translation support. Data and advice on hydrocarbon exploration was provided by I. Nissim & M. Livnat, and Marxan software support and GIS data from N. Levin. We thank the two anonymous reviewers and the editor for their constructive comments.

Author Contributions

Conceived and designed the experiments: TM HPP SK. Performed the experiments: TM DE. Analyzed the data: TM HPP DE EB SK. Wrote the paper: TM HPP DE EB SK.

References

1. Knight AT, Cowling RM (2007) Embracing Opportunism in the Selection of Priority Conservation Areas. Conservation Biology 21: 1124–1126.

2. Knight AT, Cowling RM, Rouget M, Balmford A, Lombard AT, et al. (2008) Knowing but not doing: selecting priority conservation areas and the research-implementation gap. Conservation Biology 22: 610–617.

3. Weeks R, Russ GR, Bucol AA, Alcala AC (2010) Incorporating local tenure in the systematic design of marine protected area networks. Conservation Letters 3: 445–453.

4. Biggs D, Abel N, Knight AT, Leitch A, Langston A, et al. (2011) The implementation crisis in conservation planning: could "mental models" help? Conservation Letters 4: 169–183.

5. Douvere F (2008) The importance of marine spatial planning in advancing ecosystem-based sea use management. Marine Policy 32: 762–771.

6. Naidoo R, Balmford A, Ferraro PJ, Polasky S, Ricketts TH, et al. (2006) Integrating economic costs into conservation planning. Trends in Ecology & Evolution 21: 681–687.

7. Norse EA (2008) EBM perspective: Ocean zoning is inevitable. Washington, DC: Marine Conservation Biology Institute. Available http://mcbi.marine-conservation.org/publications/pub_pdfs/norse_MEAM_2008.pdf. Accessed 2013 October 20.

8. Butt N, Butt N, Beyer HL, Bennett JR, Biggs D (2013) Biodiversity Risks from Fossil Fuel Extraction. Science 342: 425–426.

9. Shaffer B (2011) Israel—New natural gas producer in the Mediterranean. Energy Policy 39: 5379–5387.

10. Tagliapietra S (2013) Towards a New Eastern Mediterranean Energy Corridor? Review of Environment, Energy and Economics-Re3. Available http://dx.doi.org/10.7711/feemre3.2013.02.002. Accessed 2013 August 24.

11. Douvere F, Ehler CN (2009) New perspectives on sea use management: Initial findings from European experience with marine spatial planning. Journal of Environmental Management 90: 77–88.

12. Barr LM, Possingham HP (2013) Are outcomes matching policy commitments in Australian marine conservation planning? Marine Policy 42: 39–48.

13. Pressey RL, Bottrill MC (2008) Opportunism, Threats, and the Evolution of Systematic Conservation Planning. Conservation Biology 22: 1340–1345.

14. Redfern JV, McKenna MF, Moore TJ, Calambokidis J, Deangelis ML, et al. (2013) Assessing the Risk of Ships Striking Large Whales in Marine Spatial Planning. Conservation Biology 27: 292–302.

15. Casale P (2011) Sea turtle by-catch in the Mediterranean. Fish and Fisheries 12: 299–316.

16. Inger R, Attrill MJ, Bearhop S, Broderick AC, James Grecian W, et al. (2009) Marine renewable energy: potential benefits to biodiversity? An urgent call for research. Journal of Applied Ecology 46: 1145–1153.

17. NOAA (2009) Setting an new course: shipping lane shift helps mariners steer clear of whales. SanctuaryWatch, National Oceanic and Atmospheric Administration. Washington D.C. pp.6–7. Available http://sanctuaries.noaa.gov/news/pdfs/sanctuarywatch/sw_1209.pdf Accessed 2014 January 15.

18. Agardy T (2010) Ocean Zoning: Making Marine Management More Effective: Earthscan. UK, London. pp.207.

19. Ehler C, Douvere F (2009) Marine spatial planning: a step-by-step approach toward ecosystem-based management. Intergovernmental Oceanographic Commission and Man and the Biosphere Programme. IOC Manual and Guides No. 53, ICAM Dossier No. 6. Paris: UNESCO.

20. Day JC (2002) Zoning—lessons from the Great Barrier Reef Marine Park. Ocean & Coastal Management 45: 139–156.

21. Fernandes L, Day J, Lewis A, Slegers S, Kerrigan B, et al. (2005) Establishing Representative No-Take Areas in the Great Barrier Reef: Large-Scale Implementation of Theory on Marine Protected Areas. Conservation Biology 19: 1733–1744.

22. Boyes SJ, Elliott M, Thomson SM, Atkins S, Gilliland P (2007) A proposed multiple-use zoning scheme for the Irish Sea: An interpretation of current legislation through the use of GIS-based zoning approaches and effectiveness for the protection of nature conservation interests. Marine Policy 31: 287–298.

23. DSEWPC (2013) Commonwealth marine reserves. Commonwealth Marine Reserves Branch, Parks Australia, Department of Sustainability, Environment, Water, Population and Communities Kingston, Tasmania. Available http://www.environment.gov.au/marinereserves/index.html Accessed 2014 February 22.

24. Cao W, Wong MH (2007) Current status of coastal zone issues and management in China: A review. Environment International 33: 985–992.

25. Klein CJ, Steinback C, Watts M, Scholz AJ, Possingham HP (2009) Spatial marine zoning for fisheries and conservation. Frontiers in Ecology and the Environment 8: 349–353.

26. Foley MM, Halpern BS, Micheli F, Armsby MH, Caldwell MR, et al. (2010) Guiding ecological principles for marine spatial planning. Marine Policy 34: 955–966.

27. Watts ME, Ball IR, Stewart RS, Klein CJ, Wilson K, et al. (2009) Marxan with Zones: Software for optimal conservation based land- and sea-use zoning. Environmental Modelling & Software 24: 1513–1521.

28. Grantham HS, Agostini VN, Wilson J, Mangubhai S, Hidayat N, et al. (2013) A comparison of zoning analyses to inform the planning of a marine protected area network in Raja Ampat, Indonesia. Marine Policy 38: 184–194.

29. Agostini VN, Margles SW, Schill SR, Knowles JE, Blyther RJ (2010) Marine Zoning in Saint Kitts and Nevis: A Path Towards Sustainable Management of Marine Resources. Arlington, Virginia: The Nature Conservancy. pp.39.

30. European Commission (2011) Exploring the potential of Maritime Spatial Planning in the Mediterranean Sea – Country reports: Israel. Directorate-General for Maritime Affairs and Fisheries, MARE.E.1 "Maritime Policy in the Baltic, North Sea and Landlocked Member States", European Commission, Brussels, Belgium.

31. Diamant A, Tuvia AB, Baranes A, Golani D (1986) An analysis of rocky coastal eastern Mediterranean fish assemblages and a comparison with an adjacent small artificial reef. Journal of Experimental Marine Biology and Ecology 97: 269–285.

32. Spanier E, Pisanty S, Tom M, Almog-Shtayer G (1989) The fish assemblage on a coralligenous shallow shelf off the Mediterranean coast of northern Israel. Journal of Fish Biology 35: 641–649.

33. Goren M, Galil BS (2001) Fish Biodiversity in the Vermetid Reef Of Shiqmona (Israel). Marine Ecology 22: 369–378.

34. Golani D, Reef-Motro R, Ekshtein S, Baranes A, Diamant A (2007) Ichthyofauna of the rocky coastal littoral of the Israeli Mediterranean, with reference to the paucity of Red Sea (Lessepsian) migrants in this habitat. Marine Biology Research 3: 333–341.

35. Stern N (2010) The impact of invasive species on the soft bottom fish communities in the eastern Mediterranean [PhD thesis]. Tel-Aviv University. Tel-Aviv, Israel.

36. Levit Y (2012) The influence of the depth gradient on the status of alien fish in the Mediterranean coast of Israel. [PhD thesis (in Hebrew)]. Tel-Aviv University. Tel-Aviv, Israel.

37. Lipsky G (2012) Sunken ships in the Levant: Spot of attraction for various species or invasive species paradise? [MSc thesis (in Hebrew)]. Tel-Aviv University. Tel-Aviv, Israel.

38. Edelist D (2013) Fishery management and marine invasion in Israel. [PhD thesis]. University of Haifa. Haifa, Israel.

39. ESRI (2008) ArcMap 9.3 Geographical Information System Software. ESRI Inc.

40. Amante C, Eakins BW (2009) ETOPO1 1 Arc-Minute Global Relief Model: Procedures, Data Sources and Analysis. NOAA Technical Memorandum NESDIS NGDC-24. Colorado, USA. Available http://www.ngdc.noaa.gov/mgg/global/ Accessed 2013 March 20.

41. Tognelli MF, Silva-Garcia C, Labra FA, Marquet PA (2005) Priority areas for the conservation of coastal marine vertebrates in Chile. Biological Conservation 126: 420–428.

42. Clark MR, Tittensor DP (2010) An index to assess the risk to stony corals from bottom trawling on seamounts. Marine Ecology 31: 200–211.

43. Golani D, Öztürk B, Başusta N (2006) Fishes of the eastern Mediterranean: Turkish Marine Research Foundation.

44. Mazor T, Levin N, Possingham HP, Levy Y, Rocchini D, et al. (2013) Can satellite-based night lights be used for conservation? The case of nesting sea turtles in the Mediterranean. Biological Conservation 159: 63–72.

45. Scheinin A (2010) The Population of Bottlenose Dolphins (Tursiops truncatus), Bottom Trawl Catch Trends and the Interaction between the Two along the Mediterranean Continental Shelf of Israel Haifa. [PhD thesis]. Haifa University. Haifa, Israel.

46. The Israel Nature and Parks Authority (2012) Data - Ecological Geographical Information Centre, Department of Telecommunication and Information Systems. Jerusalem, Israel.

47. IUCN (2013) The IUCN Red List of Threatened Species. Version 2013.2. Available http://www.iucnredlist.org. Accessed 2013 November 27.

48. Kark S, Levin N, Grantham HS, Possingham HP (2009) Between-country collaboration and consideration of costs increase conservation planning efficiency in the Mediterranean Basin. Proceedings of the National Academy of Sciences 106: 15368–15373.

49. Lieberknecht L, Ardron JA, Wells R, Ban NC, Lötter M, et al. (2010) Addressing Ecological Objectives through the Setting of Targets. In: Ardron JA., Possingham HP, and Klein CJ (eds), Marxan Good Practices Handbook, Version 2. Pacific Marine Analysis and Research Association, Victoria, BC, Canada. pp.24–38.

50. Edelist D, Scheinin A, Sonin O, Shapiro J, Salameh P, Rilov G, (2013) Israel: Reconstructed estimates of total fisheries removals in the Mediterranean, 1950–2010. Acta Adriatica 54: 253–264.

51. Mazor T, Giakoumi S, Kark S, Possingham HP (2014) Large-scale conservation planning in a multinational marine environment: cost matters. Ecological Applications 24:1115–1130.

52. Israel Department of Fisheries (2012) Trawling routes along Israel's Mediterranean coast. 1:100000. Israel Ministry of Agriculture and Rural development - Department Of Fisheries and Aquaculture. Haifa, Israel.

53. State of Israel Ministry of Energy and Water resources (2012) Oil and Gas Section Ownership in Petroleum Rights. Haifa, Israel. Available www.energy.gov.il. Accessed 2014 December 20.

54. Varshavsky A (2012) Current Status of Offshore Oil and Gas Exploration in Israel. The Ministry of Energy and Water Resources of Israel. Haifa, Israel. Available www.energy.gov.il. Accessed 2013 September 4.

55. Ball IR, Possingham HP, Watts M (2009) Marxan and relatives: Software for spatial conservation prioritization. In: Moilanen A, Wilson KA, Possingham H, P., editors. Spatial conservation prioritization: Quantitative methods and computational tools. Oxford, UK. Oxford University Press. pp.185–195.

56. McDonald RI (2009) The promise and pitfalls of systematic conservation planning. Proceedings of the National Academy of Sciences 106: 15101–15102.

57. Pressey RL, Cabeza M, Watts ME, Cowling RM, Wilson KA (2007) Conservation planning in a changing world. Trends in Ecology & Evolution 22: 583–592.

58. Warman LD, Sinclair ARE, Scudder GGE, Klinkenberg B, Pressey RL (2004) Sensitivity of Systematic Reserve Selection to Decisions about Scale, Biological Data, and Targets: Case Study from Southern British Columbia. Conservation Biology 18: 655–666.

59. Levin N, Coll M, Fraschetti S, Gajewski J, Gal G, et al. (In Press) Review of biodiversity data requirements for systematic conservation planning in the Mediterranean Sea. Marine Ecology Progress Series.

60. Sivas DA, Caldwell MR (2008) New Vision for California Ocean Governance: Comprehensive Ecosystem-Based Marine Zoning. Stan Envtl LJ 27: 209–216.

61. Game ET, Meijaard E, Sheil D, McDonald-Madden E (2013) Conservation in a wicked complex world; challenges and solutions. Conservation Letters: 7: 271–277.

62. Hirsch PD, Adams WM, Brosius JP, Zia A, Bariola N, et al. (2011) Acknowledging Conservation Trade-Offs and Embracing Complexity. Conservation Biology 25: 259–264.

63. Halpern BS, Klein CJ, Brown CJ, Beger M, Grantham HS, et al. (2013) Achieving the triple bottom line in the face of inherent trade-offs among social equity, economic return, and conservation. Proceedings of the National Academy of Sciences 110: 6229–6234.

64. Halpern BS, Gaines SD, Warner RR (2004) Confounding effects of the export of production and the displacement of fishing effort from marine reserves. Ecological Applications 14: 1248–1256.

65. Roberts CM, Hawkins JP, Gell FR (2005) The role of marine reserves in achieving sustainable fisheries. Philosophical Transactions of the Royal Society B: Biological Sciences 360: 123–132.

66. EIA (2013) Overview of oil and natural gas in the Eastern Mediterranean region. EIA US Energy Information Administration. US Department of Energy, Washington, DC. Available http://www.eia.gov/countries/analysisbriefs/Eastern_Mediterranean/eastern-mediterranean.pdf Accessed 2013 October 3.

67. Makino A, Klein CJ, Beger M, Jupiter SD, Possingham HP (2013) Incorporating Conservation Zone Effectiveness for Protecting Biodiversity in Marine Planning. PLOS ONE 8: e78986.

68. Gomez Gesteira JL, Dauvin JC, Salvande Fraga M (2003) Taxonomic level for assessing oil spill effects on soft-bottom sublittoral benthic communities. Marine Pollution Bulletin 46: 562–572.

69. Lee L-H, Lin H-J (2013) Effects of an oil spill on benthic community production and respiration on subtropical intertidal sandflats. Marine Pollution Bulletin 73: 291–299.

70. Rooker JR, Kitchens LL, Dance MA, Wells RJD, Falterman B, et al. (2013) Spatial, Temporal, and Habitat-Related Variation in Abundance of Pelagic Fishes in the Gulf of Mexico: Potential Implications of the Deepwater Horizon Oil Spill. PLOS ONE 8: e76080.

71. Davies AJ, Roberts JM, Hall-Spencer J (2007) Preserving deep-sea natural heritage: Emerging issues in offshore conservation and management. Biological Conservation 138: 299–312.

72. Yahel R, Angert N (2012) Nature conservation policy in the Mediterranean, marine reserves as biodiversity and environmental conservation tool [in Hebrew]. Israel Nature and Parks Authority. Jerusalem, Israel.

73. Weeks R, Russ GR, Bucol AA, Alcala AC (2010) Shortcuts for marine conservation planning: The effectiveness of socioeconomic data surrogates. Biological Conservation 143: 1236–1244.

74. Fenberg PB, Caselle JE, Claudet J, Clemence M, Gaines SD, et al. (2012) The science of European marine reserves: Status, efficacy, and future needs. Marine Policy 36: 1012–1021.

12

A Biodiversity Indicators Dashboard: Addressing Challenges to Monitoring Progress towards the Aichi Biodiversity Targets Using Disaggregated Global Data

Xuemei Han[1,2]*, Regan L. Smyth[1], Bruce E. Young[1], Thomas M. Brooks[1,3,4,5], Alexandra Sánchez de Lozada[1], Philip Bubb[6], Stuart H. M. Butchart[7], Frank W. Larsen[8,9], Healy Hamilton[1], Matthew C. Hansen[10], Will R. Turner[9]

1 NatureServe, Arlington, Virginia, United States of America, 2 Department of Environmental Science and Policy, George Mason University, Fairfax, Virginia, United States of America, 3 International Union for Conservation of Nature, Gland, Switzerland, 4 World Agroforestry Center, International Center for Research in Agroforestry, University of Philippines, Los Baños, Laguna, Philippines, 5 School of Geography and Environmental Studies, University of Tasmania, Hobart, Australia, 6 United Nations Environment Programme World Conservation Monitoring Centre, Cambridge, United Kingdom, 7 BirdLife International, Cambridge, United Kingdom, 8 European Environment Agency, Copenhagen, Denmark, 9 Conservation International, Arlington, Virginia, United States of America, 10 Department of Geographical Sciences, University of Maryland, College Park, Maryland, United States of America

Abstract

Recognizing the imperiled status of biodiversity and its benefit to human well-being, the world's governments committed in 2010 to take effective and urgent action to halt biodiversity loss through the Convention on Biological Diversity's "Aichi Targets". These targets, and many conservation programs, require monitoring to assess progress toward specific goals. However, comprehensive and easily understood information on biodiversity trends at appropriate spatial scales is often not available to the policy makers, managers, and scientists who require it. We surveyed conservation stakeholders in three geographically diverse regions of critical biodiversity concern (the Tropical Andes, the African Great Lakes, and the Greater Mekong) and found high demand for biodiversity indicator information but uneven availability. To begin to address this need, we present a biodiversity "dashboard" – a visualization of biodiversity indicators designed to enable tracking of biodiversity and conservation performance data in a clear, user-friendly format. This builds on previous, more conceptual, indicator work to create an operationalized online interface communicating multiple indicators at multiple spatial scales. We structured this dashboard around the Pressure-State-Response-Benefit framework, selecting four indicators to measure pressure on biodiversity (deforestation rate), state of species (Red List Index), conservation response (protection of key biodiversity areas), and benefits to human populations (freshwater provision). Disaggregating global data, we present dashboard maps and graphics for the three regions surveyed and their component countries. These visualizations provide charts showing regional and national trends and lay the foundation for a web-enabled, interactive biodiversity indicators dashboard. This new tool can help track progress toward the Aichi Targets, support national monitoring and reporting, and inform outcome-based policy-making for the protection of natural resources.

Editor: Julia A. Jones, Oregon State University, United States of America

Funding: This work was supported by awards from the John D. and Catherine T. MacArthur Foundation (URL: http://www.macfound.org/) to NatureServe addressing "Dashboard assessments: proof-of-concept and baselines" (Grant No. 11-98252-000-INP), "Moving graphical presentation of biodiversity conservation monitoring indicators from a static proof-of-concept to a dynamic prototype" (Grant No. 12-100096-000-INP), and the bridge grant between the two (Grant No. 12-102962-000-INP). The funders supported the concept development in study design phase, but had no role in data collection and analysis, decision to publish, or preparation of the manuscript.

Competing Interests: The authors have declared that no competing interests exist.

* Email: xuemei_han@natureserve.org

Introduction

Resource monitoring has long been recognized as a cornerstone of biodiversity and conservation science [1,2,3]. In 2010, at the 10th Conference of the Parties of the Convention on Biological Diversity (CBD), 193 nations agreed to twenty "Aichi Biodiversity Targets", and in doing so committed to updating their National Biodiversity Strategies and Action Plans and developing monitoring programs to assess progress [4]. The Aichi Targets rely upon indicators to report progress towards reducing pressure on biodiversity, maintaining and improving the state of biodiversity, implementing conservation actions to ameliorate biodiversity loss, and providing benefits to human well-being [4]. Many other initiatives and multilateral agreements call for similar indicator-based biodiversity monitoring, including (a) the United Nations Millennium Development Goal #7 [5] and the draft new Sustainable Development Goals [6]; (b) intergovernmental treaties that provide mechanisms for national action and international cooperation, such as the Ramsar Convention on Wetlands [7] and the Convention on Migratory Species [8], (c) science-policy interfaces such as the Intergovernmental Platform on Biodiversity and Ecosystem Services [9]; and (d) partnerships or networks in support of the above mentioned bodies, such as the Biodiversity Indicators Partnership [10,11,12], and the Group on Earth Observations Biodiversity Observations Network Working Group #9 [13].

Monitoring called for by these programs is essential both to document biodiversity change over time [14], to shed light onto key ecological processes [15], and to measure the success or failure of conservation interventions through counterfactual analysis [16,17,18,19]. However, most existing monitoring programs have been designed primarily at localized scales, and often produce information that is disaggregated, heterogeneous, and non-standardized when considered at national or regional scales [20]. Monitoring requirements for measuring conservation performance, of the kind necessary to track the Aichi Targets, require data that transcend the fine temporal, spatial, and organizational scales commonly addressed in current literature [15].

Documentation of conservation impacts and biodiversity response must be accomplished in ways that are scientifically defensible, at appropriate temporal and spatial scales, and simple enough to inform decision-making by the diverse group of individuals and organizations working at the intersection of science and policy. Mounting global evidence shows that biodiversity loss is continuing at alarming rates [21,22], yet currently, two thirds of national reports submitted to the CBD lack evidence-based measures to illustrate changes in the status of biodiversity [23]. National capacity is often insufficient to measure many indicators of interest using on-the-ground methods, particularly in developing countries [24]. Even when national data are available, a lack of standardization across countries can make regional assessment difficult or impossible [20].

To better understand the challenges to effective biodiversity monitoring at national and regional scales, and how finer-scale (e.g. national) data might be integrated into a framework for global monitoring of biodiversity status and trends, we surveyed local conservation experts working in areas of high conservation value on monitoring and capacity needs. Building from the needs identified in those workshops, we then developed the concept for a biodiversity indicators dashboard using indicators derived from global data sets and constructed a dashboard prototype. This is the first operationalized dashboard to date that communicates multiple biodiversity indicators at multiple scales, and directly serves the global need to monitor progress towards Aichi Targets. Full development of the biodiversity indicators dashboard will encompass: (1) identification of appropriate indicators, (2) proof of concept using global data, (3) building the technological infrastructure necessary to host the dashboard, (4) designing the visual interface for multiple platforms (i.e. web and mobile users), and (5) creating systems to support the integration of finer-scale (regional and national) data. Here, we address in detail steps 1 and 2 of the dashboard design, laying the foundation for a web-based tool freely available to all with an interest in biodiversity conservation. A prototype of the tool is now available to the international conservation community at http://dashboarddev.natureserve.org, with steps 3–5 being implemented in an on-going iterative process.

Methods and Results

1. Study Area

We considered three geographically diverse areas with exceptional biodiversity value, that confront a high degree threat and that receive significant investment by international conservation agencies (Figure 1) [25,26]. The Tropical Andes region encompasses the eastern slope of the Andes, containing eight watersheds of headwater rivers (Japura, Putumayo, Rio Maranon, Ucayali, Guapore, Madre de Dios/Beni, Amazon, Magdalena) across Venezuela, Colombia, Ecuador, Peru, and Bolivia. The Great Lakes region of Africa includes five major watersheds (Lake Victoria, Upper Nile, Lake Tanganyika, Lake Malawi/Nyasa, Turkana/Omo) across Ethiopia, South Sudan, Kenya, Uganda, Democratic Republic of Congo, Rwanda, Burundi, Tanzania, Zambia, Malawi, and Mozambique. The Greater Mekong region encompasses the entire Mekong River Basin, spanning China, Myanmar, Vietnam, Lao P.D.R., Cambodia, and Thailand [25].

We delineated regional boundaries for the Tropical Andes, African Great Lakes and Greater Mekong regions using hydrological basins derived from HydroSHEDS and compiled by the UN-FAO, [27,28,29]. We performed analyses at both this regional scale, and at the national scale for the 22 countries that these three regions overlap (including areas outside the focal watershed boundaries).

2. Challenges to Biodiversity Monitoring and Capacity Needs at Regional and National Scales

We conducted seven consultation workshops in the three study regions between September 2011 and August 2012 to (1) better understand the challenges to effective biodiversity monitoring at national and regional scales, (2) identify gaps in current monitoring capacity and potential mechanisms for filling those gaps, and (3) begin to explore mechanisms for integrating local and national monitoring data into future regional and national biodiversity indicators. In total, 260 individuals from 20 countries attended at least one of the workshops, with broad representation from the public, civil-society, and academic sectors. Invitees included those with professional responsibility for National Biodiversity Strategies and Action Plans for monitoring progress towards Aichi Targets, and managers and technical experts responsible for designing and conducting biodiversity monitoring programs at multiple scales.

At each workshop, we solicited multiple-choice feedback on two issues: 1) the spatial scales of monitoring that participants required to guide their work (regional, national, sub-national, watershed, and/or site scales); and 2) the status of monitoring of selected biodiversity indicators for pressure, state, response, and benefits at the national scale, with answer options of "Monitored", "Limited Monitoring" (monitoring that has been conducted in some areas but not systematically done across the country), "Not Monitored", or "Unknown". Of the 260 workshop participants, 132 (51%) submitted answers to these written questionnaires, of which 39% came from the public sector, 45% from civil-society, and 16% from the academic sector. We also recorded and categorized responses to open-ended questions addressing (1) the utility of tracking biodiversity indicators derived from existing global data with a dashboard approach and (2) national challenges in developing sustainable biodiversity monitoring.

To identify the preferred scales of monitoring, we tabulated the frequency of the scales that participants indicated were important. To quantify the existing capacity for monitoring in each of the targeted countries, we calculated a score based on the perceived monitoring status for each biodiversity indicator. The score is scaled 0 (not monitored) to 1 (monitored), and equals $P_1 + 0.5P_2$,

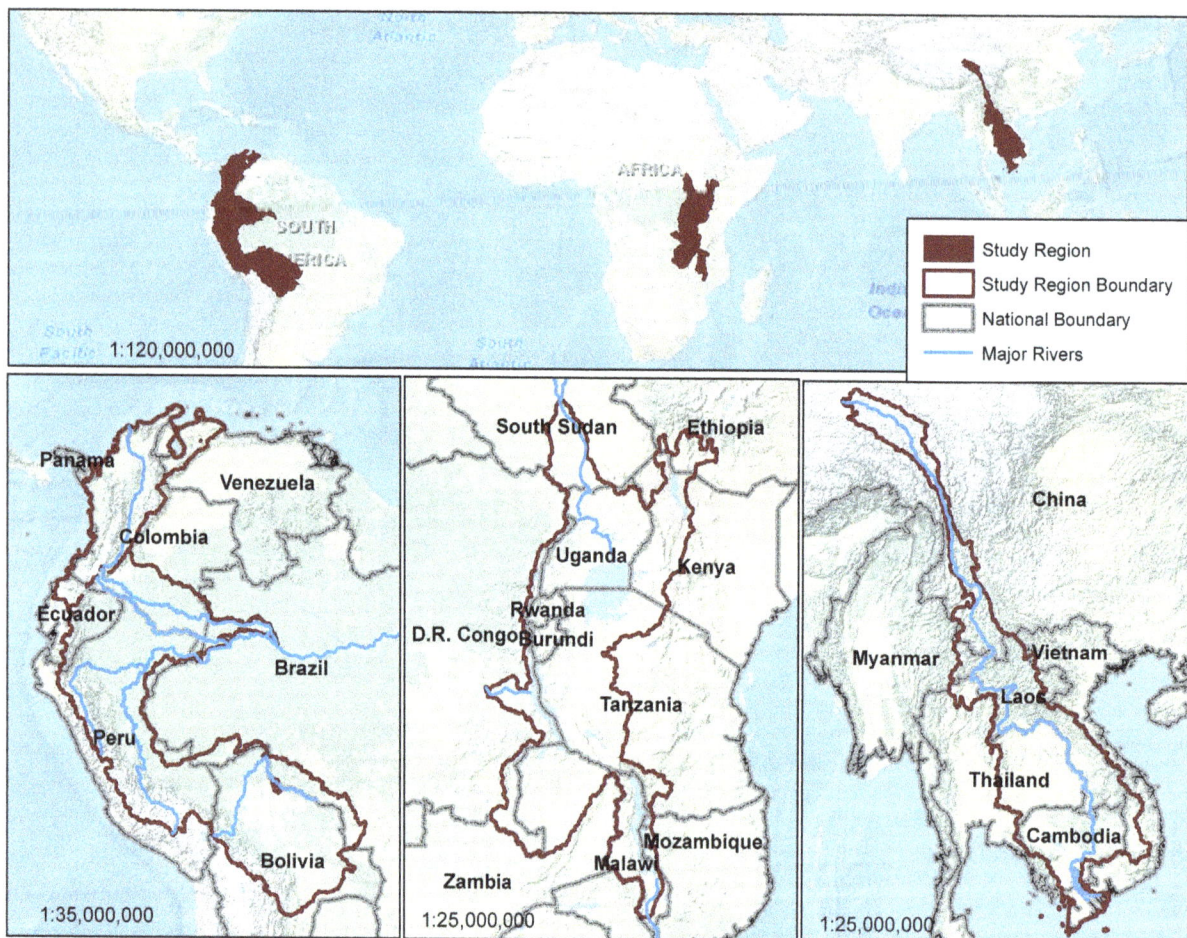

Figure 1. Study area regions. From left to right: the Tropical Andes, the African Great Lakes, and the Greater Mekong.

with P_1 the percent of respondents who answered "monitored" and P_2 the percent of respondents who answered "limited monitoring". We used ANOVA to explore differences of monitoring status between regions, and a repeated-measures ANOVA to examine differences in monitoring status among indicators.

Responses to the questionnaire indicate a strong demand for reliable information on the state of, and pressures facing, biodiversity. Regarding scales of monitoring, participants were most interested in analyses carried out at the site (82%) and national levels (76%), followed by watershed (71%), sub-national (68%) and regional levels (65%).

Our questionnaires revealed significant differences in the degree to which indicators are currently monitored ($p<0.001$), with hydrologic measures (average score $= 0.40$) and species extinction risk (0.57) less frequently monitored than deforestation (0.72) and protected area coverage (0.79) (Figure 2). While there were no differences in the average score across regions ($p = 0.88$), the status of monitoring differed widely among nations. Of the 22 countries, those with the highest overall scores for existing monitoring were Colombia (0.875), Malawi (0.875), and Thailand (0.75). Countries with very limited monitoring include South Sudan and the D. R. Congo (both ≤0.25).

Among the open-ended questions, a third of survey respondents from all sectors expressed high interest in using the dashboard approach, and employing appropriate subsets of global scale data, as a means to gather and share information to assess biodiversity status and threats, assess and improve conservation impacts, and inform policy, planning, and decision-making. Supporting capacity building, promoting stakeholder participation and dissemination of information were also frequently cited by survey respondents as potential benefits of this effort (Figure 3).

Across regions, the challenges to effective monitoring (Figure 4) include the lack of personnel, technology, and financial support for data collection and management (45%), and limited information accessibility and interoperability (40%). Emphasis varies among regions, with African respondents stressing the need for support in data management (24%), and Andean respondents more concerned about scientific standards and methods (25%) and conservation expertise and analysis (39%).

3. Creation of a Biodiversity Indicators Dashboard to Support Monitoring Needs

3.1 The Dashboard Concept. To address the challenges to biodiversity monitoring at regional and national scales identified by the survey, we envision the creation of a biodiversity "dashboard" – a visualization of biodiversity indicators designed

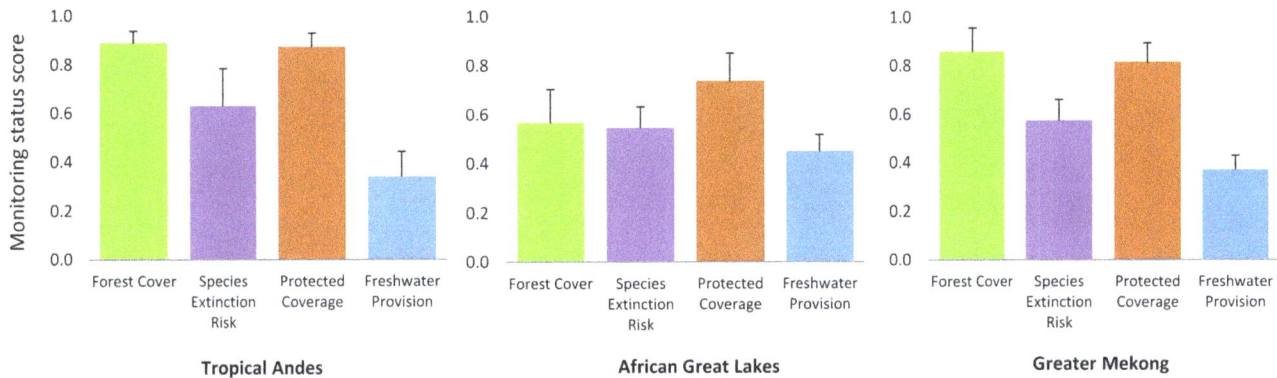

Figure 2. Monitoring status of the indicators, as reported by national experts via questionnaire responses. The mean score and its standard error for each indicator are shown by region. Number of respondent is 36 for Tropical Andes, 46 for African Great Lakes, and 50 for Greater Mekong.

to enable ongoing tracking of biodiversity status and trends and present biodiversity monitoring and conservation performance information in a clear, user-friendly, and unified format that facilitates iterative adaptive management. Using biodiversity indicators of the type developed by Butchart et al. (2010) [22] as the foundation, the dashboard provides a means to disseminate information, promote stakeholder participation, support capacity building, and allow users to better understand the relationship between conservations actions and impact. The utility of the biodiversity indicators dashboard in meeting these needs was confirmed by responses to our survey (Figure 3).

Originating as a business performance monitoring tool, dashboard visualizations are an information management and reporting instrument that has seen increasing use in a variety of contexts to communicate complicated information on current status and historical trends to broad audiences [30,31]. From the World Bank Atlas of Global Development [32] to commercial products used to track stock performance and guide financial investment (e.g., [33]), dashboards distill complicated data by tracking key indicators, usually via a combination of charts and maps. This information is typically served on websites or mobile applications and updated regularly (e.g., annually for World Bank indicators, minute-by-minute for financial markets).

Dashboards have been proposed and employed in various biological and resource management contexts. For example, the CITES Trade Data Dashboard [34] allows users to explore

patterns in species exploitation across space, time, and taxonomic affiliation through a dynamic interface [35]. Dashboards also support fisheries management by providing a framework to better visualize relationships among fish populations, socio-economics, and exploitation [36,37].

If a dashboard is to be useful for decision makers, the indicators chosen must present information critical to influencing the decisions to be made. We used the Pressure-State-Response-Benefit (PSRB) framework to guide selection of indicators, following Sparks et al. 2011 [14]. This is derived from the causal-chain Pressure-State-Response and Driver-Pressure-State-Impact-Response frameworks, widely used for reporting on the state of the environment [22,23,38,39,40,41,42,43], and one that has been used by the CBD Ad-Hoc Technical Expert Group of the CBD to guide indicator development for the CBD [44] and recommended for communicating biodiversity indicators [42]. The core elements in PSRB as applied in the dashboard assessments are pressure on biodiversity, its drivers, (e.g., habitat destruction, climate change, invasive species), the state of species and ecosystems (e.g., species extinction risk, animal and plant populations, ecosystem integrity), conservation action or policy responses (e.g., protected areas establishment and management, investment in biodiversity conservation) and benefit to human well-being from the social, economic and cultural impacts of conservation (e.g., maintenance of hydrological functions, climate change mitigation, maintenance of indigenous cultures). By

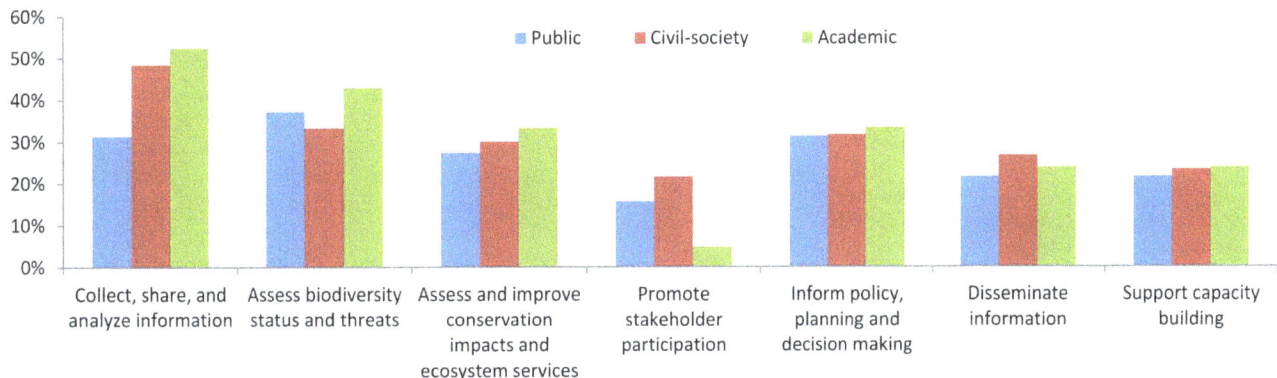

Figure 3. Perceived benefits of using global data within a dashboard approach, by sector. Number of respondent is 51 for public sector, 60 for civil-society, and 21 for academic sector.

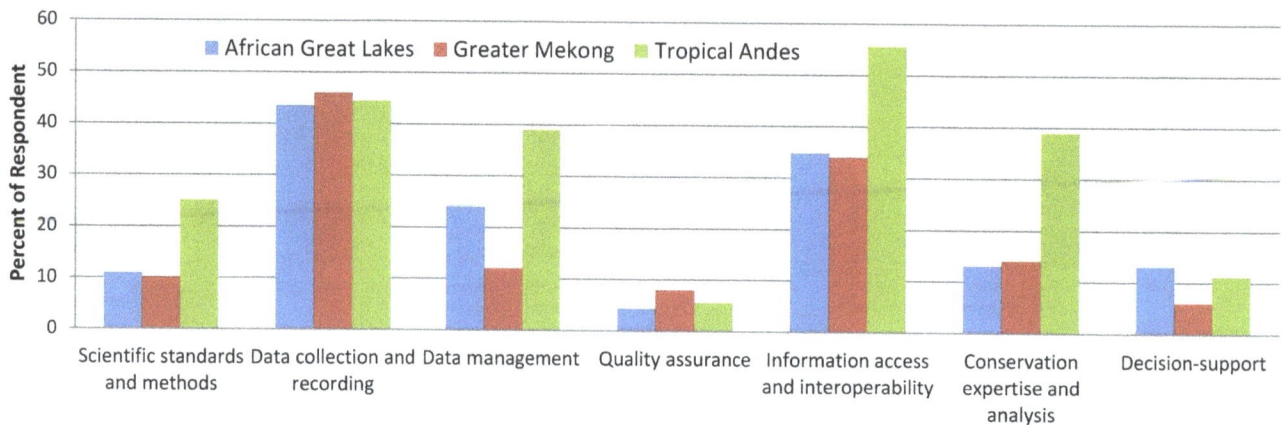

Figure 4. Perceived challenges to biodiversity monitoring by region. Number of respondent is 36 for Tropical Andes, 46 for African Great Lakes, and 50 for Greater Mekong.

viewing dashboard indicators together within the PSRB framework, users can begin to understand interactions between indicators [14].

3.2 Methods for Constructing a Biodiversity Indicators Dashboard using Global Data. We selected one indicator from each PSRB component, with consideration for the availability of global datasets, the degree to which the indicator contributes to evaluating progress made towards the Aichi Targets, the feasibility of trend estimates, and the likely availability of analogous data generated locally for future integration into the dashboard. The four selected indicators are examples of the types of data the biodiversity indicators dashboard can be used to track. At this stage in development of the dashboard, the chosen indicators are not intended to address causal relationships; however, as additional indicators are added, the PSRB approach will facilitate exploration of causal links between indicators. We selected forest cover loss as the pressure indicator, species extinction risk as the state indicator, protected area coverage of key biodiversity areas (KBAs) as the response indicator, and freshwater provisioning to downstream human populations as the benefit indicator. We used data from the first decade of the 21st century to represent current status and provide an initial baseline, or reference point, against which future trends can be assessed. For all but the benefit indicator (freshwater provision), existing data from either previous time steps (i.e. species extinction risk and percent protection of KBAs) or later time steps (i.e. forest cover loss) supporting the tracking of trends.

Global data were disaggregated to provide regional and national indicator values (Table 1). For each indicator, we mapped current condition, charted and mapped trends over time, and generated tabular summaries. All spatial analyses were performed using ArcGIS 10.1 [45] and all statistical calculations were performed in R [46].

a. Pressure Indicator: Forest Loss

The forest loss indicator is derived from the Global Forest Monitoring Project [47,48], which estimated forest cover in 2000 and forest cover loss between 2000 and 2005 using MODIS data [49] calibrated with Landsat [50] imagery, at 18.5-km resolution. Values represent the percent forest cover within each pixel, with forest cover defined as areas with at least 25% cover of trees at least 5 meter in height. The deforestation measure, Gross Forest Cover Loss (GFCL), represents a unidirectional change in forest cover, calculated from the percent forest loss between 2000 and 2005. For each analysis unit (e.g., region, nation) we calculated a

mean value for forest cover in 2000 and mean GFCL between 2000 and 2005. We then derived the average annual rate of GFCL for 2000–2005 for each analysis unit, presented as the annual percent forest loss from the 2000 baseline.

While, to our knowledge, the GFCL data provide the best globally consistent spatial representation of deforestation to date, the data are limited in that they do not incorporate information on forest gain from restoration, natural regrowth, and plantation. They also do not address finer resolution forest degradation, as some other regional mapping products do [51,52,53].

b. State Indicator: Species Extinction Risk

The Red List Index (RLI) is a measure of trends in survival probability (the inverse of extinction risk) for sets of species. It is based on the numbers of species within each IUCN Red List category (i.e., Extinct (EX), Extinct in the Wild (EW), Critically Endangered (CR), Endangered (EN), Vulnerable (VU), Near Threatened (NT), Least Concern (LC)) and the changes in these numbers over time resulting from genuine improvement or deterioration in status between assessments [22,54,55,56,57,58].

For this indicator, we used the first and last comprehensive Red List assessment (when all species of a taxonomic group were assessed) for each of three vertebrate groups (1988 and 2008 for birds, 1996 and 2008 for mammals, and 1980 and 2004 for amphibians, noting that the 1980 assessment for amphibians was based on a retrospective assessment), following Butchart et al. (2004, 2005, 2007 and 2010) [22,54,55,56]. We identified all species falling partially or completely within each region and each country using 2010 spatial distribution data for each species [58] (Table 2). For each region and country, we calculated the RLI for each taxonomic group individually and for all taxonomic groups together. This standardized RLI varies between 1 (all species LC) and 0 (all species EX or EW). Following Butchart et al. 2004, 2005 [55,56], species undergoing genuine Red List category changes between assessments contributed to RLI trends only if the driving process of the change (i.e. threat or conservation action) operated within the relevant country or region. For each vertebrate group, we calculated the annual change in aggregate extinction risk by dividing the difference in RLI between the last and first assessment by the number of intervening years. Data Deficient and Not Evaluated species were excluded from this calculation and the annual change value across all taxonomic groups is computed using the mean time difference between assessments for the three groups.

Table 1. Biodiversity indicators summary and data sources.

Framework Component	Pressure/Driver	State	Response	Benefit (Impact)
Indicator	Forest coverage and rate of gross forest cover loss	Red List Index	Protected area coverage in key biodiversity areas	Quality-weighted freshwater provision from natural ecosystems to downstream human population
Aichi Target	Target 5: Loss of habitats is at least halved by 2020	Target 12: Extinctions of known threatened species has been prevented by 2020	Target 11: At least 17% of terrestrial...areas, especially important areas for biodiversity and ecosystem services are conserved by 2020	Target 14: Ecosystems that provide essential services are restored and safeguarded by 2020
What does it show	Spatial data represents the percent of forest cover for each 18.5 km pixel in 2000 and percent gross forest cover loss (i.e., deforestation) from 2000 to2005. Tabular FAO data summarize forest land use coverage and "net forest cover change" by country in 2005 and 2010.	An index of aggregate survival probability of species that occur in the given spatial unit. Values range from 1 (all species Least Concern) to 0 (all Extinct).	Mean percent area of key biodiversity areas covered by protected areas	Quality-weighted delivery of clean freshwater from natural habitats to downstream human populations per unit area
Data Source	Forest cover for year 2000 and gross forest cover loss 2000–2005 through Global Forest Monitoring Project	IUCN Red List assessment for: - Amphibians (1980, 2004) - Birds (1988, 2008) - Mammals (1996, 2008)	- World Database on Protected Areas (UNEP-WCMC) (2010) - Global KBAs as represented by Important Bird and Biodiversity Areas (IBAs) and Alliance for Zero Extinction (AZE) sites	- World WaterGAP 2 model runoff map - Hydrological drainage direction - Landscan Global Population Database - GlobCover land cover
Time frame	2000–2005	1980–2008	1950–2010	2010
Limitation and caveat	- Resolution is too coarse (18.5 km) to detect deforestation in small areas. - The gross forest cover loss data shows deforestation only, not taking account afforestation - Forest degradation was not quantified	- Differing assessment dates requires interpolation and extrapolation to estimate aggregate trends - Because of the heterogeneous distribution of species, regional extinction risk can skew national indicator values - The proportion of a species' range within a given analysis unit is not considered - Red List categories are necessarily broad classes of extinction risk, so the RLI is moderately sensitive	- The WDPA omit recently decreed protected areas - The WDPA does not currently document management effectiveness - Key biodiversity areas for taxa other than birds that are not endemic to single sites have only been identified in some countries	- Values are relative, not absolute - Only baseline (2010) data currently available; not able to estimate trend - Spatial resolution is too coarse (2,5921km^2 pixels) to estimate freshwater provision in small areas

c. Response Indicator: Protected Area Coverage of Key Biodiversity Areas

For the Response indicator, we calculated the mean percentage area of key biodiversity areas (sites contributing significantly to the global persistence of biodiversity [59]) falling within protected areas for each analysis unit [60]. We used the World Database on Protected Areas for 2010 [61] to delimit protected areas. Within our study area, key biodiversity areas include 757 Important Bird & Biodiversity Areas (IBAs; the subset of key biodiversity areas important for birds[62]), and 139 Alliance for Zero Extinction sites (AZE; the subset of key biodiversity areas holding effectively the entire populations of highly threatened species, i.e., CR and EN species>95% restricted to single sites [63,64]). We calculated the percentage of each KBA that overlaps protected area boundaries

Table 2. Number of species recorded and analyzed to derive Red List Index (only extant species that are not Data Deficient were included).

	Number of all assessed species				Number of species with changed Red List category			
	Overall	Mammals	Birds	Amphibians	Overall	Mammals	Birds	Amphibians
Assessment year	2008	2008	2008	2004	1980–2008	1996–2008	1988–2008	1980–2004
Tropical Andes	5357	880	2978	1499	179	13	29	137
African Great Lakes	2483	613	1534	336	30	10	17	3
Greater Mekong	2499	642	1479	378	142	41	71	30

and subsequently generated a national mean. We used the year of establishment to generate time series graphs for each year from 1950 to 2010, assigning an establishment date to those protected areas lacking establishment date by randomly sampling from known dates of designation of protected areas in the same country, and then bootstrapping following the methods of Butchart et al. [60]. We plotted the mean with 95% confidence intervals based on uncertainty arising from the missing data [60]. We also calculated the annual rate of change in protection of key biodiversity areas between 1980 and 2010. The annual rate of change is thus from a time period comparable to that calculated for the State indicator (the Red List Index).

d. Benefit Indicator: Freshwater Provision

Freshwater provision data were developed by Larsen et al. [65,66] using spatially explicit maps of runoff from the global hydrological water model WaterGAP [67], hydrological drainage directions [27,68], downstream human population density [69], and global land cover data (used to weight flow estimates by a quality coefficient, based on information from previous studies [70,71,72,73]). Estimated quality-weighted freshwater provision, reported as a freshwater flow index, was calculated for 2,592 km^2 hexagonal grid cells [66]. Using this grid, we calculated a mean value for each analysis unit. Because the freshwater provision data is currently only available for a single time step (2010), we cannot yet calculate trends.

3.3 Results of Biodiversity Indicators Dashboard with Global Data. The indicators are presented as a series of maps and charts visualized to support a web-enabled biodiversity indicators dashboard (Table 3, Figures 5–7). A web prototype of the dashboard, displaying the results discussed here, is available at http://dashboarddev.natureserve.org.

Together, the dashboard graphics present a picture of the current status (Figure 5) and trends (Figure 6) in biodiversity. Baseline data show large geographical variation in the status of forest cover. National forest coverage is the lowest in Kenya (7.47%), and the highest in D. R. Congo (71.18%) and Lao P.D.R. (69.06%) (Table 3 and Figure 5A). The baseline Red List Index reveals high species extinction risk in Tropical Andes for all taxa (0.89), with variations among countries and among taxonomic groups (Table 3). The conservation response, measured as the average percentage of key biodiversity areas under protection, varies somewhat among regions (44% in the Tropical Andes, 63% in the African Great Lakes region, and 49% in the Greater Mekong as of 2010) but the dashboard shows larger differences at national levels, with lows in Mozambique (20%), Ethiopia (25%), Peru (25%), Vietnam (34%) and Myanmar (35%), and highs in Burundi (100%), Malawi (88%), Venezuela (79%) and Thailand (73%) (Table 3 and Figure 5C). Similarly, baseline data for freshwater provision show large national differences (Figure 5D), with Burundi, Rwanda, Vietnam, and China standing out as areas of high importance.

The trend of forest loss is documented as ongoing in all regions evaluated, with national rates of loss lowest in Peru (0.08%/yr) and D. R. Congo (0.12%/yr), and highest in Kenya (1.2%/yr) (Table 3 and Figure 6A). The Red List Index indicates the worsening status of species, with a decline in Red List Index observed for all nations between 1980 and 2008 (Table 2). Rates of decline were highest in the Tropical Andes, largely driven by amphibians (1.36×10^{-3}/yr), and in the Mekong, due to both mammals (1.29×10^{-3}/yr) and amphibians (0.98×10^{-3}/yr). Protection of key biodiversity areas increased in all regions since 1980, with some regional variation in the rate of increase (0.20% in the African Great Lakes, 0.86% in the Tropical Andes, and 1.43% in the Greater Mekong). Nationally, rates of increase in the

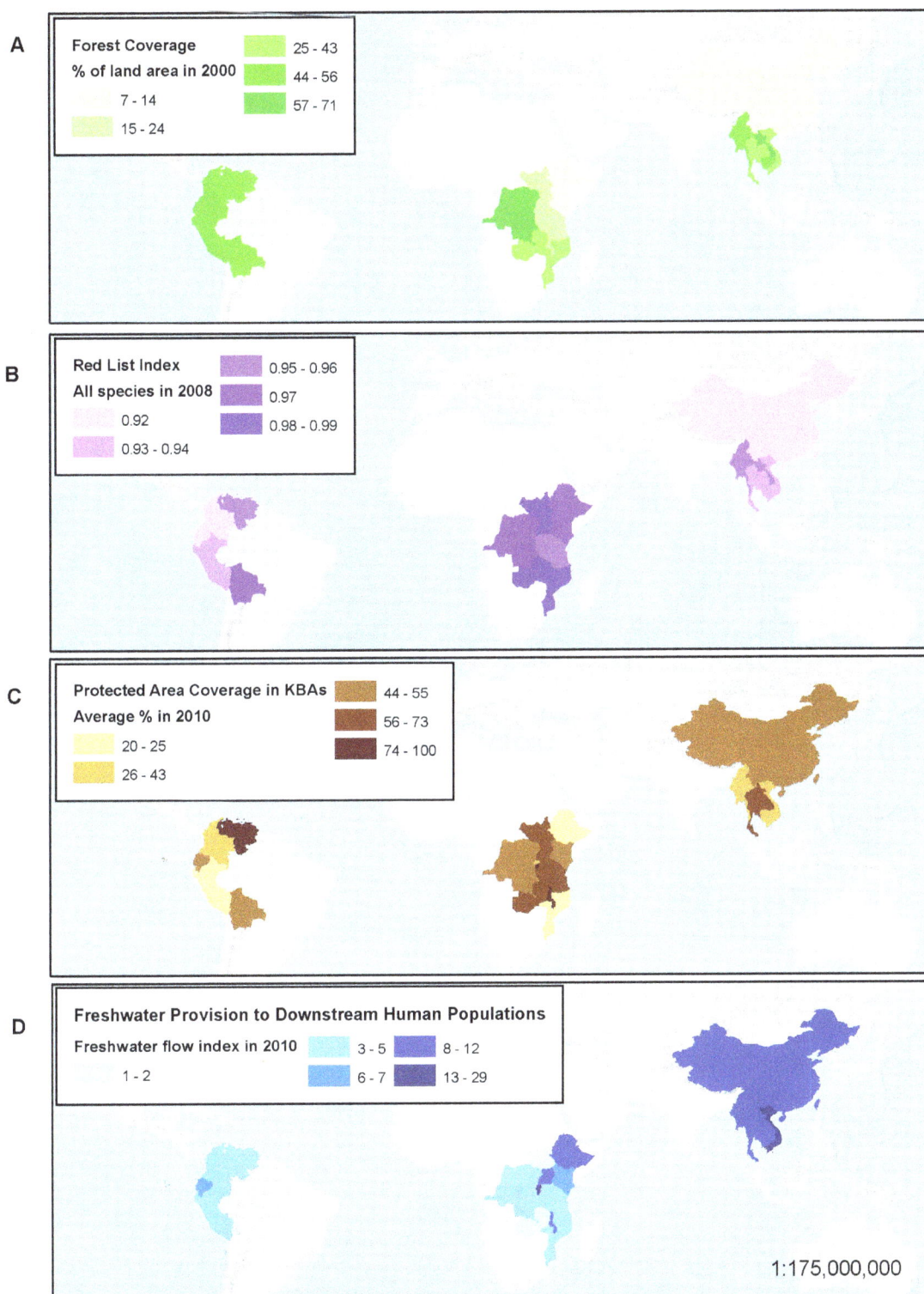

Figure 5. Dashboard indicator baseline results. Results for (A) Forest Cover (2000); (B) Red List Index a measure of change in extinction risk (2008); (C) Protected Area Coverage of Key Biodiversity Areas (2010); and (D) Freshwater Provision (2010).

protection of key biodiversity areas ranged from lows of zero in Burundi, Malawi, Mozambique and Rwanda, and highs of 2.09%, 1.54%, and 1.42% in Lao P.D.R., China, and Cambodia (Table 3 and Figure 6C).

Discussion

1. National and Regional Monitoring Challenges

As approaches to biodiversity management shift towards data-intensive and science-driven methods [74,75], addressing gaps in

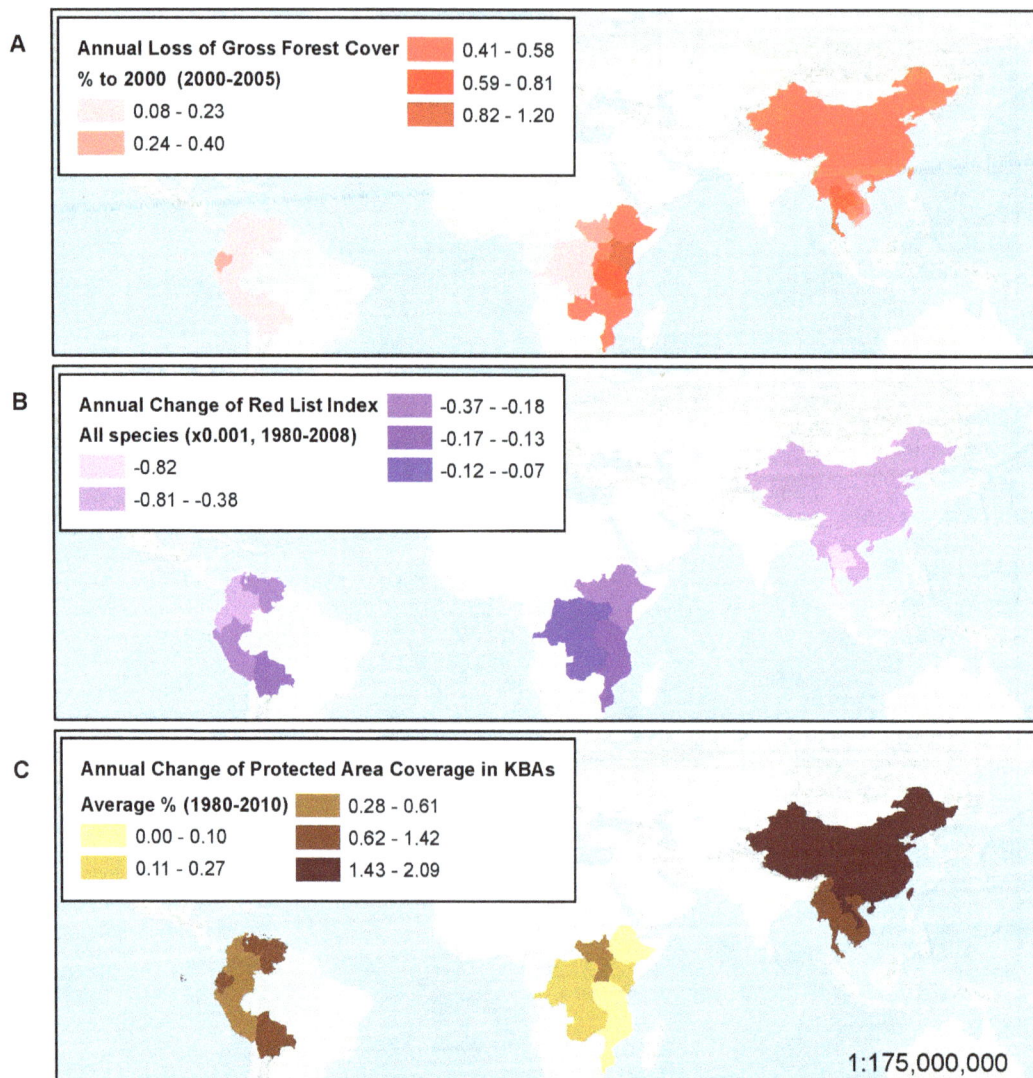

Figure 6. Dashboard indicator trend results. Annual rate of (A) Gross Forest Cover Loss (2000–2005); (B) Change in Red List Index as a measure of extinction risk (change for all species of mammals, birds, and amphibians; 1980–2008); and (C) Change of Protected Area Coverage of Key Biodiversity Areas (1980–2010).

capacity for information generation and dissemination has become increasingly important [76,77,78,79,80]. The prevailing and widely recognized challenges to addressing these gaps include sustaining financial and human resources for on-the-ground monitoring [76,81], overcoming cultural and technical barriers associated with data generation [82,83,84] and information sharing [74,85,86], adopting scientific standards for monitoring and data analysis [79,82,84,87], and developing indicator sets that can effectively inform policy and decision-making processes [10,40,88,89]. Our survey of regional biodiversity experts reaffirms these challenges in generating, managing, and sharing biodiversity information (Figure 4), demonstrates the strong demand for access to biodiversity status and trend data at multiple spatial scales, and indicates that our proposed biodiversity indicators dashboard could be an effective tool to address varied conservation needs (Figure 3).

Recognizing that national and local indicator data are often limited or non-existent, the survey respondents affirmed the value of deriving indicators from global datasets as an intermediate

measure necessary to meet current demand. Moving forward, the respondents noted that it will be necessary to augment and validate globally-derived measures with national and local monitoring results, and doing so will require both cost-effective participatory monitoring protocols that ensure sustainable data collection and well-designed standards that ensure data interoperability (Figure 4). A lack of baseline data was the most frequently mentioned monitoring challenge in our survey. The few studies that have systematically evaluated the availability of indicators for monitoring biodiversity targets [90,91] support our findings that biodiversity indicators, particularly indicators of state and benefit, are deficient (Figure 2). Unstable political situations, lack of financial support, and the low priority of biodiversity monitoring culturally or in national development strategies all can impair continuous and systematic data collection. At the same time, the barriers to information access and interoperability, prevent the information that does exist from fully informing conservation efforts. Biodiversity data are generated and kept by different agencies in a fragmented manner. Within our area of study, civil society and

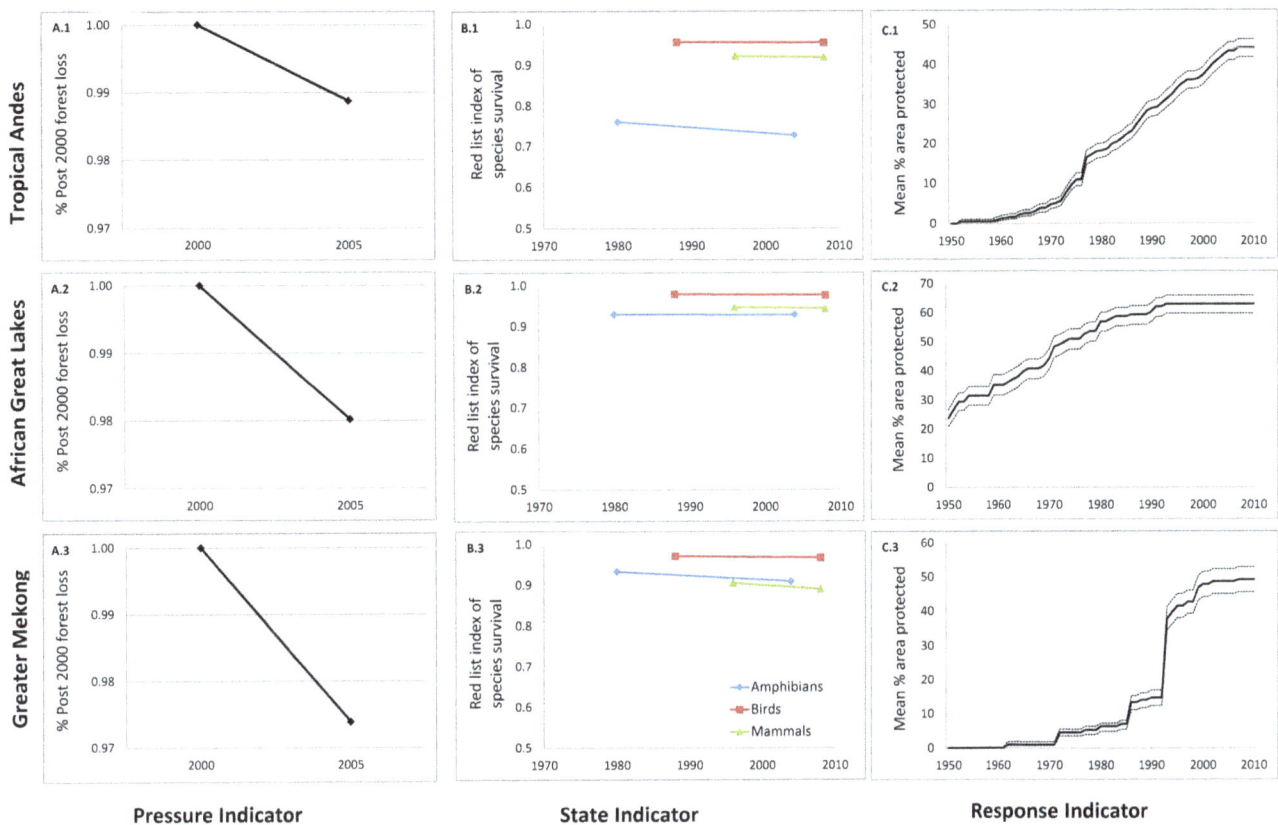

Figure 7. Dashboard indicator trend graphs by region. A.1 – A.3 chart gross forest loss as a percent of forest cover in 2000; B.1-B.3 chart change in Red List Index for mammals (green), birds (red), and amphibians (blue); and C.1-C3 chart change in protected area coverage of key biodiversity areas (1950–2010) with solid lines indicating the mean percent protected across all sites, and dashed line indicating the 95% confidence intervals [60].

academics (such as the Wildlife Conservation Society in Cambodia, NatureKenya in Kenya, NatureUganda in Uganda) accumulate a wide range of site-level monitoring data maintained as project-based resources, while national level monitoring data are typically held by specific government divisions who may or may not share that information with other government entities, much less outside organizations. These challenges are well-known and efforts to address them are being made at global (e.g., Biodiversity Indicators Partnership [12]), regional (e.g., ASEAN Centre for Biodiversity [92], Red Amazónica de Información Socioambiental Georreferenciada [93], Streamlining European Biodiversity Indicators [94], Conservation of Arctic Flora and Fauna [95]), and national levels (e.g., National Biodiversity Data bank in Uganda [96], National Biodiversity Database System in Vietnam). However, those efforts do not cover all countries and regions, and ready access to reliable and geographically consistent biodiversity indicator information at multiple scales remains a confirmed need.

2. Towards a Dashboard

The biodiversity indicators dashboard, as explained here, is designed to address the unmet needs expressed in our survey of biodiversity experts by laying the foundation for better accessibility and interpretability of existing biodiversity trend data within a framework that enhances monitoring capacity and promotes data interoperability and sharing. Many of these indicators are widely used at global scales, but until now have rarely been reported at national scales. While previous studies have demonstrated the

utility of biodiversity indicators in monitoring conservation status and trends [14,22], our biodiversity indicators dashboard is the first operationalized online interface that communicates multi-dimensional indicators with spatial representation. By providing easy access to indicator information at national and local scales, it complements global efforts such as the Biodiversity Indicators Partnership [12] and facilitates reporting for Aichi National Biodiversity Strategies and Action Plans (NBSAPs). The intuitive graphics and ease of data access that flow from the dashboard are intended to engage and enhance partnerships at all levels – international, regional, national, and local.

As a data visualization tool, the biodiversity indicators dashboard is designed so that a quick examination communicates the overall status of biodiversity conservation, important trends and patterns, and previously hidden challenges. For example, the graphics and values for the state indicator, such as those presented in Table 3 and Figure 7, communicate a high extinction risk in the Tropical Andes, driven largely by the extinction risk of amphibian species. This finding is consistent with other recent studies [97,98] and highlights the importance of addressing threats to amphibian species if biodiversity is to be maintained.

By using data and methods that are globally consistent, the dashboard facilitates direct comparison of baseline and trends across regions and nations in the three continents targeted for this first stage of dashboard developments. Regional patterns are readily evident, such as the higher rate of decline of the Red List Index in the tropical Andes or the enormous importance of fresh water provisioning in the Greater Mekong countries (Figure 5).

Table 3. Baseline and trend results for all indicators by country and region.

| Country/Region | Indicators – Baseline | | | | | | | | Indicators – Annual Rate of Change | | | | |
| | Forest Coverage % (2000) | Red List Index of species survival | | | | Protection Coverage of KBAs % (2010) | Freshwater Flow Index | GFCL % (2000–2005) | Red List Index of species survival (x 0.001) | | | | Protection Coverage of KBAs % (1980–2010) |
		3 taxa (2008)	Mammal (2008)	Bird (2008)	Amphibian (2004)				3 taxa (1980–2008)	Mammal (1996–2008)	Bird (1988–2008)	Amphibian (1980–2004)	
Bolivia	56.26	0.97	0.96	0.98	0.90	51.51	1.13	0.21	−0.16	−0.19	−0.06	−0.62	1.03
Colombia	56.06	0.92	0.93	0.97	0.77	42.15	4.82	0.22	−0.38	−0.35	−0.04	−1.09	0.59
Ecuador	50.1	0.92	0.93	0.97	0.73	51.77	6.56	0.29	−0.44	−0.39	−0.03	−1.57	1.24
Peru	52.23	0.94	0.93	0.97	0.82	25.07	2.89	0.08	−0.23	−0.21	−0.05	−0.91	0.61
Venezuela	53.95	0.95	0.95	0.98	0.80	79.34	3.3	0.23	−0.24	−0.32	−0.02	−1.15	1.05
Tropical Andes	*64.04*	*0.89*	*0.92*	*0.95*	*0.73*	*44.10*	*2.06*	*0.23*	*−0.51*	*−0.24*	*−0.07*	*−1.36*	*0.86*
Burundi	19.67	0.96	0.95	0.97	0.88	100	29.13	0.51	−0.07	0.00	−0.10	0.00	0
D.R. Congo	71.18	0.97	0.95	0.98	0.96	55	3.25	0.12	−0.09	−0.27	−0.07	0.00	0.17
Ethiopia	14.27	0.96	0.92	0.98	0.91	24.55	6.76	0.58	−0.20	−0.21	−0.22	0.00	0.1
Kenya	7.47	0.97	0.95	0.98	0.95	54.60	5.71	1.2	−0.18	−0.23	−0.18	−0.10	0.27
Malawi	19.79	0.98	0.98	0.98	0.96	88.42	9.8	0.52	−0.13	−0.20	−0.14	0.00	0
Mozambique	43	0.98	0.97	0.98	0.98	19.70	3.71	0.57	−0.13	−0.17	−0.15	0.00	0
Rwanda	18.92	0.95	0.93	0.97	0.86	38.57	27.48	0.54	−0.10	−0.18	−0.09	0.00	0
South Sudan	22.18	0.99	0.97	0.99	1.00	64	1.39	0.39	−0.18	−0.32	−0.17	0.00	0.45
Tanzania	24.3	0.95	0.93	0.97	0.84	60.97	4.66	0.66	−0.14	−0.17	−0.11	−0.23	0.07
Uganda	21.45	0.98	0.96	0.99	0.95	61.61	8.24	0.45	−0.11	−0.18	−0.11	0.00	0.61
Zambia	38.52	0.98	0.97	0.99	1.00	59.66	2.09	0.47	−0.09	−0.16	−0.09	0.00	0.19
Africa Great Lakes	*25.07*	*0.96*	*0.94*	*0.98*	*0.93*	*62.89*	*7.15*	*0.40*	*−0.14*	*−0.27*	*−0.14*	*−0.05*	*0.20*
Cambodia	49.81	0.94	0.88	0.96	0.96	42.53	8.91	0.57	−0.57	−1.79	−0.37	−0.19	1.42
China	13.17	0.92	0.91	0.96	0.79	53.00	11.36	0.45	−0.42	−0.88	−0.14	−1.03	1.54
Lao P.D.R.	69.06	0.95	0.88	0.98	0.95	62.80	8.51	0.54	−0.40	−1.42	−0.21	−0.33	2.09
Myanmar	50.22	0.95	0.90	0.97	0.97	35.38	9.52	0.55	−0.54	−1.20	−0.47	−0.23	0.92
Thailand	31.77	0.95	0.89	0.96	0.96	73.41	9.37	0.81	−0.82	−1.57	−0.72	−0.56	1.19
Vietnam	38.27	0.94	0.87	0.96	0.91	33.68	19.44	0.4	−0.42	−1.45	−0.16	−0.77	0.99
Greater Mekong	*44.03*	*0.94*	*0.89*	*0.97*	*0.91*	*49.21*	*6.34*	*0.53*	*−0.46*	*−1.29*	*−0.19*	*−0.98*	*1.43*

Stark differences in the efforts of neighboring countries at safeguarding key biodiversity areas, such as for Mozambique and Tanzania, are also easily discernible by a non-scientific audience (Figure 5).

To develop the dashboard concept, we used indicators derived from global data to bypass the many obstacles in obtaining consistent and comprehensive regional and national data. While the reporting of these global indicators at the national level provides new and valuable information, there are limitations in using global data to represent national indicators. While advances in remote sensing technology provide an unprecedented opportunity to gain temporally repetitive information [99], currently available estimates of land cover change can differ substantively among sources (i.e., FAO forest resource assessment [100], Global Forest Cover Loss mapping [48], ESA Global Land Cover and GlobCover products [101,102]) and are often coarse in resolution (342 km^2 and 2592 km^2 respectively for the deforestation and freshwater provision datasets used here). With regard to protected areas, global datasets tend to omit recently decreed areas and often fail to capture important differences in on-the-ground management. Similarly, the use of global distributions and Red List categories may not adequately reflect the conservation status of a given species in a particular region, and because global distribution data is only available for terrestrial vertebrates, the results do not reflect the status of many other components of biodiversity.

The shortcomings of indicators derived from global data can be addressed by integrating nationally and locally derived data into the final dashboard design. Despite the numerous obstacles, data are being generated at a variety of scales deemed useful by survey respondents, both through governmental efforts [92,96,103,104], regional consortiums [92,93,94,105,106,107], and site-specific projects [108,109]. The next stage of the biodiversity indicators dashboard development focuses on building an effective data sharing mechanism the promotes shared identifiers to link data from these different sources [83], digital architecture to coordinate data flow and ensure data ownership, and promoting consent and trust among data contributors. With continued development, we envision the biodiversity indicators dashboard as an interactive, web-accessible platform that can facilitate national reporting towards biodiversity targets while allowing for the integration of localized data to support the type of site-scale monitoring deemed important to survey participants. The dashboard framework has also been designed so that over time, the indicators discussed here can be supplemented with other metrics capturing complementary aspects of each of state, pressures, responses and benefits (e.g.

population trends, agricultural intensity, environmental legislation and additional ecosystem service measures).

By serving these data as a web-accessible dashboard, we can put information on status and trends in biodiversity within easy access of users and organizations from all sectors and backgrounds, and facilitate more informed decision-making, enable exploration of patterns among variables, and support the tracking of progress towards conservation goals. The maps presented here and information contained in Table 3 can be depicted in a dashboard format via various means, including as mapped values (Figures 5 and 6), mouse-over boxes displaying the numerical values associated with those maps, tabular data summaries by nation or indicator accessible via interactive menus, and charts of trends that users could generate either by region (Figure 7) or nation.

In agreeing to the Aichi Targets, the nations of the world implicitly committed to developing the data necessary to effectively monitor progress towards meeting biodiversity goals. The challenges in reporting towards those goals are many, but we believe the dashboard approach, as outlined here, provides a valuable framework that can facilitate and advance the type of reporting required by the conservation community. Starting with global indicators and expanding to incorporate additional national and site-scale data identified as important by conservation practitioners on the ground, the biodiversity indicators dashboard can serve as a tool to track progress towards Aichi Targets, support national monitoring and reporting, and inform outcome-based policy-making in the realm of conservation.

Acknowledgments

We thank A Chenery, D. Stanwell-Smith, J Tordoff, A. Ajagbe, C. Josse, the Biodiversity Indicator Partnership, and ASEAN Biodiversity Center for assistance on consulatation workshops; A. S. L. Rodrigues, T. Evans, R. Hoft, and C. Josse for comments on the indicator data and manuscript preparation; the survey respondents; and the many thousands of individuals and organisations who contribute to the compilation of IUCN Red List assessments, the World Database on Protected Areas, and the indentification and documentation of Important Bird and Biodiverseity Areas and Alliance for Zero Extinction Sites.

Author Contributions

Conceived and designed the experiments: TMB XH BEY ASL. Performed the experiments: XH ASL TMB BEY. Analyzed the data: XH ASL SHMB FWL. Contributed reagents/materials/analysis tools: PB SHMB FWL MCH WRT. Wrote the paper: XH RLS BEY TMB ASL PB SHMB FWL HH MCH WRT.

References

1. Holland GJ, Alexander JSA, Johnson P, Arnold AH, Halley M, et al. (2012) Conservation cornerstones: Capitalising on the endeavours of long-term monitoring projects. Biological Conservation 145: 95–101.

2. Lindenmayer DB, Likens GE (2010) The science and application of ecological monitoring. Biological Conservation 143: 1317–1328.

3. Lovett GM, Burns DA, Driscoll CT, Jenkins JC, Mitchell MJ, et al. (2007) Who needs environmental monitoring? Frontiers in Ecology and the Environment 5: 253–260.

4. Diversity CoB (2010) Conference of Parties Decision X/2.Strategic Plan for Biodiversity 2011–2020.

5. United Nations Millennium Development Goals. Available: http://www.un.org/millenniumgoals/. Accessed 2013 January 22.

6. Rio +20 United Nations Conference on Sustainable Development. Available: http://www.uncsd2012.org/. Accessed 2013 January 22.

7. The Ramsar Convention on Wetlands. Available: http://www.ramsar.org/. Accessed 2013 January 22.

8. Convention on Migratory Species. Available: http://www.cms.int/. Accessed 2013 January 22.

9. Intergovernmental Platform on Biodiversity and Ecosystem Services. Available: www.ipbes.net. Accessed 2013 January 22.

10. Walpole M, Almond REA, Besancon C, Butchart SHM, Campbell-Lendrum D, et al. (2009) Tracking Progress Toward the 2010 Biodiversity Target and Beyond. Science 325: 1503–1504.

11. Walpole M, Herkenrath P (2009) Measuring progress towards the 2010 target and beyond: an international expert workshop on biodiversity indicators. Oryx 43: 461.

12. Biodiversity Indicator Partnership (BIP). (2008) Available: http://www.bipindicators.net/.

13. Group on Earth Observations Biodiversity Observations Network (GEO BON). (2008) Available: www.earthobservations.org. Accessed 2013 Janurary 22.

14. Sparks TH, Butchart SHM, Balmford A, Bennun L, Stanwell-Smith D, et al. (2011) Linked indicator sets for addressing biodiversity loss. Oryx 45: 411–419.

15. Pimm SL (1991) The balance of nature?: ecological issues in the conservation of species and communities. Chicago: University of Chicago Press. xiii, 434 p. p.

16. Ferraro PJ, Pattanayak SK (2006) Money for nothing? A call for empirical evaluation of biodiversity conservation investments. Plos Biology 4: 482–488.

17. Pattanayak SK, Wunder S, Ferraro PJ (2010) Show Me the Money: Do Payments Supply Environmental Services in Developing Countries? Review of Environmental Economics and Policy 4: 254–274.

18. Sutherland WJ, Pullin AS, Dolman PM, Knight TM (2004) The need for evidence-based conservation. Trends in Ecology & Evolution 19: 305–308.

19. Ferraro PJ (2009) Counterfactual thinking and impact evaluation in environmental policy. In: Birnbaum M, Mickwitz P, editors. Environmental program and policy evaluation. pp. 75–84.

20. Roberts D, Moritz T (2011) A framework for publishing primary biodiversity data INTRODUCTION. Bmc Bioinformatics 12.

21. Butchart SHM, Baillie JEM, Chenery AM, Collen B, Gregory RD, et al. (2010) National Indicators Show Biodiversity Progress Response. Science 329: 900–901.

22. Butchart SHM, Walpole M, Collen B, van Strien A, Scharlemann JPW, et al. (2010) Global Biodiversity: Indicators of Recent Declines. Science 328: 1164–1168.

23. Bubb P, Chenery A, Herkenrath P, Kapos V, Mapendembe A, et al. (2011) National indicators, monitoring and reporting for the strategic plan for biodiversity 2011–2020. In: UNEP-WCMC, editor. A Report by UNEP-WCMC with IUCN and ECNC for the UK Department for Environment Food and Rural Affairs (Defra) ed. Cambridge.

24. Chandra A, Idrisova A (2011) Convention on Biological Diversity: a review of national challenges and opportunities for implementation. Biodiversity and Conservation 20: 3295–3316.

25. MacArthur Foundation (2011) Conservation & Sustainable Development: international programs strategic framework 2011–2020. Chicago, IL: MacArthur Foundation.

26. Myers N, Mittermeier RA, Mittermeier CG, da Fonseca GAB, Kent J (2000) Biodiversity hotspots for conservation priorities. Nature 403: 853–858.

27. Lehner B, Verdin K, Jarvis A (2008) New Global Hydrography Derived From Spaceborne Elevation Data. Eos, Transactions American Geophysical Union 89: 93–94.

28. HydroSHEDS WWF, USGS (2011) http://www.worldwildlife.org/hydrosheds for general information; http://hydrosheds.cr.usgs.gov for data download and techinical information. March 20, 2008 ed.

29. FAO GeoNetwork (2009) Hydrological Basins derived from Hydrosheds.

30. Eckerson WW, ebrary Inc. (2011) Performance dashboards measuring, monitoring, and managing your business. 2nd ed. Hoboken, N.J.: Wiley.

31. Few S, Safari Tech Books Online. (2006) Information dashboard design. Sebastopol, Calif.: O'Reilly.

32. The World Bank. Available: World Bank Open Data. Accessed 16 February 2013.

33. CNN Money. Available: Accessed 26 February 2014.

34. Convention on International Trade in Endangered Species of Wild Fauna and Flora. Available: CITES trade data dashboards. Accessed 16 February 2014.

35. Chalakkal K (2010) 3. Trade in Endangered Species. Yearbook of International Environmental Law 21: 301–310.

36. Tessier E, Pothin K, Chabanet P, Fleury P-g, Bissery C, et al. (2011) Définition d'Indicateurs de performance et d'un Tableau de bord pour la Réserve Naturelle Marine de La Réunion (RNMR) (Definition of performance indicators and a dashboard to the Natural Marine Reserve of La Reunion (RNMR)). Ifremer, Plouzane, France.

37. Clua E, Beliaeff B, Chauvet C, David G, Ferraris J, et al. (2005) Towards multidisciplinary indicator dashboards for coral reef fisheries management. Aquatic living resources 18: 199–213.

38. Carr ER, Wingard PM, Yorty SC, Thompson MC, Jensen NK, et al. (2007) Applying DPSIR to sustainable development. International Journal of Sustainable Development and World Ecology 14: 543–555.

39. European Environmental Agency. (2007) Available: The DPSIR framework used by the EEA. Accessed Jan 2 2013.

40. Mace GM, Baillie JEM (2007) The 2010 biodiversity indicators: Challenges for science and policy. Conservation Biology 21: 1406–1413.

41. Maxim L, Spangenberg JH, O'Connor M (2009) An analysis of risks for biodiversity under the DPSIR framework. Ecological Economics 69: 12–23.

42. Tscherning K, Helming K, Krippner B, Sieber S, Paloma SGY (2012) Does research applying the DPSIR framework support decision making? Land Use Policy 29: 102–110.

43. UNEP-WCMC (2009) International Expert Workshop on the 2010 Biodiversity Indicators and Post-2010 Indicator Development. Cambridge, UK: UNEP-WCMC.

44. Convention on Biological Diversity (2011) Report of the Ad Hoc Technical Expert Group on Indicators for the Strategic Plan for Biodiversity 2011–2020. High Wycombe, United Kingdom.

45. ESRI (2011) ArcGIS Desktop: Release 10. Redlands, CA: Environmental Systems Research Institute.

46. R Core Team (2013) R: A language and environment for statistical computing.. Vienna, Austria: R Foundation for Statistical Computing.

47. Hansen MC, Stehman SV, Potapov P (2010) Global Forest Monitoring Project. In: Geographic Information Science Center of Excellence tSDSU, editor.

48. Hansen MC, Stehman SV, Potapov PV (2010) Quantification of global gross forest cover loss. Proceedings of the National Academy of Sciences of the United States of America 107: 8650–8655.

49. NASA's Earth Observing System. Available: MODIS: Moderate-resolution Imaging Spectroradiometer. Accessed 2013 June 17.

50. NASA, USGS. Available: LANDSAT Missions. Accessed 2013 June 17.

51. Asner GP, Clark JK, Mascaro J, Galindo Garcia GA, Chadwick KD, et al. (2012) High-resolution mapping of forest carbon stocks in the Colombian Amazon. Biogeosciences 9: 2683–2696.

52. Asner GP, Knapp DE, Balaji A, Paez-Acosta G (2009) Automated mapping of tropical deforestation and forest degradation: CLASlite. Journal of Applied Remote Sensing 3.

53. Asner GP, Powell GVN, Mascaro J, Knapp DE, Clark JK, et al. (2010) High-resolution forest carbon stocks and emissions in the Amazon. Proceedings of the National Academy of Sciences of the United States of America 107: 16738–16742.

54. Butchart SHM, Akcakaya HR, Chanson J, Baillie JEM, Collen B, et al. (2007) Improvements to the Red List Index. Plos One 2.

55. Butchart SHM, Stattersfield AJ, Baillie J, Bennun LA, Stuart SN, et al. (2005) Using Red List Indices to measure progress towards the 2010 target and beyond. Philosophical Transactions of the Royal Society B-Biological Sciences 360: 255–268.

56. Butchart SHM, Stattersfield AJ, Bennun LA, Shutes SM, Akcakaya HR, et al. (2004) Measuring global trends in the status of biodiversity: Red list indices for birds. Plos Biology 2: 2294–2304.

57. Hoffmann M, Hilton-Taylor C, Angulo A, Bohm M, Brooks TM, et al. (2010) The Impact of Conservation on the Status of the World's Vertebrates. Science 330: 1503–1509.

58. IUCN (2010) IUCN Red List of Threatened Species.

59. Eken G, Bennun L, Brooks TM, Darwall W, Fishpool LDC, et al. (2004) Key biodiversity areas as site conservation targets. Bioscience 54: 1110–1118.

60. Butchart SHM, Scharlemann JPW, Evans MI, Quader S, Arico S, et al. (2012) Protecting Important Sites for Biodiversity Contributes to Meeting Global Conservation Targets. Plos One 7.

61. IUCN, UNEP (2010) The World Database on Protected Areas (WDPA). In: UNEP-WCMC, editor. Cambridge, UK.

62. BirdLife Important Bird Area (IBA) Programme (2011) Sites- Important Bird Areas (IBAs). BirdLife International.

63. Extinction AfZ (2010) 2010 AZE Update.

64. Ricketts TH, Dinerstein E, Boucher T, Brooks TM, Butchart SHM, et al. (2005) Pinpointing and preventing imminent extinctions. Proceedings of the National Academy of Sciences of the United States of America 102: 18497–18501.

65. Larsen FW, Londono-Murcia MC, Turner WR (2011) Global priorities for conservation of threatened species, carbon storage, and freshwater services: scope for synergy? Conservation Letters 4: 355–363.

66. Larsen FW, Turner WR, Brooks TM (2012) Conserving Critical Sites for Biodiversity Provides Disproportionate Benefits to People. Plos One 7.

67. Alcamo J, Doll P, Henrichs T, Kaspar F, Lehner B, et al. (2003) Development and testing of the WaterGAP 2 global model of water use and availability. Hydrological Sciences Journal-Journal Des Sciences Hydrologiques 48: 317–337.

68. U.S.Geological Survey (2000) HYDRO1k Elevation Derivative Database In: Sci. CfEROa, editor. Sioux Falls, SD, USA.

69. Oak Ridge National Laboratory (2006) Landscan Global Population Database Oak Ridge, TN.

70. Balmford A, Rodrigues ASL, Walpole M, ten Brink P, Kettunen M, et al. (2008) Review on the economics of biodiversity loss: scoping the science. European Commission.

71. Brauman KA, Daily GC, Duarte TK, Mooney HA (2007) The nature and value of ecosystem services: An overview highlighting hydrologic services. Annual Review of Environment and Resources. pp. 67–98.

72. Bruijnzeel LA (2004) Hydrological functions of tropical forests: not seeing the soil for the trees? Agriculture Ecosystems & Environment 104: 185–228.

73. Dudley N, Stolton S (2003) Running Pure: The importance of forest protected areas to drinking water. World Bank/WWF Alliance for Forest Conservation and Sustainable Use.

74. Hampton SE, Strasser CA, Tewksbury JJ, Gram WK, Budden AE, et al. (2013) Big data and the future of ecology. Frontiers in Ecology and the Environment 11: 156–162.

75. Kelling S, Hochachka WM, Fink D, Riedewald M, Caruana R, et al. (2009) Data-intensive Science: A New Paradigm for Biodiversity Studies. Bioscience 59: 613–620.

76. Danielsen F, Mendoza MM, Alviola P, Balete DS, Enghoff M, et al. (2003) Biodiversity monitoring in developing countries: what are we trying to achieve? Oryx 37: 407–409.

77. Danielsen F, Mendoza MM, Tagtag A, Alviola P, Balete DS, et al. (2003) On participatory biodiversity monitoring and its applicability - a reply to Yoccoz, et al. and Rodriguez. Oryx 37: 412–412.

78. Rodriguez JP (2003) Challenges and opportunities for surveying and monitoring tropical biodiversity - a response to Danielsen, et al. Oryx 37: 411–411.

79. Yoccoz NG, Nichols JD, Boulinier T (2001) Monitoring of biological diversity in space and time. Trends in Ecology & Evolution 16: 446–453.

80. Yoccoz NG, Nichols JD, Boulinier T (2003) Monitoring of biological diversity - a response to Danielsen, et al. Oryx 37: 410–410.

81. Jones JPG, Collen B, Atkinson G, Baxter PWJ, Bubb P, et al. (2011) The Why, What, and How of Global Biodiversity Indicators Beyond the 2010 Target. Conservation Biology 25: 450–457.

82. Bertzky M, Stoll-Kleemann S (2009) Multi-level discrepancies with sharing data on protected areas: What we have and what we need for the global village. Journal of Environmental Management 90: 8–24.
83. Page RDM (2008) Biodiversity informatics: the challenge of linking data and the role of shared identifiers. Briefings in Bioinformatics 9: 345–354.
84. Scholes RJ, Mace GM, Turner W, Geller GN, Jurgens N, et al. (2008) Ecology - Toward a global biodiversity observing system. Science 321: 1044–1045.
85. Chavan VS, Ingwersen P (2009) Towards a data publishing framework for primary biodiversity data: challenges and potentials for the biodiversity informatics community. Bmc Bioinformatics 10.
86. Editorial (2005) Let data speak to data. Nature 438: 531–531.
87. Magurran AE, Baillie SR, Buckland ST, Dick JM, Elston DA, et al. (2010) Long-term datasets in biodiversity research and monitoring: assessing change in ecological communities through time. Trends in Ecology & Evolution 25: 574–582.
88. Rands MRW, Adams WM, Bennun L, Butchart SHM, Clements A, et al. (2010) Biodiversity Conservation: Challenges Beyond 2010. Science 329: 1298–1303.
89. Rudd MA, Beazley KF, Cooke SJ, Fleishman E, Lane DE, et al. (2011) Generation of priority research questions to inform conservation policy and management at a national level. Conservation Biology 25: 476–484.
90. Biala K, Conde S, Delbaera B, Jones-Walters L, Torre-Marin A (2012) Streamlining European biodiversity indicators 2020: building a future on lessons learnt from the SEBI 2010 process. Copenhagen, Danmark: European Environment Agency.
91. Chenery A, Plumpton H, Brown C, Walpole M (2013) Aichi targets passport. Cambridge, UK: Biodiversity Indicators Partnership.
92. ASEAN Center for Biodiversity (ACB). (2005) Available: http://www.aseanbiodiversity.org/. Accessed 2013 Janurary 22.
93. Red Amazónica de Información Socioambiental Geoffeferenciada (RAISG). (1996) Available: Amazonian Network of Georeferenced Socio-Environmental Information, http://raisg.socioambiental.org/. Accessed 2013 Janurary 22.
94. Streamlining European Biodiversity Indicators (SEBI). Available: http://biodiversity.europa.eu/topics/sebi-indicators. Accessed 2013 Janurary 22.
95. Conservation of Arctic Flora and Fauna (CAFF). Available: Accessed 2013 Janurary 22.
96. Uganda National Biodiversity Data Bank. Available: Accessed 2013 Janurary 22..
97. Stuart SN, Chanson JS, Cox NA, Young BE, Rodrigues ASL, et al. (2004) Status and trends of amphibian declines and extinctions worldwide. Science 306: 1783–1786.
98. Young BE, Lips KR, Reaser JK, Ibanez R, Salas AW, et al. (2001) Population declines and priorities for amphibian conservation in Latin America. Conservation Biology 15: 1213–1223.
99. Turner W, Spector S, Gardiner N, Fladeland M, Sterling E, et al. (2003) Remote sensing for biodiversity science and conservation. Trends in Ecology & Evolution 18: 306–314.
100. UN-FAO (2010) Global Forest Resources Assessment. FOOD AND AGRI-CULTURE ORGANIZATION OF THE UNITED NATIONS,.
101. European Space Agency (ESA) (2010) GlobCover 2009 (Global Land Cover Map). 21st December 2010 ed.
102. European Space Agency (ESA) (2008) Global Land Cover Product (2005–06).
103. National Ecological Observatory Network. (2010)Available: NEON, www.neoninc.org. Accessed 2013 Janurary 22.
104. Hopkin M (2006) Ecology - Spying on nature. Nature 444: 420–421.
105. The Circumpolar Biodiversity Monitoring Programme (CBMP). Available: http://caff.is/monitoring. Accessed 2013 Janurary 22.
106. Natural Heritage Program. (1974) Available: http://www.natureserve.org/visitLocal/. Accessed 2013 Janurary 22.
107. Stein BA, Kutner LS, Adams JS (2000) Precious heritage: the status of biodiversity in the United States. New York: Oxford University Press. xxv, 399 p. p.
108. TEAM Network. Available: Tropical Ecology Assessment & Monitoring Network: Early Warning System for Nature. http://www.teamnetwork.org/. Accessed 2013 Janurary 22.
109. WCS Conservation Support. Available: Status and Impact Monitoring, http://wcslivinglandscapes.com/WhatWeDo/StatusImpactMonitoring.aspx. Accessed 2013 Janurary 22.

13

Ecosystem Services and Opportunity Costs Shift Spatial Priorities for Conserving Forest Biodiversity

Matthias Schröter[1,2]*, **Graciela M. Rusch**[2], **David N. Barton**[2], **Stefan Blumentrath**[2], **Björn Nordén**[2]

1 Environmental Systems Analysis Group, Wageningen University, Wageningen, The Netherlands, **2** Norwegian Institute for Nature Research (NINA), Trondheim/Oslo, Norway

Abstract

Inclusion of spatially explicit information on ecosystem services in conservation planning is a fairly new practice. This study analyses how the incorporation of ecosystem services as conservation features can affect conservation of forest biodiversity and how different opportunity cost constraints can change spatial priorities for conservation. We created spatially explicit cost-effective conservation scenarios for 59 forest biodiversity features and five ecosystem services in the county of Telemark (Norway) with the help of the heuristic optimisation planning software, Marxan with Zones. We combined a mix of conservation instruments where forestry is either completely (non-use zone) or partially restricted (partial use zone). Opportunity costs were measured in terms of foregone timber harvest, an important provisioning service in Telemark. Including a number of ecosystem services shifted priority conservation sites compared to a case where only biodiversity was considered, and increased the area of both the partial (+36.2%) and the non-use zone (+3.2%). Furthermore, opportunity costs increased (+6.6%), which suggests that ecosystem services may not be a side-benefit of biodiversity conservation in this area. Opportunity cost levels were systematically changed to analyse their effect on spatial conservation priorities. Conservation of biodiversity and ecosystem services trades off against timber harvest. Currently designated nature reserves and landscape protection areas achieve a very low proportion (9.1%) of the conservation targets we set in our scenario, which illustrates the high importance given to timber production at present. A trade-off curve indicated that large marginal increases in conservation target achievement are possible when the budget for conservation is increased. Forty percent of the maximum hypothetical opportunity costs would yield an average conservation target achievement of 79%.

Editor: Runguo Zang, Chinese Academy of Forestry, China

Funding: MS received funding from the European Research Council under grant 263027 ('Ecospace'), and from the Research Council of Norway (Yggdrasil grant). GMR, DNB and SB received funding from the European Union's Seventh Programme for research, technological development and demonstration under grant agreement No. 244065 (POLICYMIX project (http://policymix.nina.no)). BN was supported by the project ECOSERVICE (Approaches for integrated assessment of forest ecosystem services under large scale bioenergy utilization) financed by the Research Council of Norway, RCN grant no 233641/E50. The funders had no role in study design, data collection and analysis, decision to publish, or preparation of the manuscript.

Competing Interests: The authors have declared that no competing interests exist.

* Email: matthias.schroter@wur.nl

Introduction

The ecosystem services (ES) concept comprises multiple contributions of ecosystems to human well-being [1], and has increasingly been used to raise awareness about the benefits that people derive from ecosystems [2,3]. Considering ES when making decisions about the use of ecosystems could provide additional, anthropocentric arguments to support either management aimed at sustainable use of ecosystems or biodiversity conservation [4]. However, there is a still unresolved debate about to what extent components of biodiversity correspond with ES provision [4–7] and about the extent to which considering ES in decision making matches with biodiversity conservation objectives. Furthermore, accounting for ES within conservation planning is a fairly new practice [8–11]. In a conservation decision-making context, ES can be seen as benefits of conservation (many cultural and regulating services), or in the case of extractive provisioning services as an opportunity cost of conservation since their use may become restricted [8]. Trade-offs between extractive provisioning services, such as clear-cutting timber harvest, and other ES [12]

and biodiversity protection [11,13–15] require choices to be made on whether and where to protect an area. However, certain management systems restrict timber production and might thus allow for a synergy between an extractive provisioning service and other ecosystem services [16,17] as well as some aspects of biodiversity conservation [17–21]. This leads to the crucial question within cost-effective conservation planning on how multiple-use areas, in which extractive exploitation is restricted, can potentially contribute to biodiversity conservation [22–24]. Cost-effective conservation means minimizing opportunity costs in terms of foregone commodity production [25]. As some conservation targets are compatible with a certain level of use [26], and since the opportunity costs of setting aside areas can be potentially high, a mixture of fully protected areas and areas allowing for partial use is likely to render more cost-effective and less conflictive conservation solutions, and may open opportunities for overall higher levels of biodiversity protection.

Spatial considerations play an integral role in the assessment of cost-effectiveness of conservation as the spatial configurations of important habitats [27] and of opportunity costs of conservation do not necessarily coincide [28]. A 'policyscape' may be defined as the spatial configuration of a mix of policy instruments [29], which aims at conserving biodiversity and ES at an aggregated spatial level. This framing suggests that there is an optimal and complementary spatial allocation of different types of instruments across a space containing all possible combinations of conservation values and opportunity costs within a study area. The spatial configuration of the policyscape has important practical implications for decision-making. For instance, it opens opportunities to evaluate disproportionate economic burdens between administrative units.

In this study, we suggest ways of creating cost-effective policyscapes. We address a mix of instruments that combines non-use (strict protection) and partial use (forestry restricted) for the conservation of forest biodiversity and ES in the county of Telemark (Norway). Indicators of the state of forests in Norway show a decline of certain species populations, especially of species associated to old-growth forest and species whose habitats are threatened by current forestry practices [14,30]. There is a need to modify and adapt current conservation policies to help secure portions of unprotected biodiversity as well as to halt the processes that lead to forest biodiversity loss [14,30,31]. One approach is to increase protected forest areas in Norway, particularly within the ecological zones that are most favourable for forestry production [31]. Currently, new nature reserves in Norway are mostly implemented through voluntary forest conservation schemes that are based on a negotiation between forest owners and conservation authorities in Norway [32]. The exploration of different policyscapes for conservation of biodiversity and ES can give guidance to support such conservation efforts.

We used the conservation planning software Marxan with Zones [33] for near-optimal selection of areas for cost-effective policyscapes on a county level. Some experience has been developed in applying (earlier versions of) Marxan to conservation optimisation with ES [8,11,34–37]. However, to our knowledge integrated targeting of both biodiversity and multiple ES within a policyscape with different levels of protection has not been systematically studied before.

We addressed the following specific questions. We first analysed how optimal conservation outcomes differ between two scenarios that either take into account biodiversity only (scenario 1) or a set of ES next to biodiversity (scenario 2). The outcome of both scenarios was measured in terms of spatial configuration, area protected, conservation target achievement, and opportunity costs.

Second, we assessed the trade-off between biodiversity and ES conservation goals and timber production. We analysed this relationship by constructing a production possibility frontier (PPF) [25], while considering timber production as a private good and the sum of biodiversity features and other ES as public goods. These public goods are either spared from timber production in the case of full protection or jointly produced with the private good in the case of partial protection. We compared current instrument targeting, i.e. the effectiveness of current reserves to achieve conservation targets set in our scenario, to a 'benchmark' defined as the cost-effective policyscape traced by the PPF [38,39].

Third, we explored differences in conservation burden across administrative units. For this purpose, we calculated the expected opportunity costs of an optimal conservation outcome for each municipality in Telemark. Significant differences in conservation burden across municipalities would suggest potential efficiency gains with concomitant distributional consequences, which could justify considering the introduction of a conservation instrument such as ecological fiscal transfer schemes [40].

Methods

Study area

Telemark is a county in southern Norway with an area of 15,300 km^2 and a population of about 170,000 people [41], concentrated mainly in the south-eastern part of the county. The climate varies across the region with temperate conditions in the south-east (Skien, average temperature January $-4.0°C$, July $16.0°C$, 855 mm annual precipitation) and alpine conditions in the north-west (Vinje, January $-9.0°C$, July $11.0°C$, 1035 mm) [42]. The southern part of Telemark is mainly covered by forest exploited by forestry activities as well as by large inland lakes, with few towns and a small agricultural area (247 km^2, i.e. about 1.6% of the land area) [41]. The northern part is characterised by treeless alpine highland plateaus covered by bogs, fens and heathlands [43]. The forest landscape in Telemark is characterized by coniferous and boreal deciduous forest [43]. Important forest ecosystem services include moose hunting, free range sheep grazing and timber production [44]. In addition, forests of Telemark sequester and store considerable amounts of carbon, prevent snow slides and provide opportunities for recreational hiking and residential amenities [44]. In 2011, conservation areas comprised about 5.1% in national parks, 4.6% in landscape protection areas (both types cover mainly highland plateaus), as well as 1.7% in nature reserves [41]. As a result of forestry activities, the status of biodiversity in forests of Telemark shows relatively low values compared to other ecosystems and regions within Norway [14]. We conducted our analysis for the forest area within Telemark, however, as forest field mapping is lacking for a small south-eastern part of the county [45], this area was excluded from the analysis.

Principle of Marxan with Zones

Marxan with Zones [33] builds on a heuristic optimisation algorithm that incorporates key principles of systematic conservation planning, including comprehensiveness, cost-effectiveness and compactness of the reserve system [46]. Marxan with Zones enables to consider zones with different levels of protection and thus spatial differences in costs, thereby allowing for planning and evaluation of policyscapes that include full and partial protection. Marxan with Zones requires a series of inputs, which are specified below.

Data input Marxan with Zones

ES and biodiversity features and conservation targets. Depending on the scenario, a total of 59 (scenario 1, biodiversity) and 64 (scenario 2, biodiversity and ES) input features were used, respectively. Table 1 provides an overview of all features.

We included five key ES of importance within a Norwegian context for which spatial models have been developed (Table 1) [44]. We specifically included biodiversity features which are characteristic of old-growth, largely undisturbed forest and which are not maintained under current commercial forestry practices. We included 40 types of old-growth forest, to a large extent remnants of previously high-graded forests, occurring across a range of vegetation zones, climate zones and productivity conditions to represent the ecological variability across the county (Text S1 for details). Six proposed forest corridors of national importance that connect existing reserves [47] were included as a spatial indicator of conditions enabling species dispersal between habitats [48]. Forest habitats of particular conservation importance on a national level in Norway [49,50] were also included. Three classes of priority habitats for conservation (very important,

Table 1. Features, targets, fraction of targets to be achieved across the two zones (non-use and partial use), and contribution (effectiveness) of the partial zone in meeting respective targets.

Feature name	Feature target (%)	Fraction non-use (%)	Fraction partial (%) (contribution in %)
Wilderness-like areas (ES)	100	100	0 (0)
Recreational hiking (ES)	20	50	50 (100)
Carbon storage (ES)	10	50	50 (25)
Carbon sequestration (ES)	5.57	75	25 (25)
Snow slide protection (ES)	100	0	100 (100)
Old-growth forest types (40)	50	75	25 (50)
Corridors (6)	50	50	50 (50)
Priority habitats for conservation (very important)	100	100	0 (0)
Priority habitats for conservation (important)	100	100	0 (0)
Priority habitats for conservation (locally important)	50	100	0 (0)
Hollow deciduous trees	100	100	0 (0)
Late successional forests with deciduous trees	100	100	0 (0)
Logs	100	100	0 (0)
Old trees	100	100	0 (0)
Rich ground vegetation	100	100	0 (0)
Snags	100	100	0 (0)
Trees with nutrient-rich bark	100	100	0 (0)
Trees with pendant lichens	100	100	0 (0)
Recently burned forest	100	100	0 (0)
Stream gorges	100	100	0 (0)

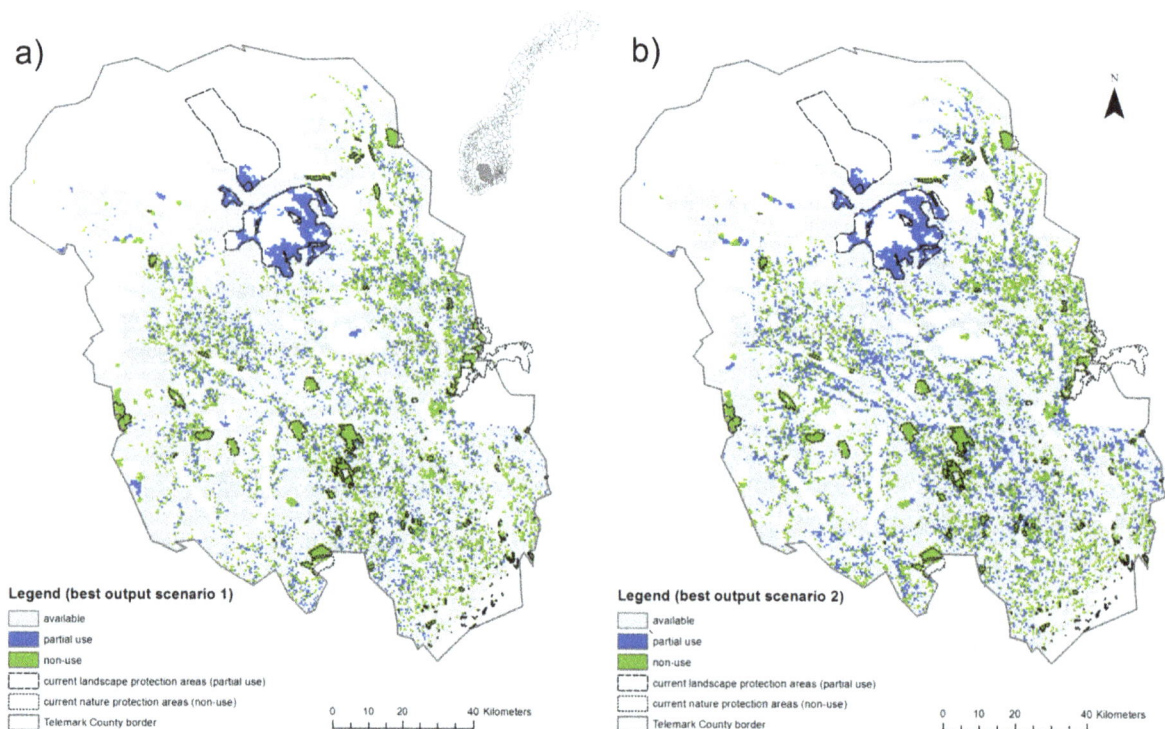

Figure 1. Best solution of the reserve network for scenario 1 (a) and scenario 2 (b). Scenario 1, considers biodiversity conservation criteria only; scenario 2, both biodiversity and ecosystem services criteria. Grey, areas available for forestry; blue, areas in the partial use zone and green, areas in the non-use zone. Current reserves are demarcated in dashed lines. Map inlay shows the location of Telemark within Norway (grey).

important and locally important) were taken from the Norwegian Environmental Agency's database (Naturbase) [51]. In addition, we included ten types of important forest habitats (Table 1) from a Norwegian Forest and Landscape Institute database (MiS) [52].

Marxan with Zones requires setting quantitative conservation feature targets that reflect the proportion of the abundance of each feature to be protected. Targets were based on expert judgments and, wherever possible, on interpretation of policy documents (Table 1, and Text S1 for details). In order to verify targets an expert workshop was organised (Text S1). Written consent to participate in this study was obtained from the participants of the expert workshop.

The policyscape – definition of zones, zone targets, zone contributions. Two types of area protection were included in our analysis, namely a non-use and a partial use zone. Non-use referred to nature reserves, where forestry is completely restricted, i.e. "use" refers to forestry activities. The partial use zone was an 'umbrella' zone covering three different current forms of protection where forestry is partially restricted, namely landscape protection areas, mountain forest ('fjellskog'), and outdoor recreation areas ('friluftsområder') (Text S1). All current nature reserves in Telemark [51] were 'locked-in' as non-use zones and all current landscape protection areas were 'locked-in' as partial use zones, which means that spatial units overlapping with these areas were selected for the respective zone in each run of Marxan.

Marxan with Zones allows for distribution of the targets across zones. Zone targets were defined according to an own expert judgement about how well the non-use and partial use areas were compatible with the persistence of the respective feature. Zone targets (Table 1) were discussed, reviewed and as far as possible confirmed during the expert workshop (Text S1).

Marxan with Zones allows for differentiation of how effective zones are in order to achieve targets (zone contribution). We considered the effectiveness of partial use areas as "the relative contribution of actions to realizing conservation objectives" [53]. We assumed that non-use areas are fully effective to reach the targets of all features (100% contribution). There is growing, but yet inconclusive knowledge on how low impact logging could be compatible with biodiversity conservation [17–21,54,55]. This means that effectiveness of partial use areas is highly uncertain, and may affect features differently. Zone contributions were thus discussed and as far as possible confirmed during the expert workshop. In a sensitivity analysis we further explored the consequences of changing the zone contribution of the partial use zone (File S1).

Planning units. The forest area in Telemark was divided into 43.513 grid planning units of 25 ha size (500 m×500 m). This resolution was suitable in terms of time and computing capacity, and considered relevant for land-use planning. Property sizes in Norwegian forests vary widely from as little as 0.1 ha to several hundred hectares [32] and as such are not a good guide to setting the size of the planning unit.

Opportunity costs of conservation. Foregone timber harvest was selected as an indicator of opportunity costs of conservation since harvest activities are constrained by different forms of protection [25]. We used a net revenue (stumpage value) forest model to determine opportunity costs (Text S1). In non-use areas opportunity costs were set to 100%, while in partial-use areas, we estimated that restrictions would account for 25% of the stumpage value. This estimate was based on different logging restrictions [56] which ranged from 15% (landscape protection area), to 20% (outdoor recreation area) and 30% (mountain forest).

Analyses

Marxan with Zones was run 20 times with the parameters described above (for further parameter adjustments see Table S1 and Table S2). The software was run for both scenarios to determine the best solution and the selection frequency of each planning unit over all runs, which ranged from 0 (never chosen) to the maximum of 20 (chosen in each run) and indicated importance of a particular planning unit to achieve the overall conservation targets [57]. Marxan with Zones input files, including spatial information on all conservation features, can be found in the supporting information for scenario 1 (File S2) and scenario 2 (File S3).

Comparison of scenarios. We used selection frequency of planning units to determine how the policyscapes of both scenarios differed spatially. Selection frequency of each planning unit to each of the two zones in scenario 1 (biodiversity only) was subtracted from selection frequency in scenario 2 (biodiversity and ES) to determine the difference. To compare the spatial configuration of the policyscapes, we calculated Pearson's correlation coefficient between the selection frequency of each scenario for the partial and the non-use zone. We calculated Cohen's Kappa on the selection frequency of each planning unit as a measure of agreement between the scenarios for each zone. To compare the two scenarios in absolute terms we calculated a number of statistics, including total costs, number of planning units without protection, planning units in the partial and non-use zone and average target achievement.

Trade-off between conservation target achievement and timber harvest. The PPF was identified by running a series of cost constraints for scenario 2. Cost constraints are a restricting condition that defines an upper limit of costs when selecting planning units. We started by running the scenario with no cost constraints and close to 100% average target achievement, and recorded the total unconstrained cost. We then introduced cost constraints at different levels (80%, 60%, 40%, 20%, 10%, 5%, 1%) of the total unconstrained cost in consecutive runs (see Table S4 for parameter details). The value of timber production (horizontal axis in the PPF) was determined as the total sum of stumpage value across all planning units in the study area minus the opportunity cost of the best solution of each run. The vertical axis in the PPF was determined as the average percentage of target achievement for all biodiversity and ES features. To assess the opportunity costs of conservation and the conservation target achievement of the current existing reserve network, we used an overlay analysis (r.stats in GRASS GIS).

Conservation burden across Telemark. To determine the conservation burden among the municipalities in Telemark, the expected opportunity cost for each municipality was calculated as the summed expected value of opportunity costs:

$$C_e = \sum \begin{cases} \frac{f_{n_i}}{20} * C_i \text{ if } n_i \geq f_{p_i} \\ \frac{f_{p_i}}{20} * 0.25 * C_i \text{ if } f_{n_i} < f_{p_i} \end{cases} \text{ for } i = 1, \dots, 43513 \quad (1)$$

where C_e is the expected opportunity cost, f_{ni} is the selection frequency of non-use areas for planning unit i, f_{pi} is the selection frequency of partial use areas for planning unit i and C_i is the opportunity cost of planning unit i. The denominator 20 stands for the number of runs in our case and the factor 0.25 specifies the harvest restriction in the partial use areas.

This analysis was run on scenario 2 with first, no cost constraint and, second, a medium cost constraint of 60% of the maximum costs needed to achieve close to 100% of the average targets. Opportunity costs per municipality were determined with zonal statistics in ArcMap for both expected opportunity cost layers and for current reserves. Municipalities were ranked according to relative opportunity costs, i.e. opportunity costs divided by municipal forest area. To analyse the spatial shift of the conservation burden across municipalities, Spearman's rank correlation coefficient was calculated between the current situation and the unconstrained scenario, as well as between the 60% cost constraint and the unconstrained scenario.

Results

Incorporating ES in the policyscape for biodiversity conservation

Incorporating ES into the policyscape changed the absolute sum of area in the two zones, the opportunity costs (Table 2) as well as the spatial configuration of the policyscape (Figures 1 and 2). When considering ES, the sum of partial use areas increased by 36.2% and the sum of non-use-areas by 3.2% compared to the scenario that only considered biodiversity. Opportunity costs were 6.6% higher in scenario 2 than in scenario 1. As an illustration of a policyscape, Figure 1 shows the best solution per scenario for scenario 1 (a) and scenario 2 (b). Selection frequencies of planning units for both scenarios can be found in Figure S1. The differences in selection frequencies are shown in Figure 2 for the partial (a) and non-use zone (b). A positive difference means higher selection frequency in the policyscape of scenario 2 than in scenario 1, while a negative difference indicates a lower selection frequency in the policyscape of scenario 2 than in scenario 1. Comparison of the spatial configuration of the policyscapes of both scenarios led to the following results. Pearson's correlation coefficient between selection frequencies of sites in the non-use zone was r = 0.90, while for the partial use zone, it was r = 0.58. This indicates that relatively larger differences can be expected in the partial use zone than in the non-use zone when ES were considered, which partly rests upon the fact that ES can, in contrast to most of the biodiversity features in this study, partly be protected in this zone. Cohen's Kappa statistics was K = 0.577 (sig≤0.0001) for the non-use zone and K = 0.398 (sig ≤0.0001) for the partial use zone. These results imply 'moderate agreement' in non-use and 'fair agreement' in partial use zone, respectively [58], which supports the observation of a relatively larger agreement between non-use areas in the different spatial configurations of the policyscapes.

Trade-offs between conservation and timber production: Production possibility frontier (PPF)

The PPF shows a concave curve representing the trade-off between timber production and conservation of biodiversity and non-forestry related ES (Figure 3). Creating a reserve network to achieve the conservation targets comes at a cost of timber production. The marginal increase in conservation target achievement is initially high when the current constraint on conservation cost is relaxed (i.e. moving left in Figure 3). This marginal conservation gain decreases more rapidly after having passed a cost constraint of about 40% of the total cost required to achieve 100% of the overall conservation target. The current policyscape (black square) lies under the PPF curve, meaning that more cost-effective policyscape configurations than the current one are possible. This means that higher average target achievement could hypothetically be realised at current levels of timber production, or that the same target could be achieved at lower costs. At the same time, the location of the current policyscape shows a strong preference of decisions towards timber production. Consequently, the conservation targets we set in our scenario are

barely met by the current reserve system (average achievement 9.1%).

While Figure 3 shows the average target achievement of all 64 features, Figure 4 shows the development of target achievement along changing opportunity cost constraints for single, exemplary features (for all features see Table S3). Some features meet high targets at low (20%) cost constraints (carbon sequestration and one type of low productive old-growth forest), which means that these features did not constrain the solution to a high degree. Some conservation features decreased at higher rates than the average (e.g., one type of high productive forest and recently burned forest). Such features are more costly to be comprehensively conserved in a compact reserve network.

Distribution of the conservation burden of cost-effective conservation areas

The creation of the policyscape for conservation of biodiversity and ES formed the basis for determining the 'conservation burden' across municipalities of Telemark (Table 3, spatial distribution in Figure S2). Conservation burdens across municipalities were slightly shifted in a (hypothetical) scenario with no cost constraint in which approximately 100% of the average target could be achieved compared to the current situation. For instance, while Porsgrunn ranked 6th in terms of the conservation burden of the current policyscape, it ranked 1st in the policyscape of with no cost constraints. The Spearman's correlation coefficient between the current situation and the scenario with unconstrained costs was r = 0.67. The Spearman's correlation coefficient between a 60% cost constraint and the unconstrained scenario was r = 0.46, which means that spatial priorities for conservation, and thus conservation burdens, shift with the level of the opportunity cost constraint.

Discussion

A policyscape for conservation of biodiversity and ES

The use of spatial planning tools that simultaneously consider conservation of biodiversity and ES in a cost-effective way is a fairly new approach, facilitated by recent advancement in computational science. This approach provides a range of opportunities [8,10], but still presents challenges in operationalization. Considering ES within biodiversity conservation could be beneficial for incorporating sustainable use of ecosystems [4] when achieving overall conservation goals in land use planning (land sharing), compared to a land use strategy that separates conservation and provision of ES (land sparing). A land sharing principle was included in our study in the partial use zone, which partly allows for the development of synergies between ES, biodiversity and timber production and which complements strict protection zones in policyscapes analysed in this study. In our analysis, we had to rely on expert-backed assumptions when describing the effects of the partial use zone on conservation. This is due to inconclusive knowledge on how restricted logging affects particular elements of biodiversity and ES [17–21,55]. Our study suggests that in forest areas of Telemark the configuration of a policyscape for conservation changes when ES were incorporated (scenario 2) compared to considering only biodiversity conservation criteria (scenario 1). This change was twofold and included a change in total areas assigned to the two protection zones and a change in the spatial configuration of selected sites. Including ES resulted in an increase in the size of the reserve network, a result that is in line with previous studies [8,36] in that when optimizing for cost-effective representation of conservation targets more areas with lower opportunity costs that contribute to target achievements of both biodiversity and ES are selected.

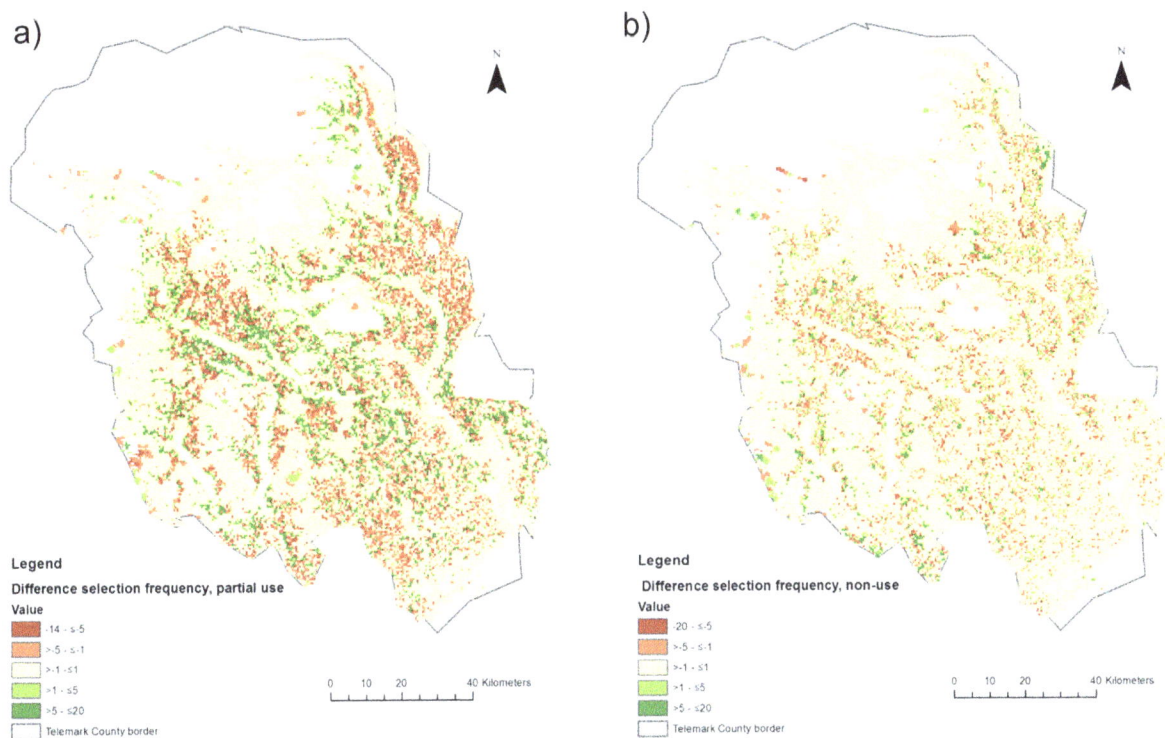

Figure 2. Differences in selection frequency of sites for partial (a) and non-use (b) areas. The maps show the difference of scenario 2 (biodiversity and ES features) versus scenario 1 (biodiversity only). A positive difference means higher selection frequency in scenario 2 than in scenario 1.

In contrast to former studies, we used different levels of protection, which enabled us to also specify the change in the policyscape in terms of the spatial distribution of the different zones. Including ES resulted in a strong increase in partial use areas (+36.2%), which was partly expected due to the fact that ES features were considered to be protected for a relatively larger proportion in partial use zones than biodiversity features (Table 1). The difference in spatial configurations of the policyscapes of the two scenarios can partly be explained by relatively low degrees of pairwise spatial overlaps between some ES and the biodiversity features (File S4). It also depends, for instance, on various combinations of biodiversity and ES features on cost-effective sites and proximity of suitable combinations to existing reserves. The difference in spatial configuration leads to different spatial prioritisations of sites to preserve in both zones and thus would have important implications for regional and local decision making.

Trade-off between commercial timber production and conservation of biodiversity and ES

Including ES next to biodiversity into a conservation scenario reflects different values [4,59] and as such could lead to more informed policy decisions. In our conservation scenario we thus treated ES of public interest representing partly intangible values (regulating and cultural services) as conservation features with an own target. While in the ES discourse, ES are often treated as generally beneficial [4], here we shed light on potential specific trade-offs among ES and between ES and biodiversity conservation priorities. We included timber production in our analysis, a provisioning service that contributes to private economic benefits, and assessed the form of the trade-off curve (PPF) between timber production on the one hand and cultural and regulating services and biodiversity on the other. The existence of a trade-off on a system level was expected based on our assumption that outside

Table 2. Summary statistics describing the difference between scenario 1 (considering biodiversity conservation criteria only) and 2 (considering biodiversity and ecosystem services) in terms of opportunity costs, area in the different zones and average conservation target achievement.

Statistics	Scenario 1	Scenario 2	Difference 2 vs. 1 in %
opportunity costs (billion NOK)	1.912	2.038	+6.6
without protection (no. of planning units of 25 ha)	32,183	30,279	−5.9
partial use area (no. of planning units of 25 ha)	4,661	6,349	+36.2
non-use (no. of planning units of 25 ha)	6,669	6,885	+3.2
average conservation target achievement (%)	99.86	99.23	−0.6

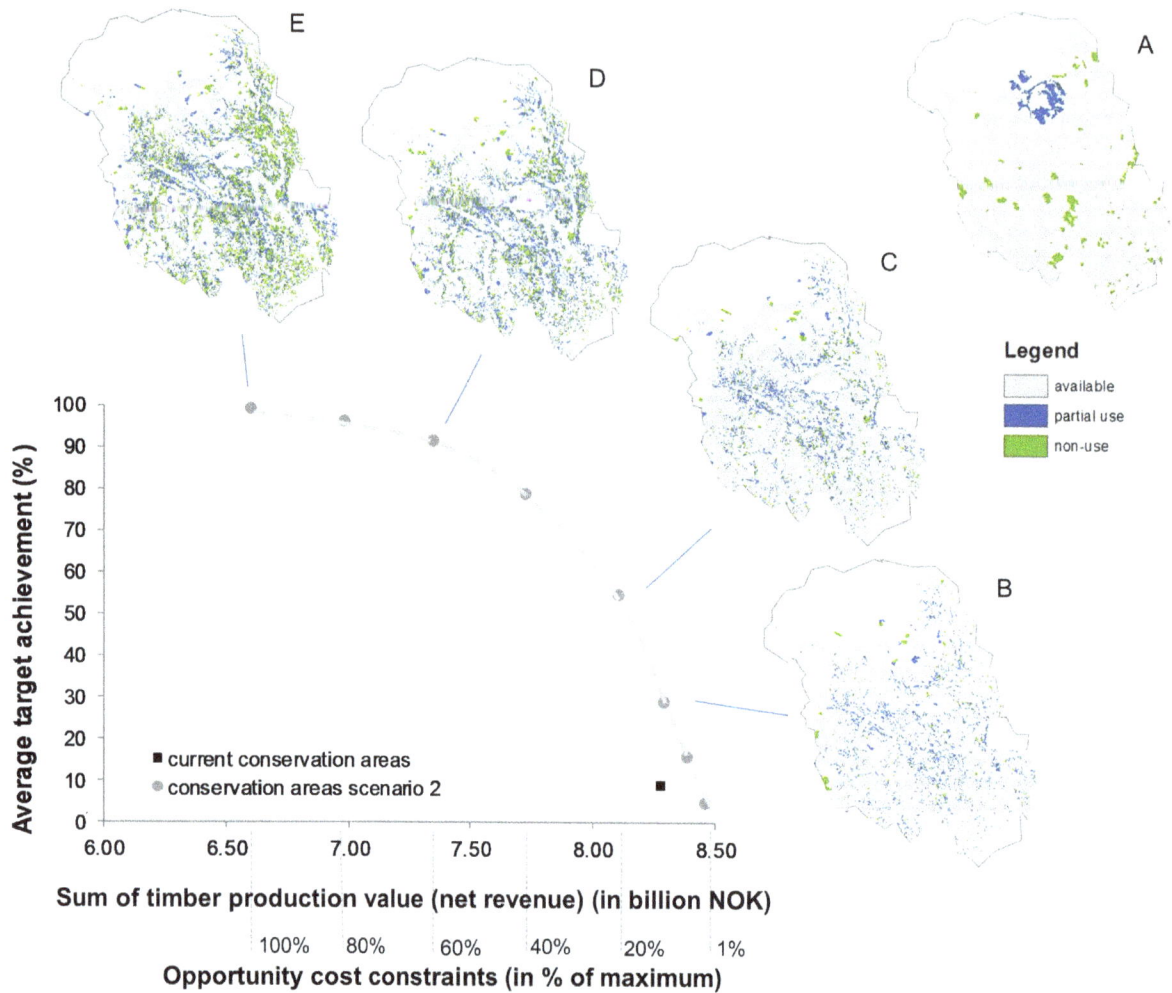

Figure 3. Forest conservation-timber production possibility frontier (PPF). Note that the x-axis (sum of timber production value) starts at 6.00 billion NOK. The maps indicate current reserve network (A) and selected (B–E) available, partial and non-use areas when current reserves are not locked-in. The spatially explicit solutions (policyscapes) are shown as maps on the trade-off between net revenues from timber production and average conservation target achievement, along a range of opportunity costs constraints.

the two conservation zones, elements of biodiversity and ES would not be conserved. This assumption might seem strong, but can be defended by the fact that the dominant form of forest management in Norway is characterised by large-scale clear-cutting [60].

From the PPF, we derive two broad policy conclusions. First, the currently designated nature reserves and landscape protection areas achieved a very low proportion (9.1%) of the conservation targets we set in our scenario. This is partly because the conservation network has not been initially designed to meet the conservation targets we defined in our study. For instance, while attention has been given to rare and threatened forest types [31], we did not assign different conservation targets to the different old-growth forest types, which might in practice be of different importance for forest biodiversity conservation. The result is, however, in agreement with the relatively little forest area that is currently allocated to conservation [31] due to low conservation budgets and conflicts. Further, our findings support the observation of a biased representation of protected areas towards high altitudes and lower opportunity cost areas [61]. This pattern, as well as the under-representation of productive forest in the current conservation network, have also been found for Norway [29,31,62]. Our present scenario was deliberately designed to

include high productive forest, which partly explains the low target achievement of the current conservation network.

Second, the PPF analysis also provides insights for policy-makers regarding balancing private and public interests. It is a societal choice to determine the level of production of either timber or biodiversity and regulating and cultural ES. The PPF illustrates the high importance given to timber production at present. At the same time, it shows that the relationship between gains in conservation and opportunity costs is not linear. This means that high marginal improvements in conservation can be obtained with relatively smaller increases in costs when a low opportunity cost constraint is relaxed. Thus, with relatively little investment, e.g. spending 40% of the maximum opportunity costs, on average 79% of the scenario targets could be achieved under the assumptions applied in this study. However, inspection of the PPF curve also reveals that lowering the cost constraint reduces the probability of achieving conservation targets for certain habitats (e.g. recently burned forests, high productive forests) within the reserve network. In contrast, carbon sequestration reaches high proportions of the target at low cost which indicates that carbon sequestration can be seen as a co-benefit of protecting biodiversity and other ES, assessed at the scale of all prioritised full and partial

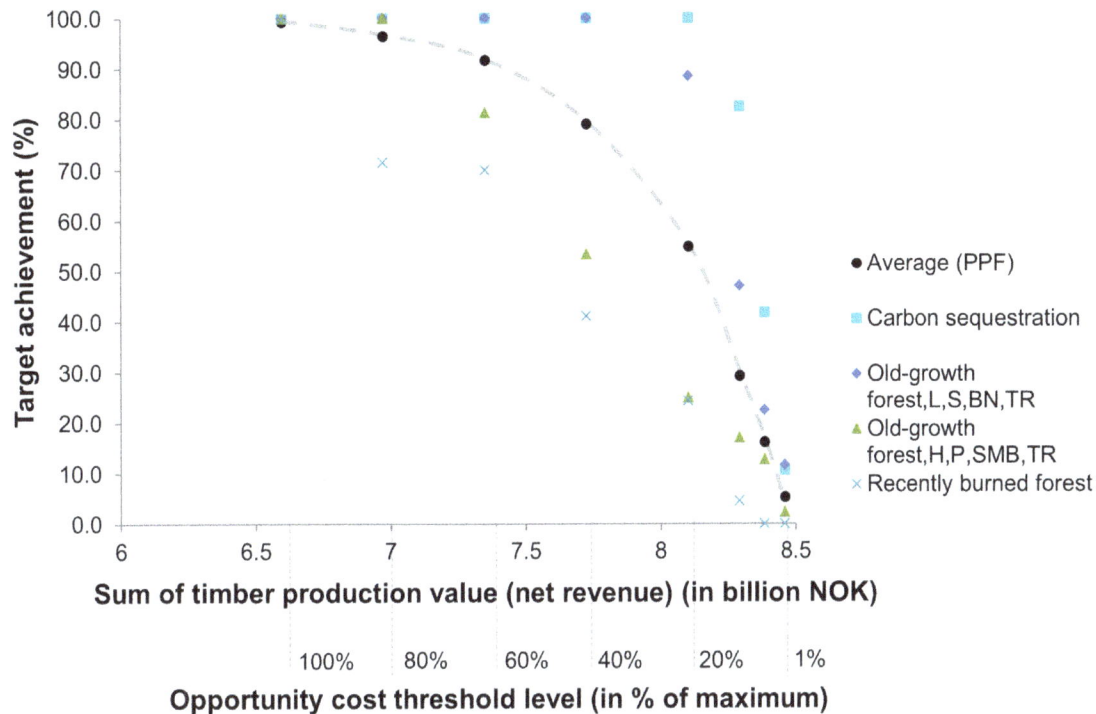

Figure 4. Forest conservation-timber production possibility frontier (PPF) for single, exemplary features. Old-growth forest L, S, BN, TR = impediment and low productivity, spruce dominated, boreonemoral zone, oceanic-inland transition zone. Old-growth forest H, P, SMB, TR = high & very high productivity, pine dominated, South & Mid- boreal zone, oceanic-inland transition zone.

protection areas across the study area. This is the inverse logic of the current international debate (i.e. REDD+), where carbon sequestration is targeted to be protected while (unmeasured) biodiversity is a (hoped for) co-benefit [63], but is in agreement with findings of process-based models in recent studies [17].

Uncertainties in creating the conservation scenario

We encountered several challenges in creating the conservation scenario. The choice of conservation features is a crucial factor that determines the outcome of the site prioritisation. Operationalizing biodiversity conservation requires quantifiable and obtainable indicators [64,65]. Given restrictions on data availability, we believe that our choice of biodiversity surrogates represents a first step for planning the maintenance of biodiversity in Norwegian forest ecosystems.

Despite the "inevitable subjectivity" in setting conservation targets [66], there is some experience in setting targets for biodiversity conservation [65,67]. However, setting explicit targets for ES when determining spatial priorities has seldom been done [68]. Current studies using Marxan for ES conservation have pointed out the need for experimentation, explicitly stated assumptions and expertise in setting targets given the absence of this information [8,11,36,37], particularly because ES targets influence the size of the reserve network [35]. A systematic sensitivity test of target levels was, however, out of scope of this current study. ES targets may vary considerably because alternative means are available for substituting forest ES depending on location. Preferences for recreational hiking can shift outside the forest towards mountainous areas. In some areas, feasible technical substitutes for snow slide prevention by forests are available. Since different interests and values are reflected in ES, a systematic stakeholder involvement could provide more insight on target levels for each conservation feature. In a future

study, sensitivity analyses could be run based on integrated consultation of forest owners. Because Marxan is a regional level policy-support tool its suitability to be used for conservation planning at the property level is restricted. For example, once priority areas have been identified in a regional planning exercise, local authorities in collaboration with the local forest association try to reach agreement with several adjacent property owners [32]. The conservation outcome is the result of multiple negotiations to achieve a single voluntary nature reserve, the final spatial configuration of which does not depend on the result of a near-optimal site prioritisation software. However, Marxan with Zones could be run iteratively on different agreement configurations to show how marginal conservation burden and target achievement are shifted to other locations, for instance when particular forest owners have declined to agree with an area which would in the first place have been prioritised. Scenario analyses in Marxan with Zones could help planners evaluate the cost-effectiveness of local level conservation decisions, in light of the portfolio of other options, instead of negotiating about one or a few sites at a time.

Another uncertainty in conservation planning lies in the underlying opportunity costs [69]. While we did not test this uncertainty in our analysis, we point out that the advent of forest harvesting for bioenergy could be a 'game changer' as it would probably change expected returns to forestry and thus change the spatial distribution of opportunity costs.

Partial use areas, where extractive resource exploitation is restricted, can host high levels of biodiversity [17,18,26,55] and integrating such areas in conservation networks may improve overall conservation effectiveness by reducing costs and conflicts between different economic activities [53]. A combination of non-use and partial-use areas may also help to maintain a landscape that enables processes such as colonization and forest succession, particularly if non-use areas are small. The determination of

Table 3. Absolute and relative conservation burden per municipality in the current situation, with a cost constraint of 60% and with no cost constraint.

Municipality	Forest area in planning units (km²)	Total opportunity costs[1] (million NOK)			Relative opportunity costs (NOK per km² forest area)			Total additional burden[2] (million NOK)	Relative additional burden[2] (NOK/km²)	Ranks relative opportunity costs (NOK/km²) (largest to smallest)			
		Current	60% cost con-straint	No cost con-straint	Current	60% cost con-straint	No cost con-straint			Current	60% cost con-straint	No cost con-straint	Addi-tional burden
Porsgrunn	175.5	3.2	30.0	60.0	18,457	170,677	341,874	56.8	323,417	6	4	1	1
Bamble	318.8	13.4	110.9	105.0	42,011	347,859	329,518	91.6	287,507	3	3	2	3
Notodden	818.8	14.7	39.0	254.3	17,945	47,655	310,558	239.6	292,613	7	15	3	2
Sauherad	316.5	13.1	52.3	95.3	41,404	165,259	301,208	82.2	259,804	4	5	4	4
Kragerø	341.8	5.8	15.3	88.4	16,979	44,866	258,777	82.6	241,797	8	16	5	5
Nome	412.8	54.2	150.9	105.7	131,320	365,660	256,155	51.5	124,835	1	2	6	10
Drangedal	1050.8	26.2	63.4	265.6	24,970	60,353	252,817	239.4	227,846	5	12	7	7
Bø	239.3	1.5	25.3	56.8	6,122	105,791	237,347	55.3	231,225	14	6	8	6
Skien	582.5	7.0	54.8	138.1	11,996	94,157	237,166	131.2	225,169	11	7	9	8
Siljan	130.5	9.7	57.6	22.2	74,231	441,457	169,989	12.5	95,758	2	1	10	13
Nissedal	855.3	12.5	57.5	110.6	14,630	67,191	129,361	98.1	114,731	9	11	11	11
Tokke	712.0	0.4	57.5	89.6	527	80,752	125,781	89.2	125,255	18	9	12	9
Kviteseid	662.8	0.9	56.6	72.7	1,433	85,374	109,691	71.7	108,258	17	8	13	12
Tinn	880.0	10.4	44.2	91.9	11852	50,223	104,416	81.5	92,564	12	14	14	15
Fyresdal	1147.5	5.0	31.2	113.7	4336	27,190	99,098	108.7	94,762	15	18	15	14
Hjartdal	649.8	5.6	36.2	57.0	8584	55,755	87,702	51.4	79,118	13	13	16	16
Seljord	577.3	1.4	25.2	47.0	2442	43,680	81,465	45.6	79,023	16	17	17	17
Vinje	939.8	12.2	64.4	68.9	13025	68,492	73,306	56.6	60,281	10	10	18	18

[1]Calculated as foregone net stumpage value.
[2]Calculated as the difference between opportunity costs for the case of no cost constraint and the current opportunity costs.

effectiveness of zones to achieve a conservation target has been identified as a major challenge for conservation planning given limited availability of knowledge [34,70]. For the sake of simplicity, we assumed a 100% effectiveness to protect biodiversity and ES for the non-use zone, given that this is the highest level of protection that can be achieved. We acknowledge, however, that considering a lower effectiveness level would most probably have led to a larger network of protected areas. In face of natural dynamics and disturbances, effectiveness of conservation areas should be monitored in terms of representativeness and persistence [66,71]. Because of the uncertainty about the probability of biodiversity persistence in the partial use zone, we explored the consequences of changing the zone contribution for the partial use zone as input in Marxan for 46 biodiversity features (Figure A1 in File S1). With a lower zone contribution, Marxan with Zones tended to select more planning units in the non-use and less in the partial use zone despite considerably lower opportunity costs of the partial use zone; a result that is in line with the findings by Makino et al. [53] in a study of partial protection zones in a marine environment in Fiji.

Assessing regional level implications of site prioritisation for ES and biodiversity: conservation burden

Decision-making about cost-effective area allocation to protect biodiversity and ES takes place at various levels of governance that may justify the design of new policy instruments. Cost-effective selection of priority sites for conservation can guide measures directed to land owners, for instance by consultation with land owners of selected priority sites on whether they would agree to convert forestry land into voluntary nature reserves, as is the current practice in Norway [32]. While land owners voluntarily entering conservation agreements in Norway are generally compensated for their private opportunity cost [32] accumulated loss of forestry activity in a region may, on the one hand, result in unequal public conservation burdens, particularly across different municipalities. Large protected areas may lead to foregone business opportunities, loss of tax income and additional expenses for municipal governments. On the other hand, protected areas can also provide positive externalities to others, through tourism opportunities and protection of biodiversity more generally. Local governments can be compensated for costs of conservation by state-to-municipal "ecological fiscal transfers" [40], an instrument that has been implemented in Brazil and Portugal, and is currently being considered in several European countries [72]. Ecological fiscal transfers have mainly been based on compensation scaled by area. Proposals to scale ecological fiscal transfers using criteria reflecting the effectiveness of conservation in a municipality have generally been limited by the availability of spatially representative data on biodiversity. We have demonstrated how the creation of cost-effective policyscapes could be used to determine distributional effects of additional conservation efforts.

Conclusions

Marxan with Zones provides a spatially explicit way to include different types of ES and biodiversity conservation criteria to study a policyscape for cost-effective conservation. We have shown that, in the case of Telemark, including a number of ES shifts priority sites for conservation and increases the area of both a partial use and a non-use zone, compared to a situation where only biodiversity conservation criteria are considered. Conservation of a number of regulating and cultural ES leads to additional conservation efforts, in terms of higher opportunity costs and a larger area protected. We show how carbon sequestration can be viewed as a side-benefit of the protection of other ES and biodiversity in the context of the current Kyoto-based setting of national targets. This is opposite to current thinking about biodiversity as a hoped-for side-benefit of climate mitigation measures under REDD+. The current conservation situation in Telemark clearly prioritises timber production against the protection of biodiversity and ES, and relatively large marginal increases in conservation target achievement could be reached with modest additional investments in terms of compensation for foregone timber production. Our analysis also shows potential differences in conservation burden among municipalities in Telemark, opening the debate on policy instruments such as ecological fiscal transfers that support county-level cost-effective conservation through stimulation of local conservation efforts.

Although the integration of partial use areas into conservation could provide opportunities to increase cost-effectiveness in conservation, significant work is needed to document effectiveness of different levels of protection on particular conservation features. Despite the high level of uncertainty, a policy mix of conservation measures appears to have the potential to contribute to address the complexity of cost-effective conservation problems.

Conservation targets for many aspects of biodiversity and especially ES are currently absent. Conservation planning could be better operationalised with more knowledge on stakeholder preferences about the importance of ES as well as with more ecological knowledge on area size needed to preserve a biodiversity feature.

Our analysis should not be understood as a concrete regional management plan, but rather as an exploratory analysis to provide insights about the current forest conservation situation, about which conservation outcomes could be achieved at which opportunity costs levels. In practice, selection of protected areas is often based on other criteria and motives than cost-effective, comprehensive site prioritisation [61]. Decision makers could use the results of this study to encourage disproportional conservation efforts at local level that achieve cost-effective, near optimal solutions to a conservation problem of multiple biodiversity and ES features. For this to happen, decision makers have to decide to what extent additional information, such as mapping of ES, could be integrated into land-use planning [73]. We have shown how ES mapping, conservation benchmarking and distributional impact analysis using conservation planning tools could inform decision-making and support compensation of land owners' and local governments' conservation efforts.

Acknowledgments

We thank Lars Hein and Lucie Vermeulen for useful comments on an earlier draft. We thank the advisory board members of the POLICYMIX project as well as all participants of the expert workshop held in Oslo on 7 May 2014. We express our sincere gratitude to two reviewers whose constructive remarks helped to improve the paper.

Author Contributions

Conceived and designed the experiments: MS GMR DNB SB BN. Performed the experiments: MS SB. Analyzed the data: MS GMR DNB SB BN. Contributed reagents/materials/analysis tools: MS GMR DNB SB BN. Contributed to the writing of the manuscript: MS GMR DNB SB BN.

References

1. Haines-Young R, Potschin M (2010) Proposal for a Common International Classification of Ecosystem Goods and Services (CICES) for integrated environmental and economic accounting. New York, USA: European Environment Agency.

2. Carpenter SR, Mooney HA, Agard J, Capistrano D, DeFries RS, et al. (2009) Science for managing ecosystem services: Beyond the Millennium Ecosystem Assessment. Proc Natl Acad Sci U S A 106: 1305–1312.

3. Larigauderie A, Prieur-Richard A-H, Mace GM, Lonsdale M, Mooney HA, et al. (2012) Biodiversity and ecosystem services science for a sustainable planet: the DIVERSITAS vision for 2012–20. Curr Opin Environ Sustain 4: 101–105.

4. Schröter M, van der Zanden EH, van Oudenhoven APE, Remme RP, Serna-Chavez HM, et al. (2014) Ecosystem services as a contested concept: a synthesis of critique and counter-arguments. Conserv Lett: doi:10.1111/conl.12091.

5. Mace GM, Norris K, Fitter AH (2012) Biodiversity and ecosystem services: a multilayered relationship. Trends Ecol Evol 27: 19–26.

6. Faith DP (2012) Common ground for biodiversity and ecosystem services: The "partial protection" challenge [v1; ref status: indexed, http://f1000r.es/QPrmmt]. F1000Research 2012 1.

7. Reyers B, Polasky S, Tallis H, Mooney HA, Larigauderie A (2012) Finding Common Ground for Biodiversity and Ecosystem Services. Bioscience 62: 503–507.

8. Chan KMA, Hoshizaki L, Klinkenberg B (2011) Ecosystem services in conservation planning: Targeted benefits vs. co-benefits or costs? PLoS ONE 6: e24378.

9. Egoh BN, Paracchini ML, Zulian G, Schägner JP, Bidoglio G (2014) Exploring restoration options for habitats, species and ecosystem services in the European Union. J Appl Ecol 51: 899–908.

10. Egoh B, Rouget M, Reyers B, Knight AT, Cowling RM, et al. (2007) Integrating ecosystem services into conservation assessments: A review. Ecol Econ 63: 714–721.

11. Chan KMA, Shaw MR, Cameron DR, Underwood EC, Daily GC (2006) Conservation Planning for Ecosystem Services. PLoS Biol 4: e379.

12. Bennett EM, Peterson GD, Gordon LJ (2009) Understanding relationships among multiple ecosystem services. Ecol Lett 12: 1394–1404.

13. Faith D (in press) Ecosystem services can promote conservation over conversion and protect local biodiversity, but these local win-wins can be a regional disaster. Aust Zool: doi:10.7882/az.2014.7031.

14. Certain G, Skarpaas O, Bjerke J-W, Framstad E, Lindholm M, et al. (2011) The Nature Index: A General Framework for Synthesizing Knowledge on the State of Biodiversity. PLoS ONE 6: e18930.

15. Anderson BJ, Armsworth PR, Eigenbrod F, Thomas CD, Gillings S, et al. (2009) Spatial covariance between biodiversity and other ecosystem service priorities. J Appl Ecol 46: 888–896.

16. Chhatre A, Agrawal A (2009) Trade-offs and synergies between carbon storage and livelihood benefits from forest commons. Proceedings of the National Academy of Sciences 106: 17667–17670.

17. Pichancourt J-B, Firn J, Chadès I, Martin TG (2014) Growing biodiverse carbon-rich forests. Global Change Biology 20: 382–393.

18. Persha L, Agrawal A, Chhatre A (2011) Social and Ecological Synergy: Local Rulemaking, Forest Livelihoods, and Biodiversity Conservation. Science 331: 1606–1608.

19. Götmark F (2013) Habitat management alternatives for conservation forests in the temperate zone: Review, synthesis, and implications. For Ecol Manag 306: 292–307.

20. Lindenmayer DB, Franklin JF, Fischer J (2006) General management principles and a checklist of strategies to guide forest biodiversity conservation. Biol Conserv 131: 433–445.

21. Nordén B, Paltto H, Claesson C, Götmark F (2012) Partial cutting can enhance epiphyte conservation in temperate oak-rich forests. For Ecol Manag 270: 35–44.

22. Bengtsson J, Angelstam P, Elmqvist T, Emanuelsson U, Folke C, et al. (2003) Reserves, Resilience and Dynamic Landscapes. Ambio 32: 389–396.

23. Hanski I (2011) Habitat Loss, the Dynamics of Biodiversity, and a Perspective on Conservation. Ambio 40: 248–255.

24. Daily GC, Ceballos G, Pacheco J, Suzán G, Sánchez-Azofeifa A (2003) Countryside Biogeography of Neotropical Mammals: Conservation Opportunities in Agricultural Landscapes of Costa Rica. Conserv Biol 17: 1814–1826.

25. Hauer G, Cumming S, Schmiegelow F, Adamowicz W, Weber M, et al. (2010) Tradeoffs between forestry resource and conservation values under alternate policy regimes: A spatial analysis of the western Canadian boreal plains. Ecol Model 221: 2590–2603.

26. Eigenbrod F, Anderson BJ, Armsworth PR, Heinemeyer A, Jackson SF, et al. (2009) Ecosystem service benefits of contrasting conservation strategies in a human-dominated region. Proc R Soc Biol Sci Ser B 276: 2903–2911.

27. Nalle DJ, Montgomery CA, Arthur JL, Polasky S, Schumaker NH (2004) Modeling joint production of wildlife and timber. J Environ Econ Manage 48: 997–1017.

28. Murdoch W, Polasky S, Wilson KA, Possingham HP, Kareiva P, et al. (2007) Maximizing return on investment in conservation. Biol Conserv 139: 375–388.

29. Barton DN, Blumentrath S, Rusch GM (2013) Policyscape–A Spatially Explicit Evaluation of Conservation in a Policy Mix for Biodiversity Conservation in Norway. Soc Nat Resour 26: 1185–1201.

30. Kålås JA, Viken Å, Henriksen S, Skjelseth S (2010) The 2010 Norwegian Red List for Species. Trondheim: Norwegian Biodiversity Information Centre Norway.

31. Framstad E, Økland B, Bendiksen E, Bakkestuen V, Blom H, et al. (2002) Evaluering av skogvernet i Norge. NINA fagrapport. Oslo: NINA. 146.

32. Skjeggedal T, Gundersen V, Harvold KA, Vistad OI (2010) Frivillig vern av skog - evaluering av arbeidsformen (Norwegian: voluntary forest conservation - an evaluation of the approach). Oslo: NIBR/NINA.

33. Watts ME, Ball IR, Stewart RS, Klein CJ, Wilson K, et al. (2009) Marxan with Zones: Software for optimal conservation based land- and sea-use zoning. Environ Model Softw 24: 1513–1521.

34. Reyers B, O'Farrell P, Nel J, Wilson K (2012) Expanding the conservation toolbox: conservation planning of multifunctional landscapes. Landsc Ecol 27: 1121–1134.

35. Egoh BN, Reyers B, Rouget M, Richardson DM (2011) Identifying priority areas for ecosystem service management in South African grasslands. J Environ Manag 92: 1642–1650.

36. Egoh BN, Reyers B, Carwardine J, Bode M, O'Farrell PJ, et al. (2010) Safeguarding Biodiversity and Ecosystem Services in the Little Karoo, South Africa. Conserv Biol 24: 1021–1030.

37. Izquierdo AE, Clark ML (2012) Spatial Analysis of Conservation Priorities Based on Ecosystem Services in the Atlantic Forest Region of Misiones, Argentina. Forests 3: 764–786.

38. Barton DN, Faith DP, Rusch GM, Acevedo H, Paniagua L, et al. (2009) Environmental service payments: Evaluating biodiversity conservation trade-offs and cost-efficiency in the Osa Conservation Area, Costa Rica. J Environ Manag 90: 901–911.

39. Rusch GM, Barton DN, Bernasconi P, Ramos-Bendaña Z, Pinto R (2013) Best practice guidelines for assessing effectiveness of instruments on biodiversity conservation and ecosystem services provision. POLICYMIX Technical Brief, Issue No 7. Oslo: NINA.

40. Ring I, May P, Loureiro W, Santos R, Antunes P, et al. (2011) Ecological Fiscal Transfers. In: I Ring and C Schröter-Schlaack, editors. Instrument Mixes for Biodiversity Policies POLICYMIX Report, Issue No 2/2011. Leipzig: Helmholtz Centre for Environmental Research - UFZ. 98–118.

41. SSB (2012) Statistisk årbok 2012. Oslo, Kongsvinger: SSB.

42. Meteorological Institute (2012) Monthly normal values. Oslo: Meteorological Institute.

43. Moen A (1999) National Atlas of Norway: Vegetation. Hønefoss: Norwegian Mapping Authority.

44. Schröter M, Barton DN, Remme RP, Hein L (2014) Accounting for capacity and flow of ecosystem services: A conceptual model and a case study for Telemark, Norway. Ecol Indic 36: 539–551.

45. NFLI (2010) Arealressurskart AR5. Ås: National Forest and Landscape Institute (NFLI, Skog og Landskap).

46. Margules CR, Sarkar S (2007) Systematic conservation planning. Cambridge [etc.]: Cambridge University Press. 270 p.

47. Framstad E, Blumentrath S, Erikstad I., Bakkestuen V (2012) Naturfaglig evaluering av norske verneområder: Verneområdenes funksjon som økologisk nettverk og toleranse for klimaendringer. NINA rapport 888. Oslo: NINA.

48. Opdam P, Steingröver E, van Rooij S (2006) Ecological networks: A spatial concept for multi-actor planning of sustainable landscapes. Landsc Urban Plann 75: 322–332.

49. Directorate for Nature Management (2007) Kartlegging av naturtyper - Verdisetting av biologisk mangfold. DN håndbok 13. Trondheim: Directorate for Nature Management.

50. Gjerde I, Baumann C (2002) Miljøregistrering i skog - biologisk mangfold. Ås: Norwegian Institute for Forest Research.

51. Norwegian Environmental Agency (2013) Naturbase. Trondheim: Norwegian Environmental Agency (Miljødirektoratet).

52. NFLI (2013) Miljøregistrering i skog (Register of important forest habitats), Telemark. Ås: National Forest and Landscape Institute (NFLI, Skog og Landskap).

53. Makino A, Klein CJ, Beger M, Jupiter SD, Possingham HP (2013) Incorporating Conservation Zone Effectiveness for Protecting Biodiversity in Marine Planning. PLoS ONE 8: e78986.

54. Faith D (1995) Biodiversity and regional sustainability analysis. Canberra: CSIRO.

55. Fisher B, Edwards DP, Larsen TH, Ansell FA, Hsu WW, et al. (2011) Cost-effective conservation: calculating biodiversity and logging trade-offs in Southeast Asia. Conserv Lett 4: 443–450.

56. Søgaard G, Eriksen R, Astrup R, Øyen B-H (2012) Effekter av ulike miljøhensyn på tilgjengelig skogareal og volum i norske skoger. Rapport fra Skog og Landskap 2/2012. Ås: National Forest and Landscape Institute (NFLI, Skog og Landskap).

57. Wilson KA, Possingham HP, Martin TG, Grantham HS (2010) Key Concepts. In: J. A Ardron, H. P Possingham and C. J Klein, editors. Marxan Good Practices Handbook, Version 2. Victoria, BC, Canada: Pacific Marine Analysis and Research Association. 18–23.

58. Landis JR, Koch GG (1977) The Measurement of Observer Agreement for Categorical Data. Biometrics 33: 159–174.

59. Chan KMA, Satterfield T, Goldstein J (2012) Rethinking ecosystem services to better address and navigate cultural values. Ecol Econ 74: 8–18.

60. Granhus A (2014) Miljøhensyn ved hogst og skogkultur. In: S Tomter and L. S Dalen, editors. Bærekraftig skogbruk i Norge. Ås: Norwegian Forest and Landscape Institute. 90–99.

61. Joppa LN, Pfaff A (2009) High and Far: Biases in the Location of Protected Areas. PLoS ONE 4: e8273.

62. Framstad E, Blindheim T, Erikstad L, Thingstad PG, Sloreid SE (2010) Naturfaglig evaluering av norske verneområder. NINA rapport 535. Oslo: NINA.

63. Venter O, Laurance WF, Iwamura T, Wilson KA, Fuller RA, et al. (2009) Harnessing Carbon Payments to Protect Biodiversity. Science 326: 1368.

64. Sarkar S, Margules C (2002) Operationalizing biodiversity for conservation planning. J Biosci 27: 299–308.

65. Carwardine J, Klein CJ, Wilson KA, Pressey RL, Possingham HP (2009) Hitting the target and missing the point: target-based conservation planning in context. Conserv Lett 2: 4–11.

66. Margules CR, Pressey RL (2000) Systematic conservation planning. Nature 405: 243–253.

67. Margules CR, Pressey RL, Williams PH (2002) Representing biodiversity: Data and procedures for identifying priority areas for conservation. J Biosci 27: 309–326.

68. Luck GW, Chan KMA, Klein CJ (2012) Identifying spatial priorities for protecting ecosystem services [v1; ref status: indexed, http://f1000r.es/T0yHOY].F1000Research 2012.

69. Carwardine J, Wilson KA, Hajkowicz SA, Smith RJ, Klein CJ, et al. (2010) Conservation Planning when Costs Are Uncertain. Conserv Biol 24: 1529–1537.

70. Chape S, Harrison J, Spalding M, Lysenko I (2005) Measuring the extent and effectiveness of protected areas as an indicator for meeting global biodiversity targets. Philos Trans R Soc Lond B Biol Sci 360: 443–455.

71. Gaston KJ, Charman K, Jackson SF, Armsworth PR, Bonn A, et al. (2006) The ecological effectiveness of protected areas: The United Kingdom. Biol Conserv 132: 76–87.

72. Schröter-Schlaack C, Ring I, Koellner T, Santos R, Antunes P, et al. (2014) Intergovernmental fiscal transfers to support local conservation action in Europe. The German Journal of Economic Geography 58: 98–114.

73. European Commission (2014) Mapping and Assessment of Ecosystems and their Services. Technical Report – 2014-080. Brussels.

14

Global Invasion of *Lantana camara*: Has the Climatic Niche Been Conserved across Continents?

Estefany Goncalves[1,2⌓], **Ileana Herrera**[1*⌓], **Milén Duarte**[3,4], **Ramiro O. Bustamante**[3,4*],
Margarita Lampo[1*], **Grisel Velásquez**[1], **Gyan P. Sharma**[5], **Shaenandhoa García-Rangel**[2]

1 Centro de Ecología, Instituto Venezolano de Investigaciones Científicas, Caracas, Venezuela, 2 Departamento de Estudios Ambientales, Universidad Simón Bolívar, Caracas, Venezuela, 3 Departamento Cs. Ecológicas, Facultad de Ciencias, Universidad de Chile, Santiago, Chile, 4 Instituto de Ecología y Biodiversidad, Facultad de Ciencias, Universidad de Chile, Santiago, Chile, 5 Department of Environmental Studies, University of Delhi, Delhi, India

Abstract

Lantana camara, a native plant from tropical America, is considered one of the most harmful invasive species worldwide. Several studies have identified potentially invasible areas under scenarios of global change, on the assumption that niche is conserved during the invasion process. Recent studies, however, suggest that many invasive plants do not conserve their niches. Using Principal Components Analyses (PCA), we tested the hypothesis of niche conservatism for *L. camara* by comparing its native niche in South America with its expressed niche in Africa, Australia and India. Using MaxEnt, the estimated niche for the native region was projected onto each invaded region to generate potential distributions there. Our results demonstrate that while *L. camara* occupied subsets of its original native niche in Africa and Australia, in India its niche shifted significantly. There, 34% of the occurrences were detected in warmer habitats nonexistent in its native range. The estimated niche for India was also projected onto Africa and Australia to identify other vulnerable areas predicted from the observed niche shift detected in India. As a result, new potentially invasible areas were identified in central Africa and southern Australia. Our findings do not support the hypothesis of niche conservatism for the invasion of *L. camara*. The mechanisms that allow this species to expand its niche need to be investigated in order to improve our capacity to predict long-term geographic changes in the face of global climatic changes.

Editor: Jose Luis Gonzalez-Andujar, Instituto de Agricultura Sostenible (CSIC), Spain

Funding: This project received financial support from Instituto Venezolano de Investigaciones Científicas, Venezuela (to IH); Instituto de Ecología y Biodiversidad-Universidad de Chile, Chile (project ICM P05-002 to ROB), Centre of Excellence for Invasion Biology (CIB), University of Stellenbosch, South Africa for the Post Doctoral funding (to GPS) and a seed and research grant -University of Delhi, India (to GPS). UniSig-Instituto Venezolano de Investigaciones Científicas provided the computers used in modeling. The funders had no role in study design, data collection and analysis, decision to publish, or preparation of the manuscript.

Competing Interests: The authors have declared that no competing interests exist.

* Email: herrera.ita@gmail.com (IH); rbustama@uchile.cl (ROB); mlampo@ivic.gob.ve (ML)

⌓ These authors contributed equally to this work.

Introduction

The West Indian Lantana, *Lantana camara* L., is considered among the most harmful invasive species in the world [1,2]. Its ability to form dense monospecific stands through the high reproductive capacity and allelopatic exclusion of other plant species can significantly reduce the productivity of agricultural systems and negatively impact the biodiversity of the invaded regions [3]. Although its natural range extends from Mexico to Brazil, the species has established populations in more than 60 countries worldwide [4] causing large economic losses in some of these. For example, in Australia losses associated with the introduction and expansion of this weed has been estimated in the order of $2.2 million per annum [5].

Mechanical and chemical management are currently used for the eradication and control of *L. camara*. These options, however, are often costly and inefficient because invaded areas tend to be vast with limited access [6,7]. On the other hand, biological control agents seem insufficient for reducing the abundance of *L. camara* to manageable levels (*e.g.* [8]). Thus, as with many invasive plant species, prevention is the first recommendation to limit the expansion of *L. camara*. The efficient implementation of preventive actions to stop the arrival and establishment of invasive species relies on the correct identification of potentially suitable areas.

Species distribution models (SDMs) are powerful tools for predicting the potential distribution of invasive plants (*e.g.* [9–11]). Based on spatial correlations between species occurrence and environmental variables, SDMs identify sets of variables associated with the presence of invasive species to project their requirements onto the geographic space [12–14]. One fundamental assumption underlying SDMs is the principle of niche conservatism, which states that species tend to preserve their ancestral niches requirements over time and space [15–18]. In the context of biological invasions, a niche is conserved whenever the species occupies the same environmental conditions in its native and invaded ranges [14,18]. If, on the contrary, the environmental conditions where the species occurs differ between the native and the invaded ranges, the species' niche is considered to have shifted. Recent studies have suggested that niche conservatism does not occur in all invasive species [19–21]. Based on a comprehensive study including 50 invasive plants species, Petitpierre *et al.* [21] found that 15% of the invasive plants species evaluated did not

Table 1. Biodiversity databases used to obtain occurrences of *L. camara*.

Database	Description
GBIF	Global Biodiversity Information Facility: Network for the exchange of biodiversity data through the internet. Compiles data from specimens deposited in museums, herbaria, and published in atlases and references worldwide. Web page: http://data.gbif.org/welcome.htm
REMIB	"Red Mundial de Información sobre Biodiversidad": It is an interagency network that shares biological information. It consists of research centers nodes that host the digitized scientific collections. Web page: http://www.conabio.gob.mx/remib_ingles/doctos/remibnodosdb.html
TROPICOS	TROPICOS is a Botanical Information System. It's a public-access botanical database with records of millions of specimens, images and references by the Department of Bioinformatics of the Missouri Botanical Garden. Here we used the "Geographical Search" tool. Web page: http://www.tropicos.org/SpecimenGeoSearch.aspx
PRECIS	Information Computerized Pretoria: This is a digital database on the state of biodiversity in southern Africa by the South African Biodiversity Institute (SANBI). It includes data from herbaria and published. Web page:
Zimbabwe	Data collection of scientific research on plants in Zimbabwe. The data are collected and compiled by the authors of the website: Hyde, M.A., Wursten, B.T. & Ballings, P. Web page: http://www.zimbabweflora.co.zw/
AVH	Australia's Virtual Herbarium: It is a compilation of occurrence data and specimens from several herbaria in Australia. Web page : http://chah.gov.au/avh/avhServlet?task=simpleQuery

conserve their niche during invasion. Although this represents a small proportion of all invasive species, it highlights the need to test the niche conservatism assumption before using SDMs. When the assumption of niche conservatism fails, SDMs underestimate the potential distribution of invasive plants. Alternative hypotheses must be formulated to account for the observed niche shift and new methods are required to increase predictive value of these models [22].

Several authors have predicted current and future distributions of *L. camara*, under possible scenarios of global change [23–26]. Some foresee a contraction of its distribution at global scale [24], while others expect expansion in some particular regions (*e.g.* Australia and China) [25,26]. All of these studies assume that the niche of this species has been conserved, although this premise has not been tested. Some characteristics of *L. camara* and its invasion history could favor niche shifts due to adaptive evolution. First, the species complex history of introductions (see [27]) has probably involved several genetic bottlenecks. Second, its polyploid condition and ability to hybridize are characteristics that promote rapid evolutionary changes [28,29]. Third, more than 600 ornamental varieties have been produced as a result of artificial selection and hybridization [6,29]. It is now widely recognized that naturalized populations of *L. camara* are morphologically different than populations in its native range [30]. Therefore, it is possible for naturalized populations to differ genetically from the original ancestral populations, to the extent that their niches are no longer exact.

Here, we tested the niche conservatism hypothesis for *L. camara* using a niche dynamics analysis [21,31]. Estimated niches were projected onto the geographic space to identify potentially suitable areas for *L. camara*, and to evaluate the possible implications of a niche shift on its potential distribution in three invaded regions, Australia, Africa and India (*e.g.* [19]).

Materials and Methods

Study species

Lantana camara (Verbenaceae, a perennial evergreen shrub native to the Neotropics, was introduced into Europe from Brazil as an ornamental plant in the 17th century [32]. For the next 100 years after its initial introduction, this species was extensively exported from Europe to Africa, Asia, America and Oceania.

Although it established populations in several countries, *L. camara* only became invasive throughout tropical, subtropical and warm temperate areas [33]. This species currently occupies millions of hectares in South Africa, Australia and India, and continues to expand (reviewed by [27]). *L. camara* has several traits that explain its high invasiveness: it is autocompatible; it is pollinated by different groups of insects (*e.g.* butterflies and honey bees); it has a high seed output with birds dispersing the seeds over long distance; it forms large seedbanks and has high potential of vegetative reproduction; it is a fire-tolerant and it has a high phenotypic plasticity. Also, this species frequently outcompetes native flora (for review see [3]).

Despite its cosmopolitan distribution, the taxonomic status of *L. camara* has not been resolved yet. It is considered a species complex, *L. camara sensu lato*, consisting of several taxa that are morphologically and ecologically very similar to the initial description of the species [29]. For the purpose of this study, *L. camara sensu lato* will be hereafter referred as *L. camara*.

Occurrence data

The occurrences of *L. camara* from their native and introduced ranges were obtained from several sources (Table 1). To filter occurrence data, we selected records collected after to 1950 that included a detailed description of the locality. A total of 896 occurrences were used: 167 from Australia, 96 from Africa, 84 from India and 549 from its native range in America (Figure 1; Table S1). We defined its native range as the geographic area between 24°N (Mexico) and 24°S (Southern Brazil), and constrained our analysis to Australia, Africa and India, where the species has invaded and caused major impact [27].

Environmental data

We chose 12 from the 20 environmental variables available in the WorldClim dataset [34]. This selection was based on natural history data of *L. camara* [3] and the contribution of these variables to a previous test model. The remaining eight variables were not included because they had no discriminatory power (low contribution); for example, variables for which the species showed wide tolerance (*i.e.* altitude) or those with most values outside the physiological tolerance of *L. camara* (*i.e.* minimum temperature of coldest month). Also, to minimize redundancy due to potential multicollinearity among variables, we omitted highly correlated

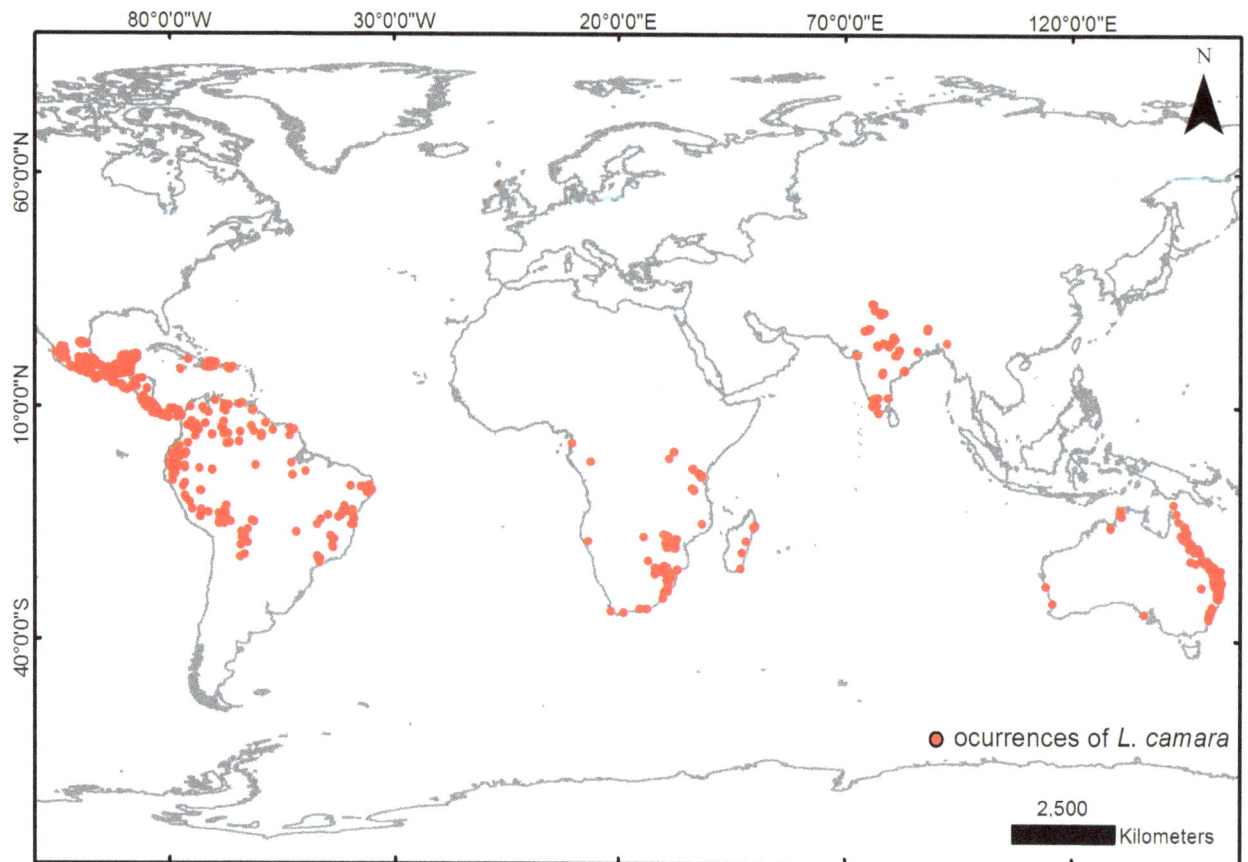

Figure 1. Filtered occurrences of *L. camara* used for this study.

variables. Using a cross-correlation analysis in software *R* [35], we estimated the Pearson's correlation coefficient between pairs of variables, and when >0.9, we selected from the pair that variable more relevant for the ecology of the species. Thus, the initial set of 12 variables was further reduced to six: i) annual mean temperature (BIO1), ii) maximum temperature of the warmest month (BIO5), iii) annual precipitation (BIO12), iv) precipitation of the driest month (BIO14), v) precipitation seasonality (BIO15) and, vi) precipitation of the warmest quarter (BIO18). Raster layers (resolution = 2.5 arc-min~25 km^2) for these climatic variables were obtained from WorldClim dataset (http://www.worldclim.org/) [34].

Niche analysis

We used PCA-env analyses [31] to assess the similarity between niches. This procedure allowed us to evaluate the hypothesis of niche conservatism between native and invaded ranges. Three PCA were conducted to compare the native niche of *L. camara* with its niches estimated for each invaded region: Australia, Africa and India. For each PCA, we used the first two axes to define the environmental space. The environmental space was divided into 100×100 cells, and the occurrence points were converted into densities of occurrences, o_{ij}, using a kernel smoothing function. Then, 10,000 randomly generated points (*i.e.* pseudo-absences) were used to estimate the density of available environments, e_{ij}, in each cell of the environmental space. Based on the values of o_{ij} and e_{ij} an occupancy index, z_{ij}, was estimated. This metric allowed for the unbiased comparison of occurrence densities, when environments were not equally available. Finally, the values of z_{ij} were

plotted on the environmental space to delimit the climatic niche occupied by *L. camara* in their native and invaded ranges.

We used four approximations to compare invaded niches with native niche [36]: (i) qualitatively changes in the niche center, cells with the highest values of z_{ij}; (ii) niche overlap D; (iii) niche equivalence, the correspondence between an observed D and that expected by randomly reallocating the occurrences from both entities on both ranges, and (iv) the niche similarity, the comparison between an observed D and that expected by randomly reallocating the occurrences in only one of the ranges. For the equivalence test, the null hypothesis (*i.e.* niches are not identical) is rejected if $p < 0.05$. For the similarity test, in contrast, a p value > 0.05 indicated that niches were no more similar than expected by chance.

Additionally, we indentified niche zones within the environmental space by overlapping the native and invasive niches according to Petitpierre *et al.* [21]: (i) unfilled (*U*), the zone on the native niche not shared with the invaded niche; (ii) overlap (*O*), the zone shared between native and invasive niches; and (iii) expansion (*E*), the zone on the invaded niche not shared with the native niche. While the *O* values measured the proportion of niche conserved, the *E* values estimated the proportion niche expanded. The unfilled zone (*U*) assesses the fraction of niche not yet occupied by the species in the invaded range.

Species distribution models

We used species distribution models (SDMs) to predict potential suitable areas for the invasion of *L. camara*. One limitation of these models is that they do not distinguish if a particular

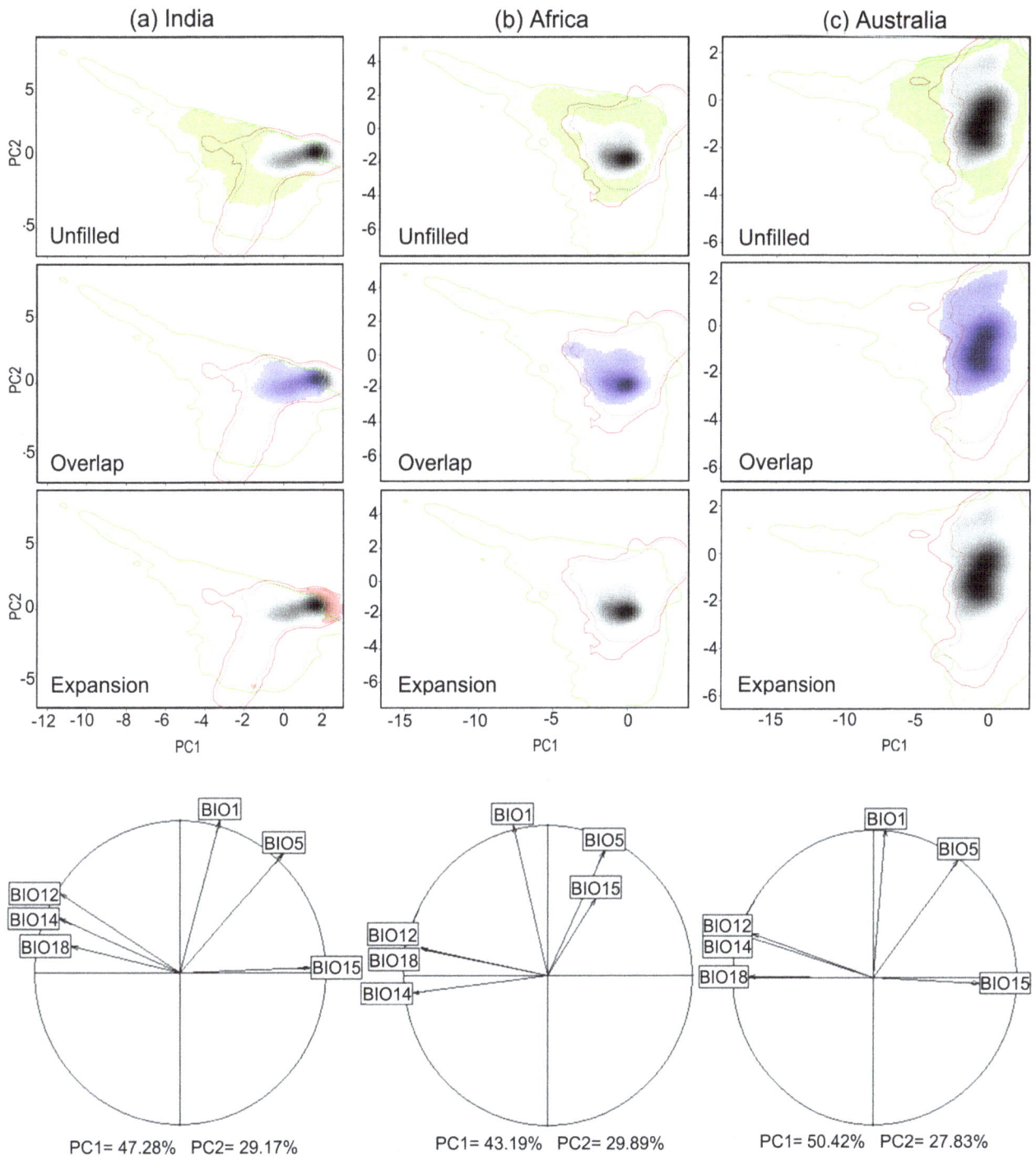

Figure 2. Niche dynamics of *Lantana camara*: from native to invaded ranges. The contour lines delineate the available niche in its native range (green) and in its invaded range (red) in India (column a), Australia (column b) and Africa (column c). The solid and dashed contour lines illustrate, respectively, 100% and 50% of the available (background) environment. The colored areas correspond to the unfilled zone (green; line 1), the overlap zone (blue; line 2), or the expansion zone (red; line 3) resulting from overlaying the native niche with the invaded niche. The last line shows the correlation circles, which indicate the weight of each bioclimatic variable on the niche space defined by the first two principal component axes. The predictor climatic variables are BIO1 (annual mean temperature), BIO5 (temperature of warmest month), BIO12 (annual precipitation), BIO14 (precipitation of driest month), BIO15 (precipitation seasonality), BIO18 (precipitation of warmest quarter).

occurrence is associated with a high abundance (source) or a low abundance population (sink). Therefore, an area classified as "suitable" corresponds to an area with high establishment risk of the species, but not necessarily with a high invasion risk [37].

However, for exotic plants with high invasive potential, as *L. camara*, the risk of establishment can be considered equivalent to of the risk of invasion.

Table 2. Niche dynamics values estimated using climatic conditions in the native and invaded regions.

Niche comparisons	Overlap (*O*)	Unfilled (*U*)	Expansion (*E*)
Native *vs.* Australia	1	0.17	0
Native *vs.* Africa	1	0.47	0
Native *vs.* India	0.79	0.23	0.20

We generated potential distributions of *L. camara* in their native and invaded regions using MaxEnt (*v.* 3.3.3.k) [38]. Using a maximal entropy function, this software estimates the probability of occurrence based on the environmental characteristics of the habitats where the species is present [38]. We choose use the MaxEnt's default settings after contrasting several models, and corroborating that this selection gave the best model based on the Akaike information criterion (AIC). For each study region, presence data was used to construct a set of 10 candidate models. Then, we selected the best four models (one for each region) based in the AIC, to generate potential distributions of *L. camara* in its native region (native-to-native distributions), and in each of the invaded regions (invaded-to-invaded distributions). We then projected the model for the native region onto each invaded region to generate three additional native-to-invaded distributions (native-to-Africa, native-to-Australia, and native-to-India). Whenever the niche analyses indicated expansion (*E*>0) or niche shift (niche similarity<expected), we projected this modified niche onto the rest of the invaded regions (modified-niche-to-invaded). The latter illustrates the effect of possible niche changes in the potential distribution of *L. camara* in the invaded regions. All potential distributions were determined using a threshold value equivalent to the 10th percentile of the probability of occurrence. We overlaid the predicted native-to-invaded distributions with the predicted invaded-to-invaded distributions to estimate the unfilled area (*U*),

the overlap area (*O*), and the expansion area (*E*). This similar procedure was used to overlay the modified-niche- to-invaded distribution with the invaded-to-invaded distribution [21].

Model evaluation and validation

To evaluate model accuracy we used a cross-validation method. For each region, 90% of the occurrence points were set as training data, and the remaining 10% as test data. To assess the model's accuracy in predicting the species' presence in a particular grid cell, we used the area under the curve (AUC) of the receiver operating characteristic (ROC) estimated by MaxEnt for the training and test data sets. Using an R package developed by B. Petitpierre, we also estimated the Boyce Index, *i.e.* threshold-independent accuracy estimator based on the Spearman rank-correlation coefficient between the predicted points and the predicted areas for both data sets. For latter analysis, we randomly selected 10,000 pixels for each model prediction to reduce computational time. The omission rates of occurrences in each invaded region were also calculated to assess how accurately the native-to invaded distributions predicted the occurrence points on the invaded regions.

Climatic analogy between native and invaded ranges

To assess the risk of extrapolating species distributions to regions with substantially different climates [39], we evaluated the climatic

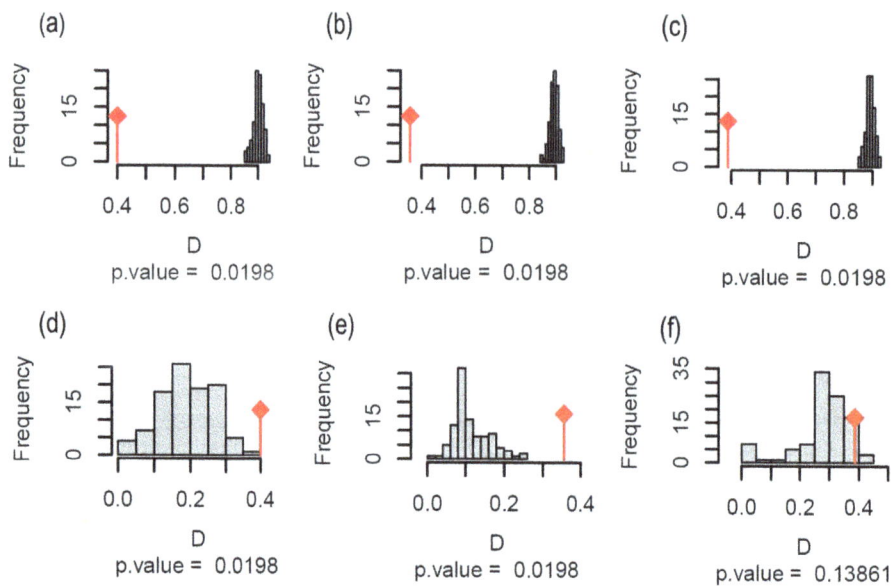

Figure 3. Statistical tests for niche comparisons between native and invaded regions. Observed frequencies for the niche overlap index (*D*) in relation to the expected *D* for *p* = 0.05. The first line shows the tests for niche equivalency (a, b and c) and the second line for niche similarity (d, e, and f). The first column compares niches between the native range and Australia (a and d), the second column between the native range and Africa (b and e) and the third column between the native range and India (c and f).

Table 3. Evaluation index values (AUC and Boyce Index) and omission rate for the obtained models.

Model	$AUC_{training}$	AUC_{test}	$B_{training}$	B_{test}
Native	0.811	0.717	0.998	0.899
Australia	0.976	0.944	0.977	0.669
Africa	0.968	0.962	0.934	0.576
India	0.823	0.802	0.959	0.610

Values for training and test evaluation are shown.

analogy between native and invaded regions using the Multivariate Environmental Similarity Surface (MESS) analysis in MaxEnt [40]. We identified the "most dissimilar" climatic variable and its geographic location, and then we compared the values of these climatic variables for the non-analogous regions by a *Student-t* test.

Results

Niche analysis

The PCA-env analyses of native and invaded regions showed changes in the size of the niches and in the position of the areas with the highest values of z_{ij} (*i.e.* highest density of occurrences) (Figure 2). In all cases, the native region had a greater niche breadth than any of the invaded regions. In India, the shift on the highest density of occurrences surpasses the limits of the native niche (Figure 2 a), while in Africa and Australia, this shift occurs

within these limits (Figures 2 b and c, respectively). In the correlation circle, the arrow directions show that, in India, the niche moved towards lower values of precipitation (BIO18), and greater seasonality of the latter (BIO 15) (Figure 2 a). In Africa towards colder climates (BIO1) (Figure 2 b), and in Australia, the niche moved towards lower temperatures (BIO1) (Figure 2 c). These plots also identified presence of unfilled niches in all invaded regions and niche expansion only in India, where the ~20% of climatic conditions occupied in India are not available in the native region (Table 2; Figure 2).

The niche equivalency tests confirmed that niches from the three invaded regions are not identical to the native niche (Figure 3 a–c). The niche similarity tests showed that the niches of *L. camara* in Australia and Africa are more similar to the niche of the native region than would be expected by chance (Australia: $D = 0.3$, $p = 0.02$; Africa: $D = 0.4$, $p = 0.02$; Figure 3 d–e). In Australia and Africa, *L. camara* only occupies areas with similar

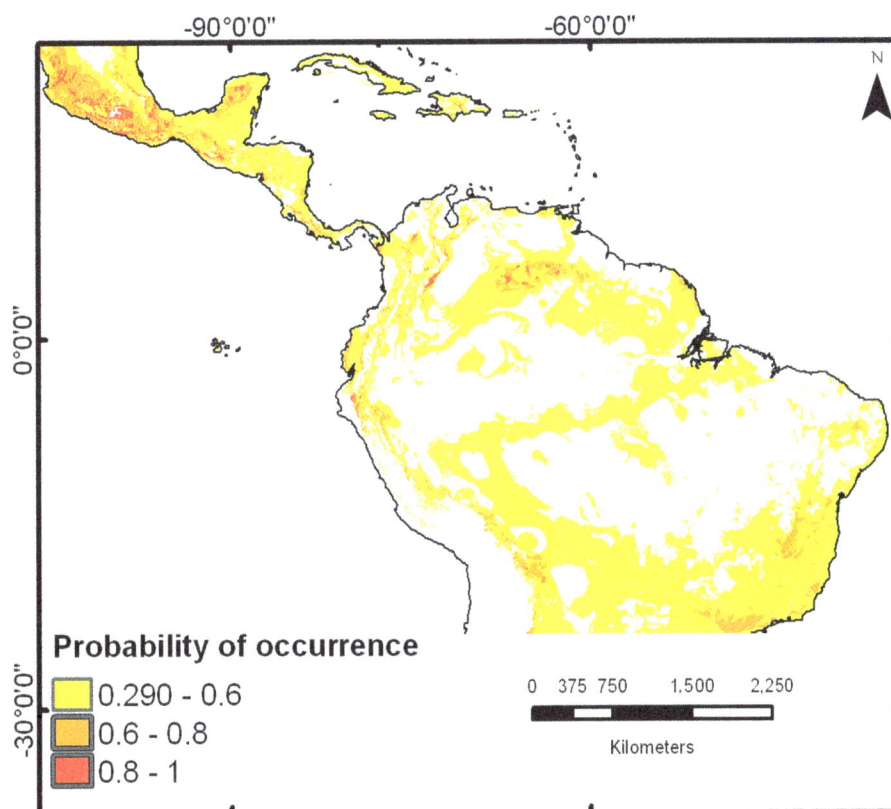

Figure 4. Potential geographic distribution of *L. camara* in its native region. Predictions are based on current occurrences in the Neotropics and the climatic data from the places where this plant inhabits there.

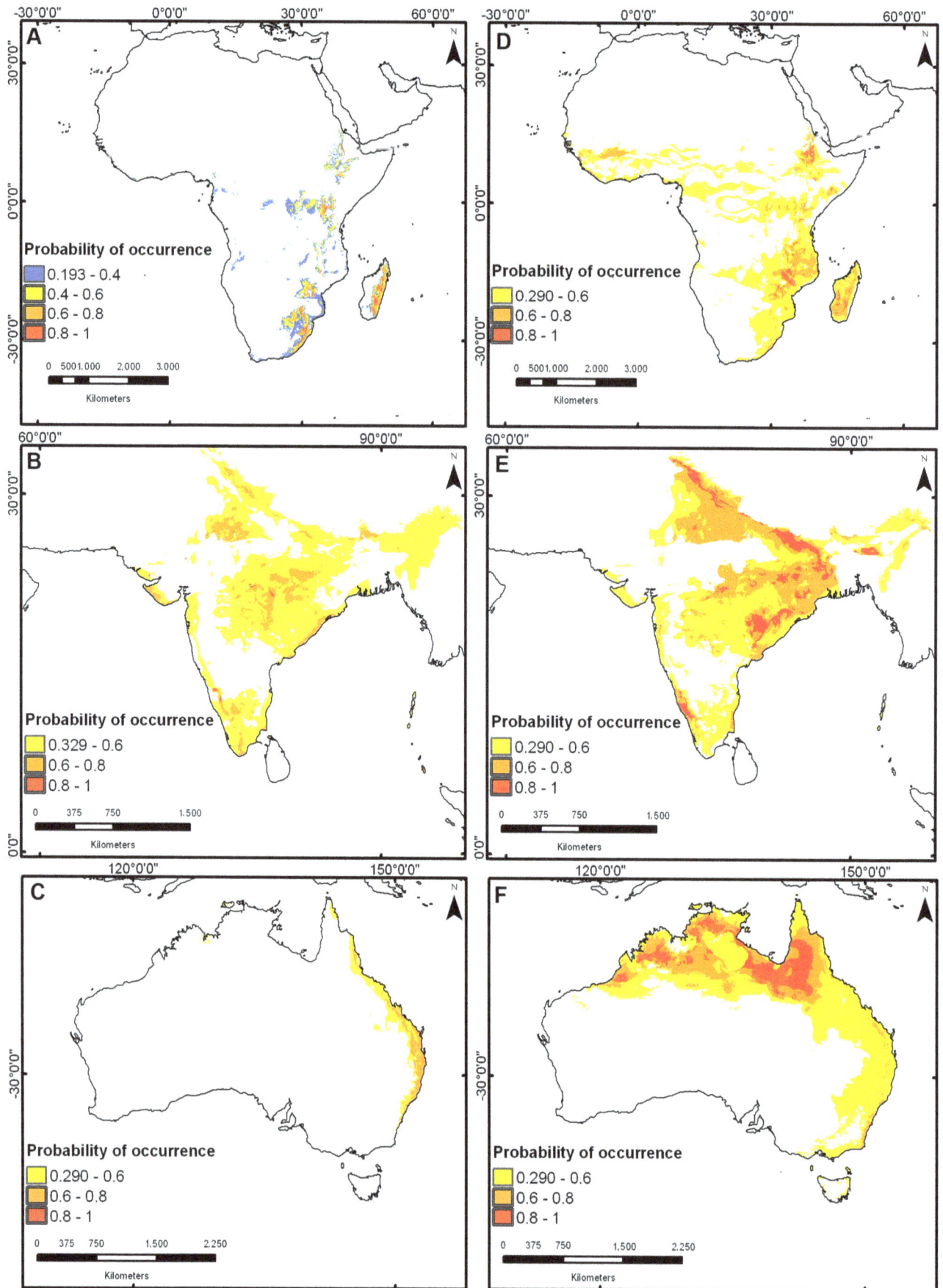

Figure 5. Potential geographic distributions of *L. camara* **in each invaded region.** The first line shows the predicted geographic distributions in Africa (a, d), the second line in India (b, e) and the third in Australia (c, f). The first column illustrates predictions based on invaded-to-invaded projections (a, b, c), and the second column based on native-to-invaded projections (d, e, f).

climatic condition to those found in its native range. In contrast, the climate niches in India and the native region are not more similar than expected by chance ($D = 0.3$, $p = 0.4$, Figure 3 f). This indicates that in India the occupation of the species does not follow a pattern expected by native niche requirements and seems to be random. In India, *L. camara* occupies various climatic conditions, some of which are similar to the climatic condition in its native range, while others are different.

Species distribution models

The selected model showed proper fit of the data. The AUC and the continuous Boyce index were high for the training or test data sets (Table 3). The projection of native model onto all invaded regions had an omission rate of 10% of occurrence points, indicating an adequate prediction. Most of these omissions occurred in India. Similarly, the model training with the Indian range and their extrapolation to Australia and Africa had a low omission rate (14%) (data not shown). These results suggest that both models are informative and did fairly well predictions.

The native potential distribution showed the highest probabilities of presence across southwest of Mexico, the lower slopes of the Andean cordillera in Colombia, and some savannas and evergreen forests in southern Venezuela – all characterized by dry and warm climates – showed the highest presence probabilities (Figure 4). In general, native-to-invade models generated wider distributions than invade-to-invade models (Figure 5 a–f). In the invaded-to-invaded distribution the higher probability of presence occurred in Cape Town, South Africa, center of India and all of the east coast of Australia (Figure 5 a–c). In the native-to-invaded distribution (Figure 5 d–f), the higher probability of presence occurred in Madagascar, around the Malawi Lake in Africa, in northeast India and northern Australia. No climatically suitable areas for *L. camara* were identified in central or southern Australia.

Australia and Africa were the regions with the highest unfilled areas; more than 70% of the predicted distribution for *L. camara* in both regions are not occupied yet (Table 4). India, on the contrary, was the region with the highest value of overlap area; 68.3% of the predicted distribution is occupied by the species. Expansion areas were small for Australia and Africa (<3%). Only in India *L. camara* occupied areas not predicted by its distribution in the native range (Table 4). When the India model was projected onto Australia or Africa (modified-niche-to-invaded), the unfilled area in both regions (*i.e.* new potential distribution areas) increased in 6.1% in Australia and 24.3% in Africa (Table 5; Figure 6).

Climatic analogy between native and invaded ranges

According to the MESS analysis, India was the only region with climatic conditions non-analogous to those observed in the native region (Figure 7). In India, 34% of the occurrences are in locations where maximum temperatures of the warmest month (BIO 5) reach in average 43°C, a value that is significantly higher (*t*-test, $t = 87.159$, $p<0.001$, see Figure 8) than the average of maximum temperature of the warmest month in the native region (35°C).

Discussion

For the first time the hypothesis of niche conservatism is evaluated for the *L. camara* invasion. Our results demonstrate that even though the niches occupied by *L. camara* in Africa and Australia are subsets of its native niche in the Neotropics, in India this species' niche shifted significantly towards warmer climates, with temperatures that frequently exceed the maxima recorded in its native region. The presence of *L. camara* in novel climatic conditions indicates that its niche has not been conserved throughout the process of invasion, therefore suggesting a greater capacity to invade new regions than previously thought.

Niche shift has been documented in several invasive plant species [20,21]. In Australia, a continent where biological invasions are common, 19 invasive plant species are known to have shifted into novel biomes not present in their native range [20]. In theory, niche shifts may conceal one of two mechanisms. First, the species could find new suitable conditions in the invaded regions that are absent from their native range (*i.e.* non-analog climate) but lie within their tolerance ranges or fundamental niche [41]. This mechanism involves a shift only in the realized niche (*i.e.* filling a pre-adapted niche) and whether it should be considered as a true niche shift is still controversial [21,41]. Secondly, the species can undergo genetic changes that allow it to adapt to conditions outside their tolerance ranges, changing its fundamental niche [42]. Identifying the underlying mechanism using SDMs is difficult, and has prompted a recent discussion on whether these analyses should be constrained to niche shifts between analog climates exclusively [21], or include non-analog climates [41]. Either way it is not possible to distinguish whether a species evolved or had pre-adaptations to the conditions in the invaded region using a correlative approach (*i.e.* SDMs) [43]. Thus, we do not know the contribution of these mechanisms to the niche shift observed in India. This species could have filled a niche space absent in its native range but for which it was pre-adapted, or it could have evolved adaptations to the new climate encountered in India. A recent study suggests that many invasive

Table 4. Percent overlap, unfilled and expansion areas in Australia, Africa and India obtained by overlaying the native-to-invaded distributions with the invaded-to-invaded distributions in each invaded region.

Distribution comparisons	Overlap area	Unfilled area	Expansion area
Native *vs.* Australia	10.6% (~3.0×10⁵ km²)	89.2% (~2.8×10⁶ km²)	0.2% (~5.3×10³ km²)
Native *vs.* Africa	26.8% (~1.9×10⁶ km²)	70.2% (~6.8×10⁶ km²)	3.0% (~2.1×10⁵ km²)
Native *vs.* India	68.3% (~8.8×10⁵ km²)	19.8% (~1.9×10⁶ km²)	11.9% (~2.6×10⁵ km²)

Percentages were adjusted by the predicted total area for *L. camara* in each invaded region.

Table 5. Percent unfilled and expansion areas in Australia and Africa obtained by overlaying the modified-niche-to-invaded distributions with the invaded-to-invaded distributions in both invaded region.

Distribution comparisons	Unfilled area	Expansion area
Expanded *vs.* Australia	90.0% (~2.7×10^6 km^2)	6.1% (~1.7×10^5 km^2)
Expanded *vs.* Africa	78.7% (~7.7×10^6 km^2)	24.3% (~2.4×10^6 km^2)

Percentages were adjusted by the predicted total area for *L. camara* in both invaded regions.

plants have evolved co-adaptations to new environmental conditions in their introduced ranges [44]. Ray *et al.* [45] found that individuals of *L. camara* in India were originated from genetically differentiated native allopatric populations that gradually homogenized. In addition, several ornamental varieties have been produced since its introduction in India through hybridization and artificial selection [29,46]. Thus, it is possible for mixtures of different genetic pools to have increased the species' ability to evolve adaptations to novel climates (*i.e.* [47]). However, genetic characterization of populations, reciprocal transplant experiments or a mechanistic modeling approach at a global-scale are required to differentiate between filling a pre-adapted niche and rapid evolution of *L. camara* in India.

The different directions of niche change observed among continents (*i.e.* expansion in India *vs.* contraction in Australia and Africa) may be attributed to contrasting scenarios encountered during the early stages of invasion. Although the first introductions of *L. camara* in these three regions were relatively contemporary (1807–1858), its invasion appeared to have been faster in India [27]. There, *L. camara* extends over 13 million ha whereas in the other two continents it occupies less than 5 million ha [27]. One possible explanation is that the initial introductions in India [27] occurred in highly suitable habitats, as suggested by our model (see Figures 4 b and e), while in Australia and Africa [27] the species arrived to less suitable habitats (see Figures 4 a, c, d and f). A difficult and slow establishment in Australia and Africa could have delayed subsequent invasion phases to leave no sufficient time for

L. camara to colonize its entire climatic niche, a phenomenon known as the colonization-lag non-equilibrium [48]. Non-equilibrium distributions can be generated by local community or demographic processes that prevent the full occupancy of suitable habitats. For instance, the large extensions of dense forests in Africa and Australia could have acted as a barrier to the dispersal of *L. camara* by inhibiting its growth through light competition [3]. Empirical examination of dispersal capacity, extinction-colonization dynamics and a more precise assessment of the habitat suitability when colonization occurred are necessary to test the non-equilibrium hypothesis. Alternatively, the observed niche contractions could have resulted from genetic bottlenecks during early invasion (*i.e.* [49–51]) that reduced the genetic variability of *L. camara*, and its ability to invade its entire niche. Although this is a possible explanation, the number and origin of founders involved in the invasion of Australia and Africa are not known.

Finally, our results highlight the influence that the choice of geographic scale may have on the ability a particular study has to test the hypothesis of niche conservatism. If we had restricted the exotic range to Australia or Africa (*e.g.* [23]) omitting India, we would have missed the evidence that *L. camara* could expand its niche and occupy novel climatic conditions. Biological invasions are a global problem and, thus, a global-scale approach is necessary to test the underlying mechanisms –biogeographic, demographic and evolutionary– involved in this process.

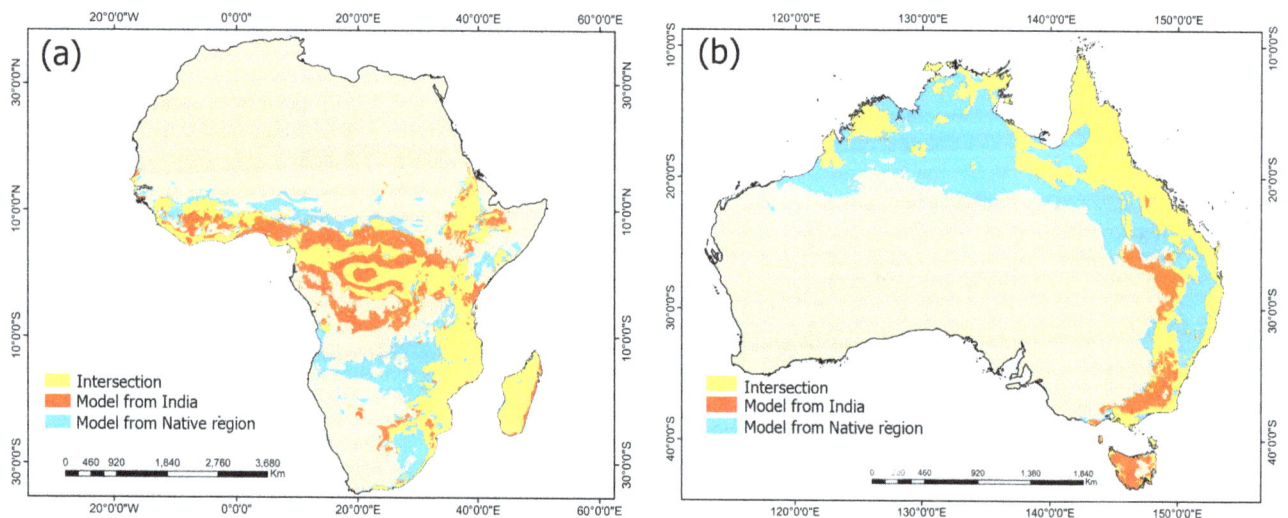

Figure 6. New potentially invadable areas resulting from the observed niche shift in India. Overlay of the potential geographic distributions in Africa (a) and Australia (b) estimated from the native niche and the modified niche in India. The orange areas identify new vulnerable areas. They correspond to locations predicted as unsuitable according to its current distribution in its native range, but as suitable based on its current distribution in India.

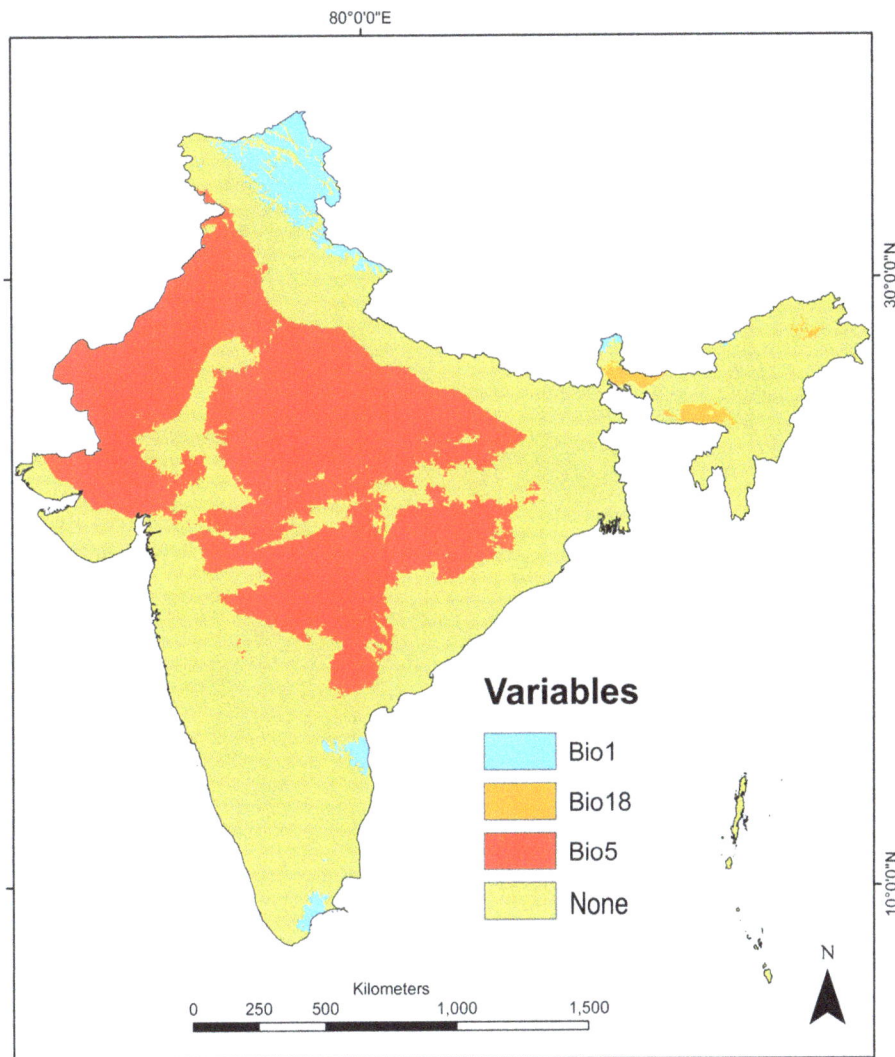

Figure 7. Climatic analogy between the native range of *L. camara* **and its invaded range in India.** Using the multivariate environmental similarity surface (MESS) we identified BIO1 (annual mean temperature), BIO18 (precipitation of the warmest quarter) and BIO5 (temperature of the warmest month) as the most dissimilar variables. The red, blue and yellow areas identify locations in India with values for BIO5, BIO1 and BIO18, respectively, outside the observed value range in its native region.

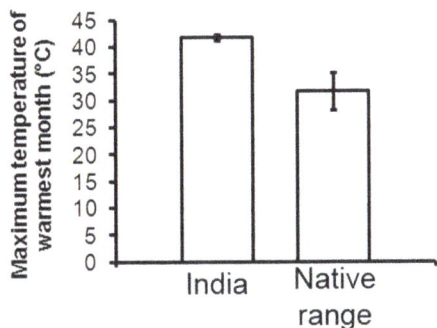

Figure 8. Differences in maximum temperature of the warmest month between the native range of *L. camara* **and its invaded range in India.** The mean values and their standard errors were estimated using the temperature of the warmest month at each location where this plant is present in India and in its native range.

Niche shift and invasion potential of L. camara

Our capacity to predict long-term changes in the geographic distribution of *L. camara* appears to be hindered by its ability to invade novel environmental conditions. The presence of *L. camara* in warmer climates in India suggests that this plant could invade similar habitats in other regions. In Africa, potentially vulnerable areas include the Democratic Republic of the Congo, Cameroon and Centro-African Republic, where several Reserves and Nationals Parks are located. In Australia, southern Queensland and southeastern Victoria may also be invaded by *L. camara*. While some authors already highlighted the vulnerability of some of these regions [23–26], here we add Canberra, Melbourne and their surroundings, and Tasmania to the list of potentially invasible areas (see Figure 5b). In light of possible niche shifts, predictions under climate-change scenarios must be done with caution (*e.g.* [24]).

Acknowledgments

We are grateful to B. Petitpierre, who facilitated the R scripts to perform the PCA analyses, and to anonymous referees for their valuable comments on previous versions of this manuscript.

Author Contributions

Conceived and designed the experiments: IH ROB EG. Performed the experiments: EG MD IH GV. Analyzed the data: EG IH MD ROB ML SGR. Contributed reagents/materials/analysis tools: IH ROB GPS. Wrote the paper: EG IH MD ROB ML GPS SGR.

References

1. GISIN (2013) Global Invasive Species Information Network, providing free and open access to invasive species data. Available: http://www.gisin.org. Accessed 2013 Dec 18.
2. Lowe S, Browne M, Boudjelas S, De Poorter M (2000) 100 of the world's worst invasive alien species: A selection from the global invasive species database. Gland: The Invasive Species Specialist Group (ISSG) a specialist group of the Species Survival Commission (SSC) of the World Conservation Union (IUCN).
3. Sharma GP, Raghubanshi AS, Singh JS (2005) *Lantana* invasion: An overview. Weed Biol Manag 5: 157–165.
4. Parsons WT, Cuthbertson EG (2001). Parsons WT, Cuthbertson EG (2001) Noxious weeds of Australia, 2nd edn. Melbourne: CSIRO Publishing.
5. Johnson S (2008) Review of the declaration of Lantana species in New South Wales. Orange: Department of Primary Industries New South Wales.
6. Day M, Wiley C, Playford J, Zalucki M (2003) Lantana: Current management, status and future prospects. ACIAR Monograph 102. Canberra: Australian Centre for International Agricultural Research.
7. Ensbey R (2003) Managing Lantana. NSW Agriculture, Orange. Available: http://www.weeds.org.au/WoNS/lantana/docs/29_NSW_Ag_Fact.pdf. Accessed 2013 Jan 18.
8. Baars JR, Heystek F (2003) Geographical range and impact of five biocontrol agents established on *Lantana camara* in South Africa. BioControl 48: 743–759.
9. Ficetola GF, Thuiller W, Miaud C (2007) Prediction and validation of the potential global distribution of a problematic alien invasive species - the American bullfrog. Divers Distrib 13: 476–485.
10. Muñoz A-R, Real R (2006) Assessing the potential range expansion of the exotic monk parakeet in Spain. Divers Distrib 12: 656–665.
11. Rouget M, Richardson DM, Nel JL, Le Maitre DC, Egoh B, et al. (2004) Mapping the potential ranges of major plant invaders in South Africa, Lesotho and Swaziland using climatic suitability. Divers Distrib 10: 475–484.
12. Colwell RK, Rangel TF (2009) Colloquium Papers: Hutchinson's duality: The once and future niche. Proc Natl Acad Sci 106: 19651–19658.
13. Elith J, Leathwick JR (2009) Species Distribution Models: Ecological explanation and prediction across space and time. Annu Rev Ecol Evol Syst 40: 677–697.
14. Guisan A, Thuiller W (2005) Predicting species distribution: offering more than simple habitat models. Ecol Lett 8: 993–1009.
15. Holt R, Gaines M (1992) Analysis of adaptation in heterogeneous landscapes: Implications for the evolution of fundamental niches. Evol Ecol 6: 433–447.
16. Peterson AT, Soberón J, Sanchez-Cordero V (1999) Conservatism of ecological niches in evolutionary time. Science 285: 1265–1267.
17. Prinzing A (2001) The niche of higher plants: evidence for phylogenetic conservatism. Proc R Soc Lond B Biol Sci 268: 2383–2389.
18. Wiens JJ, Ackerly DD, Allen AP, Anacker BL, Buckley LB, et al. (2010) Niche conservatism as an emerging principle in ecology and conservation biology. Ecol Lett 13: 1310–1324.
19. Broennimann O, Treier UA, Müller-Schärer H, Thuiller W, Peterson AT, et al. (2007) Evidence of climatic niche shift during biological invasion. Ecol Lett 10: 701–709.
20. Gallagher RV, Beaumont LJ, Hughes L, Leishman MR (2010) Evidence for climatic niche and biome shifts between native and novel ranges in plant species introduced to Australia. J Ecol 98: 790–799.
21. Petitpierre B, Kueffer C, Broennimann O, Randin C, Daehler C, et al. (2012) Climatic niche shifts are rare among terrestrial plant invaders. Science 335: 1344–1348.
22. Gallien L, Douzet R, Pratte S, Zimmermann NE, Thuiller W (2012) Invasive species distribution models – how violating the equilibrium assumption can create new insights. Glob Ecol Biogeogr 21: 1126–1136.
23. Taylor S, Kumar L, Reid N (2012) Impacts of climate change and land-use on the potential distribution of an invasive weed: a case study of Lantana camara in Australia. Weed Res 52: 391–401.
24. Taylor S, Kumar L, Reid N, Kriticos DJ (2012) Climate Change and the Potential Distribution of an Invasive Shrub, *Lantana camara* L. PLoS One7: e35565.
25. Van Oosterhout E, Clark A, Day MD, Menzies E (2004) *Lantana* control manual: Current management and control options for *Lantana* (*Lantana camara*) in Australian State of Queensland. Department of Natural Resources, Mines and Enegry, Brisbane, Australia. Available: http://www.nrm.qld.gov.au/pests/wons/Lantana. Accessed 2013 Jan 18.
26. Lüi XR (2011) Quantitative risk analysis and prediction of potential distribution areas of common lantana (*Lantana camara*) in China. Comput Ecol and Softw 2: 60–65.
27. Bhagwat SA, Breman E, Thekaekara T, Thornton TF, Willis KJ (2012) A battle lost? Report on two centuries of invasion and management of *Lantana camara* L. in Australia, India and South Africa. PLoS One7: e32407.
28. Sanders RW (1987) Taxonomic significance of chromosome observations in Caribbean species of *Lantana* (Verbenaceae). Am J Bot 74: 914–920.
29. Sanders RW (2006) Taxonomy of *Lantana* sect. *Lantana* (Verbenaceae): I. correct application of *Lantana camara* and associated names. SIDA 22: 381–421.
30. Smith LS, Smith DA (1982) The naturalised *Lantana camara* complex in eastern Australia. Queensland Bot Bull 1: 1–26.
31. Broennimann O, Fitzpatrick MC, Pearman PB, Petitpierre B, Pellissier L, et al. (2012) Measuring ecological niche overlap from occurrence and spatial environmental data. Glob Ecol Biogeogr 21: 481–497.
32. Howard RA (1969) A check list of cultivar names used in the genus *Lantana*. Arnoldia 29: 73–109.
33. Swarbrick JT, Willson BW, Hannan-Jones MA (1995) The biology of australian weeds 25. *Lantana camara* L. Plant Prot Q 10: 82–82.
34. Hijmans RJ, Cameron SE, Parra JL, Jones PG, Jarvis A (2005) Very high resolution interpolated climate surfaces for global land areas. Int J Climatol 25: 1965–1978.
35. R-Development-Core-Team (2010) R: A language and environment for statistical computing, Vienna, Austria.
36. Warren DL, Glor RE, Turelli M (2008) Environmental niche equivalency versus conservatism: quantitative approaches to niche evolution. Evolution 62: 2868–2883.
37. Bradley BA (2012) Distribution models of invasive plants over-estimate potential impact. Biol Invasions 15: 1417–1429.
38. Phillips SJ, Anderson RP, Schapire RE (2006) Maximum entropy modeling of species geographic distributions. Ecol Model 190: 231–259.
39. Peterson A, Soberón J, Pearson RG, Anderson R, Martinez-Meyer E, et al. (2011) Ecological niches and geographic distributions. Oxford: Princeton University Press.
40. Elith J, Kearney M, Phillips S (2010) The art of modelling range-shifting species. Methods Ecol Evol 1: 330–342.
41. Webber BL, Le Maitre DC, Kriticos DJ (2012) Comment on "Climatic niche shifts are rare among terrestrial plant invaders". Science 338: 193–193.
42. Mukherjee A, Williams DA, Wheeler GS, Cuda JP, Pal S, et al. (2011) Brazilian peppertree (*Schinus terebinthifolius*) in Florida and South America: evidence of a possible niche shift driven by hybridization. Biol Invasions 14: 1415–1430.
43. Tingley R, Vallinoto M, Sequeira F, Kearney MR (2014) Realized niche shift during a global biological invasion. PNAS 111: 10233–10238.
44. Maron JL, Vila M, Bommarco R, Elmendorf S, Beardsley P (2004) Rapid evolution of an invasive plant. Ecol Monogr 74: 261–280.
45. Ray A, Quader S, Loudet O (2013) Genetic diversity and population structure of "*Lantana camara*" in India indicates multiple introductions and gene flow. Plant Biol doi:10.1111/plb.12087.
46. Stirton CH (1977) Some thoughts on the polyploid *Lantana camara* L, (Verbenaccae). In Proceedings of the Second National Weeds Conference, Stellenbosch, South Africa. Cape Town: Balkema. 321–340.
47. Ellstrand N, Schierenbeck K (2000) Hybridization as a stimulus for the evolution of invasiveness in plants? Proc Natl Acad Sci 97: 7043–7050.
48. De Marco P, Diniz JAF, Bini LM (2008) Spatial analysis improves species distribution modelling during range expansion. Biol Lett 4: 577–580.
49. Baker AJ, Moeed A (1987) Rapid genetic differentiation and founder effect in colonizing populations of common mynas (*Acridotheres tristis*). Evolution: 525–538.
50. Dlugosch KM, Parker IM (2008) Founding events in species invasions: genetic variation, adaptive evolution, and the role of multiple introductions. Mol Ecol 17: 431–449.
51. Easteal S (1989) The effects of genetic drift during range expansion on geographical patterns of variation: a computer simulation of the colonization of Australia by *Bufo marinus*. Biol J Linn Soc Lond 37: 281–295.

15

Diversity, Distribution and Nature of Faunal Associations with Deep-Sea Pennatulacean Corals in the Northwest Atlantic

Sandrine Baillon[1]*, Jean-François Hamel[2], Annie Mercier[1]

1 Department of Ocean Sciences, Memorial University, St. John's, Newfoundland and Labrador, Canada, **2** Society for the Exploration & Valuing of the Environment (SEVE), St. Philips, Newfoundland and Labrador, Canada

Abstract

Anthoptilum grandiflorum and *Halipteris finmarchica* are two deep-sea corals (Octocorallia: Pennatulacea) common on soft bottoms in the North Atlantic where they are believed to act as biogenic habitat. The former also has a worldwide distribution. To assist conservation efforts, this study examines spatial and temporal patterns in the abundance, diversity, and nature of their faunal associates. A total of 14 species were found on *A. grandiflorum* and 6 species on *H. finmarchica* during a multi-year and multi-site sampling campaign in eastern Canada. Among those, 7 and 5 species, respectively, were attached to the sea pens and categorized as close associates or symbionts. Rarefaction analyses suggest that the most common associates of both sea pens have been sampled. Biodiversity associated with each sea pen is analyzed according to season, depth and region using either close associates or the broader collection of species. Associated biodiversity generally increases from northern to southern locations and does not vary with depth (~100–1400 m). Seasonal patterns in *A. grandiflorum* show higher biodiversity during spring/summer due to the transient presence of early life stages of fishes and shrimps whereas it peaks in fall for *H. finmarchica*. Two distinct endoparasitic species of highly modified copepods (families Lamippidae and Corallovexiidae) commonly occur in the polyps of *A. grandiflorum* and *H. finmarchica*, and a commensal sea anemone frequently associates with *H. finmarchica*. Stable isotope analyses (δ^{13}C and δ^{15}N) reveal potential trophic interactions between the parasites and their hosts. Overall, the diversity of obligate/permanent associates of sea pens is moderate; however the presence of mobile/transient associates highlights an ecological role that has yet to be fully elucidated and supports their key contribution to the enhancement of biodiversity in the Northwest Atlantic.

Editor: Erik V. Thuesen, The Evergreen State College, United States of America

Funding: This work was supported by Discovery grants from the Natural Sciences and Engineering Research Council of Canada (NSERC). The funders had no role in study design, data collection and analysis, decision to publish, or preparation of the manuscript.

Competing Interests: The authors have declared that no competing interests exist.

* Email: sbaillon@mun.ca

Introduction

Corals form one of the most complex biological habitats of the deep sea, offering a variety of microhabitats that serve as feeding, shelter, foraging and spawning sites to other species [1–4]. Deep-sea corals occur as unitary forms (i.e. composed of a single polyp) or colonial forms (i.e. composed of many polyps), and can be sparsely distributed or form fields, large thickets and even reefs that may stretch 300 m high and several kilometres wide [1,2,5]. A good understanding of deep-sea corals and their associated fauna, i.e. the organisms that live in or on the corals [1], is essential to evaluate the importance of these unique deep-sea ecosystems and to implement adequate measures for their conservation [6].

Studies of the associated fauna have shown that biodiversity around deep-sea corals can be comparable to that of tropical coral reefs and that main associates include crustaceans, molluscs, echinoderms, cnidarians, sponges, polychaetes and fishes [3,7–9]. A review catalogued 983 invertebrate species associated with 74 species of deep-sea corals; 114 of the associates were characterized as symbionts (living in a close relationship with the coral host) of which 53% were parasites (detrimental to the host) and 47% were commensals (having no impact on the host) [7]. Deep-sea corals

feed on zooplankton and phytodetritus, based on analyses of δ^{13}C and δ^{15}N [10,11] as indicators of food sources and trophic levels, respectively [12]. However, to our knowledge, trophic relationships between deep-sea corals and their associated species have not been explicitly studied. So far, more studies have examined the fauna associated with hard corals than soft corals. We are aware of only one previous work on deep-sea octocorals in the Northwest Atlantic, which reported a total of 114 associates on 2 gorgonian species [1]. Additional information exists for soft corals (excluding Pennatulacea and Helioporacea) from other regions, with a total of 59 symbionts (83% listed as commensals and 17% as parasites) catalogued on 42 octocorals [13]. Sea pens (order Pennatulacea) are typically not afforded the attention of other deep-sea corals [7,13] even though they are very common and have been identified as vulnerable organisms in both shallow and deep environments [4,14–16]. Moreover, sea pens can be collected whole, allowing precise determination/quantification of faunal species living in, on or around them, which is not always the case with larger or more fragile branching corals (e.g. gorgonians) for which analyses of colony fragments is often the rule.

Sea pens can be considered "structural" species due to their extension above the seafloor [17] and have been suggested to create complex biohabitats [8]. However, so far no clear evidence has been provided to support their role as a biogenic habitat, although one study reported the presence of adult fish in large sea pen fields [18]. According to Etnoyer et al. [19], the majority of the species forming biogenic habitats exhibit complex morphology (e.g. branches) and a sufficient size to provide substrate or refuge for other species. Sea pens do not correspond to this definition but have nevertheless been shown to serve as biogenic substrate for different species [8,20,21] and to act as nursery habitat for fish larvae [3]. Moreover, sea pens can cover extensive areas in the deep sea, and are sometimes found in high densities [22], occurring on mud or sand flats, where they could provide an important structural biohabitat to other organisms [23] in relatively featureless environments.

Buhl-Mortensen et al. [8] noted that there seemed to be few species associated with sea pens, indicating that this observation was plausibly due to a lack of data, and only mentioned the association between the ophiuroid *Asteronyx loveni* and the sea pen *Funiculina quadrangularis* [8]. Other associates have been described, including a copepod parasite in *Anthoptilum grandiflorum* [24] in the Labrador Sea (1210 m depth) and a polychaete living between the polyps of *Funiculina quadrangularis* [20] along the Swedish coast (300 m depth). More associated species have been found in, on or around shallow-water sea pens, including different parasitic copepods on various host species [24–27], the gametophyte of an algae living inside the tissues of *Ptilosarcus gurneyi* [28], and the hydrozoan *Eudendrium ramosum* on *Virgularia mirabilis* [21]. At least 5 symbionts were reported on *Ptilosarcus gurneyi* [29], and a porcellanid crab was found between the leaves of *Pteroeides esperi* [30].

Pennatulacean corals are common on the continental slope of eastern Canada, where 16 species have been inventoried [4,31]. The present study focuses on two of the most common ones: *Anthoptilum grandiflorum* (Anthoptilidae) and *Halipteris finmarchica* (Halipteridae) which were recently found to act as essential larval fish habitat [3]. *A. grandiflorum* exhibits a cosmopolitan distribution, with confirmed occurrence in the North and South Atlantic, North and South Pacific, Indian and Antarctic Oceans [32] while *H. finmarchica* is restricted to the North Atlantic [33]. Both species are present from 100 to >2000 m [22]. The main goal of this study was to better define their role and importance as biogenic substrate or habitat with the following objectives: (1) determine the diversity and abundance of their associated species, with an emphasis on spatial and temporal patterns; (2) characterize the dominant symbiotic relationships; and (3) elucidate trophic interactions between the most common associates and their hosts.

Materials and Methods

Collection

Samples of *Anthoptilum grandiflorum* (from 98–1347 m) and *Halipteris finmarchica* (from 256–1333 m) were obtained in 2006 and 2007 as by-catch from annual research surveys (Multispecies Surveys and Northern Shrimp Research Surveys), and the At-Sea Observer Program, along the continental slope of eastern Canada (Fig. 1, Tables S1 and S2) which were all led by Fisheries and Oceans Canada (DFO). The DFO surveys followed a stratified random sampling design with a Campellen 1800 trawl towed for 15 minutes on approximately 1.4 km (gear opened and closed at depth). For more information on the At-Sea Observer Program see Wareham et al. [34]. The sampling area can be divided into 5

regions: Laurentian Channel (LC), Grand Banks (GB), Flemish Cap (FC), North Newfoundland (NNL) and Labrador (LB, Fig. 1, Table 1). Additional samples collected in April and May of 2009 and 2010 were used to determine the consistent presence of some associated suspected to be particularly abundant during the spring months. Colonies of *A. grandiflorum* and of *H. finmarchica* were frozen at −20°C on board the vessels.

Processing of Samples

Colonies to be analysed were selected haphazardly among all samples from a given site. When less than three colonies were sampled at a site, all the colonies were analysed. When more than three colonies were available, a minimum of three colonies were analysed, more if needed, in order to reach a minimum of 20% of the colonies sampled at each site. Few exceptions occurred when samples were unavailable or damaged. Overall, samples of *A. grandiflorum* examined included 185 colonies (measuring 15–83.9 cm) in 2006–2007 (Table S1) and 60 colonies (19.8–76.8 cm) in 2009–2010 (Table S2). Samples of *H. finmarchica* consisted of 92 colonies (17.2–148.6 cm) in 2006–2007 (Table S1) and 12 colonies (15.8–94.0 cm) in 2009–2010 (Table S2). Colonies were thawed in filtered seawater before measuring colony length (from the peduncle to the tip of the sea pen), polyp diameter (n = 10) and density in the three rachis sections, coined lower, middle and upper section as in previous studies on sea pens [35–37]. Colonies were subsequently inspected under a stereomicroscope (Nikon SMZ1500) coupled to a digital camera (Nikon DXM1200F) to isolate and identify associated species. The position of each associate along the central axis was recorded (peduncle, lower, middle and upper sections of the rachis). After extraction from the sea pens, samples of associated species were preserved in 100% ethanol for DNA analyses or dried for 48 h at 60°C for isotopic analyses.

Identification of the Associated Species

While there is no explicit or universal definition of faunal associates or associated species, the terms typically refer to species that find living space, shelter and/or food in or around a given substrate, habitat or species. Here, they were divided into three categories: (1) endobionts (living inside the tissues of the sea pen), (2) ectobionts (or epibionts, living attached to the surface of the sea pen) and (3) free-living. The latter were found unattached to the sea pen but trapped between the polyps, evoking a close association at the moment of sampling. Whenever there was doubt that a specimen might be a by-catch species, it was omitted from the analysis. It is important to note that free-living associates may be lost during sampling, leading to an underestimation of their importance. Studies have sometimes considered only the associates living inside or attached to the corals [1]. Therefore, analyses were conducted on all three categories (all associates) as well as on categories 1 and 2 only (close associates/symbionts). Associated species were grouped according to their morphology and identified to the lowest possible taxonomic level. For the dominant associates, measures of length (e.g. copepod) or basal diameter (e.g. sea anemone) were recorded.

A total of 93 samples of associates were processed by the Canadian Centre for DNA Barcoding (University of Guelph, Canada) for genetic identification. They were analyzed using standard polymerase chain reaction (PCR) and DNA sequencing protocols [38,39]. Identifications were made by running the sequences against the BOLD and BLAST databases.

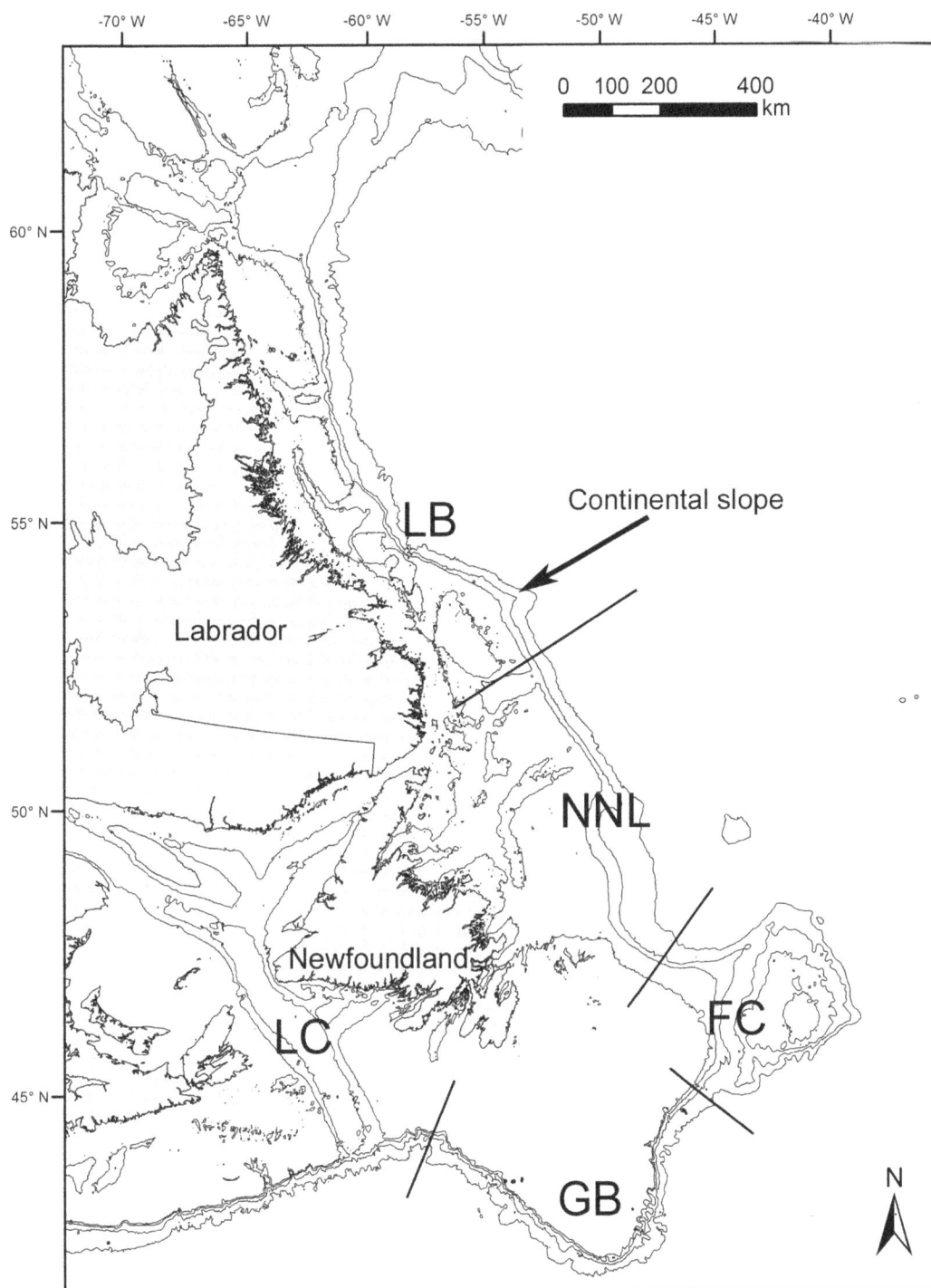

Figure 1. Map showing the five geographic regions where colonies of the sea pens *Anthoptilum grandiflorum* **and** *Halipteris finmarchica* **were collected along the continental slope.** LC: Laurentian Channel, GB: Grand Banks, FC: Flemish Cap, NNL: North Newfoundland, LB: Labrador.

Distribution of the Associated Species

The prevalence of associates (percentage of sea pen colonies harbouring a given species) was determined for pooled associates and for the three categories separately (endobiont, ectobiont, free-living; described above). The mean yield (MY) was defined as the mean number of associates per colony (ind colony^{-1}) considering all sea pens examined, and the mean exact yield (MEY) was defined as the mean number of associates colony^{-1} considering only sea pens harbouring this associated species. Both measures were extrapolated to obtain total yields for the associates (MYtot and MEYtot), both overall and within each category of associate. The MY for a site (site mean yield [SMY] or site mean exact yield [SMEY]) was defined as the number of associates found in that site divided by the number of sea pen colonies examined for that site

Table 1. Number of colonies sampled in the different geographic regions.

	LC	GB	FC	NNL	LB
Anthoptilum grandiflorum	34	35	56	12	31
Halipteris finmarchica	11	33	25	1	18

LC: Laurentian Channel, GB: Grand Banks, FC: Flemish Cape, NNL: North Newfoundland, LB: Labrador.

(as individuals colony^{-1}). All parameters, i.e. prevalence, MY, MEY were also separately determined for the most common (major) associated species.

Specificity of the Lamippidae and Corallovexiidae

Complementary data were obtained from histological sections of polyps of *A. grandiflorum* colonies infested by *L. bouligandi* that were preserved in 4% formaldehyde (n = 12). Polyp samples were prepared using standard histology protocols [40]. They were dehydrated in an ethanol series (70–100%), embedded in paraffin, sectioned (6–10 μm) and stained with haematoxylin and eosin. They were examined under a light microscope (Nikon Eclipse 80i) coupled to a digital camera (Nikon DXM1200F) and analyzed using the imaging software Simple PCI (v. 6.0).

To determine the effect of *Lamippe bouligandi* on the fecundity of *A. grandiflorum*, the density and Feret diameter of oocytes were determined in 5 polyps harbouring a copepod and 5 polyps without copepods sampled in a given colony. The measures were limited to the upper section of the colony to avoid the variation of fecundity along the rachis (increase of the fecundity from the lower to the upper section [37]).

Trophic Interactions

Due to putative regional variations in carbon and nitrogen signatures of pennatulaceans [10], only samples from the Laurentian Channel sampled in 2007 were used for isotopic analysis; this location/date yielded several colonies with enough copepods to allow comparisons. Analyses of stable isotopes were conducted according to Sherwood et al. [10] on 16 samples of associates (2 *L. bouligandi*, 3 undescribed Corallovexiidae and 5 *S. nexilis*) and on their hosts (2 *A. grandiflorum* and 4 *H. finmarchica*). Briefly, dried samples were ground to powder and treated with 5% (v/v) HCl to remove carbonates, then rinsed three times with de-ionised water and dried again for 24 h at 60°C. Between 0.6 and 2.3 mg of sample was placed into 10×10 mm ultralight Sn capsules. Due to the small size of the copepods, specimens from a given colony were pooled to obtain the minimum weight necessary. The analyses were carried out using a Carlo Erba 1500 elemental analyser connected via a ConFlo-II interface to a FinniganTM MAT 252 isotope ratio mass spectrometer in the Department of Earth Sciences at Memorial University. The carbon and nitrogen isotopic values are provided using the standard δ-notation: $\delta X = [(R_{sample}/R_{standard}) - 1] \times 10^3$, where X corresponds to ^{13}C or ^{15}N and R is $^{13}C/^{12}C$ and $^{15}N/^{14}N$, respectively.

As per Sherwood et al. [10] a proxy for particulate organic matter (POM) was used in the form of sedimentary organic matter (SOM) from the LC sampled at 268–531 m between October and December 1990 [41]. Data for pelagic and benthic invertebrates were not available for LC. However, previous data from offshore NNL were used [42] including amphipods and euphasiids for the

pelagic invertebrates, and shrimps (*Pandalus borealis* and *Pasiphae multidentata*) and snow crab for benthic invertebrates to situate the sea pens in the food web.

Trophic level (TL) was estimated from the $\delta^{15}N$ values using the following equation [43]: $TL_{consumer} = [(\delta^{15}N_{consumer} - \delta^{15}N_{base})/\Delta\delta^{15}N] + TL_{base}$ where $\delta^{15}N_{consumer}$ corresponds to the $\delta^{15}N$ of the taxa considered, while $\delta^{15}N_{base}$ and TL_{base} correspond to the value of the baseline of the trophic web considered, and $\Delta\delta^{15}N$ is the trophic fractionation for $\delta^{15}N$ (average 3.8‰ for polar and deep-sea studies [44]). Here, the base value was determined as per Gale et al. [45] using zooplankton as the primary consumer (TL_{base} = 2.3, $\delta^{15}N_{base}$ = 9).

In addition, gastro-vascular contents of the sea anemone *Stephanauge nexilis* (an associate of *H. finmarchica*, see results) were extracted and preserved in 100% ethanol for DNA analyses. Eight samples were processed for DNA identification as outlined above.

Data Analysis

Rarefaction curves [46] were used to compare species richness of faunal associates between sea pen host species using BioDiversity Pro software (Natural History Museum, London/Scottish Association of Marine Sciences). Rarefaction analysis allows an estimation of the number of species expected ($E_{(Sn)}$) for a specific number of individuals observed (n) removing the influence of the sample effort [47]. The evenness (or equitability, indicating whether or not species are represented by a similar number of individuals) of the assemblage of species was determined for both sea pens using the Shannon–Wiener diversity index: $H' = -\sum_{i=1}^{S} (N_i/N) \times \log(N_i/N)$ [47] where S is the total number of taxa, N the total number of individuals, N_i the number of individuals of the ith taxa. Biodiversity (rarefaction curve, expected number of species and the Shannon–Wiener diversity) was determined separately for all associates and for close associates (endobionts and ectobionts only).

Principal component analyses (PCA) were used to determine the influence of season and region on the species distribution at the studied sites. Data were pooled per site and a log(x+1) transformation was applied to the faunal abundance values [47]. This transformation allows the consideration of both the most abundant and rarer species. The general repartition of the associated species, their diversity and the repartition of the most common associates were analysed according to sea pen colony length, colony section, depth, region (Fig. 1; Laurentian Channel, Grand Banks, Flemish Cape, North Newfoundland, Labrador) and season. Additionally, sea pen morphometry (polyp density and polyp diameter) was used to analyze the fine scale distribution of the most common associated species. According to the parameter considered, linear regression and one-way ANOVA or t-test were used, after verifying assumptions of normality and homogeneity of variances. Post-hoc pairwise analysis (Student-Newman test) was conducted as appropriate. When assumptions were not met even after transforming the data, Spearman correlation and Kruskal-Wallis or Mann-Whitney tests were used, followed by Dunn's tests as appropriate. The number and distribution of ectobionts and free-living associates among seasons, depths and regions precluded the statistical analysis for these associates alone. Therefore, the analyses of seasonal, bathymetric and regional variations were carried out using MEYtot and biodiversity index. Due to the sample size, analysis of the influence of depth on the yield was carried out only when more than 10 colonies with associated species were sampled in the same region for a specific season. Therefore analyses were limited for *A. grandiflorum* to LC-spring

Table 2. Species found on the sea pens *Anthoptilum grandiflorum* and *Halipteris finmarchica* during the present study (2006–2007).

Species	Number of individuals	Prevalence (%)	Type of association	Life stage	Link to pictures
On *Anthoptilum grandiflorum*					
Actinopterygii					
Scorpaeniformes					
Sebastes spp.	150	17.1	Free-living	Larvae	[3]
Myctophiformes					
Benthosema glaciale	1	0.4	Free-living	Larvae	[3]
Perciformes					
Lycodes esmarkii	1	0.4	Ectobiont	Egg	[3]
Crustacea					
Copepoda					
Lamippe bouligandi	1458	66.2	Endobiont	Adult	Fig. 4
Unidentified Copepoda	4	1.7	Free-living	Adult	Fig. 2F
Decapoda*					
Acanthephyra pelagica	2	0.4	Free-living	Larvae	Fig. 2A
Pandalus montagui	3	0.9	Free-living	Larvae	Fig. 2A
Unidentified Decapoda	7	1.7	Free-living	Larvae	Fig. 2A
Amphipoda					
Unidentified Amphipoda	3	1.3	Free-living	Adult	—
Nematoda					
Unidentified Nematoda	2	0.4	Free-living	Adult	—
Unidentified species					
Unidentified sp. 1	6	2.6	Endobiont	Egg	Fig. 2C
Unidentified sp. 2	2	0.9	Endobiont	Egg	Fig. 2E
Unidentified sp. 3	6	2.1	Endobiont	Egg	Fig. 2B
Unidentified sp. 4	1	0.4	Endobiont	Egg	Fig. 2D
Unidentified sp. 5	1	0.4	Endobiont	?	—
On *Halipteris finmarchica*					
Actinopterygii					
Scorpaeniformes					
Sebastes spp.	17	4.3	Free-living	Larvae	[3]
Cnidaria					
Actinaria					
Stephanauge nexilis	28	16.0	Ectobiont	Adult	Fig. 5
Hydrozoa					
Unidentified Hydrozoa	1	1.1	Ectobiont	Adult	—
Crustacea					
Copepoda					
Undescribed Corallovexiidae	112	29.8	Endobiont	Adult	Fig. 6
Unidentified Lamippidae	7	7.5	Endobiont	Adult	Fig. 2G
Unidentified species					
Unidentified sp 7	10	4.3	Endobiont	?	Fig. 2H, I

* 6 more larvae were found in April 2009 with a third species identified as *Pasiphaea multidentata*.

(n = 28), FC-fall/winter (n = 39), GB-spring (n = 12), GB-fall (n = 24) and LB-summer (n = 20); while only GB-fall (n = 11) was used for *H. finmarchica*. For the influence of depth on biodiversity, all data irrespective of region and season were used and data were pooled per range of depth (100-m interval) to determine $E_{(S15)}$. Comparison of biodiversity between seasons (fall vs. spring) was done on samples from GB and LB for *A. grandiflorum* and GB for *H. finmarchica*. Due to the difference in the number of associates found in the different regions/seasons,

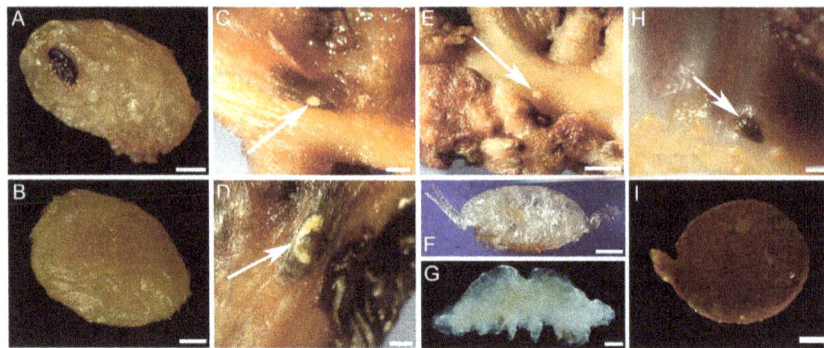

Figure 2. Associates of *Anthoptilum grandiflorum*: **(A) decapod larva, (B) unidentified sp. 3, (C) unidentified sp. 1, (D) unidentified sp. 4, (E) unidentified sp. 2, (F) unidentified copepod.** Unidentified sp. 1 to 4 correspond to potential egg mass. Associates of *Halipteris finmarchica*: (G) unidentified Lamippidae, (H and I) unidentified sp. 7. Scale bar in A = 200 μm, B, F and H = 500 μm, C and D = 2 mm, E = 4 mm, I = 100 μm. Species numbers linked to Table 2.

different expected number of species were used ($E_{(S170)}$, $E_{(S150)}$, $E_{(S120)}$ and $E_{(S20)}$, respectively).

Results

Species Identification and Diversity

A total of 1647 individuals belonging to 14 species (7 scored as close associates or symbionts) were found on the 175 colonies of *A. grandiflorum* examined and a total of 189 individuals belonging to 6 species (5 close associates) occurred on the 43 colonies of *H. finmarchica* (Table 2, Fig. 2, 3, 4 and 5). Seven species associated with *A. grandiflorum* were classified as free-living, 1 as ectobiont and 6 as endobionts, whereas 1 free-living associate, 2 ectobionts and 3 endobionts were found on *H. finmarchica*. On the 93 samples prepared for genetic identification, 52.7% were successfully sequenced. Partial COI sequences with all meta-data are registered in the Barcode of Life Data Systems [48], project SBDSC, and deposited in GenBank (Table S3). This analysis allowed identification down to species for fish and shrimp larvae. While no precise identification was obtained for the other specimens, higher taxonomic levels were determined.

The free-living species included fish larvae (*Sebastes* spp. and *Benthosema glaciale* [3]), shrimp larvae (*Acanthephyra pelagica*, *Pandalus montagui*), amphipods, copepods and nematodes. The ectobionts included one occurrence of one egg of the fish *Lycodes esmarkii* attached to the tissues of one colony of *A. grandiflorum* [3], several sea anemones *Stephanauge nexilis* and a hydrozoan colony found on the naked upper section (exposed skeleton) of

colonies of *H. finmarchica*. Finally the endobionts included parasitic copepods (*Lamippe bouligandi* on *A. grandiflorum*, an undescribed Corallovexiidae and an unidentified Lamippidae both found on *H. finmarchica*) and 6 unidentified species (including 4 putative egg masses on *A. grandiflorum*; Table 2).

Analysis of Close Associates

When only the close associates were considered (endobionts and ectobionts), the values of $E_{(S150)}$ and evenness were lower for *A. grandiflorum* than *H. finmarchica* ($E_{(S150)}$: 2.45 and 4.00, H': 0.07 and 0.84, respectively). Rarefaction curves did not reach the asymptote. Overall, 97.9% of the individuals found on the two sea pens belonged to 3 species. The most common (89.3% of the associates) occurred on *A. grandiflorum* and was identified as *Lamippe bouligandi*, a parasitic copepod living inside the tissues of the polyp column (Fig. 3A). The next two most common species were found on *H. finmarchica*: a sea anemone (representing 1.7% of the associates) found attached to the central axis, showing 96% DNA similarity with Hormathiidae and identified as *Stephanauge nexilis* (Fig. 4A), and a parasitic copepod (6.8% of the associates; Fig. 5A) living inside the polyp, in the space typically hosting reproductive cells. The latter was identified as a copepod from the family Corallovexiidae based on the presence of nauplii (characteristic of crustacean) and its general morphology. The parasitic copepod found in *H. finmarchica* presents lateral extensions (5 or 6 pairs of pereionites) consistent with the Corallovexiidae described by Stock [49]. The male of the undescribed Corallovexiidae, which was always found close to the female, surrounded by eggs/nauplius, differs from previous descriptions. However, only 10 species have so far been described, and it is likely that variation in the shape of males exist. Finally, a genetic similarity of ~85.5% was obtained between the undescribed Corallovexiidae and *L. bouligandi* (family: Lamippidae) suggesting that the two species belong to different families. Given the localisation of these copepods in their hosts, they were considered endobionts.

Principal component analysis (PCA) on the close associates of *A. grandiflorum* revealed that the copepod *L. bouligandi* was the main contributor to the first principal component (PC1: 94.0%) and the unidentified sp. 1 to the second principal component (PC2: 4.0%). For *H. finmarchica* the main contributor to the first principal component was the undescribed Corallovexiidae (PC1: 65.5%) and the sea anemone *S. nexilis* for the second component (PC2: 21.4%). No clear grouping was visible on the PCAs for any sea pen.

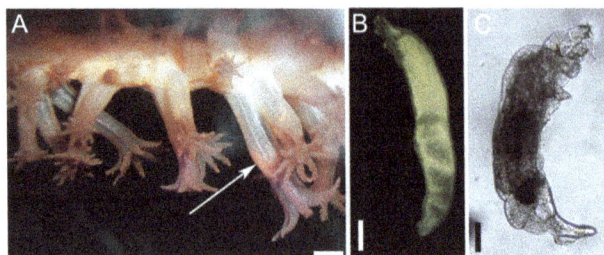

Figure 3. *Lamippe bouligandi*, **a parasitic copepod living inside the polyps of** *Anthoptilum grandiflorum*: **(A) in situ view of the copepod (arrow) through the transparent polyp wall, (B) a female, (C) a male.** Scale bar in A = 1 mm, in B = 500 μm and in D = 100 μm.

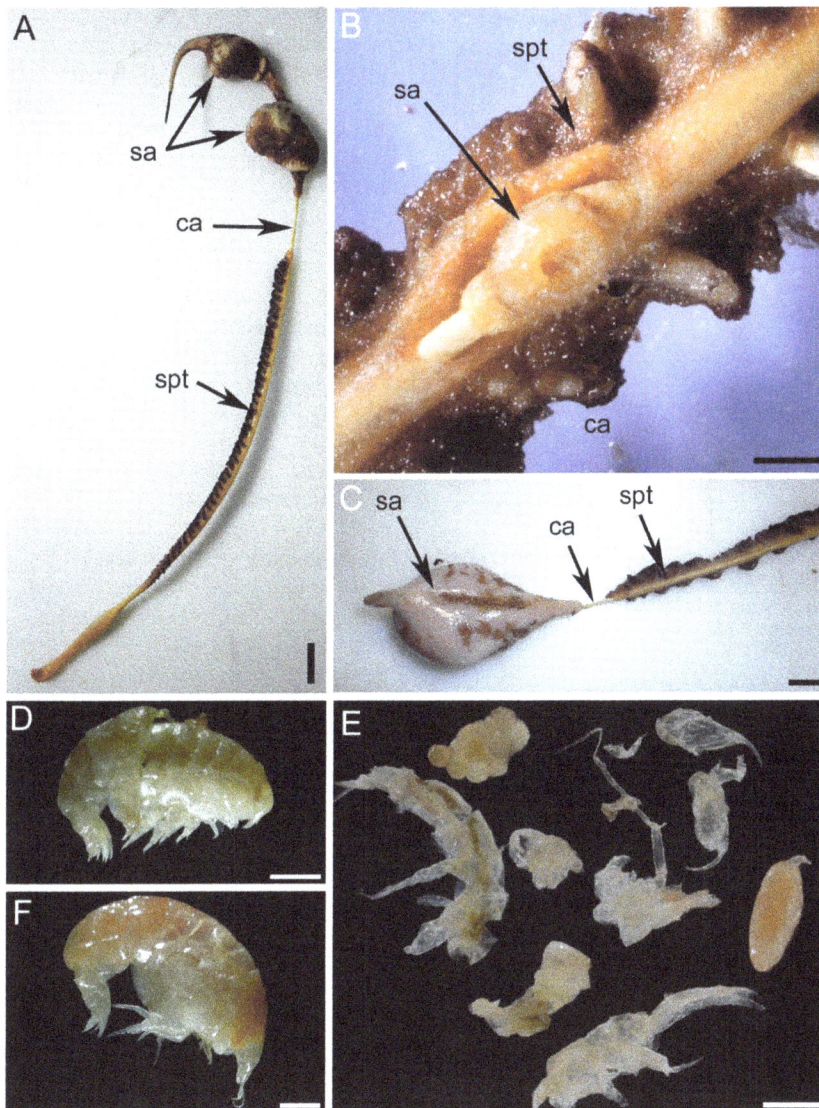

Figure 4. The sea anemone *Stephanauge nexilis* **using the central axis of** *Halipteris finmarchica* **as a substrate: (A) general view of a colony of** *H. finmarchica* **harbouring two sea anemones in the upper section, (B) a small sea anemone surrounded by sea pen tissues, (C) dorsal view of the sea anemone found on the upper section of the sea pen colony.** Gastro-vascular contents were found: (D and F) amphipod, (E) mix of prey including amphipods, halocyprids, egg mass, and unidentified food item extracted from one sea anemone. sa: sea anemone, ca: central axis, spt: sea pen soft tissues. Scale bar in A = 2 cm, in B and D = 2 mm and C and F = 1 cm, E = 1 mm.

Seasonal analyses showed a higher diversity of species associates with *A. grandiflorum* in spring/summer ($E_{(S200)}$ = 3.25) than in fall ($E_{(S200)}$ = 1.66) while the MEYtot showed no variation among seasons (H = 4.04, df = 3, P = 0.258). The opposite trend was observed for *H. finmarchica* with a lower diversity in spring/summer ($E_{(S40)}$ = 2.3) compared to fall ($E_{(S40)}$ = 4). However, the MEYtot showed no seasonal variation (H = 2.71, df = 2, P = 0.258).

Regional analyses showed different biodiversity associated with *A. grandiflorum* among regions (Fig. 6A); however, no pattern was visible and no variation of the MEYtot was detected (H = 8.95, df = 4, P = 0.062). A general southward decrease emerged for the biodiversity associated with *H. finmarchica* among regions (Fig. 6D) while no variation of the MEYtot occurred (H = 1.65, df = 3, P = 0.648).

There was no influence of depth on the biodiversity associated with either sea pen species (*A. grandiflorum*: r^2 = .11, $F_{(1,6)}$ = 0.60, P = 0.474, log-transformed data; *H. finmarchica*: r^2 = 0.12, $F_{(1,4)}$ = 0.42, P = 0.563) or their MEYtot (*A. grandiflorum*: r_s = −0.06, P = 0.429, log-transformed data; *H. finmarchica*: r_s = −0.21, P = 0.187).

Analysis of All Associates

Values of $E_{(S170)}$ when all species found considered were lower for *A. grandiflorum* (~5 expected species) than *H. finmarchica* (~6 expected species). The rarefaction curve for *A. grandiflorum* did not reach an asymptote while the curve for *H. finmarchica* showed a steeper increase of the number of species towards an asymptote. However, when the rarest species (with only one observation) were removed, the rarefaction curve of both species

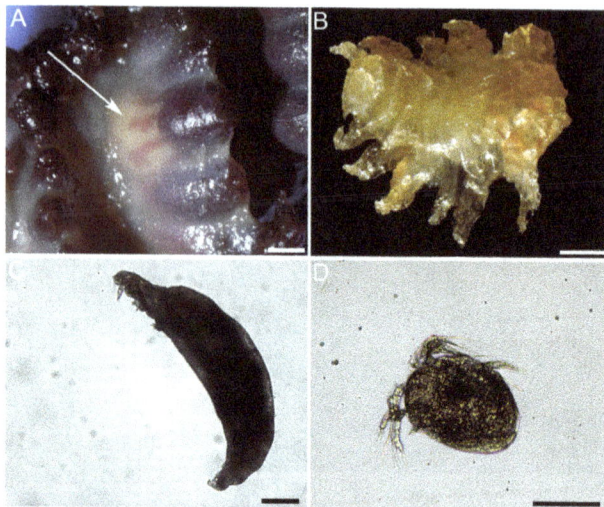

Figure 5. Undescribed copepod species belonging to Corallo-vexiidae living inside the polyps of *Halipteris finmarchica*: **(A) row of polyps including a polyp infested with a copepod (arrow), (B) female copepod, (C) male copepod and (D) nauplius larvae.** Scale in A = 1 mm, in B = 500 μm, in C and D = 100 μm.

reached an asymptote. Evenness was lower for *A. grandiflorum* (H′ = 0.44) than for *H. finmarchica* (H′ = 1.12).

Larvae of redfish, *Sebastes* spp., were the fourth most common species found during this study (representing 9.3% of associates); they were present on both sea pens (for more details see Baillon et al. [3]). In addition to fish larvae, 12 shrimp larvae were found in April 2006 and April 2007 on *A. grandiflorum*; they were identified as *Acanthephyra pelagica* (DNA: 99% certainty) and *Pandalus montagui* (DNA: 100% certainty). Six shrimp larvae were also found on four colonies of *A. grandiflorum* in April 2009, one of them identified as *Pasiphaea multidentata* (DNA: 100% certainty).

Principal component analysis (PCA) on the associated species of *A. grandiflorum* revealed that the copepod *L. bouligandi* was the main contributor to the first principal component (PC1: 69.1%) and the fish larvae to the second principal component (PC2: 22.5%). Two groupings were visible (Fig. 7A) corresponding, for the first, to the colonies harbouring fish larvae (April-May in the LC region, Fig. 1) and, for the second group, to all other samples in various regions/months. PCA on the associated species of *H. finmarchica* showed that the undescribed Corallovexiidae was the main contributor to the first principal component (PC1: 56.0%) and fish larvae and the sea anemone *S. nexilis* to the second principal component (PC2: 24.1%). However, no specific group-ings emerged (Fig. 7B). Therefore, to account for the influence of fish larvae on the repartition of the study sites, the remaining analyses were conducted considering both regions and seasons (spring/summer vs. fall/winter).

Seasonal analyses inside specific regions showed that the diversity of species associated with *A. grandiflorum* was higher in spring/summer than in fall in GB ($E_{(S120)spring}$ = 7.0 > $E_{(S120)fall}$ = 4.5) and LB ($E_{(S150)spring}$ = 3.0 > $E_{(S150)fall}$ = 2.0). However, the MEYtot did not show any significant seasonal variations at any site (GB: U = 139.0, P = 0.596; LB: U = 94.5, P = 0.826). The associates of *H. finmarchica* showed a lower diversity in spring than fall in GB ($E_{(S20)spring}$ = 2.0 < $E_{(S20)fall}$ = 3.8) but no significant difference in MEYtot was observed in GB (U = 27.0, P = 0.565).

Regional analyses within the various seasons revealed that the associated biodiversity of *A. grandiflorum* exhibited a general northward decrease in fall and spring/summer (Fig. 6B and C) while the MEYtot showed no significant variation among regions (spring/summer: $F_{(2,91)}$ = 2.82, P = 0.065; fall: H = 4.72, df = 3, P = 0.193; log-transformed data). In fall, *H. finmarchica* showed the same biodiversity of associates in FC and GB ($E_{(S15)}$ = 3.94 and 3.48, respectively; Fig. 6E) as well as the same MEYtot (U = 37.0, P = 0.925), while in summer colonies showed a higher biodiversity of associates in LC than FC and LB (Fig. 6F) but no regional differences in MEYtot ($F_{(2,22)}$ = 1.49, P = 0.247; log-transformed data).

No significant influence of depth was found on the biodiversity of associates for either sea pen host (*A. grandiflorum*: r^2 = 0.21, $F_{(1,8)}$ = 1.83, P = 0.218, *H. finmarchica*: r^2 = 0.03, $F_{(1,5)}$ = 0.14, P = 0.721). No bathymetric variation in MEYtot was found either, except a decrease of MEYtot with depth in GB during the fall for *A. grandiflorum* (Table 3).

Species Distribution on the Hosts

All faunal associates were found on the rachis section of the host colonies. At least one of the associates was found on 75.9% of *A. grandiflorum* and 46.6% of *H. finmarchica* colonies. Across regions, prevalence proportion varied between 58.3% (NNL) and 96.8% (LB) for *A. grandiflorum* and between 23.8% (FC) and 90.0% (LC) for *H. finmarchica* (Table 4). For both species the endobionts were the most common (prevalence on *A. grand-iflorum* = 72.3%; on *H. finmarchica* = 38.6%) across geographic regions. They were principally represented by *L. bouligandi* (98.9%) in *A. grandiflorum* and by the undescribed Corallovex-iidae (87.5%) in *H. finmarchica*.

The yield of associates (as MEY) on *A. grandiflorum* was significantly greater for endobionts (9.3±0.9 ind colony^{-1}) than for ectobionts (1.0±0.0 ind colony^{-1}) and for free-living species (3.2±0.7 ind colony^{-1}; H = 42.83, df = 2, P<0.001). No significant differences were found in the MEY of each category of associate on *H. finmarchica* (endobiont: 4.0±0.9 ind colony^{-1}; ectobiont: 1.9±0.5 ind colony^{-1}; free-living: 4.3±2.3 ind colony^{-1}; H = 4.64, df = 2, P = 0.099). Comparisons between the two sea pens showed that they harboured the same number of ectobionts (U = 30.0, P = 0.121) and free-living associates (U = 84.5, P = 0.401) whereas *A. grandiflorum* hosted a significantly higher number of endobionts than *H. finmarchica* (U = 1707.0, P<0.001).

Endobionts were present in all the sections of the rachis in both sea pen species. A significant increase of the endobiont MEY occurred from the lower to the upper section of *A. grandiflorum* colony (H = 95.50, df = 2, P<0.001), while the endobionts in *H. finmarchica* showed a significantly higher MEY in the middle section than in the two other sections (middle> lower = upper; H = 12.39, df = 2, P = 0.002). For both sea pens, when removing the most common associate (i.e. *L. bouligandi* and the undescribed Corallovexiidae), no significant differences were found among sections for other associates (*A. grandiflorum*: H = 3.10, df = 2, P = 0.212; *H. finmarchica*: $F_{(2,6)}$ = 1.5, P = 0.296). In *H. finmarchica*, the sea anemone *S. nexilis* and a hydrozoan (ectobionts) were always attached directly to the central axis in the upper section of the colonies. *A. grandiflorum* showed a significant increase of the MEYtot with colony length (r_s = 0.16, P = 0.036) while no variation was noted for *H. finmarchica* (r_s = 0.05, P = 0.735). Analyses per category of associate showed an increase of the MEY with colony length for free-living associates (r_s = 0.36, P = 0.007) of *A. grandiflorum* while no variation occurred for other categories in either sea pen species.

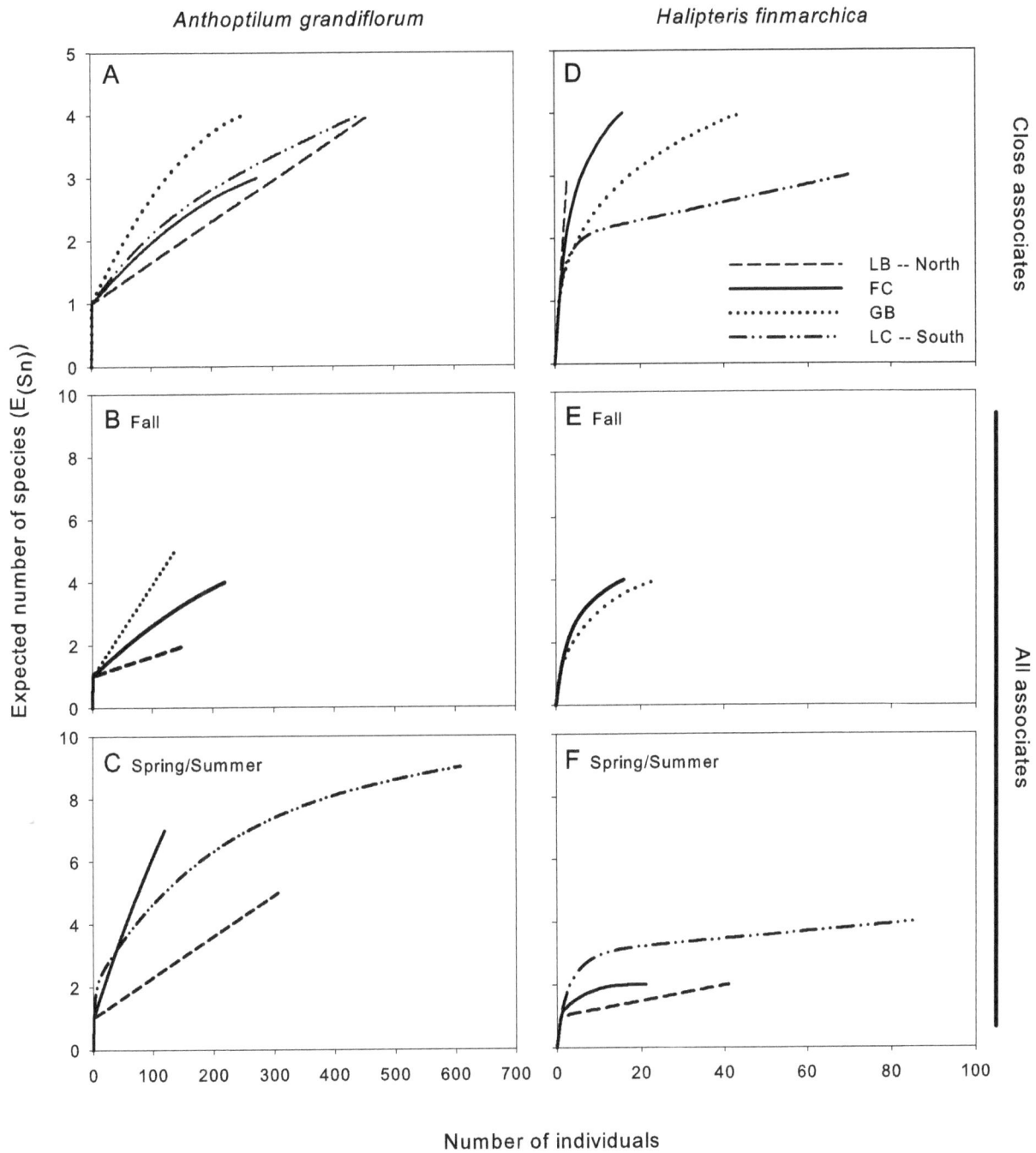

Figure 6. Rarefaction curves for *Anthoptilum grandiflorum* **(left panels) and** *Halipteris finmarchica* **(right panels), based on close associates only (top two panels) or all associates in spring and fall (bottom four panels).** (A) Close associates of *A. grandiflorum*; (B) all the associated fauna of *A. grandiflorum* in the fall and (C) in spring/summer. (D) Close associates of *H. finmarchica*; (E) all the associated fauna of *H. finmarchica* in the fall and (F) in spring/summer. LC: Laurentian Channel, GB: Grand Banks, FC: Flemish Cap, NNL: North Newfoundland, LB: Labrador.

Relationship between Hosts and Dominant Associates

Lamippe bouligandi **in** *Anthoptilum grandiflorum.* A total of 1126 females and 23 males of the copepod *L. bouligandi* (MEYtot = 9.4±0.9 copepods colony^{-1}) were recorded from 118 colonies (15–84 cm) of *A. grandiflorum* (prevalence of 71.1%) from all five geographic regions under study. Eggs and nauplius larvae of *L. bouligandi* were found in association with 36 females (3.2%) in 18 sea pen colonies (10.8%) sampled year-round. Females mainly occurred singly in a polyp; whereas males were

always paired with a female. The female copepods measured 5.06±0.07 mm (Fig. 3B) while the males were smaller at 1.39±0.17 mm (Fig. 3C). Two females occurred in the same polyp on 25 occasions (in 18 sea pen colonies) while larger groups of 3–4 females were found in only 4 polyps distributed on 3 colonies sampled year-round. No seasonal pattern emerged to explain the pairings/groupings. Infestation was between 0.1 and 19.1% of the polyps in an affected colony (i.e. 1–50 polyps). Overall, most (57.6%) of the colonies had less than 2% of polyps

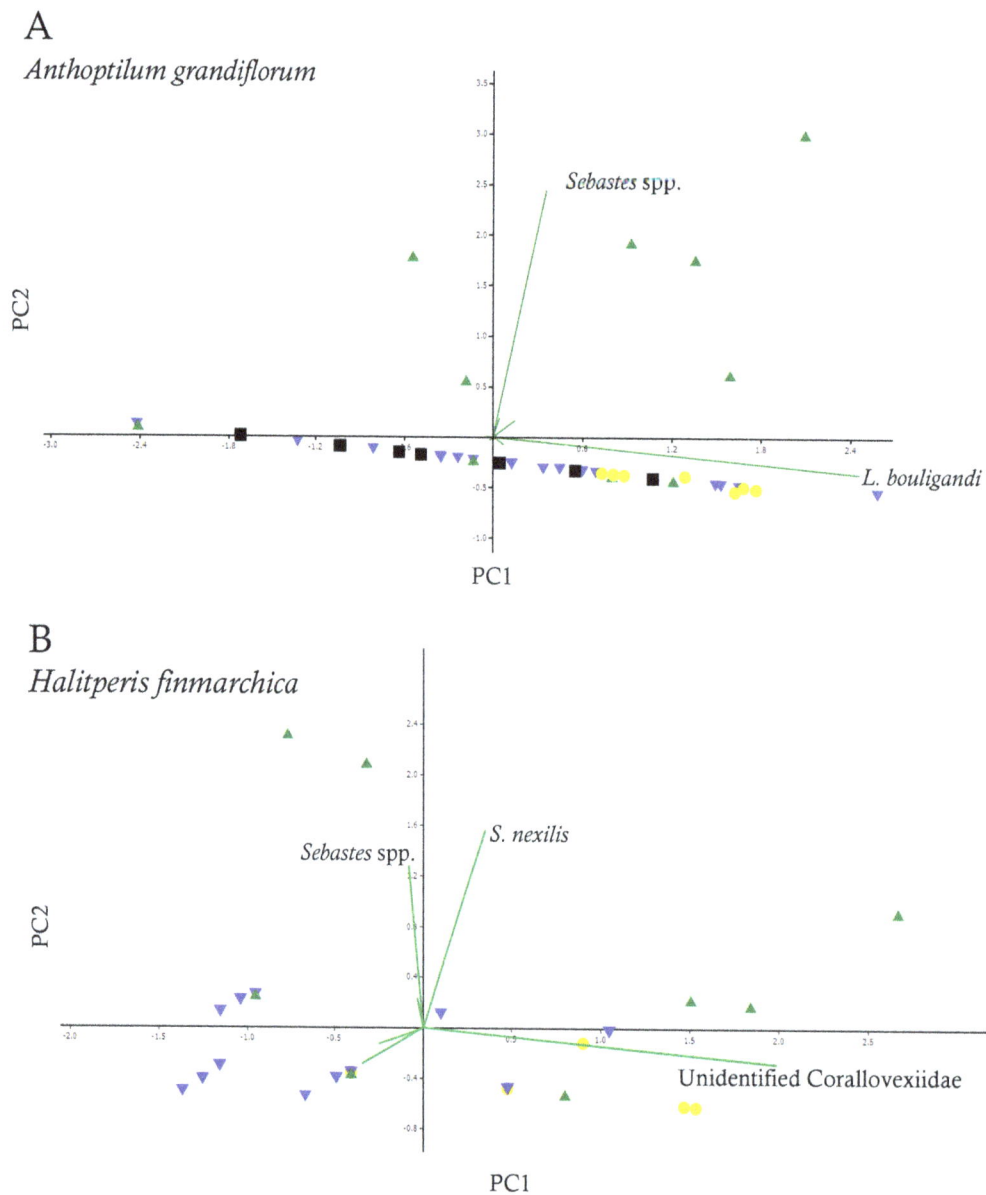

Figure 7. Principal component analyses and biplots based on the associated species of *Anthoptilum grandiflorum* **(A, B) and** *Halipteris finmarchica* **(C, D).** Green up triangle: spring; yellow circle: summer; blue down triangle: fall; black square: winter.

Table 3. Influence of increasing depth on total mean yield (MEYtot) of all associates on colonies of *Anthoptilum grandiflorum* and *Halipteris finmarchica* in the different geographic regions (only sites with more than 10 colonies harbouring associated species were used).

Species	Region	Depth range (m)	n	Spring/summer	Fall/winter
A. grandiflorum	LC	301–488	28	$r^2 = 0.07$, $F_{(1,27)} = 1.81$, $P = 0.190$	
	GB	98–603	13/24	$r^2 = 0.03$, $F_{(1,11)} = 0.34$, $P = 0.570$	$r^2 = 0.41$, $F_{(1,23)} = 15.06$, $P < 0.001$
	FC	273–1208	39		$r_s = -0.06$, $P = 0.734$
	LB	176–883	20	$r^2 = 0.02$, $F_{(1,19)} = 0.34$, $P = 0.569$	
H. finmarchica	GB	579–1333	11		$r^2 = 0.01$, $F_{(1,10)} = 0.05$, $P = 0.825$

LC: Laurentian Channel, GB: Grand Banks, FC: Flemish Cape, LB: Labrador, n: number of sea pen colonies analysed. Empty cells correspond to regions without enough data available for analyses.

Table 4. Prevalence of associates on colonies of *Anthoptilum grandiflorum* and *Halipteris finmarchica* in the different geographic regions (as percent colonies harbouring them).

		All regions combined	LC	GB	FC	NNL	LB
A. grandiflorum	All associates	75.9	67.5	74.0	70.0	58.3	96.8
	Endobionts	72.3	52.9	72.2	70.0	58.3	96.8
	Ectobionts	0.6	0.0	0.0	3.3	0.0	0.0
	Free-living	26.0	52.9	8.3	0.0	0.0	16.1
H. finmarchica	All associates	44.7	90.9	44.7	23.8	—	55.6
	Endobionts	37.2	45.5	36.8	23.8	—	55.6
	Ectobionts	20.0	63.6	18.4	4.8	—	0.0
	Free-living	4.3	27.3	0.0	4.8	—	5.6

Data also shown separately for endobionts, ectobionts and free-living associates. LC: Laurentian Channel, GB: Grand Banks, FC: Flemish Cape, NNL: North Newfoundland, LB: Labrador.

infested and only 3.4% of the colonies had >10% of polyps infested. An average of 44.0 ± 4.7 white/yellowish oocytes were present in non-infested polyps and measured 429.1 ± 11.7 µm. The infested polyps showed a significantly lower fecundity (19.6 ± 5.1 oocytes polyp^{-1}, representing a $45 \pm 6.9\%$ decrease in relative fecundity) than the non-infested polyps (t = 3.51, df = 8, P = 0.008), and they were translucent and significantly larger (520.5 ± 18.8 µm, U = 7591.5, P<0.001).

No influence of colony length on the yield (MEY) of copepods was found ($r_s = 0.17$, P = 0.057). A significant increase in the abundance of female copepods occurred from the lower to the upper section of the rachis (H = 77.71, df = 2, P<0.001), with 60.3% of females occupying the upper section. Positive correlations were found between the abundance of copepod and both polyp density ($r_s = 0.34$, P<0.001) and polyp diameter ($r_s = 0.31$, P = 0.005). Infestation with *L. bouligandi* occurred at all sampling depths. No correlation of MEY with depth ($r_s = -0.16$, P = 0.075) and no influence of season (H = 4.06, df = 3, P = 0.255) were detected. However, significant regional differences in MEY were evidenced (H = 13.49, df = 4, P = 0.009) between LB (16.07 ± 0.95 copepods colony^{-1}) and GB (5.04 ± 0.95 copepods colony^{-1}).

Undescribed Corallovexiidae in *Halipteris finmarchica*. A total of 112 females and 2 males copepods belonging to the Corallovexiidae (MEYtot $= 4.7 \pm 1.0$ copepods colony^{-1}) were recorded inside the polyps (Fig. 5A) of 28 colonies (21–132 cm) of *H. finmarchica* (prevalence of 29.8%) from all five geographic regions under study. When a male was found, it was always paired with a female. Females measured 4.52 ± 0.51 mm (Fig. 5B) and males were smaller at 0.73 ± 0.05 mm (Fig. 5C). Females occurred at the base of the polyp where reproductive cells typically grow (Fig. 5A). No oocytes or spermatocysts were observed in the infested polyps while the surrounding non-infested polyps harboured oocytes or spermatocysts. Overall, 61.6% of the female copepods were found in association with eggs/nauplii (Fig. 5D) at various times of the year. Contrarily to *L. bouligandi* in *A. grandiflorum*, a polyp never hosted more than one female Corallovexiidae. Infestation rates varied between 0.1 and 1.6% (1–20 infested polyps) in an affected colony with only five colonies (17%) harbouring more than 5 copepods.

MEY was not influenced by colony length ($r^2 = 0.07$, $F_{1,22} = 1.56$, P = 0.225). The middle section of the rachis showed greater infestation than the upper and lower sections (H = 13.46, df = 2, P<0.001), with 50% of the corallovexiids occurring there. This copepod was present at all depths sampled. Despite a significant decrease of the MEY with depth ($r_s = 0.52$, P = 0.010), no clear

threshold was detected; i.e. there was no significant difference among 100-m depth intervals (H = 7.24, df = 7, P = 0.404). Comparison among seasons showed a higher MEY in spring than in fall (H = 7.98, df = 2, P = 0.019). No significant regional differences were evidenced ($F_{2,20} = 2.39$, P = 0.117).

***Stephanauge nexilis* on *Halipteris finmarchica*.** A total of 28 sea anemones *S. nexilis* were found attached to the central axis of *H. finmarchica*, usually in the upper section of the rachis that was devoid of soft tissues (Fig. 4A and C). However, three small individuals were found surrounded by polyps (Fig. 4B). Sea anemones had a basal diameter ranging from 0.4 to 9.9 cm (3.4 ± 0.5 cm). Between 1 and 8 sea anemones (MEYtot $= 4.7 \pm 1.0$ anemones colony^{-1}) were found on 14 colonies of *H. finmarchica* (prevalence of 15.4%).

Stephanauge nexilis was present on *H. finmarchica* colonies from all sampling depths studied (366–1125 m) with no influence of depth on the MEY ($r_s = -038$, P = 0.178). However, this association was restricted to the southern regions (85.7% in LC and GB, and 14.3% in FC). No significant seasonal difference in MEY was found (U = 12, P = 0.142).

Trophic Interactions between Hosts and Dominant Associates. Analysis of isotopic ratios in tissues of the two sea pen species collected from LC showed they had similar $\delta^{13}C$ and $\delta^{15}N$ signatures (Table 5; $\delta^{13}C$: U = 3.5, P = 0.800; $\delta^{15}N$: U = 2.0, P = 0.533). No significant differences were detected between the sea anemone *S. nexilis* and its host *H. finmarchica* despite the fact that the sea anemone had a higher $\delta^{13}C$ (~1 ‰, t = −1.36, df = 6, P = 0.224) and $\delta^{15}N$ (~1 ‰, t = −2.42, df = 6, P = 0.052). Both sea pens and the sea anemone showed the same TL (Table 5). The two associated copepods had similar $\delta^{13}C$ and $\delta^{15}N$ signatures (Table 5; $\delta^{13}C$: t = −1.12, df = 3, P = 0.344; $\delta^{15}N$: t = −1.40, df = 3, P = 0.255). They had a significantly lower $\delta^{13}C$ (~2 ‰, $F_{(4,14)} = 22.16$, P<0.001) and a significantly higher $\delta^{15}N$ (~2 ‰, $F_{(4,14)} = 10.12$, P = 0.002) than their sea pen hosts (Fig. 8). On average, copepods were approximately half a trophic level (0.4–0.6) above their hosts.

Gastro-vascular contents analysed in 8 of the sea anemones (28.5%) comprised small pelagic invertebrates: amphipods, copepods and halocyprids (based on DNA; Fig. 4D–F).

Discussion

Different measures of biodiversity exist and its estimation depends on the number of species and the respective abundance of those species [50]. When considering only the close associates,

Table 5. Carbon and nitrogen stable isotope signatures ($\delta^{13}C$ and $\delta^{15}N$), and trophic level (TL) of *Anthoptilum grandiflorum* and *Halipteris finmarchica* and their dominant associates.

	n	$\delta^{13}C$ (‰)	$\delta^{15}N$ (‰)	TL
Anthoptilum grandiflorum	2	−20.9±0.1	11.3±0.8	3.0
Lamippe bouligandi	2	−22.7±0.2	13.4±0.1	3.4
Halipteris finmarchica	4	−21.3±0.4	10.5±0.3	2.7
Stephanauge nexilis	5	−20.4±0.3	11.5±0.2	2.9
Undescribed Corallovexiidae	3	−23.3±0.3	12.8±0.3	3.3

n: number of samples analysed (mean ± SE).

biodiversity expressed as $E_{(S150)}$ showed a higher diversity for *Halipteris finmarchica* than *Anthoptilum grandiflorum*. Both species exhibited a moderate number of associated species (see below) but additional associates still remain to be found based on the rarefaction curves. When all categories of associates were considered, $E_{(S170)}$ was similar between faunal associates of *A. grandiflorum* and *H. finmarchica*; however, the rarefaction curves showed that increasing sample size would yield a greater numbers of associates for *A. grandiflorum* probably due to the higher number of free-living species found in association with this host (see below). When removing the rarest species (single occurrences), the rarefaction curves reached an asymptote, suggesting that the most common associates of both sea pens have been sampled. The Shannon-Weiner index ascribed more even abundances to the associates of *H. finmarchica* than to those of *A. grandiflorum*. Associates of *A. grandiflorum* are clearly dominated by one species, i.e. the copepod *Lamippe bouligandi*. Associates of *H.*

finmarchica comprise two common species, an undescribed Corallovexiidae (Copepoda) and the sea anemone *Stephanauge nexilis*, resulting in a slightly more even distribution. Overall, specialized copepods emerge as the predominant associates of sea pens.

In general, measures of biodiversity associated with each sea pen species showed comparable patterns of variation with depth, region and season, irrespective of whether all or only close associates were considered, with a single exception outlined below. Variations in richness of faunal associates were not observed across depths in any of the analyses. A northward decrease was generally detected, except for the close associates of *H. finmarchica*, which showed a southward decrease. The northward decrease is in accordance with previous studies reporting a general decline of biodiversity with increasing latitude [51,52]. Variation in primary productivity over large spatial scales has been proposed to generate this trend [53]. The fact that associated biodiversity

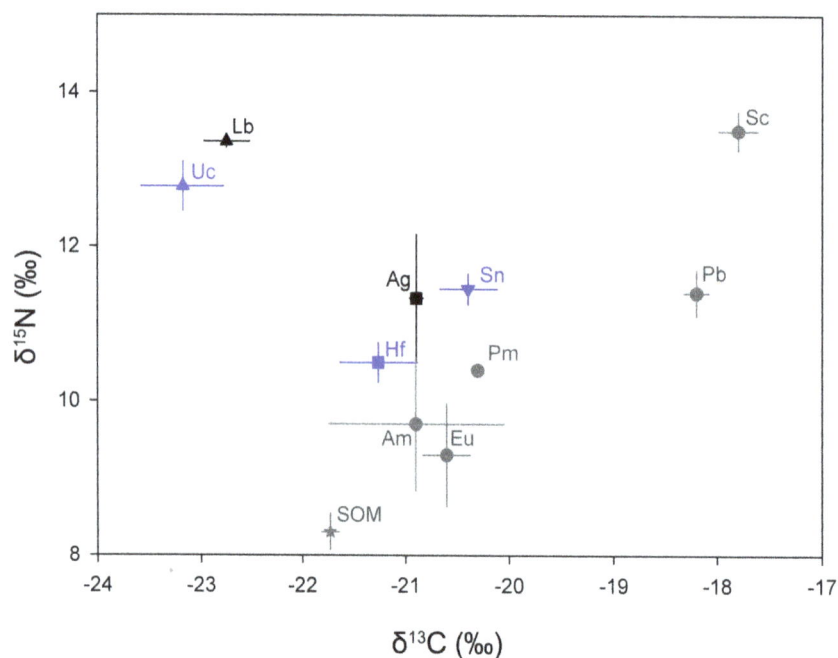

Figure 8. Stable isotope values ($\delta^{15}N$ and $\delta^{13}C$) for sea pens (Ag: *Anthoptilum grandiflorum* **and Hf:** *Halipteris finmarchica*) **and their associated species (Lb:** *Lamippe bouligandi*, **Sn:** *Stephanauge nexilis* **and Uc: undescribed Corallovexiidae).** To locate and compare the signature of the sea pens, values for other invertebrates are shown, Am: Amphipods, Eu: Euphausiids, Pm: *Pasiphae multidentata*, Pb: *Pandalus borealis* and Sc: snow crab from Sherwood & Ross [42], as well as sedimentary organic matter (SOM) from Muzuka & Hillaire-Marcel [41]. Result shown as mean ± SE (n = 2–5). Black: *A. grandiflorum* and its associates, Blue: *H. finmarchica* and this associates, Grey: other invertebrates and SOM.

showed different seasonal peaks for the two sea pens species is intriguing. The higher biodiversity in spring for *A. grandiflorum* may be explained by the presence of egg masses and early life stages of free-living species following spring reproductive events. However, no clear explanation emerges for the higher fall biodiversity associated with *H. finmarchica*.

Sea pen colonies studied here only yielded associated species on the rachis, and none on the peduncle. This is not unexpected since the peduncle is essentially buried in the sediment in both *A. grandiflorum* and *H. finmarchica*. However, a polychaete was recorded in association with the peduncle of sea pen colonies that had been maintained alive in the laboratory for a few weeks (including *A. grandiflorum* and *H. finmarchica*); the polychaete appears to be a new deep-sea species that feeds opportunistically on sea pen flesh [54]. An earlier report by Johnstone [29] described the presence of a parasitic copepod living in/on the half-buried peduncle of the shallow-water sea pen *Ptilosarcus guerneyi* from the North Pacific.

The number of associates identified in *A. grandiflorum* and *H. finmarchica* is similar to that reported in the shallow-water sea pen *P. guerneyi* [29] from Puget Sound (Northeast Pacific), suggesting that the biodiversity associated with pennatulacean octocorals may be consistent across regions and depths. It is apparently lower than that generally reported in deep-water branching corals, keeping in mind that comparison among taxonomic groups is often problematic, due to differences in methods and sampling effort. For example, 66 species have been found on seven partial colonies of the scleractinians *Madrepora oculata* and *Lophelia pertusa* sampled by trawl in the Mediterranean Sea [55]. As for octocorals in the order Gorgonacea (sea fans), 47 and 97 associated species have been found on 13 colonies/fragments of *Paragorgia arborea* and 45 colonies/fragments of *Primnoa resedaeformis*, respectively, that were either sampled by ROV or by trawl in the Northwest Atlantic [1]. The $E_{(S170)}$ of *P. arborea* and *P. resedaeformis* is around 18 and 38 expected species, respectively [1], which is 3–6 times higher than $E_{(S170)}$ in *A. grandiflorum* and *H. finmarchica* from the same geographic region. The difference in the diversity of associates likely results from the type of substrate/habitat offered by sea pens vs. sea fans, as well as from the inherently different biodiversity of their respective environments (soft vs. hard bottoms). Two different microhabitats occur in gorgonians: (1) living tissues in the young body parts of the colony and (2) exposed skeleton in the older body part of the colony [1]. The former harboured a lower biodiversity but the highest abundance of specialized associates. The greater biodiversity in the older/dead section is due to the capacity of sessile species to settle there, as also observed in dead sections of deep-sea scleractinians [56,57]. The moderate biodiversity associated with sea pens might therefore be due to the less frequent availability of exposed skeleton for other species to colonize. The central axis of sea pens is formed of collagen and calcite [58], and provides support to the colony; however, it apparently does not survive the colony's death for long since no dead skeletons were sampled here (personal observation) or reported previously. Some colonies of *H. finmarchica* showed no tissue on the older upper section where two ectobiotic species were found (sea anemone *S. nexilis* and a hydrozoan). The biodiversity in this older section was not consistently higher than elsewhere along the colony, which can be due to its small diameter and the smooth surface of the central axis, less favorable to settlement, as well as its susceptibility to erosion or to grazing predators [59].

Few ectobiotic species are reported on the living tissue of gorgonians and all are highly specialised symbionts [1]. Similarly rare ectobiotic species were identified on sea pens, none of which

were found on the living tissues of *H. finmarchica* and only one on the soft tissues of *A. grandiflorum*: an egg mass of the eelpout *Lycodes esmarkii*. Ectobiotic species are probably rare because soft corals, including sea pens, produce toxic chemicals acting as antifouling agents [60–62]. A study on the shallow-water pennatulacean *Renilla octodentata* confirmed the negative effect of those agents on the settlement of barnacle [63]. Chemicals, if present, seem to have a limited impact on colonisation by endobiotic species, which represent 87.7% of the associates recorded here. The ability of endobionts to colonize sea pen tissues might be explained by the fact that most of them are parasitic and have developed adaptations to thwart their host's defenses [64]. Overall, 38.6% of the colonies of *H. finmarchica* harboured endobionts compared to 66.7% of *A. grandiflorum*, suggesting that the former may be better protected against infestations. The rachis of *H. finmarchica* produces a larger quantity of mucus than that of *A. grandiflorum* (personal observation), which might create a barrier against settlement and mitigate infestation.

While chemical deterrents produced by corals may influence colonisation by ectobionts and endobionts, they are also known to deter predators [61,65]. Hence, corals may offer protective shelter to free-living associates. Keeping in mind that the sampling method (see below) likely underestimated the number of unattached faunal associates that derive shelter or food from sea pens, a clear difference in the number of free-living associates between the two sea pens was found. All 7 free-living associates were found on *A. grandiflorum* and only one (larvae of *Sebastes* spp.) on *H. finmarchica*. It is presumed that *A. grandiflorum* relies only on chemical defenses while *H. finmarchica* also harbours sclerites forming a calyce around the polyps (physical defense) [59]. However, the common observation of bare central axis in *H. finmarchica* suggests that this species is more often grazed than *A. grandiflorum*, possibly explaining why free-living associates might favour *A. grandiflorum*, which is predated by slow-moving species such as the sea star *Hippasteria phrygiana* [66]. Alternatively, the morphology of the two sea pens might explain this discrepancy. The elongated polyps of *A. grandiflorum* occur singly, while the polyp rows on *H. finmarchica* are fused at their base, forming ridges, as described by Williams [67]. Thus, *A. grandiflorum* is more "bushy" than *H. finmarchica*, which probably allows small invertebrates (e.g. shrimp larvae, copepods) and small vertebrates (e.g. fish larvae) to use *A. grandiflorum* for shelter and protection. The shallow-water sea pen *P. guerneyi* provides anchorage to various species against the tidal flow and a hiding place for small invertebrates, e.g. amphipods, caprellids and shrimps [29], emphasising the importance of sea pens as shelter and structural habitat. While *H. finmarchica* is a less likely shelter for free-living organisms, stomach contents of its ectobiont, the sea anemone *Stephanauge nexilis*, showed the presence of small invertebrates (free-living amphipods, copepods and halocypriods), suggesting their presence around colonies of *H. finmarchica*. The whip morphology of *H. finmarchica* may be less likely to retain small associates during sampling and lead to an underestimation of this type of association. Buhl-Mortensen and Mortensen [1] indicated that sampling of the associated species of deep-sea gorgonians by trawl led to the loss of most of the mobile crustaceans, which were sampled when using suction devices with a ROV. An additional challenge is that some free-living associates of sea pens are present only during a specific life stage and/or a specific season: three different species of shrimp larvae (*Acanthephyra pelagica*, *Pandalus montagui* and *Pasiphaea multidentata*) were found here in April/May (spring) exclusively. Previously, fish larvae of *Sebastes* spp. were also found on both species of sea pens in April and May, prompting the suggestion that sea pens act as essential fish habitat [3]. The additional presence of shrimp larvae underscores the importance of sea pens for the early life history of other species, including commercially harvested ones.

While transient free-living associates are important, the three most common associates found on both sea pens (*L. bouligandi* on *A. grandiflorum*, *S. nexilis* and undescribed Corallovexiidae on *H. finmarchica*) can be considered symbionts. *L. bouligandi* and the corallovexiid are endoparasites that spend most of their life history inside the polyp. While *A. grandiflorum* and *H. finmarchica* are sympatric species, their respective endoparasitic copepods are distinct. Lamippidae are adapted to their coral host [68] supporting the assumption that *L. bouligandi* is highly specific to *A. grandiflorum*. In contrast, Corallovexiidae are either mono-specific or found in 2 or 3 closely related coral hosts [49], suggesting that the corallovexiid in *H. finmarchica* might yet be found in other sea pens. Parasitic copepods in *A. grandiflorum* predominated in the upper rachis section, whereas those in *H. finmarchica* occurred mostly in the middle section. This trend can be explained by the greater density and larger diameter of polyps in these sections, which correspond to older polyps [59], and thus provide greater opportunity for infestation.

Both copepods had an impact on the polyps they infested: a total absence of oocytes/spermatocysts suggesting an inhibition of gametogenesis in *H. finmarchica*, and ~45% decrease in relative fecundity in *A. grandiflorum*. Parasitic copepods disrupt vitello-genesis (yolk deposition) either because they interfere with feeding or increase energy expenditure by the polyp (e.g. immune reaction). At the colony level, few polyps are infested, limiting the effect on total fecundity. Lamippidae were previously shown to increase mortality rates of sea pen hosts under stress (e.g. anoxic condition) despite their healthy appearance in optimal conditions [29]. Overall, copepods are the most common parasites identified in deep-sea octocorals [13]. Here, in addition to the two species discussed above, 7 individuals of an unidentified Lamippidae were recorded in the polyps of *H. finmarachica*. Furthermore, a copepod of the genus *Linaresia* was recently found in the polyps of a deep-sea gorgonian, *Paramuricea* sp., in the Northwest Atlantic [69].

The sea anemone *S. nexilis* found on *H. finmarchicus* is commonly reported from the Northwest Atlantic, between the Gulf of Mexico [70] and Labrador [71]. *S. nexilis* emerges as a facultative ectobiont with a low specificity for *H. finmarchica*. It is found attached to rocks, empty shells and sponges in the Gulf of Mexico [70]. The life history of this species is not known, but it can be hypothesised that it settles at the larval stage on the central axis of the sea pen and remains there due to the general absence of other suitable substrata where muddy seafloor dominates. Whether the absence of polyps around the sea anemones is a prerequisite to their settlement on sea pens, or an outcome of it, remains unclear. Some colonies of *H. finmarchica* exhibited a naked central axis without any visible ectobionts, suggesting that loss of soft tissue may precede colonization and supporting the assumption that the sea anemone is a commensal symbiont. On the other hand, a small number of sea anemones (probably newly settled) were observed to be closely surrounded by healthy tissues/polyps. Perhaps they initially settled on a small naked section of the colony and grew toward living tissues. It is not impossible that they are able to dislodge the polyps, which would correspond to a previously unreported case of parasitism.

The present study attempted to elucidate trophic relationships among sea pens and their principal associates. Previous work showed an increase of ~3.8‰ in $\delta^{15}N$ between prey and predator in polar and deep-sea environments [44]. Here, the endoparasitic copepods fell about half a trophic level above their sea pen hosts. Parasites are presumed to feed on a single source during a specific life stage [72], indicating that feeding on the host tissues should elicit a full trophic increase in $\delta^{15}N$, whereas feeding on the same food as the host should result in no difference between $\delta^{15}N$ of parasite and host [73]. The intermediate values recorded here suggest that copepods might use a mixed strategy. This hypothesis is supported by the location of the parasitic copepods inside the polyps, which suggests that they can both feed directly on sea pen tissues and feed on items ingested by the polyp. Johnstone (1969) proposed a similar hypothesis for the diet of *Lamippe* sp. associated with the shallow-water *P. gurneyi* based on its location and on the observation of orange material in its digestive tract (the color of the sea pen's tissues). Our isotopic results also confirm that the sea anemone and both sea pens feed on sedimentary organic matter in addition to small pelagic invertebrates [74]. However, the sea anemone is potentially targeting different prey based on small invertebrates found in their gastro-vascular cavity, which were not observed in the sea pen polyps, suggesting that the sea anemone is not competing directly with its host for food.

Overall, sea pens appear to have a moderate number of associated species, as previously hypothesized [8]. Nevertheless, sea pens play important roles in the life history of their associates. Some, such as parasitic copepods spend most of, possibly all, their life in association with sea pens and depend on them to survive and reproduce. The presence of the sea anemone on *H. finmarchica* confirms that sea pens offer a suitable biogenic substrate for other species. Sea pens are also important for mobile species such as fishes and shrimps that use them transiently as shelter during early life stages indicating that sea pens can be considered as biogenic habitat. However, the seasonality in these associations as well as the distribution of the sea pens (patchy occurrence of sea pen fields) emphasizes the difficulty in gaining a comprehensive understanding of their role as biogenic habitats. The sampling method used in this study (trawl by-catch) does not allow precise determination of functional interactions with free-living associates or a quantitative analysis, as some associates might be lost during sampling. However, this method is advantageous by permitting a large spatial and temporal coverage, as well as a large sample size, allowing the identification of spatiotemporal patterns which would not be possible with other sampling methods (e.g. ROV). Importantly, co-occurrences were not investigated here; only close (physical) associations. Recent studies have shown that the sea star *Mediaster bairdi* is usually found in sea pen fields in the Northwest Atlantic [66], and that lobsters often occur in association with sea pens in Norway fjords [75] suggesting that the contribution of pennatulacean corals to deep-sea biodiversity has yet to be fully elucidated.

Acknowledgments

We thank the scientific staff of Fisheries and Oceans Canada and the Canadian Coast Guard for helping us with sampling on board of the CCGS *Teleost*. We would also like to thank I. Dimitrove and K. Zipperlen for their assistance with histological processing, as well as G. Williams and two anonymous reviewers for their constructive comments.

Author Contributions

Conceived and designed the experiments: SB AM JFH. Performed the experiments: SB. Analyzed the data: SB AM. Contributed reagents/materials/analysis tools: AM. Wrote the paper: SB AM JFH.

References

1. Buhl-Mortensen L, Mortensen PB (2005) Distribution and diversity of species associated with deep-sea gorgonian corals off Atlantic Canada. In: Freiwald A and Roberts J M, editors. Cold-Water Corals and Ecosystems. Berlin: Springer-Verlag Berlin. pp. 849–879.

2. Longo C, Mastrototaro F, Corriero G (2005) Sponge fauna associated with a Mediterranean deep-sea coral bank. J Mar Biol Assoc UK 85: 1341–1352.

3. Baillon S, Hamel J-F, Wareham VE, Mercier A (2012) Deep cold-water corals as nurseries for fish larvae. Front Ecol Environ 10: 351–356.

4. Murillo FJ, Duran Munoz P, Altuna A, Serrano A (2011) Distribution of deep-water corals of the Flemish Cap, Flemish Pass, and the Grand Banks of Newfoundland (Northwest Atlantic Ocean): interaction with fishing activities. ICES J Mar Sci 68: 319–332.

5. Roberts JM, Wheeler AJ, Freiwald A (2006) Reefs of the deep: The biology and geology of cold-water coral ecosystems. Science 312: 543–547.

6. Buhl-Mortensen L, Mortensen PB (2004) Crustaceans associated with the deep-water gorgonian corals *Paragorgia arborea* (L., 1758) and *Primnoa resedae-formis* (Gunn., 1763). J Nat Hist 38: 1233–1247.

7. Buhl-Mortensen L, Mortensen PB (2004) Symbiosis in deep-water corals. Symbiosis 37: 33–61.

8. Buhl-Mortensen L, Vanreusel A, Gooday AJ, Levin LA, Priede IG, et al. (2010) Biological structures as a source of habitat heterogeneity and biodiversity on the deep ocean margins. Mar Ecol Evol Persp 31: 21–50.

9. Krieger KJ, Wing BL (2002) Megafauna associations with deepwater corals (*Primnoa* spp.) in the Gulf of Alaska. Hydrobiologia 471: 83–90.

10. Sherwood OA, Jamieson RE, Edinger EN, Wareham VE (2008) Stable C and N isotopic composition of cold-water corals from the Newfoundland and Labrador continental slope: Examination of trophic, depth and spatial effects. Deep-Sea Res I 55: 1392–1402.

11. Carlier A, Le Guilloux E, Olu K, Sarrazin J, Mastrototaro F, et al. (2009) Trophic relationships in a deep Mediterranean cold-water coral bank (Santa Maria di Leuca, Ionian Sea). Mar Ecol Prog Ser 397: 125–137.

12. Fry B (1988) Food web structure on Georges Bank from stable C, N, and S isotopic compositions. Limnol Oceanogr 33: 1182–1190.

13. Watling L, France SC, Pante E, Simpson A (2011) Biology of deep-water octocorals. Adv Mar Biol 60: 41–122.

14. NAFO (2008) Report of the NAFO SC working group on ecosystem approach to fisheries management (WGEAFM) response to fisheries commission request 9.a. Scientific Council Meeting, 22–30 October 2008, Copenhagen, Denmark. Serial No. N5592. NAFO SCS Doc. 08/24.

15. Donaldson A, Gabriel C, Harvey BJ, Carolsfeld J (2010) Impacts of fishing gears other than bottom trawls, dredges, gillnets and longlines on aquatic biodiversity and vulnerable marine ecosystems. Fisheries and Oceans Canada C.S.A.S.Ca-Canadian Science Advisory Secretariat.pp.90.

16. Hoare R, Wilson EH (1977) Observations on the behaviour and distribution of *Virgularia mirabilis* O.F. Muller (Pennatulacea) in Holyhead harbour, Anglesey. In: Keegan BF, Céidigh P and Boaden PSS, editors. Biology of Benthic Organisms Proceedings 11th European Marine Biology Symposium, Galway, 1976. Oxford: Pergamon Press pp. 329–337.

17. Troffe PM, Levings CD, Piercey GE, Keong V (2005) Fishing gear effects and ecology of the sea whip (*Halipteris willemoesi* (Cnidaria: Octocorallia: Pennatulacea)) in British Columbia, Canada: preliminary observations. Aquat Conserv-Mar Freshw Ecosyst 15: 523–533.

18. Brodeur RD (2001) Habitat-specific distribution of Pacific ocean perch (*Sebastes alutus*) in Pribilof Canyon, Bering Sea. Cont Shelf Res 21: 207–224.

19. Etnoyer P, Morgan L (2005) Habitat-forming deep-sea corals in the Northeast Pacific Ocean. In: Freiwald A and Roberts JM, editors. Berlin:Springer Berlin Heidelberg. pp. 331–343.

20. Nygren A, Pleijel F (2010) Redescription of *Imajimaea draculai* – a rare syllid polychaete associated with the sea pen *Funiculina quadrangularis*. J Mar Biol Assoc UK 90: 1441–1448.

21. Dalyell JG (1848) Rare and remarkable animals of Scotland represented from living subjects: with practical observations on their nature. London: J. Van Voorst. 270 p.

22. Baker KD, Wareham VE, Snelgrove PVR, Haedrich RL, Fifield DA, et al. (2012) Distributional patterns of deep-sea coral assemblages in three submarine canyons off Newfoundland, Canada. Mar Ecol Prog Ser 445: 235–249.

23. Tissot BN, Yoklavich MM, Love MS, York K, Amend M (2006) Benthic invertebrates that form habitat on deep banks off southern California, with special reference to deep sea coral. Fish Bull 104: 167–181.

24. Laubier L (1972) *Lamippe (Lamippe) bouligandi* sp. nov., copépode parasite d'octocoralliaire de la Mer du Labrador. Crustaceana 22: 285–293.

25. Humes AG (1978) Lichomolgid copepods (Cyclopoida), with two new species of Doridicola, from sea pens (Pennatulacea) in Madagascar. Trans Am Microsc Soc 97: 524–539.

26. Bouligand Y (1965) Notes sur la famille des Lamippidae, 3e partie. Crustaceana 8: 1–24.

27. Humes AG (1985) Cnidarians and copepods: a success story. Trans Am Microsc Soc 104: 313–320.

28. Dube MA, Ball E (1971) *Desmarestia* sp. associated with the sea pen *Ptilosarcus gurneyi* (Gray). J Phycol 7: 218–220.

29. Johnstone AK (1969) *Ptilosarcus guerneyi* (Gray) as a host for symbioses, with a review of its symbionts. Puget: University of Puget Sound. pp. 84.

30. Sankarankutty C (1961) On the porcellanid crab, *Porcellanella triloba* White (Crustacea-Anomura), a commensal on sea pen with remarks on allied species. J Mar Biol Ass India 3: 96–100.

31. Gilkinson K, Edinger E (2009) The ecology of deep-sea corals of Newfoundland and Labrador waters: biogeography, life history, biogeochemistry, and relation to fishes. Can Tech Rep Fish Aquat Sci. pp. 142.

32. Williams GC (2011) The global diversity of sea pens (Cnidaria: Octocorallia: Pennatulacea). PLoS One 6: e22747.

33. Williams GC (2013) *Halipteris finmarchica* (Sars, 1851) World Register of Marine Species. Accessed through: World Register of Marine Species athttp://www.marinespecies.org/aphia.php?p=taxdetails&id=128509

34. Wareham VE, Edinger EN (2007) Distribution of deep-sea corals in the Newfoundland and Labrador region, Northwest Atlantic Ocean. Bull Mar Sci 81(S1): 289–313.

35. Soong K (2005) Reproduction and colony integration of the sea pen *Virgularia juncea*. Mar Biol 146: 1103–1109.

36. Pires DO, Castro CB, Silva JC (2009) Reproductive biology of the deep-sea pennatulacean *Anthoptilum murrayi* (Cnidaria, Octocorallia). Mar Ecol Prog Ser 397: 103–112.

37. Baillon S, Hamel J-F, Wareham VE, Mercier A (2014) Seasonality in reproduction of the deep-water pennatulacean coral *Anthoptilum grandiflorum*. Mar Biol 161: 29–43.

38. Ivanova NV, Dewaard JR, Hebert PDN (2006) An inexpensive, automation-friendly protocol for recovering high-quality DNA. Mol Ecol Notes 6: 998–1002.

39. DeWaard JR, Ivanova NV, Hajibabaei M, Hebert PD (2008) Assembling DNA barcodes. Analytical protocols. Methods Mol Biol 410: 275–293.

40. Baillon S, Hamel J-F, Mercier A (2011) Comparative study of reproductive synchrony at various scales in deep-sea echinoderms. Deep-Sea Res I 58: 260–272.

41. Muzuka ANN, Hillaire-Marcel C (1999) Burial rates of organic matter along the eastern Canadian margin and stable isotope constraints on its origin and diagenetic evolution. Mar Geol 160: 251–270.

42. Sherwood GD, Rose GA (2005) Stable isotope analysis of some representative fish and invertebrates of the Newfoundland and Labrador continental shelf food web. Est Coast Shelf Sci 63: 537–549.

43. Nilsen M, Pedersen T, Nilssen EM, Fredriksen S (2008) Trophic studies in a high-latitude fjord ecosystem – a comparison of stable isotope analyses (δ^{13}C and δ^{15}N) and trophic-level estimates from a mass-balance model. Can J Fish Aquat Sci 65: 2791–2806.

44. Iken K, Bluhm BA, Gradinger R (2005) Food web structure in the high Arctic Canada Basin: evidence from δ^{13}C and δ^{15}N analysis. Polar Biol 28: 238–249.

45. Gale KSP, Hamel J-F, Mercier A (2013) Trophic ecology of deep-sea Asteroidea (Echinodermata) from eastern Canada. Deep-Sea Res I 80: 25–36.

46. Hurlbert SH (1971) The nonconcept of species diversity: A critique and alternative parameters. Ecology 52: 577–586.

47. Clarke KR, Warwick RM (2001) Change in Marine Communities: an approach to statistical analysis and interpretation. Plymouth Marine Laboratory, UK. 177 p.

48. Ratnasingham S, Hebert PDN (2007) BOLD: The Barcode of Life Data System (www.barcodinglife.org). Mol Ecol Notes 7: 355–364.

49. Stock JH (1975) Corallovexiidae, a new family of transformed copepods endoparasitic in reef corals with two new genera and ten new species from Curaçao. Foundation for Scientific Research in Surinam and the Netherlands Antilles 83: 1–45.

50. Pielou EC (1966) Shannon's formula as a measure of specific diversity: its use and misuse. Amer Nat 100: 463–465.

51. Rex MA, Stuart CT, Coyne G (2000) Latitudinal gradients of species richness in the deep-sea benthos of the North Atlantic. Proc Natl Acad Sci U S A 97: 4082–4085.

52. Rex MA, Etter RJ (2010) Deep-sea biodiversity: pattern and scale. Harvard University Press. 368 p.

53. Rex MA, Etter RJ, Stuart CT (1997) Large-scale patterns of species diversity in the deep-sea benthos. In: RFG. Ormond, JD. Gage and MV. Angel, editors. Marine Biodiversity: Patterns and Processes. Cambridge University Press. pp. 94–121.

54. Mercier A, Baillon S, Hamel J-F Life history and seasonal breeding of the deep-sea annelid *Ophryotrocha* sp. (Polychaeta: Dorvelleidae). Deep-Sea Res I 91:27–35.

55. Mastrototaro F, D'Onghia G, Corriero G, Matarrese A, Maiorano P, et al. (2010) Biodiversity of the white coral bank off Cape Santa Maria di Leuca (Mediterranean Sea): An update. Deep-Sea Res II 57: 412–430.

56. Mortensen PlB (2001) Aquarium observations on the deep-water coral *Lophelia pertusa* (L., 1758) (scleractinia) and selected associated invertebrates. Ophelia 54: 83–104.

57. Mortensen PB, Hovland M, Brattegard T, Farestveit R (1995) Deep water bioherms of the scleractinian coral *Lophelia pertusa* (L.) at 64°N on the Norwegian shelf: Structure and associated megafauna. Sarsia 80: 115–158.

58. Wilson MT, Andrews AH, Brown AL, Cordes EE (2002) Axial rod growth and age estimation of the sea pen, *Halipteris willemoesi* Kolliker. Hydrobiologia 471: 133–142.

59. Baillon S (2014) Characterization and role of major deep-sea pennatulacean corals in the bathyal zone of Eastern Canada. Biology. St John's: Memorial University. pp. 300.

60. Coll JC (1992) The chemistry and chemical ecology of octocorals (Coelenterata, Anthozoa, Octocorallia). Chem Rev 92: 613–631.

61. Changyun W, Haiyan L, Changlun S, Yanan W, Liang L, et al. (2008) Chemical defensive substances of soft corals and gorgonians. Acta Ecol Sin 28: 2320–2328.

62. Krug PJ (2006) Defense of benthic invertebrates against surface colonization by larvae: a chemical arms race. In: N Fusetani and AS Clare, editors. Antifouling compounds. Berlin: Springer-Verlag. pp. 1–53.

63. García-Matucheski S, Muniain C, Cutignano A, Cimino G, Faimali M, et al. (2012) Renillenoic acids: Feeding deterrence and antifouling properties of conjugated fatty acids in Patagonian sea pen. J Exp Mar Biol Ecol 416–417: 208–214.

64. Kaltz O, Shykoff JA (1998) Local adaptation in host-parasite systems. Heredity 81: 361–370.

65. Mackie AM (1987) Preliminary studies on the chemical defenses of the British octocorals *Alcyonium digitatum* and *Pennatula phosphorea*. Comp Biochem Physiol A Comp Physiol 86: 629–632.

66. Gale KSP (2013) Ecology of deep-sea Asteroidea from Atlantic Canada. Biology. St John's: Memorial University of Newfoundland. pp. 135.

67. Williams GC (1995) Living genera of sea pens (Coelenterata, Octocorallia, Pennatulacea) - Illustrated key and synopses. Zool J Linn Soc 113: 93–140.

68. Bouligand Y (1966) Recherches récentes sur les copépodes associés aux anthozoaires. In: Rees WJ, editor. The cnidaria and their evolution. New York: Academic Press. pp. 267–306.

69. de Moura Neves B, Wareham VE, Edinger E (2013) Geographic, bathymetric extension and new report of octocoral host (Cnidaria: Plexauridae) for an endoparasitic copepod (Copepoda: Lamippidae) in the north-west Atlantic. Sixth Annual Biology Graduate Student Symposium. Memorial University of Newfoundland, St John's.

70. Ammons AW, Daly M (2008) Distribution, habitat use and ecology of deepwater anemones (Actiniaria) in the Gulf of Mexico. Deep-Sea Res II 55: 2657–2666.

71. Fautin DG (2013) Hexacorallians of the World. http://geoportal.kgs.ku.edu/hexacoral/anemone2/index.cfm.

72. Lafferty KD, Allesina S, Arim M, Briggs CJ, De Leo G, et al. (2008) Parasites in food webs: the ultimate missing links. Ecol Lett 11: 533–546.

73. Iken K, Brey T, Wand U, Voigt J, Junghans P (2001) Food web structure of the benthic community at the Porcupine Abyssal Plain (NE Atlantic): a stable isotope analysis. Prog Oceanogr 50: 383–405.

74. Edwards DCB, Moore CG (2008) Reproduction in the sea pen *Pennatula phosphorea* (Anthozoa: Pennatulacea) from the west coast of Scotland. Mar Biol 155: 303–314.

75. Buhl-Mortensen P, Buhl-Mortensen L (2014) Diverse and vulnerable deep-water biotopes in the Hardangerfjord. Mar Biol Res 10: 253–267.

16

Coming to Terms with the Concept of Moving Species Threatened by Climate Change – A Systematic Review of the Terminology and Definitions

Maria H. Hällfors[1]*, Elina M. Vaara[1,4], Marko Hyvärinen[1], Markku Oksanen[2], Leif E. Schulman[1], Helena Siipi[2,3], Susanna Lehvävirta[1,5]

1 Botany Unit, Finnish Museum of Natural History, University of Helsinki, Helsinki, Finland, 2 Department of Behavioural Sciences and Philosophy, University of Turku, Turku, Finland, 3 Turku Institute for Advanced Studies, University of Turku, Turku, Finland, 4 Faculty of Law, University of Lapland, Rovaniemi, Finland, 5 Department of Environmental Sciences, University of Helsinki, Helsinki, Finland

Abstract

Intentional moving of species threatened by climate change is actively being discussed as a conservation approach. The debate, empirical studies, and policy development, however, are impeded by an inconsistent articulation of the idea. The discrepancy is demonstrated by the varying use of terms, such as *assisted migration*, *assisted colonisation*, or *managed relocation*, and their multiple definitions. Since this conservation approach is novel, and may for instance lead to legislative changes, it is important to aim for terminological consistency. The objective of this study is to analyse the suitability of terms and definitions used when discussing the moving of organisms as a response to climate change. An extensive literature search and review of the material (868 scientific publications) was conducted for finding hitherto used terms (N = 40) and definitions (N = 75), and these were analysed for their suitability. Based on the findings, it is argued that an appropriate term for a conservation approach relating to aiding the movement of organisms harmed by climate change is *assisted migration* defined as follows: *Assisted migration means safeguarding biological diversity through the translocation of representatives of a species or population harmed by climate change to an area outside the indigenous range of that unit where it would be predicted to move as climate changes, were it not for anthropogenic dispersal barriers or lack of time.* The differences between assisted migration and other conservation translocations are also discussed. A wide adoption of the clear and distinctive term and definition provided would allow more focused research on the topic and enable consistent implementation as practitioners could have the same understanding of the concept.

Editor: Paul Hohenlohe, University of Idaho, United States of America

Funding: MHH was supported by the University of Helsinki Research Fund and LUOVA – Finnish School in Wildlife Biology, Conservation and Management, EMV by World Design Capital Helsinki 2012/University of Helsinki & the International Design Foundation, MO by the Academy of Finland grant 218139, and SL by the Academy of Finland grant 126915. The funders had no role in study design, data collection and analysis, decision to publish, or preparation of the manuscript.

Competing Interests: The authors have declared that no competing interests exist.

* Email: maria.hallfors@helsinki.fi

Introduction

As the effect of climate change on biodiversity is becoming more evident through, e.g., spatial changes in species' suitable areas (e.g., [1–4]), translocation of organisms has been proposed to avoid the loss of biodiversity and to complement current conservation strategies. The idea was, to our knowledge, first proposed by Peters and Darling in 1985 [5], and nine years later termed *human-assisted dispersal* [6]. Since then, numerous other terms have been applied, including *assisted migration*, first used by Whitlock and Milspaugh in 2001 [7], *assisted colonisation* first used in 2007 [8], and *managed relocation* in 2009 [9]. In addition, the initial proposal [5] has also been articulated in various ways. Different terms have been used to refer to similar ideas, while one term may be used to denote different ideas.

The debate around the idea (see, e.g., [10]; and responses to [11]: [12–17]) has mostly focused on epistemic uncertainty, such as the possible negative effects of introduced species on a focal area, while the linguistic uncertainties involved have been neglected [18]. However, the diversity of terms and their usage predisposes the scientific discussion to confusion; see, e.g., treatments of the concepts of community and stability [19] and diversity indices [20]. Terminological confusion may lead to poor comparison of one study with another and can seriously hamper scientific development. This, in turn, perturbs public discussion and decision-making and, thus, harms efficient application [21–23], [19].

We argue that there is an evident risk for confusion as this new conservation approach is being discussed using different terms and definitions – especially since the measure is evaluated in different fields of science and society. Today, mainstream conservation aims at preserving biota within their current range and at protecting nature from human activity (e.g., [24–26]). Moving species to new areas will thus require changes in both conservation practises and regulation [27]. In the legal context, definitions often guide the interpretation of law. In some cases, too much or too little flexibility in the definitions of concepts may lead to conflicts when laws are interpreted. For example, in the USA, the legal concept of species defined in the ESA (Endangered Species Act 1973, Pub. L.

No. 93–205) has led to problems in conserving some red-listed species. Recent research has shown that the red wolf (*Canis rufus*) is a hybrid species, and as hybrids are not included in the definition of species given in the ESA, some stake-holders are trying to get the red wolf removed from the ESA listings [28]. The opinion of researchers in law, ecology, conservation biology, environmental ethics and other relevant fields, as well as the views of decision-makers and the public about the idea of moving species depends partly on how this idea is described and articulated. With a clear and concise definition the discussion could stay focused and relevant to conservation of biodiversity under climate change.

In this article we examine, through the hitherto proposed terms and definitions, the general idea of *moving organisms in response to climate change* and distinguish from it the more specific idea of *aiding the dispersal of species threatened by climate change*. We scrutinise two aspects of the original articulation of the idea: the term used to designate it and the definition of the term. The idea, the term, and the definition are interdependent as a concise definition enables communicating an idea to others, and a commonly followed terminology is essential to avoid confusion. Thus, one cannot concentrate on only one of the aspects and hope to clarify the whole concept.

Our aim is to recommend the most suitable term denoting the initial idea of *aiding the dispersal of species threatened by climate change*, and to formulate a standard definition of it, which consists of necessary and sufficient conditions to distinguish this idea from other related ones. We also describe the differences between this specific measure and other cases of translocation. We hope to provide a general, yet biologically valid, articulation of the new approach to facilitate discussion and application void of confusion caused by vague and inconsistent articulations or by definitions relating to conceptually other, however seemingly similar, ideas.

Materials and Methods

Literature review

To quantify the discussion on the proposed conservation approach and to generate data to analyse the prevailing terminology and definitions, we conducted a literature search. We used the search query ("assisted migration" OR "assisted coloni*ation" OR "managed relocation" OR "human-aided translocation" OR "assisted translocation" AND "climate change") to search for literature published in English up until the end of 2012. These terms represented our initial understanding on which might be the most commonly used terms for the idea. We included"AND climate change" since an omission of it resulted in a large number of hits that were irrelevant to this study.

To attain maximum coverage of the relevant scientific discussion we conducted the search in Google Scholar, ISI Web of Science, Scopus Elsevier, Hein Online and EBSCO (Academic Search Online). Additionally, we searched the reference lists of two review articles [29], [30]. We excluded publications that were irrelevant (i.e., did not discuss moving organisms under climate change) or did not include a specific term for the idea. We did additional searches on new terms that came up through this search, excluding some general terms that are also used in other contexts (like "translocation" or "assisted dispersal") as they proved to generate a large number of irrelevant hits (Table 1).

We acknowledge that the use of a specific term and definition in a certain publication is not independent from other publications. Quite often, the use of a term or definition in influential papers by highly-cited scientists may promote their adoption by others. Nevertheless, it depicts the actual use of the term. Moreover, all of the clauses we classified as definitions may not have been intended

as such by the authors. However, we considered them definition-like articulations and treated them as definitions in this analysis.

Analysis of terms

For the terminological analysis, we recorded all terms referring to moving species under climate change. When several terms were mentioned in a publication we chose the main one used throughout the text. In cases where several terms were used throughout the text we recorded them all (two to three). Thus, the total number of occurrences of terms is greater than the number of publications reviewed.

The approach was usually referred to using a so called 'complex term', which consists of two or more words (e.g., *managed relocation* or *facilitated dispersal*). One of the words (usually a noun) can be understood as the main term that is qualified by restrictive modifying terms (adjectives or adjectival phrases). For instance, in the complex term *assisted colonisation*, the main term is *colonisation* and the modifying term *assisted* singles it out from other instances of colonisation.

Most previous terminological discussion on this new conservation approach has focused on the main term. For example, *colonisation*, *migration*, and *introduction* have been thoroughly discussed by Hunter [31]. However, both the single words and the term as a whole are important when choosing a suitable term. We analysed the meanings of the main and modifying terms separately, as we think that in an emerging field such as this, the meanings of the words comprising a term are easily carried over from previous uses and the complex terms do not have established meanings beyond the meanings of their parts.

Analysis of definitions

In the analysis of definitions of the measure we included only peer-reviewed articles that in their title, abstract, or keywords mention a relevant term for the general measure. We used content analysis, a method that can be employed to identify patterns across qualitative data by calculating the frequency with which analysis units occur [32]. Our analysis units were single words or parts of sentences used in the definitions.

We followed the three-step view of content analysis by Miles and Huberman [33]. *Reduction* means that the data are selected and simplified by leaving out uninformative words, such as *and*, *or*, *is*. For *data display* (or *grouping*, cf. [34]) we identified similarities and differences of the analysis units and grouped them. Related words (e.g. move, movement, moving) were placed together into subgroups that were used to form larger groups that contain synonyms (e.g., threatened and endangered belong to the same group). Finally, *conclusion drawing* implies finding patterns from the previous steps: we placed groups referring to similar aspects of the approach into the same main category. For example, *moving*, *translocating* and *planting* were placed under the main category *action*.

To identify the exact meanings of the words used in the terms and definitions for the concept, and to evaluate the suitability of each word as part of the term or definition, we relied on interpretations from Oxford English Dictionary [35] and Collins Dictionary and Thesaurus [36]. E.M.V. and M.H.H. initially carried out the content analysis separately. Thereafter, they made a synthesis of their subjective views. Finally, the procedure was re-evaluated by S.L., for conflicting views.

Results and Discussion

The idea of moving species in response to climate change is discussed in various contexts, for example, in the conservation of

Table 1. Terms used in three or more publications.

Term	Times mentioned
Assisted migration	563
Assisted colonization	121
Managed relocation	94
Facilitated migration	26
Translocation	25
Human assisted migration	22
Assisted dispersal	14
Assisted translocation	8
Artificial translocation	8
Bening introduction	8
Assisted relocation	7
Managed translocation	7
Facilitated dispersal	5
Human assisted dispersal	5
Conservation introduction	4
Human assisted translocation	4
Transformative restoration	3

Other terms (used in one or two publications) are: adaptation assisted migration, assisted afforestation, assisted ecosystem migration, assisted population migration, assisted range expansion, assisted reintroduction, assisted species relocation, facilitated translocation, forestry assisted migration, human aided translocation, human assistance of dispersal, human assisted colonisation, human assisted establishment, human assisted migration management, human assisted relocation, managed migration, migration management, managed reintroduction, planned invasions process, plant refuge translocation, species rescue assisted migration, and trans situ conservation.

species facing a changing climate (cf. [5]) and in choosing the right provenances in forestry (e.g., [37]). Accordingly, the articulations of the idea differ substantially regarding what is to be moved where and why. Moreover, different authors speak about moving different kinds of units, such as individuals, populations, or species. We focus on this issue under *definition review - what* and talk about moving *species* until then for simplicity's sake.

Some authors refer to moving species *outside historic ranges* (e.g., [38]) and some *to more favourable regions* (e.g., [39]). Moving individuals beyond the range of the species is the core of many conceptualisation of the idea (e.g., [40]), although sometimes moving within the range is included in the discussion (e.g., [37]). Some authors distinguish between moving species over different distances, e.g. *assisted population migration* for "the movement of species within a species' established range" and *assisted range expansion*, for "the movement of species to areas just outside their established range"; and *assisted long-distance dispersal*, "the movement of species to areas far outside their established range" ([37]; see also [38]).

The motivation for the measure varies from a general anthropogenic threat (e.g., [11]; [40]) and an entailing need for conservation (e.g., [31]; [10]) to more specific reasons, such as managing commercial forests (e.g., [41]; [37]; [42]). IUCN [43] provides a definition of *assisted colonisation* (listing *benign introduction*, *assisted migration*, and *managed relocation* as synonyms) where the motivation for the measure is left open to include any threat to the focal species, not only climate change.

Through our literature search, we found 2983 records (Fig. 1). Of these, 868 mention moving species in connection to climate change using a specific term, and they form our data (Table S1). The data include 460 scientific peer-reviewed articles, 111 reports, 47 theses, 85 books or book chapters, and 165 popular or professional articles including published abstracts of congress presentations. The literature review established that a multitude of terms and definitions are used to describe the idea of *moving species as a response to climate change*. In the following sections we analyse the hitherto proposed definitions and terms to assess their suitability for describing the more specific idea of *aiding the dispersal of species threatened by climate change*. We discuss them by examining the modifying and the main term and the eight main categories identified in the concept analysis of the definitions.

Terminological review

Taylor and Hamilton [6] were the first to mention a specific term for the approach, in 1994. In the years 1996, 1998, and 2000 we found no specific terms referring to the measure (Fig. 2). Otherwise, up till 2006, we found one to eight publications mentioning the approach with a specific term. In 2007, it was mentioned by a term in 20 publications, and subsequently in more publications each year until the score for 2012 was 275. This steep increase in interest (Fig. 2) was probably stimulated by a combination of alarming predictions of the impacts of climate change on biodiversity and articles in high-profile scientific journals discussing the option of alleviating the impacts by moving species.

We found 40 different terms for the idea (Table 1). The most commonly used terms were *assisted migration* (mentioned 563 times; first by Whitlock and Milspaugh in 2001 [7]), *assisted colonisation* (121; by Holmes et al. in 2007 [8]), and *managed relocation* (94; by Richardson et al. in 2009 [9]) (Fig. 3).

To promote unbiased, relevant, comprehensive, and exclusive discussions and studies, the term for the approach should neither be highly value-laden nor widely used in other contexts. For example, a highly positively value-laden term might support the idea of moving species in response to climate change prematurely, without solid scientific support for the action. It is also important that the term is descriptive of the approach. The modifying term should delineate the main term, and together they should describe the focal act and communicate the action in an unbiased and unambiguous way.

The main term. *Colonisation* (used by numerous authors) means establishing colonies. In this context it implies that what is helped in moving is also helped in establishing a viable population at the new site. While this may sometimes be needed for successful conservation, in many cases dispersing the organisms would be enough. A possible problem with *colonisation* is that it might bring in negative connotations from invasion biology.

Dispersal (e.g., [44]) is central in discussions concerning climate change impacts on biodiversity and encompasses the concrete action of the new approach. Failure to disperse is the reason for the suggested need to help species move to new areas. Thus, *dispersal* would be suitable for a term describing the idea discussed here.

Introduction (e.g., [26]) is defined by the IUCN ([45] and [43]) as "the intentional or accidental dispersal by human agency of a living organism outside its historically known native range". As such, it is a much wider concept than moving organisms threatened by climate change for conservation purposes. Moreover, *introduction* may be associated with invasive alien species, which might hamper a neutral discussion. This is true also for *invasions process* [46]. *Reintroduction* [47], in a conservation context, is "movement and release of an organism inside its

PRISMA 2009 Flow Diagram

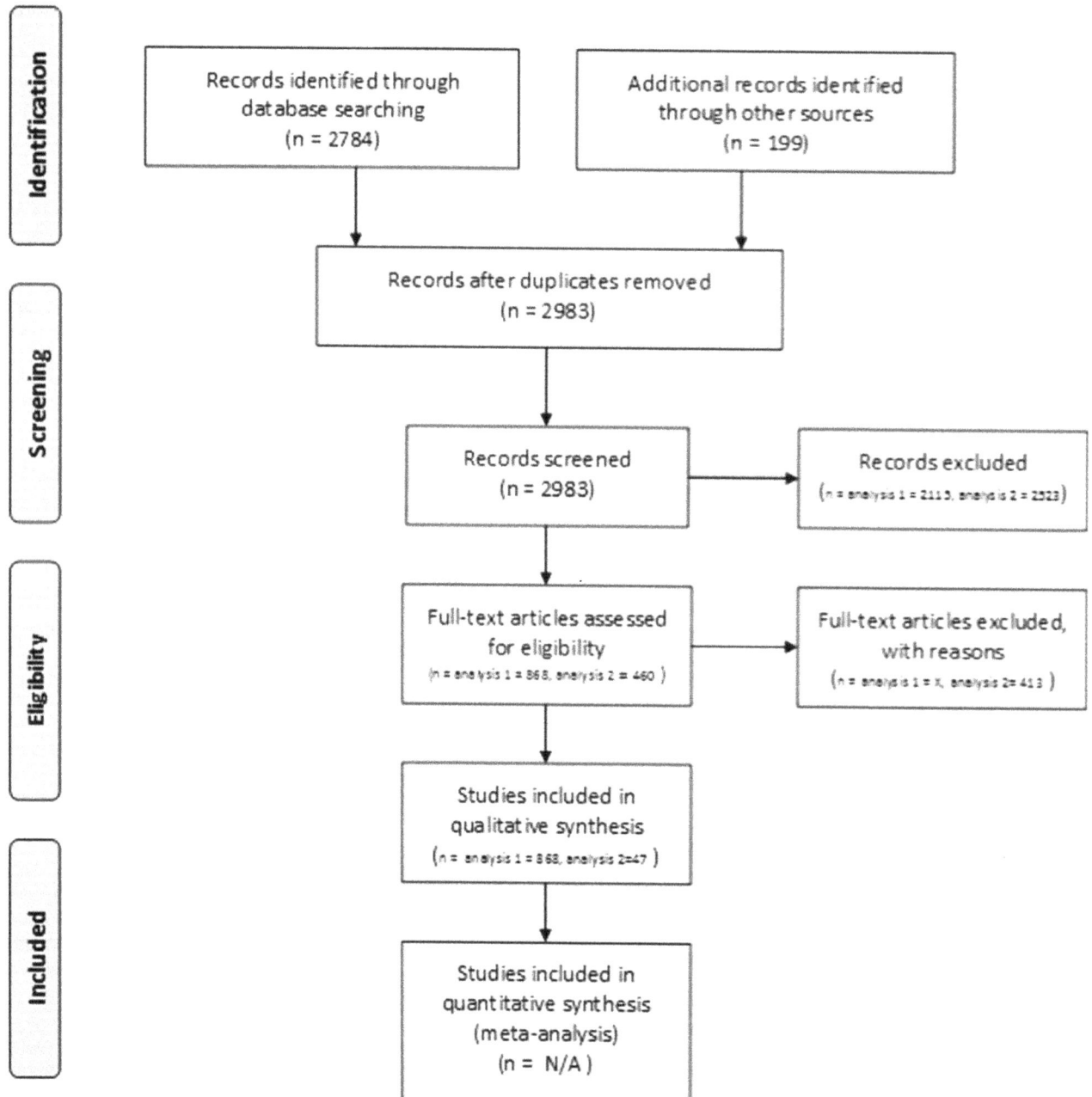

Figure 1. Work flow of the systematic review. 'Analysis 1' refers to the data used in the terminological analysis and 'Analysis 2' to the definition analysis.

indigenous range from which it has disappeared" [43]. Thus, it does not communicate the idea of moving organisms to new areas.

Migration (numerous authors) has been criticised because, in zoology, it is associated with seasonal or diurnal movements back and forth and would therefore not clearly capture the aim of establishing new populations [31], [48], [49], [40]. However, one

of the conventional meanings for the word *migration* is "extension of the distribution of a plant or animal" [35]. Hence, *migration* may be used in the term if associated with another descriptive word. *Migration management* [50] as a combined main term brings in nothing new, but puts the emphasis on *management*, making humans active managers of distribution areas. We argue

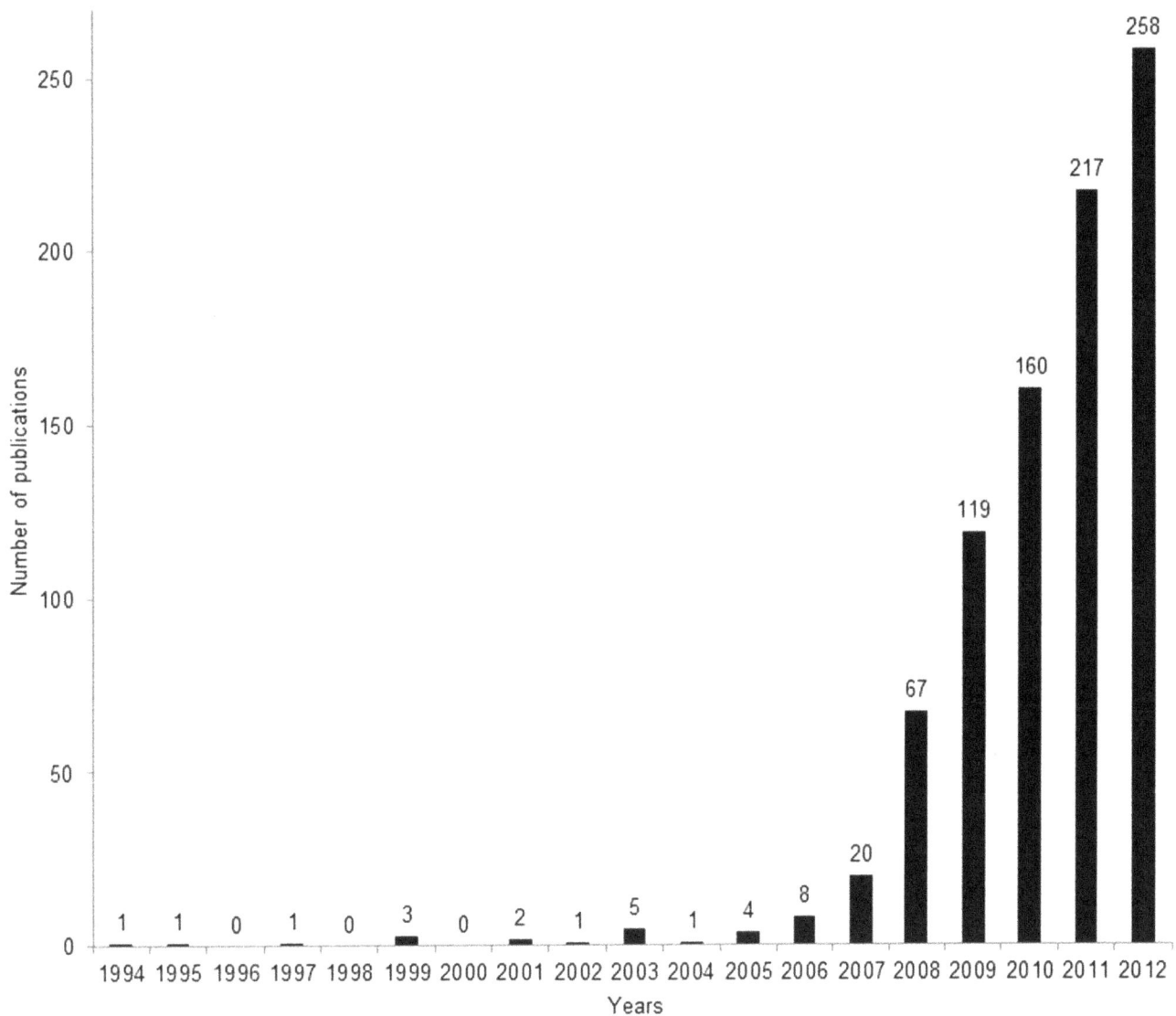

Figure 2. Number of publications mentioning a term for the measure. Number of publications per year (1994–2012) in which a term was mentioned for the measure entailing intentional human-mediated dispersal of organisms. The total number of publications mentioning a term was 868.

that the emphasis should instead be on the actual process that is being helped.

Range expansion [51] is not suitable here because expansion implies becoming larger. In many cases *range shifting* is a more appropriate description of what takes place. Other combined main terms that we found include *species dispersal* [52], *species relocation* [53], and *ecosystem migration* [54]. Although a specification of what is being moved could be useful, it may be misleading to include only one or a few units in the term (see below for a review on the *what*-part of the definition).

Relocation (numerous authors) refers to displacing individuals, but species or whole populations will not usually be actively relocated. *Translocation, introduction, relocation,* and *reintroduction* also suffer from redundancies if combined with active adjectives such as *assisted, human-aided, planned,* or *managed* since they imply human activity in themselves.

Restoration [55] and *afforestation* [56] emphasise the receiving area, not the organisms that would be moved. While these terms

are useful in the context of doing something to a degraded ecosystem, they are not descriptive of protecting threatened species by moving them.

According to the IUCN [47], [43] *translocation* [57] is an umbrella concept involving a variety of accidental or intentional "human-mediated movement[s] of living organisms from one area, with release in another" [43]. The IUCN [43] also defines a subcategory, *conservation translocation*, referring to translocating organisms specifically for conservation purposes. Thus, both *translocation* and *conservation translocation* are wider concepts that do not exclusively refer to the mitigation of a threat posed by climate. *Conservation translocation* includes situations where the organism is in danger due to other threat factors, such as land conversion. Thus, these terms are not restricted to approaches with a climate change dependent direction of the translocation.

Dispersal, colonisation and *migration* could be seen as a continuum ranging from singular *dispersal* events allowing the dispersed individuals to locally *colonise* a new site and finally

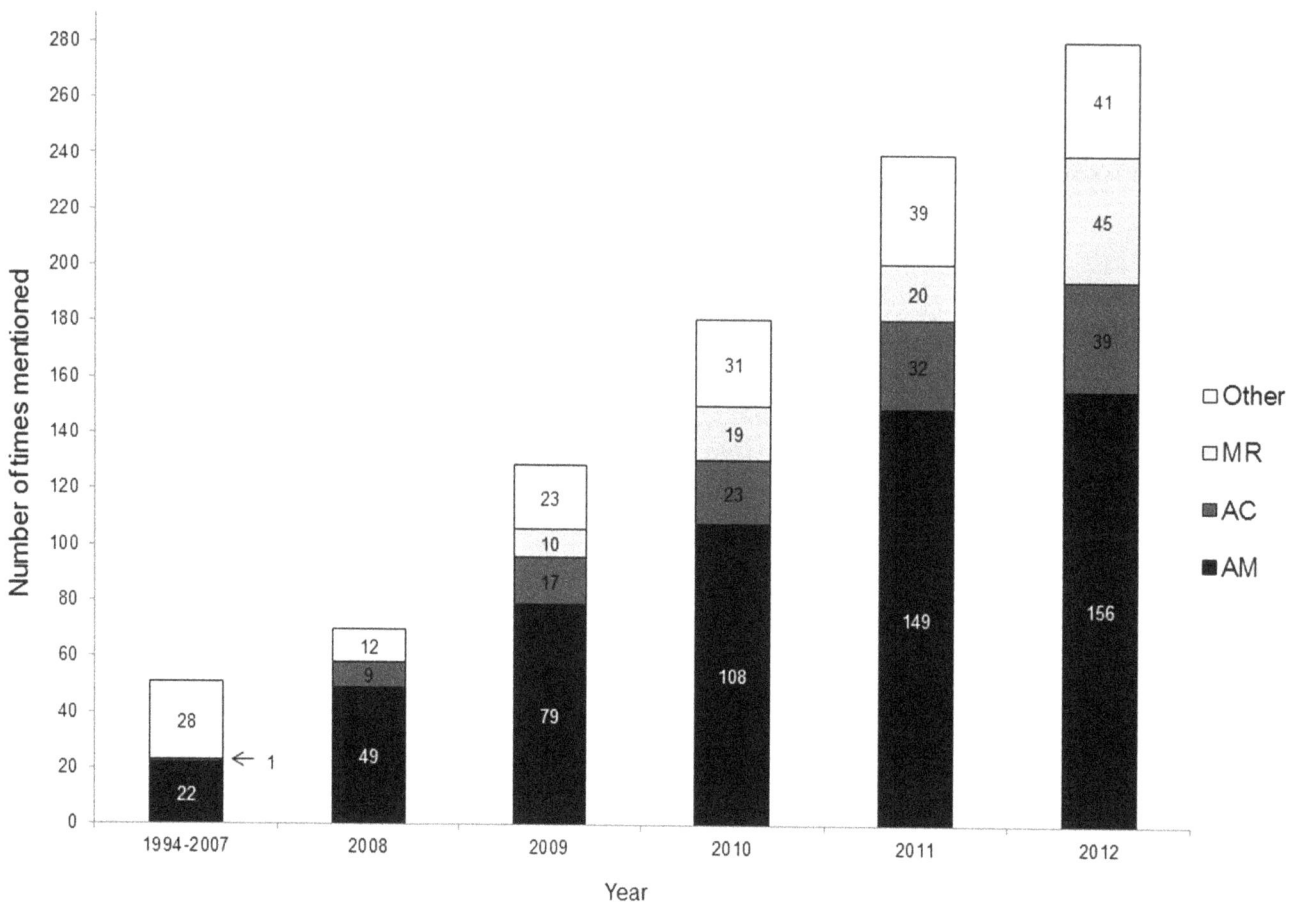

Figure 3. Number of times the three most common terms were used. Number of times the three most common terms denoting a conservation measure entailing intentional human-mediated dispersal of organisms in response to climate change were used as compared to other terms. AM = assisted migration; AC = assisted colonization; MR = managed relocation; Other = all other terms found in the literature search (N = 39; see Table 1).

resulting in *migration*, i.e., a change in distribution area. When helping organisms to move to new areas, it is essentially the *dispersal* event that is helped to enable local *colonisations* and, ultimately, *migration*. Colonisation could be helped as well, but that is not always necessary. In a directional, climate change motivated measure *dispersal* and, ultimately, *migration* is helped. As *migration* and *dispersal* thus both are descriptive words for the conservation approach, either could be a suitable main term to be used together with a modifying term.

The modifying term. *Artificial* (e.g., [58]) implies intentional human-made modifications and is related to such ambiguous terms as *unnatural*, *non-natural*, and *natural* [59–63]. In an environmental context, and especially when contrasted with *natural*, the term *artificial* may be value-laden, mostly negatively so, and may thus fail to fairly describe a conservation action.

Assisted (numerous authors) refers to helping and, hence, usually excludes accidental species introductions. *Assisted* therefore seems well suited for describing an intentional introduction. It can, however, be seen as positively value-laden to some degree. *Adaptation assisted* [64] would refer to the specific purpose of adaptation to, e.g., climatic change, and thus removes the emphasis from assisting migration. Likewise, combining *assisted* (or any other modifying term) with words such as *forestry*, *species rescue*, or *population* (as in *forestry assisted migration* or *assisted*

population migration; [65]) may be useful in specific cases, but in a general term such detail is not needed.

Benign [66] implies kindliness and a favourable outcome, and as such is positively value-laden to the extent of compromising objectivity. *Conservation* [43] indicates the purpose of an action, but is inclusive of any introductions with a conservational aim including those to areas where the organism would not disperse on itself driven by climate change.

Facilitated (e.g., [67]) does not contain the idea of the discussed approach (cf. [5–6]), but could rather refer to the already established conservation action of facilitating species movements through the construction of dispersal corridors enabling spontaneous dispersal.

Most of the modifying terms contain words that *per se* communicate human involvement. Thus, in *human-aided* [68] and *human-assisted* (e.g., [69]), *human* is redundant. *Human-assisted migration management* [50] brings human involvement into the term multiple times as *management* and even *assisted* are likely associated with human action. It also seems that *human-assisted dispersal* is already used for describing accidental dispersal of species to new areas (e.g., [70–71]).

Managed (numerous authors) has the meaning of being subject to control, guidance, and influence. It can be seen as positively or negatively value-laden, depending on one's attitudes. *Managed* also has the flavour of succeeding and coping with. It communicates that the action is handled in a well-ordered way, which indeed should be the aim when using any conservation approach. However, management often refers to a concrete and continuous intervention. Such continuous activity may not always be included in a conservation approach involving moving organisms in response to climate change: just helping their dispersal may be sufficient.

Planned [46] is a dispositional term. Planned actions follow a pre-set design, which is well in agreement with the idea of moving species. However, *planned* does not communicate the actual action of translocating: the action could be just planned but never conducted. *Transformative* [72] refers to something being rather radically altered and is thus not descriptive of the approach, since most practitioners envision minimal change in the ecosystems receiving new organisms.

Summing up, we suggest that *assisted* is the best-suited word for the first part of the term. It may be slightly positively value-laden and we acknowledge that if the measure is found unsuitable, any promotion of it conveyed by a term is undesirable. Compared to other candidates, however, *assisted* suffers from fewer downsides.

The complex term. To denote a conservation approach entailing intentional and directional moving of species threatened by climate change, the above analysis identified *dispersal* or *migration* as a suitable main term and *assisted* as the most apt modifying term. We pointed out that the essence of this approach is dispersal but that its ultimate aim is migration. Thus we find that the complex terms *assisted dispersal* and *assisted migration* are both suitable and descriptive. However, because of its wide use (Figs. 1 and 2) we propose *assisted migration* as the term for the focal conservation approach, the exact definition of which is discussed below.

Definition review

We found 84 definitions in 66 articles of the 130 peer-reviewed articles that mention a term for the approach in their title, abstract, or keywords (Fig. 1; Table S1). However, nine of the definitions (e.g. [73–74]) clearly had a different focus: they emphasised the management of the receiving *area* instead of the *unit* to be moved (species, population; Table S2). As this focus is fundamentally different from that of species-specific conservation, we included only the remaining 75 definitions from 60 publications in the analysis (TableS3).

In the content analysis, we found 485 analysis units, and classified them into 70 groups in eight main categories (see also Fig. 4 and Table S3):

1. *action* (what is done; 13 groups; 82 units);
2. *specification of action* (in what manner is something done; 9 groups; 35 units);
3. *what* (what is transferred; 6 groups; 97 units);
4. *specification of what* (8 groups; 25 units);
5. *where from* (the current location; 3 groups; 8 units);
6. *where to* (recipient area; 9 groups; 48 units);
7. *specification of where to* (13 groups; 99 units); and
8. *motivation* (9 groups; 87 units).

All definitions did not contain analysis units adhering to all of the main categories while others contained several units that were placed in the same group and/or main category. Some parts of the original definitions were split up in a way that reduced the detailed meaning of that part (see Table S3 for a description of how the definitions were divided into groups and main categories). For example, a part of a definition like "to a new area where more

favourable conditions prevail" was split up in three analysis units and was placed in major categories as follows: *area* (where); *new* (specification of where) and *where more favourable conditions prevail* (specification of where).

A definition of a term should communicate an idea in a concise and understandable manner. Ideally, a definition expresses the necessary and sufficient conditions for something to belong to the scope of the defined term, i.e., it captures all actions that the term covers and leaves out everything else [75]. With this kind of definition, everyone using the term can discuss, study, and apply the same thing, and confusion can be minimised.

Below, we evaluate how the definition should be worded to best describe this conservation approach. We conclude by constructing a definition using the articulation and categories we see necessary and descriptive for representing the idea.

Action. We constructed 13 groups referring to the *action* (Fig. 4). The definitions most commonly included some form of the word *move* (N = 32), which is neutral and, when used as a transitive verb, highlights the active nature of the method. *Move* would thus be suitable as part of the definition.

Words belonging to the group *transfer*, including *transfer* and *transport*) may refer to moving individuals or populations in their entirety, leaving nothing at the starting point. The same applies for *translocate* and *relocate*. However, *translocation* is a general concept in conservation biology referring to moving organisms (e.g., IUCN [45], [43]; and see above). Although it could be argued to bring in unnecessary confusion with related terms, the measure in question is a sort of translocation and thus the verb *translocate* could be suitable for describing the action as part of the definition of the focal approach.

Replenish implies reintroduction to a former distribution area, which is not the case with this approach. To *introduce* has a negative association from invasion ecology and would therefore not necessarily be appropriate for the definition (see review above). Although most introduced species don't have a negative impact on their new habitat, *introduction* has been defamed by the invasive ones. *Bring* could be a suitable substitute, but if used as part of a definition, it might emphasise the receiving area or the arrival phase, not what is being moved nor the entire process.

To establish refers to seeing to that the moved individuals also succeed. However, concrete actions to ensure this are not necessarily included, as a mere *release* of individuals or propagules in the new area may be sufficient. *Release*, in turn, is too specific, and refers to mobile species that can actually be released (cf. *planted*).

A *series of moves* may describe the actual application of the approach in some cases, but is unnecessarily detailed and exclusive. Similarly, to *mimic natural dispersal* and *facilitate range shift* render unnecessary complexity and imprecision. *Migrate* is unsuitable for describing the *action* in a definition since it does not convey the human action in itself, but requires an active verb (such as to *assist* the migration).

In summary, *move* and *translocate* are the most applicable words to describe the action. We propose to use *translocate*, because it is defined and established in the field of conservation (e.g., [45] and [43]). A new approach defined using this word would relate it to established translocation methodologies carried out for conservational purposes, albeit for other reasons than climate change.

Specification of action. Several definitions described the human involvement in the approach with words such as *assisted*, *human-aided*, or *mediated*. These we grouped together as referring to the idea of *assisted*. We discussed *assisted* under *Terminological review* and find it descriptive also for the definition.

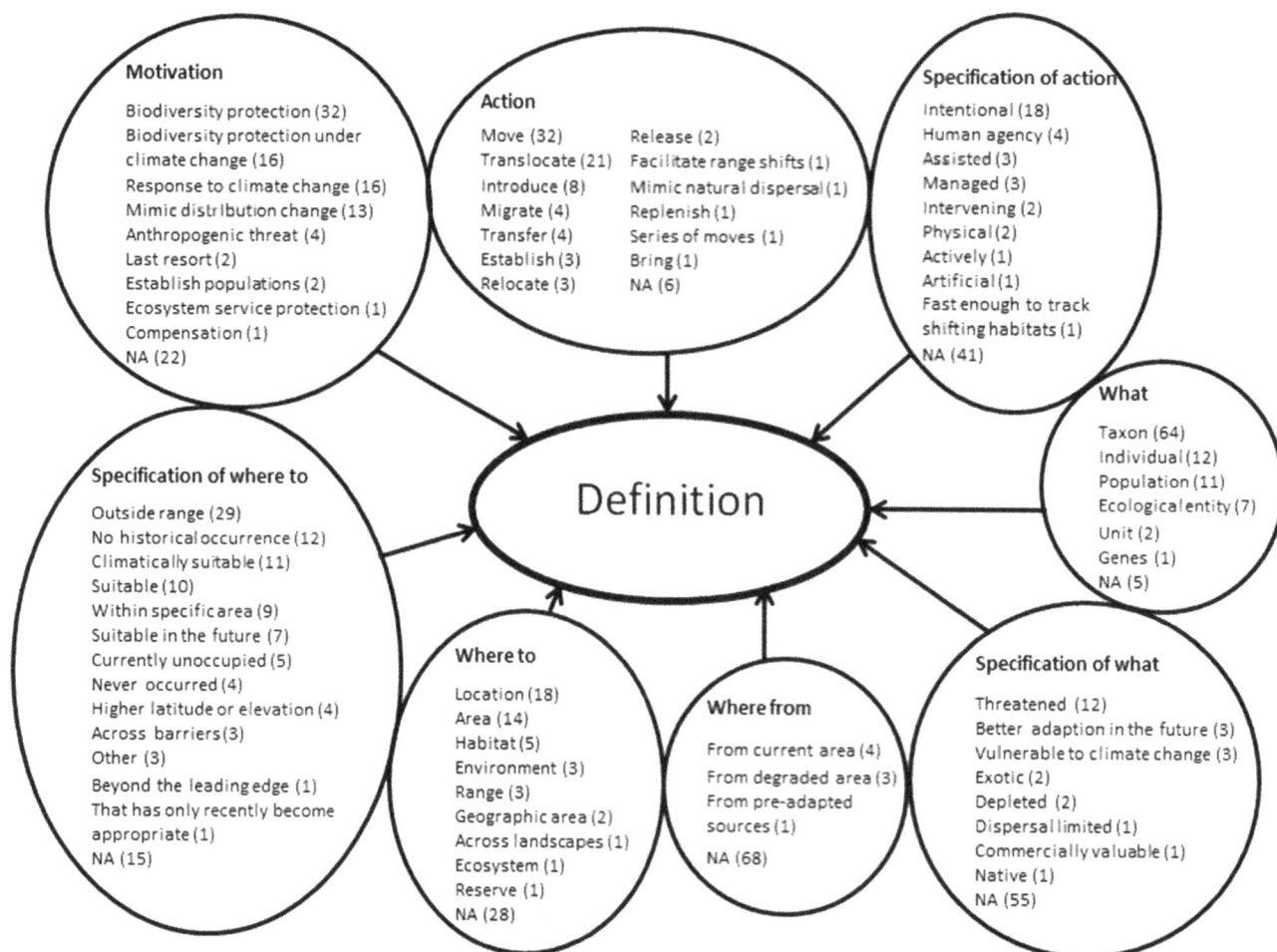

Figure 4. Main categories and groups formed from the definitions found in the literature review. The groups are divided into eight main categories (in bold; see text for further clarification). The numbers after each group refers to the total number of analysis units placed in that group. *NA* = the definition did not contain a part referable to this main category; the number denotes how many definitions lacked this part.

The approach was also described with wordings such as *by human agency*, *human intervention* or *proactive*. These refer to initiating change and a sense of human stewardship over nature, and may not clearly enough convey the adaptive rather than transformative notion of the approach.

Physical is not very informative since any moving of organisms can hardly be anything else. *Actively* is not exclusive, as also unintentional moving can be active. *Artificial* is reviewed above, and for the reasons described we do not recommend it as part of the definition. *Fast enough to track shifting habitats* emphasises the temporal dimension of the action and although it may be discussed as a detail in the implementation of the method, such level of detail seems unnecessary in a general definition.

The action was sometimes described using *purposeful* or *intentional*. This notion is apt in that it conveys the idea that something is done on purpose as opposed to accidentally and would rule out unintentional spreading of species.

Purposeful, assisted or intentional could all be useful for specifying the action. However, their suitability depends on the word describing the *action*. As we have found *translocation* to be the best word for describing it (see previous section), a specification by *purposeful*, *assisted*, or *intentional* would bring in tautology, as translocation is already established as intentional moving of species

by human agency. Thus, we do not think this part is needed in the definition.

What. Most definitions in this category referred to words that we grouped into *taxon* where the most common word was *species* (others were: *taxa*, *plants*, *animals*, *subspecies* and *ecotype*; see Fig. 4). Other groups we formed were *population*, *individual*, *ecological entity*, *genes*, and *units*.

As what is being moved is an essential part of the definition it is worthwhile considering this aspect a bit further. The threat of climate change may not appear the same for all populations within a species, since there may be both genetic and climatic differences between regions within the range of a species. Thus, this approach may be applied to only certain populations of a species, and the definition should also embrace cases where specific populations are moved *outside the population's current range*, but *within the species' current range*. Also, from a philosophical point of view, it is inappropriate to speak about moving *species*, since a species is an abstract construct used to describe patterns of recurrence in the living world or to refer to temporally delimited entities that operate within a continuous evolutionary process (e.g., [76–77]). Moreover, a species is intangible also in practice: an entire species cannot really be moved, or at least this cannot be verified. Instead, the levels at which this methodology would operate will more likely be the *individual* or *organism*, or a part of a *population*. *Genes* could be an inclusive way to describe what is being moved since they are always moved when any biological unit is moved. However, this would make the definition quite abstract and could also provoke connotations to genetic modification of organisms.

Some authors defined the unit to be moved as an *ecological entity (functional form, life form, flora, fauna, ecosystem)*. Even though moving certain parts of ecosystems may sometimes be the case, using these in the definition would divert the focus from species conservation toward conservation of ecosystem functions, which we feel should be kept separate to avoid confusion. Furthermore, the operationalization will have to happen at the level of individual organisms or their propagules, instead of broad, abstract groups.

In the group *unit*, we placed *biological units* and *focal units*, both of which allow the focus to lie case-specifically on species, populations, groups of species etc. They also allow the new area to be specified in relation to the current distribution of that particular unit (instead of, e.g., by the distribution of the species it belongs to). However, they may be vague for non-biologists.

On the basis of the aspects above, none of the descriptions seem quite suitable. Therefore, we propose to combine the best sides of the above suggestions and describe what is being translocated with *representatives of a species or population*. This wording allows the definition to include any suitable propagule and does not exclude smaller units than species.

Specification of what. In some definitions the kinds of units in focus were specified (Fig. 4). Most of them we grouped under *threatened* (e.g., *threatened, endangered* and *high priority*), while those that mentioned a characteristic of the unit as a reason for the threat were placed into *vulnerable to climate change* or *dispersal-limited*. *Exotic* is negatively value-laden due to terms such as *exotic species*, and *native* is not relevant here. *Depleted* is vague and would be more suitable if talking about reintroductions. *Commercially valuable*, in turn, underlines the economic value of the species over conservational ones.

Words such as *threatened* and *endangered* can be part of the definition but *high priority* is exclusive. However, since the translocation of species can be motivated by several reasons, including habitat destruction and over-exploitation, it is important to emphasise climate change as the main threat for the species. This is important because the recipient area is differently defined in these differing instances (see discussion under *specification of where to*). However, as *threatened* is a well-recognised category of the Red List, a rigid use of a definition containing it could result in using assisted migration only for red-listed species. In practise, the measure should be applicable also for species or populations that are not classified as threatened, but are adversely affected by climate change. In order to make a distinction between this climate change motivated directional approach and other, non-directional translocations, we suggest defining the species or population as *harmed by climate change*.

Where from. Only eight definitions mentioned from where the species should be moved, resulting in three groups: from current area (in situ, existing natural habitats, or current area of occupancy), from degraded area (habitat predicted to become unsuitable, hostile environments, or degrading ecosystems), and from pre-adapted sources. As species may be moved from different sources – e.g., directly from their natural populations or from ex situ collections when the in situ populations are too scarce to allow sufficient harvesting – the inclusion of where from in a definition is not necessary and would be too limiting.

Where to. Of the nine groups in this category, we think that *habitat*, *range*, *environment*, and *across landscapes* are too wide, while referring to a certain *geographic area* (e.g., *high latitude* or

mountain), *ecosystem*, or *reserve* is not general enough. *Area* can be interpreted as a wide entity (c.f. *distribution area*) while *location* refers to a more limited spatial entity, an exact place. Therefore we suggest using *area*. However, to properly indicate where something should be moved, most of the definitions included a *specification of where to*.

Specification of where to. This specification is a key part of the definition, since it demarcates the main difference between this and other conservation translocations. Several definitions mentioned translocating the species to where they do not exist (Fig 4). We grouped these analysis units according to their exclusiveness, from translocating to *other* areas, which does not specify the focal area at all, through *outside range* and *currently unoccupied*, further to *no historic occurrence*, and finally to the most restrictive *never occurred*. However, these are all too broad to single out translocations motivated by climate change, as they do not specify the direction of the movement and can thus imply introduction to any area outside the range of the species. Furthermore, several aspects make the use of historical species distributions as reference points ambiguous ([40]; and response [78]). Most importantly, there is no widely accepted definition of historical distribution. Various stakeholders may, hence, interpret it differently according to their needs, which in turn may lead to substantial confusion between disciplines and in practical applications. A further complication is that historical distribution is usually related to the concept of species, leading to problems when the operational unit of assisted migration is something else than a species, e.g., a locally adapted population. For example, Liu et al. [79] tested the difference in success of individuals moved outside the species' historical range vs. those moved within the range of the species. We would argue that for a specific population, which may be locally adapted, human-constructed boundaries between species and their ranges are irrelevant, and thus, operationalizing the idea in this manner leads to obscure experiments and conclusion drawing.

Many definitions specified the new area as being *suitable* (here including *favourable*) without any further specification, while others referred to areas *suitable in the future* or to an area *that has only recently become appropriate* with no reason specified. As we are trying to find a definition that could single out assisted migration from other conservation translocations, the climatic aspect is an essential criterion. Just describing the new area as *suitable* is ambiguous, since it implies any suitable area and contains neither a reason for the suitability nor a direction of the movement.

Other attributes specifying where the species should be moved were: across barriers, beyond the leading edge, to higher latitude or elevation, and within specific area that included, within natural range, to adjacent area, to contiguous environment, and to parts of the same biogeographic area. These are unnecessarily specific or too ambiguous for a general definition of the topic. For instance, using relative terms such as far outside and just outside [37] may be treacherous, as the definitions then also remain relative, hampering unambiguous operationalization and understanding.

A number of definitions specified the new place as *climatically suitable (climatically buffered* or *where climate is projected to become suitable)*, which sets the precondition that the suitability of the target site should be evaluated.

We suggest that when defining a conservation approach that aims at protecting a species or population harmed by climate change, the best description for the target area could be *outside its indigenous range where it would be predicted to move in response to climate change*. This formulation allows the definition to include the movement of organisms within the distribution area of a larger entity (e.g., populations vs. species), since *outside of its indigenous range* is related to the focal population. The description contains a precondition (*predicted to move in response to climate change*) whereby it excludes relocating a species or a population due to, e.g., land-use. It also includes the precondition of estimating where the suitable area will lie as climate changes. These estimations can be based on projections ranging from expert views to sophisticated models. In practice also other factors determining site suitability must be considered when choosing the specific sites within the climatically suitable area. However, site suitability is not a necessary part of a definition of the approach, but something requiring consideration during the planning of an actual assisted migration project.

It is important also to identify why the species or population would not reach the new area and why assisted migration is needed to ensure its continued survival. For this, a further condition is needed: where it would be predicted to move in response to climate change, *were it not for anthropogenic dispersal barriers or lack of time*. This specifies the spatial and temporal reality for why the focal species needs help: rapid climate change limits the time it has to disperse over natural barriers, and anthropogenic barriers are further dispersal obstacles.

Motivation. The underlying reason for the approach was part of almost all published definitions. The most common incentive was *biodiversity protection* (e.g. *conservation* and *reducing extinction risks*) followed by climate change, either as a *response to climate change* or more specifically *biodiversity protection under climate change*.

Also other motives were mentioned in the definitions, such as *anthropogenic threat, ecosystem service protection* or to *establish populations* in new areas. These motives imply different kinds of translocations for, e.g., conservational, cultural, or economic purposes and are not exclusive enough for this approach. Other anthropogenic threats, such as land-use, deforestation, over-harvesting, or pollution may all be good reasons for applying conservation translocations, but do not imply a movement following the direction of climate change.

To *mimic distribution change* was also suggested for explaining that the distribution area is changing. We think this verbalisation could be used if combined with climate change. However, it is not needed if the definition contains a description of *where* the biological unit is moved, which mentions climate change.

We also found some definitions containing the idea of the proposed measure as a *last resort* or as *compensation*. Although this may sometimes be the reality, it is unnecessary to delimit the application of the approach in this way, as for some species it may be the best alternative. Neither do we know whether it will be seen as a compensation for the harm done by humans to biodiversity or as way to sustain biodiversity for its instrumental value.

The motivation for the action narrows down the scope of application and thereby diminishes confusion. Therefore it is one of the necessary conditions that the definition should include. We suggest defining the motive of the measure through mentioning *safeguarding of biological diversity* in the definition. However, it is also crucial to mention climate change in order to separate the focal approach from other translocations. The motivation will become clear in the definition through *harmed by climate change* (specification of what), and *where it would be predicted to move in response to climate change* (specification of where).

Defining assisted migration. Based on the above analysis, we propose the definition "Assisted migration means safeguarding biological diversity through the translocation of representatives of a species or population harmed by climate change to an area outside the indigenous range of that unit where it would be predicted to move as climate changes, were it not for anthropogenic dispersal barriers or lack of time".

This definition follows the original idea of Peters and Darling [5], but elucidates it by bringing in the necessary conditions needed to separate it from other approaches that focus on moving organisms. Thus, only the actions that meet the definition should be regarded as assisted migration. The definition is both exclusive and inclusive. It includes translocations of threatened populations within a species' range, but it excludes translocations made to enhance economic activities (e.g., forestry) by restricting this measure to safeguarding biodiversity. It also excludes conservation translocations motivated by other threats than climate change, or targeted to areas outside the predicted climate change driven dispersal. It is important to be able to separate the motivation behind the focal action, as the motivation may be decisive as regards investment, prioritisation, and stakeholders both in research and society at large.

Assisted migration and related ideas

In Figure 5 we present the definitions for other translocation concepts and their relationships to each other and to assisted migration. *Translocation* (as defined by IUCN [45] and [43]) is an umbrella concept for several kinds of translocations, including assisted migration. In their guidelines for conservation transloca-tions, the IUCN [43] define *conservation introduction* as "the intentional movement and release of an organism outside its indigenous range" and subdivides this into two types: *assisted colonisation* and *ecological replacement*. *Conservation introduction* [66] and *assisted colonisation* [43] are defined as measures that could be used when no other options exist and the species cannot be re-enforced or translocated within its current range. However, they are not specifically related to climate change.

We argue that it is necessary to specify a type of translocation where dispersal is assisted because of a change in climate and, thus, in the suitable distribution area. We have not been able to identify any other force than climate change that would make a species' or population's current distribution area unfavourable while simultaneously making another area favourable. Anthropo-genic climate change thus requires a re-evaluation of the way we conserve biodiversity. For this, we need clear concepts. Any other translocation outside a species current range for conservation purposes should be called something else than *assisted migration*, which could then be reserved for translocations that are directional as a response to climate change. Hence, *assisted migration* should not be seen as synonymous with *assisted colonisation* but as a subcategory of it.

We noticed a discrepancy in the definition of the measure between the fields of conservation and forestry. Most of the forestry-related definitions emphasised a silvicultural viewpoint, which is not included in the original idea of assisted migration, which has to do with safeguarding biodiversity. Pedlar et al. [65] place forestry in the assisted migration debate by introducing the concepts *forestry assisted migration* and *species rescue assisted migration*. This is a movement in the right direction, since it distinguishes these two concepts, which are fundamentally different in their goal. Nevertheless, to avoid confusion, we suggest that other terms should be used for strategies seemingly similar to assisted migration but applied for purposes other than safeguard-ing biodiversity. Choosing suitable provenances in the context of agri-, silvi- or horticulture in relation to anticipated changes in climate could be called, e.g., *predictive provenancing* [80] to avoid confusion with *assisted migration*.

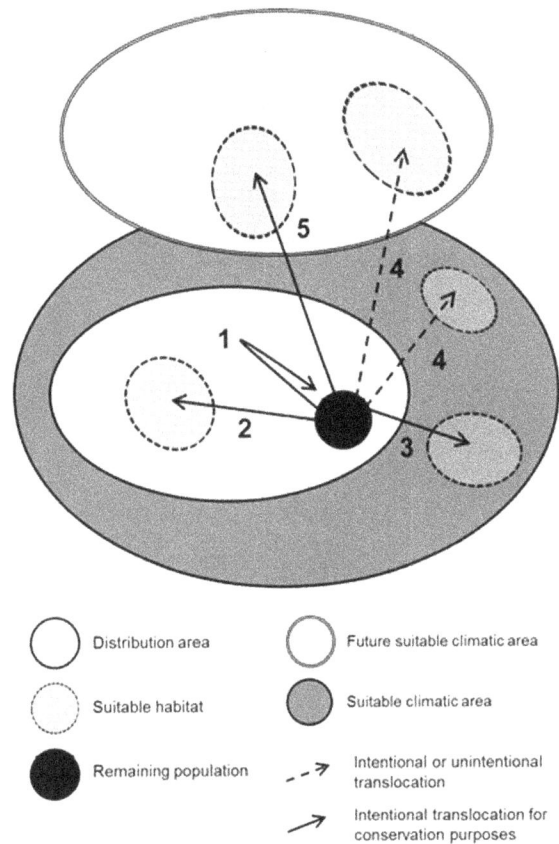

Term	Definition	Source
1. **Reinforcement**	The intentional movement and release of an organism into an existing population of conspecifics.	IUCN 2012
2. **Reintroduction**	The intentional movement and release of an organism inside its indigenous range from which it has disappeared	IUCN 2012
3. **Conservation introduction**	The intentional movement and release of an organism outside its indigenous range	IUCN 2012
4. **Introduction**	The intentional or accidental dispersal by human agency of a living organism outside its historically known native range.	IUCN 1987
5. **Assisted migration**	The safeguarding of biological diversity through the translocation of representatives of a species or population harmed by climate change to an area outside the indigenous range of that unit where it would be predicted to move as climate changes, were it not for anthropogenic dispersal barriers or lack of time	this paper

Figure 5. Definitions and a model of translocation concepts. According to the IUCN [45], *translocation* is defined as the movement of living organisms from one area with free release in another.

Conclusions

We have quantified the discussion on moving species as a response to climate change by reviewing the proposed terms and definitions for the general idea. Based on that, we propose a term and definition specifically for aiding species threatened by climate change to shift their ranges. Assisted migration is a kind of conservation translocation ([43]; Fig. 5), which can be distinguished from other conservation translocations by the following three aspects:

1. It is directional and based on a prediction of the potential future distribution of the biological unit;

2. It is limited to translocations as a way to overcome temporal or spatial dispersal limitations; and

3. It is used to mitigate threat caused directly or indirectly by anthropogenic climate change.

In these respects this measure is not only separable from other conservation translocations, but also clearly different from many kinds of species introductions discussed in the literature, where representatives of species are introduced far away from their original distribution areas, even to other continents where they would not disperse on their own. Such discussions include the invasive alien species problem (e.g., [81]), Colombian exchange [82], Pleistocene rewilding [83], and the interpretation of including the moving of polar bears from the Arctic to the Antarctic within the scope of assisted migration (cf. [84]).

A clear articulation of an idea enables better scientific operationalization. If the concept can be defined, hypotheses can be generated and tested, and the applicability of the method can be critically evaluated and, if deemed feasible, developed. Without conceptual clarity, there is a high risk for doing confused science that does not actually test what was intended. The concept can be supported by a well-constructed definition and a suitable term for the idea. All this is highly relevant in the development or restriction of a conservation strategy and methodology where scientific theory, ethical considerations, legislation, and application need to be inter-related and fluent to communicate without losing focus.

Acknowledgments

We are grateful to S. Aikio, K. Kokko, J. Kotze, T. Ryttäri, and S. Venn for discussions on various aspects of assisted migration. We thank I. Lehtimäki for assistance with preparing data tables. J. Hellmann, M. Ahteensuu and eight anonymous reviewers provided useful comments to earlier versions of the manuscript.

Author Contributions

Conceived and designed the experiments: MHH EMV SL. Performed the experiments: MHH EMV. Analyzed the data: MHH EMV MH MO LES HS SL. Contributed reagents/materials/analysis tools: MHH EMV MH MO LES HS SL. Wrote the paper: MHH EMV MH MO LES HS SL.

References

1. Devictor V, Julliard R, Couvet D, Jiguet F (2008). Birds are tracking climate warming, but not fast enough. Proceedings of the Royal Society Biological Sciences 275: 2743–8.
2. Maschinski J, Ross MS, Liu H, O'Brien J, von Wettberg EJ, et al. (2011) Sinking ships: conservation options for endemic taxa threatened by sea level rise. Climatic Change 107: 147–167.
3. Thomas CD, Cameron A, Green RE, Bakkenes M, Beaumont LJ, et al. (2004) Extinction risk from climate change. Nature 427: 145–8.
4. Thuiller W, Lavorel S, Araújo MB, Sykes MT, Prentice IC (2005) Climate change threats to plant diversity in Europe. Proceedings of the National Academy of Sciences of the United States of America 102: 8245–50.
5. Peters RL, Darling JDS (1985) The Greenhouse Effect and Nature Reserves. Bioscience 35: 707–717.
6. Taylor D, Hamilton A (1994) Impact of climatic change on tropical forests in Africa: Implications for protected area planning and management. In: Pernetta J, Leemans R, Elder D, Humphrey S (Eds.), Impacts of Climate Change on Ecosystems and Species: Implications for Protected Areas. IUCN, Gland, Switzerland.
7. Whitlock C, Millspaugh SH (2001) A paleoecologic perspective on past plant invasions in Yellowstone. Western North American Naturalist 61: 316–327.
8. Holmes B (2007) Assisted migration: Helping nature to relocate. New Scientist 196: 46–49.
9. Richardson DM, Hellmann JJ, McLachlan JS, Sax DF, Schwartz MW, et al. (2009) Multidimensional evaluation of managed relocation. Proceedings of the National Academy of Sciences of the United States of America 106: 9721–4.
10. Minteer BA, Collins JP (2010) Move it or lose it? The ecological ethics of relocating species under climate change. Ecological Applications 20: 1801–1804.
11. Ricciardi A, Simberloff D (2009a) Assisted colonization is not a viable conservation strategy. Trends in ecology & evolution 24: 248–53.
12. Fazey I, Fischer J (2009) Assisted colonization is a techno-fix. Trends in ecology & evolution 24: 476–7.
13. Ricciardi A, Simberloff D (2009b) Assisted colonization: good intentions and dubious risk assessment. Trends in Ecology & Evolution 24: 476–477.
14. Sax DF, Smith KF, Thompson AR (2009) Managed relocation: a nuanced evaluation is needed. Trends in ecology & evolution 24: 472–473.
15. Schlaepfer M, Helenbrook W, Searing K, Shoemaker K (2009) Assisted colonization: evaluating contrasting management actions (and values) in the face of uncertainty. Trends in Ecology and Evolution 24: 471–472.
16. Schwartz M, Hellmann J, McLachlan J (2009) The precautionary principle in managed relocation is misguided advice. Trends in Ecology and Evolution 24: 474.
17. Vitt P, Havens K, Kramer AT, Sollenberger D, Yates E (2010) Assisted migration of plants: Changes in latitudes, changes in attitudes. Biological Conservation 143: 18–27.
18. Kujala H, Burgman MA, Moilanen A (2013) Treatment of uncertainty in conservation under climate change. Conservation Letters 6: 73–85
19. Shrader-Frechette KS, McCoy ED (1993) Method in ecology: strategies for conservation. Cambridge University Press, Melbourne, Australia.
20. Tuomisto H (2011) Commentary: do we have a consistent terminology for species diversity? Yes, if we choose to use it. Oecologia 167: 903–911.
21. Jax K (2006) Ecological Units: Definitions and Application. The Quaterly Review of Biology 81: 237–258.
22. Peters RH (1991) A critique for ecology. Cambridge University Press, Melbourne, Australia.
23. Regan HM, Colyvan M, Burgman MA (2002) A taxonomy and treatment of uncertainty for ecology and conservation biology. Ecological Applications 12: 618–628.
24. Araújo MB, Alagador D, Cabeza M, Nogués-Bravo D, Thuiller W (2011) Climate change threatens European conservation areas. Ecology letters 14: 484–92.
25. Hannah L, Midgley GF, Lovejoy T, Bond WJ, Bush M, et al. (2002) Conservation of Biodiversity in a Changing Climate. Conservation Biology 16: 264–268.
26. Harris JA, Hobbs RJ, Higgs E, Aronson J (2006) Ecological Restoration and Global Climate Change. Restoration Ecology 14: 170–176.
27. Schwartz MW, Hellmann JJ, McLachlan JM, Sax DF, Borevitz JO, et al. (2012) Managed Relocation : Integrating the Scientific, Regulatory, and Ethical Challenges. BioScience 62: 732–743.
28. Doremus H (2010) The Endangered Species Act: Static Law Meets Dynamic World. Washington University journal of law and policy 32: 175.
29. Hewitt N, Klenk N, Smith AL, Bazely DR, Yan N, et al. (2011) Taking stock of the assisted migration debate. Biological Conservation 144: 2560–2572.
30. Loss SR, Terwilliger LA, Peterson AC (2011) Assisted colonization: Integrating conservation strategies in the face of climate change. Biological Conservation 144: 92–100.
31. Hunter ML (2007) Climate change and moving species: Furthering the debate on assisted colonization. Conservation Biology 21: 1356–1358.
32. Carley K (1993) Coding Choices for Textual Analysis: A Comparison of Content Analysis and Map Analysis. Sociological methodology 23: 75–126.
33. Miles M, Huberman A (1994) Qualitative Data Analysis: An Expanded Sourcebook, 2nd ed. Sage publications, Thousand Oaks
34. Dey I (1993) Qualitative Data Analysis. User Friendly Guide for Social Scientists. Routledge, London.
35. Oxford English Dictionary. Available: www.oed.com. Accessed 2012 March 14.
36. Collins Dictionary. Available: www.collinsdictionary.com. Accessed 2012 March 14.
37. Ste-Marie C, Nelson EA, Dabros A, Bonneau M-E (2011) Assisted migration: Introduction to a multifaceted concept. Forestry Chronicle 87: 724–730.
38. Gray LK, Hamann A (2011) Strategies for reforestation under uncertain future climates: guidelines for Alberta, Canada. PLOS ONE 6:e22977.
39. Ruhl JB (2010) Assisted Colonization: Facilitate Migration First. Science 330: 1317–1318.
40. Seddon PJ (2010) From Reintroduction to Assisted Colonization: Moving along the Conservation Translocation Spectrum. Restoration Ecology 18: 796–802.
41. Leech SM, Almuedo PL, Neill GO (2011) Assisted Migration : Adapting forest management to a changing climate. BC Journal of Ecosystem and Management 12: 18–34.
42. Ukrainetz NK, Neill GAO, Jaquish B (2011) Comparison of fixed and focal point seed transfer systems for reforestation and assisted migration : a case study for interior spruce in British Columbia. Canadian Journal of Forestry Research 41: 1452–1464.
43. IUCN (2012) IUCN Guidelines for Reintroductions and Other Conservation Translocations. Gland, Switzerland.
44. Carvalho SB, Brito JC, Crespo EJ, Possingham HP (2010) From climate change predictions to actions - conserving vulnerable animal groups in hotspots at a regional scale. Global Change Biology 16: 3257–3270.
45. IUCN (1987) Translocation of living organisms IUCN position statement. Gland, Switzerland.
46. Frascaria-Lacoste N (2010) Local versus non local : managing in the face of uncertaint. In: Proceedings 7th European Conference on Ecological Restoration Avignon. Avignon, France.
47. Beever EA, Ray C, Mote PW, Wilkening JL (2010) Testing alternative models of climate-mediated extirpations. Ecological applications 20: 164–78.
48. Kreyling J, Bittner T, Jaeschke A, Jentsch A, Steinbauer JM, et al. (2011) Assisted Colonization: A Question of Focal Units and Recipient Localities. Restoration Ecology 19: 433–440.
49. Ledig FT, Rehfeldt GE, Sáenz-Romero C, Flores-López C (2010) Projections of suitable habitat for rare species under global warming scenarios. American journal of botany 97: 970–87.
50. Bergstrom BJ, Vignieri S, Sheffield SR, Sechrest W, Carlson AA (2009) The Northern Rocky Mountain Gray Wolf Is Not Yet Recovered. BioScience 59: 991–999.
51. Hayward MW (2009) Conservation management for the past, present and future. Biodiversity and Conservation 18: 765–775.
52. Poulter B, Feldman RL, Brinson MM, Horton BP, Orbach MK, et al. (2009) Sea-level rise research and dialogue in North Carolina: Creating windows for policy change. Ocean & Coastal Management 52: 147–153.
53. Gozlan RE (2009) Biodiversity crisis and the introduction of non-native fish: Solutions, not scapegoats. Fish and Fisheries 10: 109–110.
54. Bennett C (2011) Climate change and the implications for forest fires in British Columbia. University of British Columbia, Forestry Undergraduate Essays/ Theses, 2010 winter session, FRST 497
55. Fleishman E, Chambers JC, Wisdom MJ (2009) Introduction to the Special Section on Alternative Futures for Great Basin Ecosystems. Restoration Ecology 17: 704–706.
56. Gibbon A, Silman MR, Malhi Y, Fisher JB, Meir P, et al. (2010) Ecosystem Carbon Storage Across the Grassland–Forest Transition in the High Andes of Manu National Park, Peru. Ecosystems 13: 1097–1111.

57. Lankau R, Jørgensen PS, Harris DJ, Sih A (2011) Incorporating evolutionary principles into environmental management and policy. Evolutionary Applications 4: 315–325.

58. Huntley B, Collingham YC, Green RE, Hilton GM, Rahbek C, et al. (2006) Potential impacts of climatic change upon geographical. Ibis 148: 8–28.

59. Bergin LA (2009) Latina Feminist Metaphysics and Genetically Engineered Foods. Journal of Agricultural and Environmental Ethics 22: 257–271.

60. Cooley D, Goreham G (2004) Are transgenic organisms unnatural? Ethics and the environment 9: 46–55.

61. Lee K (2003) Philosophy and Revolutions in Genetics: Deep Science and Deep Technology. Palgrave MacMillan, Houndmills.

62. Siipi H (2003) Artefacts and Living Artefacts. Environmental Values 12: 413–430.

63. Siipi H (2008) Dimensions of Naturalness. Ethics and the environment 13: 71–103.

64. Black-Samuelson S (2012) The state of forest genetic resources in Sweden. Report to FAO. 56 pp. Skogsstyrelsen, Jönköping.

65. Pedlar JH, Mckenney DW, Aubin I, Beardmore T, Beaulieu J, et al. (2012) Placing Forestry in the Assisted Migration Debate. BioScience 62: 835–842.

66. IUCN (1995) IUCN/SSC Guidelines For Re-Introductions. Gland, Switzerland.

67. Aitken S, Wang T, Smets P, Curtis-McLane S, Holliday J (2007) Adaption of forest trees to climate change. LARIX 2007: International Symposium of the IUFRO Working Group S2.02.07. p 13

68. Campbell A, Kapos V, Chenery A, Kahn S, Rashid M, et al. (2009) The linkages between biodiversity and climate change adaptation. UNEP World Conservation Monitoring Centre.

69. Nitschke CR, Innes JL (2008) Integrating climate change into forest management in South-Central British Columbia: An assessment of landscape vulnerability and development of a climate-smart framework. Forest Ecology and Management 256: 313–327.

70. Carrasco LR, Mumford JD, MacLeod A, Harwood T, Grabenweger G, et al. (2010) Unveiling human-assisted dispersal mechanisms in invasive alien insects: Integration of spatial stochastic simulation and phenology models. Ecological Modelling 221: 2068–2075.

71. Lintermans M (2004) Human-assisted dispersal of alien freshwater fish in Australia. New Zealand Journal of Marine and Freshwater Research 38: 481–501.

72. Bradley BA, Wilcove DS (2009) When Invasive Plants Disappear: Transformative Restoration Possibilities in the Western United States Resulting from Climate Change. Restoration Ecology 17: 715–721.

73. Viveros-Viveros H, Sáenz-Romero C, Vargas-Hernández JJ, López-Upton J, Ramírez-Valverde G, et al. (2009) Altitudinal genetic variation in Pinus hartwegii Lindl. I: Height growth, shoot phenology, and frost damage in seedlings. Forest Ecology and Management 257: 836–842.

74. Wang T, O'Neill GA, Aitken SN (2010) Integrating environmental and genetic effects to predict responses of tree populations to climate. Ecological applications 20: 153–163.

75. Yagisawa T (1999) Definition. In: Audi R (Ed.), The Cambridge Dictionary of Philosophy.

76. Hey J (2001) The mind of the species problem. Trends in Ecology and Evolution 16: 326–329.

77. Lidén M, Oxelmann B (1989) Species. Pattern or Process? Taxon 38: 228–232.

78. Jørgensen D (2011) What's History Got to Do with It? A Response to Seddon's Definition of Reintroduction. Restoration Ecology 19: 705–708.

79. Liu H, Feng C, Chen B, Wang Z, Xie X, et al. (2012) Overcoming extreme weather challenges: Successful but variable assisted colonization of wild orchids in southwestern China. Biological Conservation 150: 68–75.

80. Sgrò CM, Lowe AJ, Hoffmann AA (2011) Building evolutionary resilience for conserving biodiversity under climate change. Evolutionary Applications 4: 326–337.

81. McNeely J (2001) The great reshuffling: human dimensions of invasive alien species. IUCN, Cambridge, U.K.

82. Crosby AW, McNeill JR, von Mering O (2003) The Columbian exchange biological and cultural consequences of 1492. Praeger, Westport (Conn.).

83. Donlan JC, Berger J, Bock CE, Bock JH, Burney DA, et al. (2006) Pleistocene rewilding: an optimistic agenda for twenty-first century conservation. The American naturalist 168: 660–681.

84. Marris E (2008) Moving on assisted migration. Nature Reports Climate Change 2: 112–113.

17

Comprehensive Red List Assessment Reveals Exceptionally High Extinction Risk to Madagascar Palms

Mijoro Rakotoarinivo[1]*, John Dransfield[2], Steven P. Bachman[2], Justin Moat[2], William J. Baker[2]

1 Kew Madagascar Conservation Centre, Ambodivoanjo Ivandry, Antananarivo, Madagascar, **2** Royal Botanic Gardens, Kew, Richmond, Surrey, United Kingdom

Abstract

The establishment of baseline IUCN Red List assessments for plants is a crucial step in conservation planning. Nowhere is this more important than in biodiversity hotspots that are subject to significant anthropogenic pressures, such as Madagascar. Here, all Madagascar palm species are assessed using the IUCN Red List categories and criteria, version 3.1. Our results indicate that 83% of the 192 endemic species are threatened, nearly four times the proportion estimated for plants globally and exceeding estimates for all other comprehensively evaluated plant groups in Madagascar. Compared with a previous assessment in 1995, the number of Endangered and Critically Endangered species has substantially increased, due to the discovery of 28 new species since 1995, most of which are highly threatened. The conservation status of most species included in both the 1995 and the current assessments has not changed. Where change occurred, more species have moved to lower threat categories than to higher categories, because of improved knowledge of species and their distributions, rather than a decrease in extinction risk. However, some cases of genuine deterioration in conservation status were also identified. Palms in Madagascar are primarily threatened by habitat loss due to agriculture and biological resource use through direct exploitation or collateral damage. The recent extension of Madagascar's protected area network is highly beneficial for palms, substantially increasing the number of threatened species populations included within reserves. Notably, three of the eight most important protected areas for palms are newly designated. However, 28 threatened and data deficient species are not protected by the expanded network, including some Critically Endangered species. Moreover, many species occurring in protected areas are still threatened, indicating that threatening processes persist even in reserves. Definitive implementation of the new protected areas combined with local community engagement are essential for the survival of Madagascar's palms.

Editor: Francisco Moreira, Institute of Agronomy, University of Lisbon, Portugal

Funding: This work was funded by the Friends of Kew through the Threatened Plants Appeal and by the Bentham-Moxon Trust at the Royal Botanic Gardens, Kew. The funders had no role in study design, data collection and analysis, decision to publish, or preparation of the manuscript.

Competing Interests: The authors have declared that no competing interests exist.

* Email: mrakotoarinivo.rbgkew@moov.mg

Introduction

Madagascar is one of the World's most threatened biodiversity hotspots [1] because of the high endemism of its biota coupled with widespread habitat degradation, especially in humid forest areas. Despite ongoing scientific studies that have highlighted Madagascar as a place of endemic megadiversity that is facing intensifying extinction risk [2], the island's charismatic flora and fauna remain under immense pressure [3,4]. Conservation baselines are urgently required to demonstrate and strengthen the case for action on the ground.

Palms are among the most conspicuous components of the flora of Madagascar. To date, 195 species in 17 genera are recognized [5,6] with all but three being endemic to the island (98% endemism). The palm flora of Madagascar is outstandingly rich in a global context [7]. Palms inhabit mostly primary vegetation although a few species occur in disturbed areas, such as anthropogenic grassland. Consistent with global patterns of palm distribution, 90% of Madagascar palms are restricted to humid forest [8,9].

Palms are particularly vulnerable to humid forest degradation. In most species, survival and recruitment are reduced when habitat quality declines [10,11] or when habitats become fragmented [12]. The extensive degradation of Madagascar's humid forests, which have been reduced to around 25% of their original extent [13], implies that the island's humid forest-restricted biota, such as palms, are likely to be extremely threatened.

In addition to habitat loss, palms are further threatened by unsustainable, targeted exploitation by humans. Alongside grasses and legumes, palms are among the most important plant families for humans [14], providing numerous useful resources, such as materials for construction or weaving, food, medicine and ornamental plants [15]. Palms play a particularly important role in poorer countries, such as Madagascar, where they have immense economic importance at the village level [10], but they are often destructively harvested, e.g. for palm heart consumption or construction materials. In recent decades, Madagascar palms have also been targeted by plant collectors for introduction to horticultural trade [16]. These human activities place palms at greater risk of extinction than other humid forest groups that are not exploited in this way.

Extinctions at both species and population levels are of concern because unique evolutionary history and ecosystem services may be lost [17], which is particularly significant in the case of keystone groups such as palms [9]. To prevent such biodiversity loss in

Madagascar palms, a critical conservation strategy is required to focus attention on conservation priorities, to stimulate necessary actions and to raise public awareness. To take these steps, species of concern must first be properly identified based on sound taxonomy [18] so that accurate and cost-effective conservation management decisions can be made. The conservation performance of protected area networks can be improved with such information. Much of Madagascar's biodiversity is unlikely to survive unless it occurs within protected areas [19,20]. In 2003, the Madagascar government decided to increase the protected areas surface [21] from 1.7 million hectares (3%) [22] to 6 million hectare (10% of the island's surface [23]) as many unprotected areas were found to be critically important for biodiversity [24,25].

The International Union for the Conservation of Nature (IUCN) curates the IUCN Red List of Threatened Species, which is the most comprehensive, objective and authoritative data source on extinction risk in species [26–28]. Through the application of a set of five criteria (e.g. restricted range, declining population), a species can be classified according to its relative risk of extinction. In an earlier assessment of the palms of Madagascar [10], in which previous versions of the IUCN system [29,30] were applied, 113 species were identified as threatened and 18 presumed extinct. In this paper, we present a complete and updated conservation assessment of all palm species in Madagascar using the current IUCN Red List categories and criteria (version 3.1 [31]). This work builds upon a robust taxonomy for the group established in recent years [10,32,33,34] and a comprehensive database of collections and observations from recent field work [6].

The objective of this study is to produce a baseline conservation dataset for palms in Madagascar including taxonomy, species distributions, ecological factors and economic uses. We analyze this dataset to answer the following questions: 1) what is the current extinction risk to Madagascar's endemic palm species, 2) how does current status compare with the previous assessment in 1995, 3) is the existing protected area network effective for palms and 4) what are the major threats to palms?

Methods

Study area

Madagascar is a large tropical island (592,750 km^2) in the Indian Ocean [35] and is the third largest tropical island in the world, after New Guinea and Borneo. The island has a complex landscape [36] and is dominated by mountains running north-south, resulting in a central highland region above 800 m elevation. On the eastern side of the central highland is an escarpment that falls steeply away towards the Indian Ocean, whereas the western side consists of a large plain declining gently to the Mozambique Channel. Due to the impact of the southeastern tradewinds (Alizé) and the northwestern monsoon from the Equator, the eastern region is humid to perhumid, the highlands are relatively temperate, the western region is subhumid to dry, and the far south-west is subarid [37]. Consequently, the island has a great diversity of primary vegetation types, ranging from humid forest to dry spiny forest (Fig. 1a) [13]. Humid forest, the primary habitat of most palms, is restricted to the east and north-west of the island. Of the estimated 21 million inhabitants, nearly 80% live in rural areas [38] and depend on natural resources for their subsistence, contributing to the destruction of Madagascar's forests, which have declined by 40% between 1950 and 2000 alone [39].

Occurrence data

The study is based on a dataset of 2,160 georeferenced occurrence records, derived from herbarium specimens of Madagascar palms in key botanical institutions around the world: AAU, FTG, GE, K, MO, NY, P, TAN, TEF and ZT (herbarium acronyms follow [40]). Collection dates range from 1834 to 2010. Records lacking geographic coordinates on specimen labels were georeferenced using topographic maps, online gazetteers, the Madagascar gazetteer of the Missouri Botanical Garden [41] and online mapping tools such as Google Earth [42].

Of the georeferenced records, 820 (38%) postdate 1995 when the previous conservation assessment was conducted [10] of which 561 (26%) result from our own fieldwork in Madagascar since 1995 (fig. 1a). Building on the robust taxonomic baseline provided by Dransfield and Beentje [10], we have conducted targeted fieldwork in 32 sites (Table 1) across Madagascar between 1995–2010 (Fig. 1a), focusing mostly on primary forest areas far from high human density where the palm flora is rich, but poorly known (Fig. 1b). This fieldwork has substantially improved our understanding of the distribution and the populations of 152 species (78% of the total palm flora). The number of specimen records per species ranges from 1 to 85 (mean: 10 specimens per species) and 107 species are known from fewer than 5 specimen records.

All fieldwork was conducted with prior informed consent of the necessary authorities (Table 1). Permission for all fieldwork activities was obtained from the Ministry of Environment and Forests (Ministère de l'Environnement et des Forêts). Additional permissions were required depending on the status of the area visited. For National Parks and Special Reserves, additional permits were issued by Madagascar National Parks (MNP). For Système des Aires Protégées de Madagascar (SAPM) Reserves, additional fieldwork permission was sought from the specific management authority of each site (Table 1). For Local Community Forests, the local village council (Communauté de Base, COBA) was consulted on arrival. Fieldwork on private lands required permission from the land owners in advance (Table 1). Herbarium specimens from our fieldwork were deposited at the Madagascar national herbarium at Parc Botanique et Zoologique de Tsimbazaza (TAN) and the Royal Botanic Gardens, Kew (K). Additional duplicates, where available, were distributed primarily to the Missouri Botanical Garden (MO) and the Natural History Museum, Paris (P).

IUCN Red List Conservation Assessments

We conducted a complete assessment of the conservation status of all 192 endemic Madagascar palm species using the IUCN Red List categories and criteria, version 3.1 [31] with reference to the latest guidelines [43]. Assessments were independently reviewed and verified by the IUCN Palm Specialist Group Red List Authority and IUCN Red List Unit. They were subsequently published on-line on the IUCN Red List on 17 October 2012 [44]. All assessments are now accessible via the IUCN Red List web portal at www.iucnredlist.org.

Each species was classified according to one of the following IUCN categories: Extinct in the Wild (EW), Critically Endangered (CR), Endangered (EN), Vulnerable (VU), Near Threatened (NT), Least Concern (LC) or Data Deficient (DD). We used data from our palm occurrence dataset to summarise distribution, population size and threats to species in order to apply the quantitative Red List criteria. Although attempts were made to apply all five criteria in the Red List system (A, declining population; B, geographic range size and fragmentation, decline, or fluctuations; C, small population size and fragmentation, decline or fluctuations; D, very small population; E, quantitative analysis of extinction risk), as

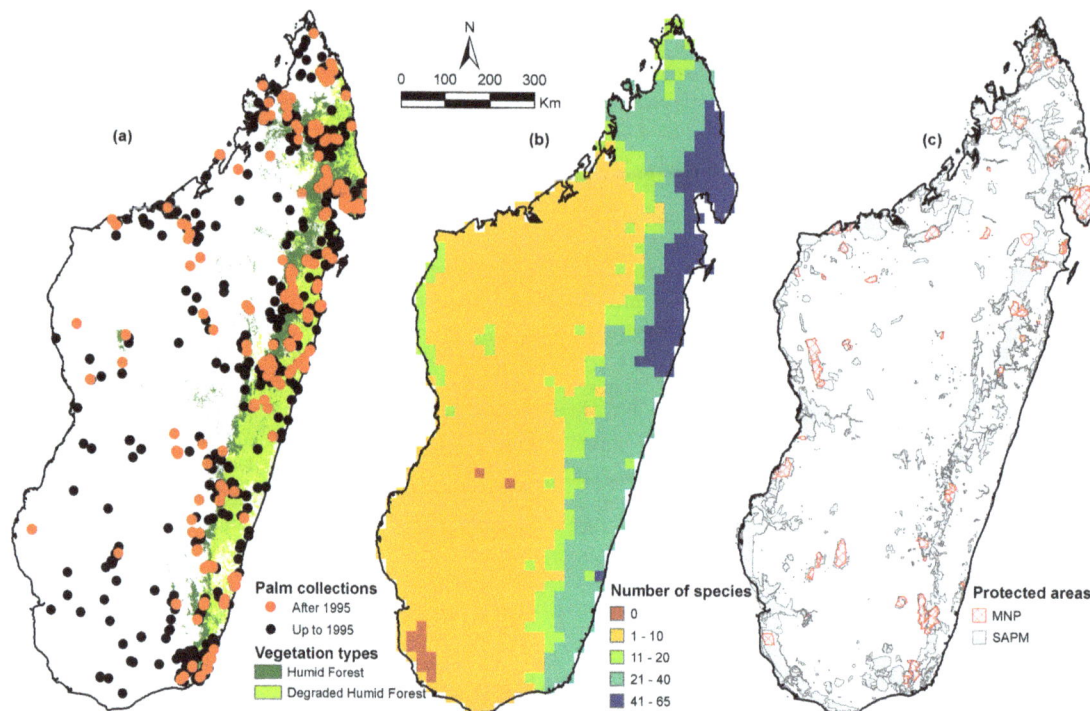

Figure 1. Palm distributions, humid forest and protected areas in Madagascar. (a) Palm specimen collection localities in Madagascar and extent of humid forest vegetation [13]. (b) Species richness of palms in Madagascar [6] illustrating predicted number of palm species across the island at a resolution of 0.2° (ca. 22 km × 22 km). (c) Protected area network in Madagascar comprising the long-standing MNP network (46 parks and reserves [50]) and the newly established SAPM (145 reserves, including those of the MNP network [23]).

recommended by IUCN, most assessments were conducted using criterion B due to the limitations of available data, a common pattern for Red List assessments of plants and some other groups [45].

The palm occurrences were carefully scrutinised for georeference precision, taxonomic identification and likelihood of a population still being extant e.g. historical collections in areas now deforested were excluded. The geographic range of each species was then quantified using two metrics, extent of occurrence (EOO) and area of occupancy (AOO) [31], both of which can be used for assessments under criterion B (restricted range species). EOO was calculated by constructing the minimum convex polygon (convex hull) around known occurrences [46,47] using the Conservation Assessment Tools extension to ArcView [48]. AOO was calculated with the same tools by overlaying a grid and interpreting known occurrences as occupied grid cells. The sum of occupied grid cells equates to the AOO value. A grid cell size of 2 × 2 km^2 was applied, as recommended by IUCN [43], where sampling effort was deemed sufficient. In some cases, larger cell sizes were used (up to 10 × 10 km^2 [25,46]) to account for inadequate sampling across the range. These larger grid cells were not scaled down to the reference scale of 2 × 2 km^2, so the assessments assume the distribution is fully saturated at the 2 × 2 km^2 reference scale [43]. In cases where a species was known from less than three unique collection sites, EOO could not be calculated and AOO alone was estimated. For species known to occur at a single locality and in a well defined habitat, AOO was estimated by considering the available suitable habitat. Satellite imagery from Google Earth [42] was used to determine suitable

areas and polygons were drawn to estimate area of occupancy (AOO).

To infer population trends, such as continuing decline or fragmentation of distribution range through time, GIS layers of the vegetation maps of Humbert & Cours-Darne [49] and Moat & Smith [13] were compared. The rate of the decline of the population of each species in the 42 years between these two baseline vegetation surveys (1965 and 2007) was then calculated from the loss of suitable habitat under its EOO and AOO.

In order to evaluate the trend in conservation status change over time, we compared the 2012 assessment [44] with the previous assessment made by Dransfield & Beentje [10]. The two assessments were based on different versions of the IUCN Red List categories and criteria. The 1995 assessment was broadly based on version 2.3 [29], whereas the 2012 assessment used version 3.1 [31]. While most Red List categories were comparable between the two assessments, the category "Rare" used by Dransfield and Beentje [10] comes from a scheme pre-dating version 2.3 and could not be related to a category in version 3.1. The category "Near Threatened" of version 3.1 was absent from version 2.3. "Not Threatened" (NotT), as used in the 1995 assessment, was regarded as equivalent to "Least Concern" in version 3.1. The change in the IUCN assessments was quantified for data sufficient species (i.e. those that had enough data to carry out a full assessment) that were assessed in both years and in comparable categories. Changes were sorted into three classes: a) no change, if the category of the species was the same in the two assessments, b) downlisted, if the assessment of extinction risk decreased, i.e. from higher to lower category (e.g. EN to VU) and

Table 1. Fieldwork locations visited by the authors.

Location	Latitude and longitude co-ordinates of sites visited	Status
Ambakireny	17.69° S 48.01° E	Local community forest
Ambatovaky	16.86° S 49.26° E	Special Reserve
Ambodivoahangy (Makira)	15.28° S 49.62° E	Local community forest
Analalava (Mahajanga)	14.76 S° 47.43° E	Local community forest
Andilamena	16.98° S 48.84° E; 16.81° S 48.68° E	Local community forest
Anosibe an'Ala	19.66° S 48.11° E	Local community forest
Betafo	20.20° S 46.50° E	Local community forest
Betampona	17.91° S 49.20° E	Special Reserve
Brickaville	18.89° S 49.12° E; 18.96° S 48.85° E	Local community forest
Daraina	13.26° S 49.59° E	SAPM Reserve (managed by Fanamby)
Fenoarivo Atsinanana	17.29° S 49.41° E	SAPM Reserve (managed by Ecole Supérieure des Sciences Agronomiques- Forêts, Antananarivo)
Fort-Dauphin	24.77° S 47.18° E	Private Land (managed by Qit Minerals Madagascar/Rio Tinto)
	24.56° S 47.20° E	SAPM Reserve (managed by Asity Madagascar/Bird Life International)
Analalava (Foulpointe)	17.71° S 49.45° E	SAPM Reserve (managed by Missouri Botanical Gardens)
Ifanadiana	21.33° S 47.71° E	Local community forest
Itremo	20.57° S 46.56° E	SAPM Reserve (managed by Kew Madagascar Conservation Centre)
Maevatanana	16.76° S 47.03° E	Local community forest
Makira	15.38° S 49.44° E; 15.28° S 49.44° E	SAPM Reserve (managed by Wildlife Conservation Society)
Manakara	21.83° S 47.90° E	Local community forest
Mananara Avaratra	15.94° S 49.54° E	Local community forest
Mangerivola	18.20° S 48.92° E	Special Reserve
Mantadia	18.88° S 48.44° E	National Park
Masoala	15.31° S 49.85° E; 15.73° S 49.96° E, 15.74° S 50.19° E, 15.77° S 50.07° E	National Park
Midongy Atsimo	23.55° S 47.08° E	National park
Soanierana Ivongo	16.68° S 49.60° E	Local community forest
Vondrozo	22.80° S 47.18° E	Local community forest

c) uplisted, if the assessment of extinction risk increased, i.e. from lower to higher category (e.g. VU to EN).

Protected area coverage

We compared the distribution of all species to the protected area network in order to assess the effectiveness and coverage of reserves for palm conservation. We used GIS layers (Fig. 1c) describing the 46 established protected areas within the MNP network [50] and the new protected area network being established by SAPM since 2011, which comprises 145 reserves, including those of MNP [23]. We assessed the relative threat status of species occurring in zero, one and two or more protected area to test the expectation that species occurring in fewer protected areas have higher threat ratings.

Threats

During the assessment process the dominant threats for each species were classified according to the IUCN Threats Classification Scheme (version 3.2) [51]. Details about the threats and the local utilization of each species were obtained from expert field observations, specimen labels and literature sources. These data were later compiled to evaluate the relative importance of major threatening processes affecting palms in Madagascar.

Results

IUCN Red List Conservation Assessments

The results of our complete assessment of the conservation status of all known Madagascar palms are summarised in Table 2 and Figure 2, with a detailed break-down given in Table 3. Of the data sufficient species (179), we found that 149 (78%) are classified as threatened (CR, EN or VU). Thirteen species were not data sufficient and were thus rated as DD. Data on the current status of these species were inadequate to complete an assessment primarily because most were known only from the type collection and have not been observed for many years. Taking into account the 13 DD species, we estimated 'lower', 'best estimate' ('mid-point') and 'upper' bounds of the percentage of threatened species [52], which were 78%, 83% and 84% respectively. The lower bound treats all DD species as unthreatened, whereas the upper bound assumes that all are threatened. The best estimate assumes that the same fraction of DD species are threatened as was found for data sufficient species. A total of 14 species were listed as NT, which

Table 2. Summary of results from the 2012 IUCN Red List Assessment of Madagascar Palms.

	Count	Percentage
IUCN Red List category		
Extinct (EX)	0	0
Extinct in the Wild (EW)	0	0
Critically Endangered (CR)	61	32
Endangered (EN)	45	23
Vulnerable (VU)	43	22
Near Threatened (NT)	14	7
Least Concern (LC)	16	8
Data Deficient (DD)	13	7
Summary Statistics		
Total Evaluated	192	100
Total Data Sufficient (CR+EN+VU+NT+LC)	179	93
Total Threatened – lower bound (CR+EN+VU)	149	78
Total Threatened – best estimate (mid-point) (CR+EN+VU+((Total Threatened/Data Sufficient) × DD))	160	83
Total Threatened – upper bound (CR+EN+VU+DD)	162	84
Total Species of Elevated Conservation Concern (CR+EN+VU+NT)	163	85
Total Not Threatened (LC+DD)	30	16

gives a total of 163 (91%) species considered to be of elevated conservation concern. Only 16 species were listed as LC.

Comparison with 1995 assessments

Comparison between the 1995 assessments and the 2012 assessments is complicated due to the change in Red List criteria from earlier versions to version 3.1 [29–31], improved knowledge of species distributions and changes to the overall taxonomy of Madagascar palms due to many new species being described after 1995. Consequently, a Red List Index approach [53,54] was not used here. Figure 3a illustrates numbers of species assessed in each of the IUCN Red List categories for categories that are

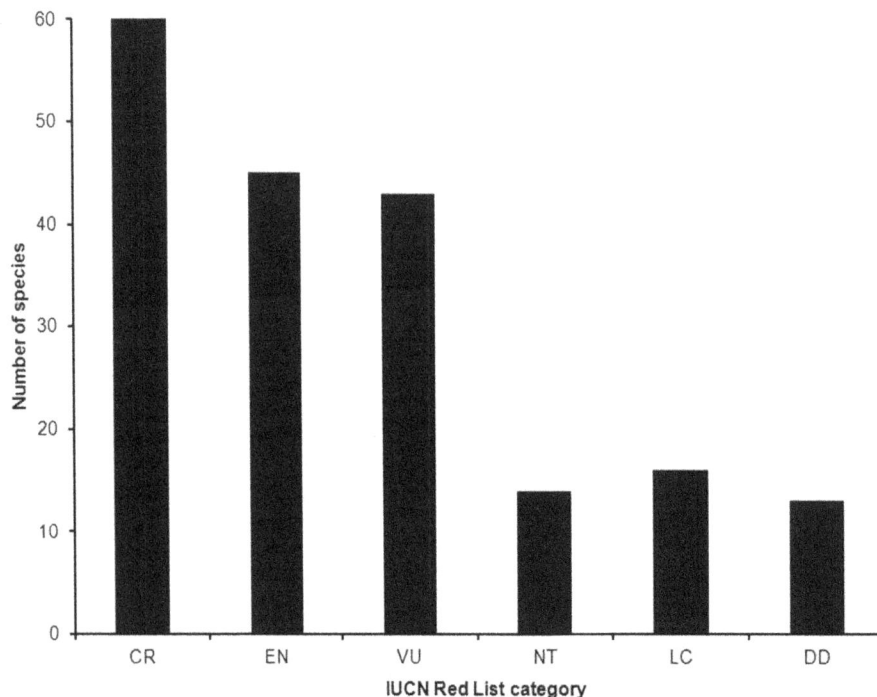

Figure 2. Summary of the 2012 IUCN Red List Assessments of Madagascar Palms (see table 2). IUCN Red List categories: Extinct in the Wild (EW), Critically Endangered (CR), Endangered (EN), Vulnerable (VU), Near Threatened (NT), Least Concern (LC), Data Deficient (DD) [31].

Table 3. The conservation status of all 192 endemic Madagascar palm species.

Species	2012 Red List Assessment	1995 Red List Assessment	EOO (km²)	AOO (km²)	Status change	Populations in protected areas: MNP only (%)	Populations in protected areas: SAPM (%)
Beccariophoenix alfredii	VU	–	8	5	Described post-1995	0	0
Beccariophoenix madagascariensis	VU	CR	15460	300	Downlisted	30	80
Bismarckia nobilis	LC	NotT	319415	17100	No change	20	60
Borassus madagascariensis	EN	VU	48872	350	Uplisted	0	40
Dypsis acaulis	EN	EW	72	8	Downlisted	30	30
Dypsis acuminum	EN	DD	12113	150	Not comparable	50	80
Dypsis albofarinosa	CR	–	4	4	Described post-1995	100	100
Dypsis ambanjae	CR	EW	1767	150	Downlisted	50	100
Dypsis ambilaensis	EN	EN	176	60	No change	0	60
Dypsis ambositrae	CR	CR	1790	150	No change	0	30
Dypsis andapae	EN	Rare	1428	35	Not comparable	70	70
Dypsis andilamenensis	CR	–	10	5	Described post-1995	0	0
Dypsis andrianatonga	VU	Rare	5909	288	Not comparable	90	90
Dypsis angusta	EN	EN	5760	123	No change	40	70
Dypsis angustifolia	EN	Rare	469	327	Not comparable	30	100
Dypsis anjae	CR	–	4	4	Described post-1995	100	100
Dypsis ankaizinensis	DD	DD	–	–	No change	–	–
Dypsis ankirindro	NT	–	14	4	Described post-1995	0	100
Dypsis antanambensis	CR	EN	6	6	Uplisted	100	100
Dypsis aquatilis	CR	EN	35	25	Uplisted	0	0
Dypsis arenarum	CR	CR	895	36	No change	0	70
Dypsis baronii	LC	NotT	239065	6075	No change	40	70
Dypsis basilonga	CR	EN	188	16	Uplisted	0	0
Dypsis beentjei	CR	EN	6	6	Uplisted	100	100
Dypsis bejofo	VU	EN	10322	150	Downlisted	60	80
Dypsis bernierana	VU	VU	14610	600	No change	50	60
Dypsis betamponensis	VU	EW	4	4	Downlisted	100	100
Dypsis betsimisarakae	VU	–	7995	512	Described post-1995	50	80
Dypsis boiviniana	EN	EN	7196	896	No change	30	60
Dypsis bonsai	VU	VU	14420	520	No change	80	90
Dypsis bosseri	EN	EW	4	1.5	Downlisted	0	90
Dypsis brevicaulis	CR	CR	380	135	No change	0	80
Dypsis brittiana	CR	–	4	4	Described post-1995	0	100
Dypsis canaliculata	CR	EW	4114	243	Downlisted	30	70
Dypsis canescens	DD	EW	–	–	Not comparable	–	–

Table 3. Cont.

Species	2012 Red List Assessment	1995 Red List Assessment	EOO (km²)	AOO (km²)	Status change	Populations in protected areas: MNP only (%)	Populations in protected areas: SAPM (%)
Dypsis carlsmithii	CR	-	2041	8	Described post-1995	50	100
Dypsis catatiana	LC	NotT	146761	3300	No change	70	80
Dypsis caudata	CR	CR	4	4	No change	100	100
Dypsis ceracea	EN	EW	11211	225	Downlisted	40	100
Dypsis commersoniana	DD	CR	-	-	Not comparable	-	-
Dypsis concinna	NT	VU	30591	2100	Downlisted	30	100
Dypsis confusa	NT	Rare	49262	1216	Not comparable	60	70
Dypsis cookei	CR	EN	6	4	Uplisted	100	100
Dypsis coriacea	NT	VU	11642	910	Downlisted	80	90
Dypsis corniculata	EN	VU	18147	1350	Uplisted	40	40
Dypsis coursii	LC	VU	3737	256	Downlisted	70	100
Dypsis crinita	NT	Rare	67442	1134	Not comparable	60	90
Dypsis culminis	EN	-	564	10	Described post-1995	0	100
Dypsis curtisii	EN	DD	7185	64	Not comparable	60	90
Dypsis decaryi	VU	VU	339	83	No change	40	40
Dypsis decipiens	VU	EN	42846	1430	Downlisted	20	40
Dypsis delicatula	VU	-	8	8	Described post-1995	40	80
Dypsis digitata	CR	CR	2334	8	No change	40	40
Dypsis dracaenoides	EN	-	4	4	Described post-1995	0	0
Dypsis dransfieldii	NT	EN	12	8	Downlisted	100	100
Dypsis elegans	CR	CR	1340	27	No change	40	70
Dypsis eriostachys	EN	EN	2395	20	No change	40	70
Dypsis faneva	EN	EN	2037	94	No change	50	90
Dypsis fanjana	EN	EN	10772	110	No change	80	80
Dypsis fasciculata	NT	VU	74598	2304	Downlisted	40	70
Dypsis fibrosa	LC	NotT	160135	4400	No change	40	60
Dypsis forficifolia	LC	NotT	29506	4500	No change	40	80
Dypsis furcata	EN	EW	7914	25	Downlisted	0	50
Dypsis gautieri	VU	-	6	3	Described post-1995	0	100
Dypsis glabrescens	EN	EN	3195	61	No change	60	70
Dypsis gronophyllum	CR	-	4	4	Described post-1995	0	0
Dypsis henrici	DD	DD	-	-	No change	-	-
Dypsis heteromorpha	DD	DD	-	-	No change	-	-
Dypsis heterophylla	NT	Rare	91693	1225	Not comparable	40	60
Dypsis hiarakae	VU	VU	17316	700	No change	70	90

Table 3. Cont.

Species	2012 Red List Assessment	1995 Red List Assessment	EOO (km²)	AOO (km²)	Status change	Populations in protected areas: MNP only (%)	Populations in protected areas: SAPM (%)
Dypsis hildebrandtii	NT	VU	22794	1700	Downlisted	40	60
Dypsis hovomantsina	CR	CR	5225	288	No change	60	100
Dypsis humbertii	VU	VU	10780	810	No change	40	80
Dypsis humilis	CR	–	4	4	Described post-1995	0	0
Dypsis ifanadianae	CR	CR	25	6	No change	0	0
Dypsis integra	EN	CR	35824	294	Downlisted	40	70
Dypsis intermedia	CR	CR	4	4	No change	100	100
Dypsis interrupta	CR	CR	378	10	No change	40	60
Dypsis jeremiei	CR	–	4	4	Described post-1995	100	100
Dypsis jumelleana	VU	VU	15117	1100	No change	40	40
Dypsis laevis	CR	CR	4	4	No change	100	100
Dypsis lantzeana	VU	VU	12141	2925	No change	50	70
Dypsis lanuginosa	CR	EW	18	18	Downlisted	50	50
Dypsis lastelliana	LC	NotT	72396	2340	No change	50	60
Dypsis leptocheilos	CR	DD	4	4	Not comparable	0	0
Dypsis ligulata	DD	EW	–	–	Not comparable	–	–
Dypsis linearis	EN	EW	153	96	Downlisted	50	50
Dypsis lokohoensis	VU	VU	6506	600	No change	40	100
Dypsis loucoubensis	CR	EN	367	100	Uplisted	100	100
Dypsis louvelii	VU	VU	8884	612	No change	20	60
Dypsis lucens	DD	EW	–	–	Not comparable	–	–
Dypsis lutea	EN	CR	1435	90	Downlisted	30	30
Dypsis lutescens	NT	NotT	51777	1700	Uplisted	20	20
Dypsis madagascariensis	LC	Rare	115274	5175	Not comparable	30	40
Dypsis mahia	CR	CR	4	4	No change	100	100
Dypsis makirae	VU	–	18	12	Described post-1995	0	80
Dypsis malcomberi	EN	VU	615	64	Uplisted	100	100
Dypsis mananjarensis	NT	VU	25568	1200	Downlisted	20	30
Dypsis mangorensis	CR	CR	4000	80	No change	40	40
Dypsis marojejyi	VU	VU	337	21	No change	100	100
Dypsis mcdonaldiana	EN	VU	3835	48-499	Uplisted	40	70
Dypsis metallica	CR	–	4	4	Described post-1995	100	100
Dypsis minuta	VU	VU	127	45	No change	80	80
Dypsis mirabilis	EN	EN	267	102	No change	70	70
Dypsis mocquerysiana	NT	VU	7596	3145	Downlisted	60	70

Table 3. Cont.

Species	2012 Red List Assessment	1995 Red List Assessment	EOO (km²)	AOO (km²)	Status change	Populations in protected areas: MNP only (%)	Populations in protected areas: SAPM (%)
Dypsis monostachya	DD	DD	492	-	No change	-	-
Dypsis montana	VU	DD	52	14	Not comparable	100	100
Dypsis moorei	EN	EN	4623	25	No change	50	50
Dypsis nauseosa	CR	CR	4295	256	No change	0	50
Dypsis nodifera	LC	NotT	162112	6400	No change	50	80
Dypsis nossibensis	CR	CR	4	4	No change	100	100
Dypsis occidentalis	VU	DD	9567	600	Not comparable	60	60
Dypsis onilahensis	VU	VU	225319	4950	No change	40	40
Dypsis oreophila	VU	VU	19830	1000	No change	30	80
Dypsis oropedionis	CR	CR	5431	120	No change	30	30
Dypsis ovobontsira	CR	CR	4	4	No change	100	100
Dypsis pachyramea	LC	VU	883	279	Downlisted	70	70
Dypsis paludosa	VU	VU	19094	1452	No change	40	70
Dypsis perrieri	VU	VU	23202	540	No change	60	80
Dypsis pervillei	CR	EW	2892	4	Not comparable	30	30
Dypsis pilulifera	VU	VU	68666	704	No change	60	90
Dypsis pinnatifrons	LC	NotT	250579	6000	No change	30	80
Dypsis plumosa	DD	-	-	-	Described post-1995	-	-
Dypsis plurisecta	DD	EW	-	-	Not comparable	-	-
Dypsis poivreana	EN	CR	289	30	Downlisted	0	100
Dypsis prestoniana	VU	VU	15208	400	No change	30	80
Dypsis procera	VU	VU	18576	756	No change	40	70
Dypsis procumbens	NT	NotT	161478	5746	Uplisted	50	70
Dypsis psammophila	EN	CR	4234	112	Downlisted	0	90
Dypsis pulchella	CR	EW	11766	147	Not comparable	0	0
Dypsis pumila	CR	VU	4	4	Uplisted	100	100
Dypsis pusilla	VU	VU	2212	342	No change	100	100
Dypsis rakotonasoloi	CR	-	6	4	Described post-1995	0	100
Dypsis ramentacea	CR	CR	4	4	No change	100	100
Dypsis reflexa	CR	-	6	4	Described post-1995	100	100
Dypsis remotiflora	CR	EW	5045	14	Not comparable	50	50
Dypsis rivularis	EN	VU	16789	457	Uplisted	60	60
Dypsis robusta	CR	-	4	4	Described post-1995	0	0
Dypsis sahanofensis	CR	EN	16653	45	Uplisted	0	0
Dypsis saintelucei	EN	CR	22453	210	Downlisted	0	80

Table 3. Cont.

Species	2012 Red List Assessment	1995 Red List Assessment	EOO (km²)	AOO (km²)	Status change	Populations in protected areas: MNP only (%)	Populations in protected areas: SAPM (%)
Dypsis sancta	CR	–	4	4	Described post-1995	100	100
Dypsis sanctaemariae	CR	CR	7	7	No change	0	0
Dypsis scandens	CR	CR	15	7	No change	0	0
Dypsis schatzii	EN	VU	108	18	Uplisted	100	100
Dypsis scottiana	VU	VU	7612	1529	No change	40	40
Dypsis serpentina	VU	VU	901	216	No change	40	80
Dypsis simianensis	EN	EN	25806	340	No change	40	70
Dypsis singularis	CR	CR	19	4	No change	50	50
Dypsis soanieranae	DD	EW	–	–	Not comparable	–	–
Dypsis spicata	LC	Rare	24904	1300	Not comparable	40	70
Dypsis tanalensis	CR	EW	8	8	Not comparable	0	0
Dypsis tenuissima	EN	EN	1242	64	No change	50	100
Dypsis thermarum	VU	Rare	77	72	Not comparable	80	100
Dypsis thiryana	VU	Rare	62521	1245	Not comparable	60	70
Dypsis thouarsiana	DD	DD	–	–	No change	–	–
Dypsis tokoravina	CR	EN	341	97	Uplisted	70	100
Dypsis trapezoidea	CR	CR	4	4	No change	0	0
Dypsis tsaratananensis	DD	DD	8	–	No change	–	–
Dypsis tsaravoasira	VU	EN	40289	891	Downlisted	70	80
Dypsis turkii	EN	–	1602	462	Described post-1995	70	70
Dypsis utilis	EN	VU	11592	440	Uplisted	70	90
Dypsis viridis	VU	VU	6893	962	No change	60	80
Dypsis vonitrandambo	CR	–	8	8	Described post-1995	100	100
Lemurophoenix halleuxii	EN	EN	1729	31	No change	50	70
Marojejya darianii	EN	CR	11080	80	Downlisted	50	80
Marojejya insignis	LC	VU	91513	2710	Downlisted	60	70
Masoala kona	EN	EN	693	36	No change	0	30
Masoala madagascariensis	CR	–	15803	77	Uplisted	40	80
Orania longisquama	LC	Rare	151841	2345	Not comparable	30	80
Orania ravaka	VU	VU	8913	220	No change	70	80
Orania trispatha	VU	CR	25198	1644	Downlisted	70	90
Ravenea albicans	EN	EN	19384	929	No change	80	90
Ravenea beentjei	CR	–	7	5	Described post-1995	0	100
Ravenea delicatula	CR	–	6	4	Described post-1995	0	0
Ravenea dransfieldii	EN	VU	37979	1856	Uplisted	80	90

Table 3. Cont.

Species	2012 Red List Assessment	1995 Red List Assessment	EOO (km²)	AOO (km²)	Status change	Populations in protected areas: MNP only (%)	Populations in protected areas: SAPM (%)
Ravenea glauca	VU	VU	9989	443	No change	100	100
Ravenea hypoleuca	CR	–	575	8	Described post-1995	0	0
Ravenea julietiae	EN	EN	34734	112	No change	40	80
Ravenea krociana	EN	VU	10241	450	Uplisted	70	70
Ravenea lakatra	VU	VU	44	9	No change	40	70
Ravenea latisecta	CR	EN	44	9	Uplisted	50	100
Ravenea louvelii	CR	EN	9	4	Uplisted	80	80
Ravenea madagascariensis	LC	Rare	137043	45300	Not comparable	60	70
Ravenea musicalis	VU	VU	4	4	No change	0	0
Ravenea nana	EN	EN	75260	220	No change	40	40
Ravenea rivularis	EN	VU	2122	144	Uplisted	20	20
Ravenea robustior	NT	Rare	312828	2200	Not comparable	60	70
Ravenea sambiranensis	LC	VU	355990	52360	Downlisted	60	60
Ravenea xerophila	VU	EN	17191	676	Downlisted	30	40
Satranala decussilvae	EN	EN	3248	86	No change	50	80
Tahina spectabilis	CR	–	4	4	Described post-1995	0	0
Voanioala gerardii	CR	CR	264	12	No change	60	80

Where available, both the 1995 and 2012 conservation assessments are given, along with EOO (extent of occurrence) and AOO (area of occupancy) from the 2012 assessment; see methods for details. The percentage of populations (geographically distinct groups [30]) recorded inside the MNP and SAPM protected area networks is also given for each species (note that the expanded SAPM network includes MNP). IUCN Red List categories: Extinct in the Wild (EW), Critically Endangered (CR), Endangered (EN), Vulnerable (VU), Near Threatened (NT), Least Concern (LC), Data Deficient (DD) [30]; two additional ratings were used in the 1995 assessment [10], Not Threatened (NotT) and Rare.

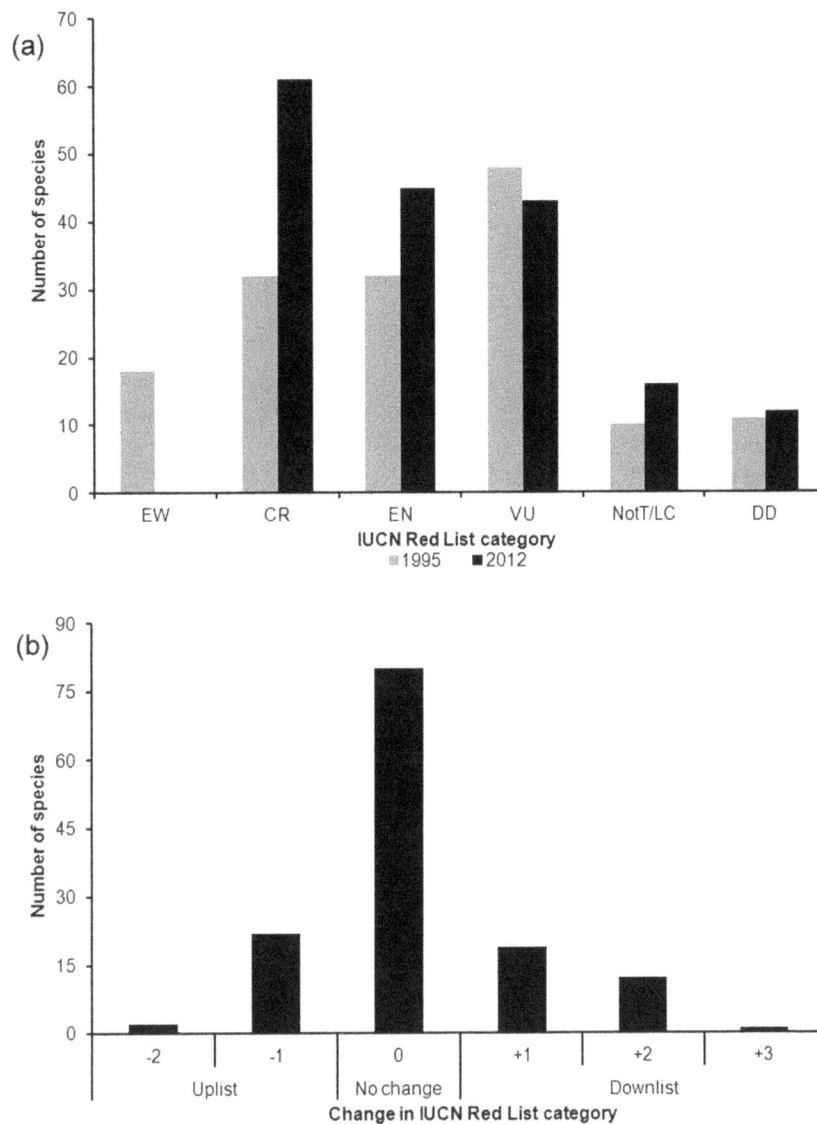

Figure 3. Comparison between the palm assessment of 1995 and 2012. IUCN Red List categories: Extinct in the Wild (EW), Critically Endangered (CR), Endangered (EN), Vulnerable (VU), Least Concern (LC), Data Deficient (DD) [31]. (a) Number of species assessed in each category (total assessments: 164 in 1995, 192 in 2012). All species assessed in each year are illustrated (see Table 3), except for those placed in categories that are not comparable (13 species assessed as "Rare" in 1995 [10] and 14 species assessed as NT in 2012 [44]; see methods). (b) Change in IUCN Red List status (see Table 3) where positive values indicate downlisting to lower extinction risk (e.g. CR to EN is a downlisting of 1 step) and negative values indicate uplisting to higher extinction risk (e.g. EN to CR is an uplisting of 1 step). Figure 3b includes 130 data sufficient species (i.e. excluding species rated as DD in either year) that were assessed in comparable categories in both 1995 [10] and 2012 [44].

comparable in both the 1995 assessment and the 2012 assessments (i.e. excluding NT and "Rare"). Numbers of species assessed as DD, LC and VU were similar in both assessments (11, 10 and 48 in 1995, 13, 16 and 43 in 2012, respectively; Fig. 3a). In contrast, numbers of species assessed as EN and CR were much higher in 2012 than 1995 (32 and 32 in 1995, 45 and 61 in 2012, respectively; Fig. 3a). There are two main reasons for this. Firstly, all 18 species assessed as EW in 1995 were assigned to other categories in 2012 based on additional information, the majority being rated as EN or CR. Five of the 18 species were assessed as DD, but insufficient evidence was found to rate any species as EW in the 2012 assessment. Secondly, 28 species were discovered and described after 1995 and were thus assessed for the first time in 2012. Of these species, only *Dypsis ankirindro* (NT) is not regarded

as threatened while the remainder are most being classified as CR (18 species). These newly discovered species are mostly known only from a single site where area of occupancy (AOO) is often low (< 4 km², e.g. *Dypsis andilamenensis, D. gronophyllum, Tahina spectabilis*) and known population sizes are small, some with less than 10 mature individuals recorded in the wild (e.g. *Dypsis humilis, D. robusta*).

Comparison of the classes of change between 1995 and 2012 reveals a contrasting pattern (Fig. 3b). Of the 130 species evaluated in both years with data sufficient, comparable assessments, most (80) species showed no change in status and more species were downlisted than uplisted. Specifically, 32 species moved from EW to CR, EN or VU, CR to EN or VU, and EN to VU or NT, and 24 species moved from EN to CR, VU to EN or

Table 4. Threatened and data deficient palm species that do not occur in the SAPM protected area network.

Species	Location	Major threats
Beccariophoenix alfredii (VU)	Betafo	Fires, harvest of seeds for horticulture
Dypsis andilamenensis (CR)	Andilamena	Habitat loss due to mining and agriculture
Dypsis ankaizinensis (DD)	Tsaratanana	Unknown
Dypsis aquatilis (CR)	Fort-Dauphin	Fire, habitat loss due to mining and agriculture
Dypsis basilonga (EN)	Andrambovato and Vatovavy	Habitat loss due agriculture, harvest of seeds for horticulture, harvest of palm heart
Dypsis canescens (DD)	Ambilobe	Unknown
Dypsis commersoniana (DD)	Fort-Dauphin	Unknown
Dypsis dracaenoides (CR)	Vondrozo	Habitat loss due to logging and agriculture
Dypsis gronophyllum (CR)	Vondrozo	Habitat loss due to logging and agriculture
Dypsis henricii (DD)	Fort-Dauphin	Unknown
Dypsis heteromorpha (DD)	Tsaratanana	Unknown
Dypsis humilis (CR)	Ambodivoahangy (Makira)	Habitat loss due to logging and agriculture
Dypsis ifanadianae (CR)	Ifanadiana	Habitat loss due to logging and agriculture, harvest of seeds for horticulture
Dypsis leptocheilos (CR)	Maevatanana	Habitat loss due to agriculture, harvest of seeds for horticulture
Dypsis ligulata (DD)	Ambilobe	Unknown
Dypsis monostachya (DD)	Maroantsetra	Unknown
Dypsis plurisecta (DD)	Maroantsetra	Unknown
Dypsis pulchella (CR)	Andilamena	Habitat loss due to mining, logging and agriculture
Dypsis robusta (CR)	Ifanadiana	Habitat loss due to agriculture
Dypsis sahanofensis (CR)	Ambositra, Anosibe an'Ala and Vatovavy	Habitat loss due to logging and agriculture
Dypsis sanctaemariae (CR)	Sainte Marie	Habitat loss due to logging and agriculture
Dypsis scandens (CR)	Ifanadiana	Habitat loss due to logging and agriculture, harvest of stems for weaving
Dypsis soanieranae (DD)	Soanierana Ivongo	Unknown
Dypsis tanalensis (CR)	Vondrozo	Habitat loss due to logging and agriculture
Dypsis trapezoidea (CR)	Vatovavy	Habitat loss due to logging and agriculture
Ravenea delicatula (CR)	Andilamena	Habitat loss due to mining, logging and agriculture
Ravenea musicalis (CR)	Fort-Dauphin	Harvest of seeds for horticulture, harvest of stems to make canoes
Tahina spectabilis (CR)	Analalava (Mahajanga)	Fire, grazing by livestock

Three remaining data deficient species are not listed here as their distributions are unknown (Dypsis lucens, D. plumosa and D. thouarsiana). Some of the locations listed here are close to protected areas (e.g. Tsaratanana, Vondrozo), but the known palm localities fall outside the protected areas boundaries.

CR, and NotT to NT. The downlisting of species is primarily due to increased knowledge of palm distributions and populations, rather than an actual change in their threat status in the wild. Changes in status may also be partly due to criteria being more rigorously applied in 2012. However, deforestation and over-exploitation of some species has resulted in the genuine decline of populations or even to local extinction (e.g. *Dypsis ambositrae, D. ifanadianae* and *Voanioala gerardii*), and has resulted in genuine uplisting of these taxa.

Protected area coverage

The expansion of protected areas in Madagascar from the older MNP network to the new SAPM network is highly beneficial to palms. Under the MNP network, 56 species were not included in protected areas. The SAPM network protects at least one population of all but 28 species, a significant improvement over the MNP Network (Tables 3 and 4). Moreover, the SAPM network protects many additional populations of threatened palm species that were not protected by the MNP network. In total, the protected area coverage of populations of 77 threatened species is increased, with an average of 42% more populations being

protected under the SAPM network than MNP for these species (Table 3).

Comparison of IUCN Red List assessments with species presence in protected areas (Fig. 4) demonstrates that species known only from outside the SAPM network are either threatened or DD (Table 4). All have small range sizes, many persist in degraded habitats and some have not been seen in the wild for several decades. The majority of the unprotected, threatened species are assessed as CR, e.g. *Dypsis ifanadianae, D. scandens* and *Ravenea musicalis*. Some occur in forested areas unconnected to the protected area network (e.g. Ambilobe, Ifanadiana, Vatovavy), whereas others occur in forest adjacent to protected area boundaries (e.g. Andilamena, Tsaratanana, Vondrozo).

Of the species that are protected within the SAPM network, 37 have been recorded in only one protected area while 124 have been documented in two or more protected areas (Fig. 4). The most important protected areas for palms are Masoala, Makira (both 43 species) and Mananara Avaratra (41 species). Marojejy, the Fandriana-Vondrozo Corridor, Manompana, Betampona and Mangerivola each contain more than 20 species. It is significant that three of these eight palm hotspots are newly designated protected areas (the Fandriana-Vondrozo Corridor, Makira,

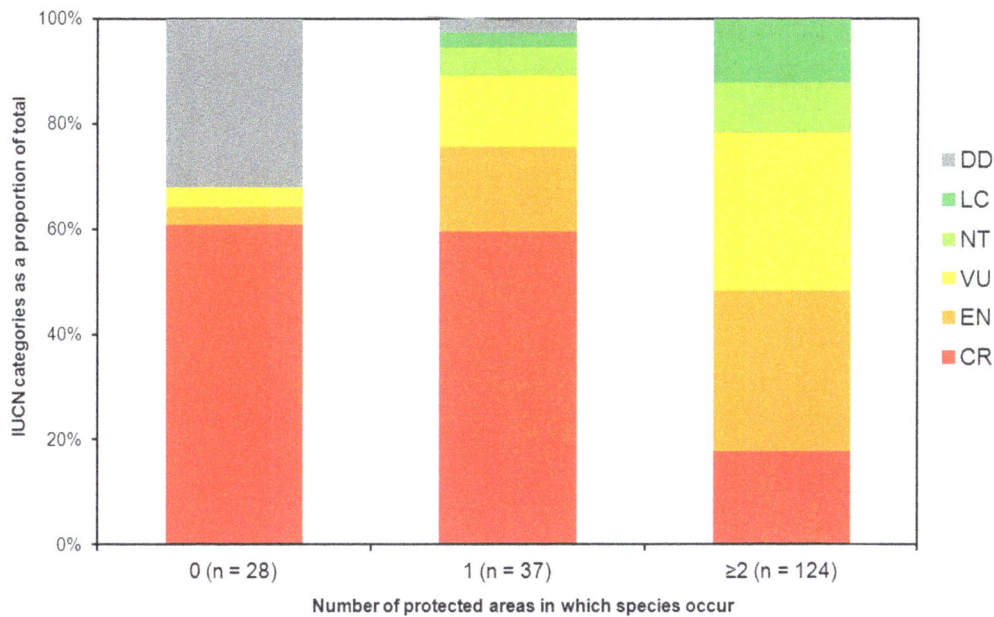

Figure 4. IUCN conservation status of palm species summarised by occurrence in protected areas (SAPM network). Of the 192 assessed species [44], 28 are not recorded from any protected area (coded as 0 in the figure), 37 species are recorded only from one protected area (coded as 1) and 124 species occur in two or more protected areas (coded as ≥2). Three data deficient species are not included as their distributions are unknown (Dypsis lucens, D. plumosa and D. thouarsiana).

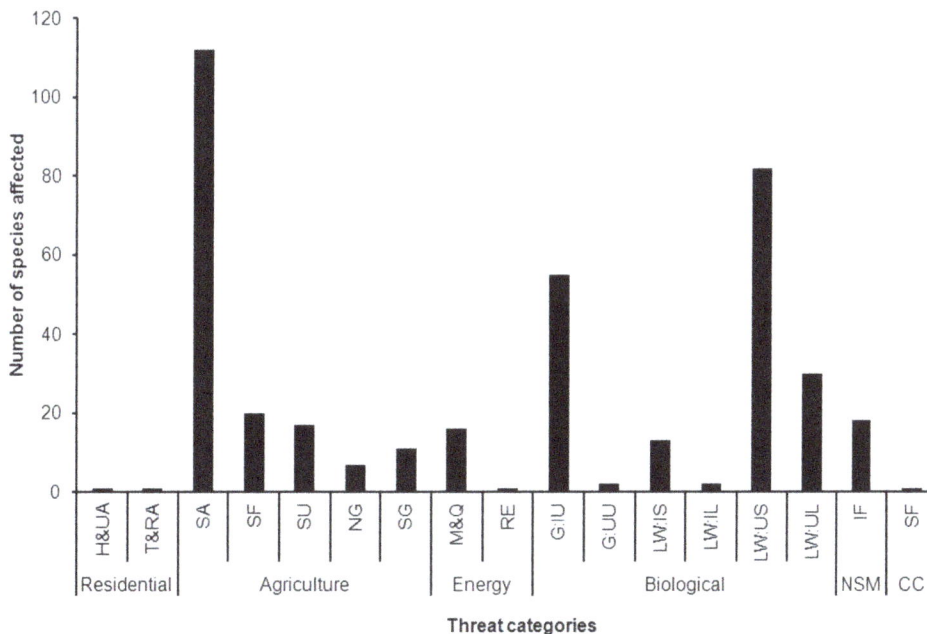

Figure 5. Major threats affecting endemic palm species in Madagascar. Bar heights reflect number of species affected by each threat, as indicated in the 2012 IUCN conservation assessment [44]. Threat categories follow the Threats Classification Scheme version 3.2 [51], using the top two levels of the hierarchy. Abbreviations: Residential & commercial development (Residential): Housing & urban areas (H&UA), Tourism & recreation areas (T&RA); Agriculture & aquaculture (Agriculture): Shifting agriculture (SA), Small-holder farming (SF), Scale unknown/unrecorded (SU), Nomadic grazing (NG), Small-holder grazing, ranching or farming (SG); Energy production & mining (Energy): Mining & quarrying (M&Q), Renewable energy (RE); Biological resource use (Biological): Gathering terrestrial plants, Intentional use (species being assessed is the target) (G:IU), Gathering terrestrial plants, unintentional use (G:UU), Logging & wood harvesting for subsistence, Intentional use: subsistence/small scale (species being assessed is the target) (LW: IS), Logging & wood harvesting, Intentional use: large scale (species being assessed is the target) (LW: IL); Logging & wood harvesting, Unintentional effects: subsistence/small scale (species being assessed is not the target) (LW: US), Logging & wood harvesting, Unintentional effects: large scale (species being assessed is not the target) (LW: UL); Natural system modifications (NSM): Increase in fire frequency (IF); Climate change & severe weather (Climate): Storm & flooding (SF).

Figure 6. Example of palm species under threat in Madagascar. (a) Anthropogenic fires in grasslands, causing decline and destruction of palm populations, such as *Dypsis decipiens* (VU), Itremo. (b) Forest clearance for slash and burn cultivation by smallholder farmers, causing habitat loss for many species, such as *Masoala kona* (EN), Ifanadiana. (c) Gathering of young leaves of *Ravenea lakatra* (VU) for production of woven hats and basketry, Masoala. (d) Destructive harvest of palm heart threatens many species such as *Dypsis saintelucei* (EN), Sainte Luce. (e) Remnant populations of species such as *Tahina spectabilis* (CR), Analalava, near Mahajanga in vegetation remnants isolated within anthropogenic landscapes, at risk from fire, grazing and other human pressures. Image credits: (a) M Rakotoarinivo, (b) WJ Baker, (c, d & e) J Dransfield.

Manompana), further emphasising the importance of the expanded SAPM network. These protected areas vary widely in extent, but all are located in the humid forested east. Nevertheless, protected areas do not guarantee low extinction risk as the majority of species that occur in protected areas are still assessed as threatened (Fig. 4), indicating that threatening processes persist in these areas.

Threats

The major threatening processes for palms in Madagascar are agriculture and biological resource use with 167 and 184 species affected by these threats respectively. More specifically the threats to palm habitats from agriculture relate to annual and perennial non-timber crop production i.e. crops planted for food, fodder, fibre, fuel or other uses, with 'shifting agriculture' listed as the scale of farming affecting the highest number of species (112) (Figs. 5

and 6; threat wordings according to IUCN Threats Classification Scheme (version 3.2) [51]). The threat from biological resource use is related to the gathering of terrestrial plants (55 species, e.g. for palm heart consumption) and logging and wood harvesting (127 species). More specifically the highest scoring threat is from logging and wood harvesting for subsistence on a large scale where the species of palm is actually not intended target, but is threatened due to collateral damage (112 species) i.e. the palms are subject to collateral damage. Other less prevalent threats relate to mining, livestock farming, fires, housing and urban development.

Discussion

As of 2013, 684 native plant species from Madagascar (out of an estimated total of ca. 13,000 [55]) were completely assessed under the IUCN categories and criteria [31] and displayed on the website of the Red List [44]. With 192 species and representing nearly 30% of the Madagascar plant species on the Red List, our study of palms is the largest and most complete IUCN conservation assessment of any plant family on the island. Recently, the IUCN Madagascar Plant Specialist Group (Groupe des Spécialistes des Plantes Malgaches, GSPM) [55] assessed ca. 3,000 plant species from 74 families and 285 genera. In addition, full assessments of smaller taxonomic groups have been completed: Pandanaceae (91 species [25]), Sarcolaenaceae (68 species) and Sphaerosepalaceae (20 species) [56], tribe Coleeae of Bignoniaceae (67 species [57]) and the genus *Delonix* (Fabaceae, 11 species; [58]). Unfortunately, none of these groups of assessments has been formally published on the IUCN Red List yet.

Our finding that as many as 83% of palm species are threatened (best estimate) indicates that the Arecaceae is among the organismal groups facing the highest risk of extinction in Madagascar. Moreover, the proportion of threatened palms in Madagascar is almost four times greater than for plants in general, estimated as 21.5% worldwide [59], and is higher than the estimate for Madagascar's flora as a whole (54%, [60]). At family level, the percentage of threatened species is variable, 49% of all endemic legumes [57], 65% for Sphaerosepalaceae and 75% for Sarcolaenaceae [56], and 81.3% for Pandanaceae [25]. It is notable that the proportions of threatened species in the Arecaceae and Pandanaceae are similar, given that they share functional similarities as woody, often arborescent monocotyledonous plants that are most species rich in humid forests. For major groups of animals in Madagascar, the proportion of threatened species varies widely, for example, 25% for amphibians [61], 37% for reptiles [60] and 94% for lemurs [62].

Comparisons between the 1995 and 2012 assessments of Madagascar palms [10] indicates that downlisting (a movement to a category of lower threat) is more frequent than uplisting, though 21 of 32 downlisted species still fall into threatened categories. These changes to a lower category come from improved knowledge of species distribution, population size, and taxonomy rather than any genuine decline of extinction risk in the wild. In contrast, almost all of our recently discovered species (species not known to science in 1995) are threatened as they typically have small range sizes and are at risk of habitat loss and direct or indirect threats from human pressure [63]. The role of the taxonomist in conservation assessment cannot be over-stated as collections-based research and knowledge both in the field and in the herbarium or museum is essential for confirming the identity of species and the distribution of their wild populations. Conservation assessment in the absence of robust taxonomy will result in inaccurate ratings and a tendency to categorise species as DD [57,64]. In our case, intensive taxonomic research and field surveys in Madagascar fundamen-tally underpin our conservation assessments and have led to the rediscovery of species previously thought to be extinct, as well as the discovery of new populations of threatened species and species new to science.

Our analysis of threats facing palms suggests that the dependency of rural people on forested lands for shifting cultivation and their continued unsustainable exploitation of wild forest products such as palms are key drivers of palm extinction risk. This applies even in remote areas where human population density is low [35]. Our analysis also reveals a novel and insidious threat to palms through the logging or harvesting of other plants at a subsistence level that causes collateral damage to palms. Unless the economic circumstances of rural communities change radical-ly, forest resources such as palms will continue to be exploited unsustainably for basic subsistence needs. Economic factors are a primary concern for the conservation of Madagascar palms, as they are for so many other organisms globally.

Time-delayed biodiversity loss [65] is an important consider-ation for Madagascar palms as many species persist locally as seedlings or juvenile plants after mature trees have been cleared with the forest. Species in decline may survive for a long time before they become extinct if a threshold in the habitat quality is maintained [66]. Without adequate protection and management, these sites are likely to be lost in the future as disturbance and fragmentation provide suitable habitat for invasive secondary species [12], which have negative impacts on native species by depressing the growth rate at various stages of the life cycle [67].

The high degree of extinction risk faced by Madagascar palms calls into question the effectiveness of previous conservation actions on the island. In a period when human population density and pressure on biodiversity are increasing, the long-term success of protected areas is at the heart of potential solutions for palm conservation. By covering 70% of the remaining humid forest in Madagascar [68], the SAPM network is expected to be considerably more effective for species protection in Madagascar compared with the previous, more limited MNP network, as the new set of reserves has been selected to include narrow range taxa [69]. Our analysis demonstrates that SAPM protects threatened palm populations much more effectively than the MNP network. Nevertheless, the SAPM network has limitations. To date, only Makira has been accorded definitive protected area status, whereas the remainder are not yet formally designated [23]. Consequently, critical protection of the forest and its biodiversity is lacking in the majority of the SAPM network.

Moreover, SAPM does not provide complete protection for Madagascar palms as several priority sites (Table 4), typically forest fragments far from protected areas, are not included. Small areas of intact habitat need to be taken into account as they often contain remnant populations of rare and endemic species [70] that are highly susceptible to environmental stochasticity and local extinction [71]. For example, a monotypic genus of massive fan palm, *Tahina spectabilis*, discovered only in 2006 [72], persists in a 160×50 m patch of forested tsingy (karst limestone) surrounded by anthropogenic grassland near Mahajanga. The protection of this forest is an urgent priority to conserve this isolated, endemic lineage. Some small fragments are included in the SAPM network, such as a ca. 2 km^2 tract of degraded coastal plain forest at Analalava (near Foulpointe), north of Toamasina, which is an outstanding palm hotspot containing 25 palm species, including one local endemic and three species known from only one other site each. This small fragment is managed locally by the Missouri Botanical Garden staff who promote the site for ecotourism [73].

Madagascar palms face exceptional levels of extinction risk by both national and global standards. The conservation of keystone species such as palms [9] is of particular importance due to the potential consequences of their extinction to other species. Humans are among the organisms that rely substantially on ecosystem services provided by palms [10,15,74,75]. The engage-ment of local communities in conservation initiatives will be critical to their success. Given the intensifying pressure from growing human populations, compounded by projected impacts of climate change on species extinction [76], there is now an urgent need for prioritised action for Madagascar palms. The rigorous IUCN conservation assessment described here provides an essential foundation for such a process.

Acknowledgments

We thank Craig Hilton-Taylor and Henk Beentje for their support and advice during the Red Listing process, and Charlotte Rajeriarison and

Stuart Cable for supervision. We acknowledge the cooperation of the IUCN SSC Madagascar Plant Specialist Group, particularly Sylvie Andriambololonera. Martin Callmander and Pete Lowry provided many useful comments on an earlier draft of the manuscript. We are grateful to the many authorities, institutions and individuals who have supported and facilitated our fieldwork.

Author Contributions

Conceived and designed the experiments: MR JD SPB JM WJB. Performed the experiments: MR JD SPB JM WJB. Analyzed the data: MR JD SPB JM WJB. Contributed reagents/materials/analysis tools: MR JD SPB JM WJB. Wrote the paper: MR JD SPB JM WJB.

References

1. Myers N, Mittermeier RA, Mittermeier CG, da Fonseca GA, Kent J (2000) Biodiversity hotspots for conservation priorities. Nature 403: 853–858. doi:10.1038/35002501.

2. Goodman SM, Benstead JP (2005) Updated estimates of biotic diversity and endemism for Madagascar. Oryx 39: 73–77 doi: 10.1017/S0030605305000128.

3. Harper GJ, Steininger MK, Tucker CJ, Hawkins F (2007) Fifty years of deforestation and forest fragmentation in Madagascar. Environ Conserv 34: 325–333. doi: 10.1017/S037689290700426.

4. Allnutt TF, Asner GP, Golden CD, Powell GV (2013) Mapping recent deforestation and forest disturbance in northeastern Madagascar. Trop Conserv Sci 6: 1–15.

5. Govaerts R, Dransfield J, Zona SF, Hodel DR, Henderson A (2012) World Checklist of Arecaceae. Facilitated by the Royal Botanic Gardens, Kew. Available: http://apps.kew.org/wcsp/. Accessed 12 June 2013.

6. Rakotoarinivo M, Blach-Overgaard A, Baker WJ, Dransfield J, Moat J, et al. (2013) Palaeo-precipitation determines palm diversity across Madagascar - a tropical biodiversity hotspot. Proc R Soc Lond B Biol Sci 280: 20123048. doi:10.1098/rspb.2012.3048.

7. Kissling WD, Eiserhardt WL, Baker WJ, Borchsenius F, Couvreur TLP, et al. (2012) Cenozoic imprints on the phylogenetic structure of palm species assemblages worldwide. Proc Natl Acad Sci USA 109: 7379–7384. doi:10.1073/pnas.1120467109.

8. Couvreur T, Forest F, Baker W (2011) Origin and global diversification patterns of tropical rain forests: inferences from a complete genus-level phylogeny of palms. BMC Biol 9: 44. doi:10.1186/1741-7007-9-44.

9. Couvreur T, Baker W (2013) Tropical rain forest evolution: palms as a model group. BMC Biol 11: 48. doi:10.1186/1741-7007-11-48.

10. Dransfield J, Beentje H (1995) The palms of Madagascar. The Royal Botanic Gardens, Kew and The International Palm Society. 475 p.

11. Fleischmann K, Edwards PJ, Ramseier D, Kollmann J (2005) Stand structure, species diversity and regeneration of an endemic palm forest on the Seychelles. Afr J Ecol 43: 291–301. doi: 10.1111/j.1365–2028.2005.00567.x.

12. Scariot A (1999) Forest fragmentation effects on palm diversity in central Amazonia. J Ecol 87: 66–76. doi: 10.1046/j.1365–2745.1999.00332.x.

13. Moat J, Smith P (2007) Atlas of the Vegetation of Madagascar. Kew Publishing. 124p.

14. Bennett BC (2011) Twenty-five economically important plant families. Encyclopedia of Life Support. Available: http://www.eolss.net/sample-chapters/c09/e6-118-03.pdf. Accessed 15 May 2013.

15. Dransfield J, Uhl NW, Asmussen CB, Baker WJ, Harley MM, et al. (2008) Genera Palmarum – the evolution and classification of palms. Kew Publishing. 744 p.

16. Dransfield J (1999) Madagascar as a source of new palm introductions. Acta Hortic 486: 21–32.

17. Isaac NJ, Turvey ST, Collen B, Waterman C, Baillie JE (2007) Mammals on the EDGE: Conservation priorities based on threat and phylogeny. PLoS ONE 2: e296. doi:10.1371/journal.pone.0000296.

18. Richard D, Evans D (2006) The need for plant taxonomy in setting priorities for designated areas and conservation management plans: a European perspective. In: Leadlay E, Jury S, editors. Taxonomy and Plant conservation. Cambridge University Press. pp. 163–176.

19. Bruner AG, Gullison RE, Rice RE, da Fonseca GA (2001) Effectiveness of parks in protecting tropical biodiversity. Science 291: 125–128. doi: 10.1126/science.291.5501.125.

20. Borrini-feyerabend G, Dudley N(2005) Les Aires Protégées à Madagascar : bâtir le système à partir de la base. World Commission on Protected Areas & International Union for Conservation of Nature. 51 p.

21. Norris S (2006) Madagascar defiant. BioScience 56: 960–960. doi:10.1641/0006–3568(2006)56[960:MD]2.0.CO;2.

22. Mittermeier RA, Hawkins F, Rajaobelina S, Langrand O (2005) Wilderness conservation in a biodiversity hotspot. International Journal of Wilderness 11: 42–45.

23. Système des Aires Protégées de Madagascar [SAPM] (2011) Atlas numerique du SAPM. Available : http://atlas.rebioma.net/index.php?option = com_wrapper&Itemid = 39. Accessed 3 June 2013.

24. Nicoll ME (2003) Forests outside protected areas In: Goodman SM, Benstead JP, editors. The Natural History of Madagascar. University of Chicago Press. pp. 1432–1437.

25. Callmander MW, Schatz GE, Lowry II PP, Laivao MO, Raharimampionona J, et al. (2007) Identification of priority areas for plant conservation in Madagascar using Red List criteria: rare and threatened Pandanaceae indicate sites in need of protection. Oryx 41: 168–176. doi: 10.1017/S0030605307001731.

26. Rodrigues ASL, Pilgrim JD, Lamoreux JF, Hoffmann M, Brooks TM (2006) The value of the IUCN Red List for conservation. Trends Ecol Evol 21: 71–76. doi:10.1016/j.tree.2005.10.010.

27. Mace G, Collar N, Gaston K, Hilton-Taylor C, Akçakaya R, et al. (2008) Quantification of extinction risk: IUCN's system for classifying threatened species. Conserv Biol 22: 1424–1442. doi: 10.1111/j.1523–1739.2008.01044.x.

28. Vié JC, Hilton-Taylor C, Pollock C, Ragle J, Smart J, et al. (2008) The IUCN Red List: a key conservation tool. IUCN, Switzerland. 13 p.

29. International Union for Conservation of Nature [IUCN] (1994) IUCN Red List Categories and Criteria version 2.3. Available: http://www.iucnredlist.org/technical-documents/categories-and-criteria/1994-categories-criteria. Accessed 29 May 2013.

30. International Union for Conservation of Nature [IUCN] (1998) 1997 IUCN Red List of Threatened Plants. World Commission on Protected Areas [WCMC] & International Union for Conservation of Nature [IUCN]. 862 p.

31. International Union for Conservation of Nature [IUCN] (2001) IUCN Red List Categories and Criteria: Version 3.1. IUCN, Gland, Switzerland and Cambridge, UK. 30 p.

32. Dransfield J, Beentje H, Britt A, Ranarivelo T, Razafitsalama J (2006) Field guide to the palms of Madagascar. Royal Botanic Gardens, Kew. 172 p.

33. Rakotoarinivo M, Trudgen MS, Baker WJ (2009) The palms of Makira protected areas. Palms 53: 125–146.

34. Rakotoarinivo M, Dransfield J (2010) New species of Dypsis and Ravenea (Arecaceae) from Madagascar. Kew Bull 65: 279–303.

35. Foibe Taosarintanin'i Madagasikara [FTM] (2006) Madagasikara et ses 22 régions. FTM. 24 p.

36. Battistini R (1996) Paléogéographie et variété des milieux naturels à Madagascar et dans les îles voisines : quelques données de base pour l'étude biogéographique de la région malgache. In: Lourenço WR, editor. Biogéographie de Madagascar, ORSTOM, Paris. pp. 1–17.

37. Jury MR (2003) The Climate of Madagascar. In: Goodman SM, Benstead JP, editors. The Natural History of Madagascar. University of Chicago Press. pp. 75–87.

38. WorldBank (2013) Population density. Available: http://data.worldbank.org/country/madagascar. Accessed 31 May 2013.

39. Allnutt TF, Ferrier S, Manion G, Powell GVN, Ricketts TH, et al. (2008) A method for quantifying biodiversity loss and its application to a 50-year record of deforestation across Madagascar. Cons Lett 1: 173–181. doi: 10.1111/j.1755–263X.2008.00027.x.

40. Thiers B (2012) Index Herbariorum, a global directory of public herbaria and associated staff. New York Botanical Garden's virtual herbarium. Available: http://sweetgum.nybg.org/ih/. Accessed 26 April 2013.

41. Schatz G, Lescot M (2003) Gazetteer to Malagasy Botanical Collecting Localities. Missouri Botanical Garden Available: http://www.mobot.org/MOBOT/Research/madagascar/gazetteer/. Accessed 15 July 2012.

42. Google (2010) Google Earth (version 5.2.1.1588). Available: http://www.google.com/earth/explore/products/. Accessed 03 November 2011.

43. International Union for Conservation of Nature [IUCN] Standards and Petitions Subcommittee (2011) Guidelines for Using the IUCN Red List Categories and Criteria. Version 9.0. Prepared by the Standards and Petitions Subcommittee. Available : http://www.iucnredlist.org/documents/RedListGuidelines.pdf. Accessed: 06 July 2013.

44. IUCN (2012) IUCN Red List of Threatened Species. Version 2013.1. Available: http://www.iucnredlist.org Accessed 10 October 2013.

45. Gaston KJ, Fuller RA (2009) The sizes of species' geographic ranges. J Appl Ecol 46: 1–9. doi: 10.1111/j.1365–2664.2008.01596.x.

46. Willis F, Moat J, Paton A (2003) Defining a role for herbarium data in Red List assessments : a case study of *Plectranthus* from eastern and southern tropical Africa. Biodivers Conserv 12: 1537–1552. doi: 10.1023/A:1023679329093.

47. Bachman S, Moat J, Hill AW, de la Torre J, Scott B (2011) Supporting Red List threat assessments with GeoCAT: geospatial conservation assessment tool. Zookeys 150: 117–126. doi: 10.3897/zookeys.150.2109.

48. Moat J (2007) Conservation assessment tools extension for ArcView 3.x, version 1.2. GIS Unit, Royal Botanic Gardens, Kew. Available: http://www.rbgkew.org.uk/gis/cats. Accessed 12 November 2012.

49. Humbert H, Cours-Darne G (1965) Notice de la carte de Madagascar : carte internationale du tapis végétal et des conditions écologiques à 1:1.000.000. Institut Français de Pondichéry. 162 p.

50. Association Nationale pour la Gestion des Aires Protégées [ANGAP] (2001) Madagascar Protected Area System Management Plan. Mye. 112 p.

51. International Union for Conservation of Nature [IUCN] (2012) Threats Classification Scheme (Version 3.2). Available: http://www.iucnredlist.org/technical-documents/classification-schemes/ threats-classification-scheme. Accessed 10 May 2013.

52. Hoffmann M, Hilton-Taylor C, Angulo A, Böhm M, Brooks TM, et al. (2010) The Impact of Conservation on the Status of the World's Vertebrates. Science 330: 1503–1509. doi: 10.1126/science.1194442.

53. Butchart SH, Akcakaya HR, Kennedy E, Hilton-Taylor C (2006) Biodiversity indicators based on trends in conservation status: strengths of the IUCN Red List index. Conserv Biol 20: 579–581. doi: 10.1111/j.1523-1739.2006.00410.x.

54. Butchart SH, Resit Akçakaya H, Chanson J, Baillie JE, Collen B, et al. (2007) Improvements to the Red List Index. PLoS ONE 2: e140. doi:10.1371/journal.pone.0000140.

55. Groupe des Spécialistes des Plantes Malgaches [GSPM] (2011) Liste rouge des plantes vasculaires endemiques de Madagascar. Sud Expert Plantes, Antananarivo. 188 p.

56. Ramananjanahary RH, Frasier CL, Lowry II PP, Rajaonary FA, Schatz GE (2010) Madagascar's endemic plant families, species Guide. Missouri Botanical Garden, Madagascar. 150 p.

57. Good TC, Zjhra ML, Kremen C (2006) Addressing data deficiency in classifying extinction risk: a case study of a radiation of Bignoniaceae from Madagascar. Conserv Biol 20: 1099–1110. doi: 10.1111/j.1523-1739.2006.00473.x.

58. Rivers MC, Bachman SP, Meagher TR, Nic Lughadha E, Brummitt NA (2010) Subpopulations, locations and fragmentation: applying IUCN red list criteria to herbarium specimen data. Biodivers Conserv 19: 2071–2085. doi: 10.1007/s10531-010-9826-9.

59. Brummitt N, Bachman S (2010) Plants under pressure a global assessment. The first report of the IUCN Sampled Red List Index for plants. Royal Botanic Gardens, Kew, UK. Available : http://www.kew.org/ucm/groups/public/documents/document/kppcont_082104.pdf. Accessed 14 September 2012.

60. International Union for Conservation of Nature (2013) The IUCN Red List of threatened species - Summary statistics. Available: http://www.iucnredlist.org/documents/summarystatistics/2013_1_RL_Stats_Table_5.pdf. Accessed 25 October 2013.

61. Andreone F, Cadle JE, Cox N, Glaw F, Nussbaum RA, et al. (2005) Species review of Amphibians extinction risks in Madagascar: conclusions from the global Amphibians assessment. Conserv Biol 19: 1790–1802. doi: 10.1111/j.1523-1739.2005.00249.x.

62. Davies N, Schwitzer C (2013) Lemur conservation status review: an overview of the Lemur Red-Listing results 2012. In: Schwitzer C, Mittermeier RA, Davies N, Johnson S, Ratsimbazafy J, et al., editors. Lemurs of Madagascar: A Strategy for their Conservation 2013–2016. IUCN SSC Primate Specialist Group, Bristol Conservation and Science Foundation, and Conservation International pp. 13–33.

63. Joppa LN, Roberts DL, Myers N, Pimm SL (2011) Biodiversity hotspots house most undiscovered plant species. P Natl Acad Sci USA 108: 13171–13176. doi:10.1073/pnas.1109389108.

64. Callmander MW, Schatz GE, Lowry II PP (2005) UICN Red List assessment and the Global Strategy for plant conservation : taxonomists must act now. Taxon 54: 1047–1050. doi: 10.2307/25065491.

65. Krauss J, Bommarco R, Guardiola M, Heikkinen RK, Helm A, et al. (2010) Habitat fragmentation causes immediate and time-delayed biodiversity loss at different trophic levels. Ecol Lett 13: 597–605. doi: 10.1111/j.1461-0248.2010.01457.x.

66. Fattorini S, Borges PAV (2012) Species-area relationships underestimate extinction rates. Acta Oecol 40: 27–30. doi:10.1038/nature09985.

67. Rojas-Sandoval J, Meléndez-Ackerman E (2012) Effects of an invasive grass on the demography of the Caribbean cactus *Harrisia portoricensis*: Implications for cacti conservation. Acta Oecol 41: 30–38. doi: 10.1016/j.actao.2012.04.004.

68. Hannah L, Dave R, Lowry II PP, Andelman S, Andrianarisata M, et al. (2008) Climate change adaptation for conservation in Madagascar. Biol Letters 4: 590–594. doi:10.1098/rsbl.2008.0270.

69. Kremen C, Cameron A, Moilanen A, Philips SJ, Thomas CD, et al. (2008) Aligning conservation priorities across taxa in Madagascar with high resolution planning tools. Nature 230: 222–226. doi: 10.1126/science.1155193.

70. Marcot BG, Raphael MG, Schumaker NH, Galleher B (2013) How big and how close? habitat patch size and spacing to conserve a threatened species. Nat Resour Model 26: 194–214. doi: 10.1111/J.1939-7445.2012.00134.X.

71. Hobbs RJ (2007) Setting effective and realistic restoration goals: key directions for research. Restor Ecol 15 : 354–357. doi: 10.1111/j.1526-100X.2007.00225.x.

72. Dransfield J, Rakotoarinivo M, Baker WJ, Bayton RP, Fisher JB, et al. (2008). A new Coryphoid palm genus from Madagascar. Bot J Linn Soc 156: 79–91. doi: 10.1111/j.1095-8339.2007.00742.x.

73. Rakotoarinivo M, Razafitsalama JL, Baker W, Dransfield J (2010) Analalava – a palm conservation hotspot in eastern Nadagascar. Palms 54: 141–151.

74. Byg A, Balslev H (2001) Diversity and use of palms in Zahamena, eastern Madagascar. Biodivers Conserv 10: 951–970. doi:10.1023/A:1016640713643.

75. Byg A, Balslev H (2003) Palm heart extraction in Zahamena, Eastern Madagascar. Palms 47: 37–44.

76. Thomas CD, Cameron A, Green RE, Bakkenes M, Beaumont LJ, et al. (2004) Extinction risk from climate change. Nature 427: 145–148. doi:10.1038/nature02121.

18

16S rRNA Gene Survey of Microbial Communities in Winogradsky Columns

Ethan A. Rundell[1], Lois M. Banta[2], Doyle V. Ward[3], Corey D. Watts[2], Bruce Birren[3], David J. Esteban[1]*

1 Department of Biology, Vassar College, Poughkeepsie, New York, United States of America, **2** Department of Biology, Williams College, Williamstown, Massachusetts, United States of America, **3** Genome Sequencing Center, Broad Institute, Cambridge, Massachusetts, United States of America

Abstract

A Winogradsky column is a clear glass or plastic column filled with enriched sediment. Over time, microbial communities in the sediment grow in a stratified ecosystem with an oxic top layer and anoxic sub-surface layers. Winogradsky columns have been used extensively to demonstrate microbial nutrient cycling and metabolic diversity in undergraduate microbiology labs. In this study, we used high-throughput 16s rRNA gene sequencing to investigate the microbial diversity of Winogradsky columns. Specifically, we tested the impact of sediment source, supplemental cellulose source, and depth within the column, on microbial community structure. We found that the Winogradsky columns were highly diverse communities but are dominated by three phyla: Proteobacteria, Bacteroidetes, and Firmicutes. The community is structured by a founding population dependent on the source of sediment used to prepare the columns and is differentiated by depth within the column. Numerous biomarkers were identified distinguishing sample depth, including Cyanobacteria, Alphaproteobacteria, and Betaproteobacteria as biomarkers of the soil-water interface, and Clostridia as a biomarker of the deepest depth. Supplemental cellulose source impacted community structure but less strongly than depth and sediment source. In columns dominated by Firmicutes, the family Peptococcaceae was the most abundant sulfate reducer, while in columns abundant in Proteobacteria, several Deltaproteobacteria families, including Desulfobacteraceae, were found, showing that different taxonomic groups carry out sulfur cycling in different columns. This study brings this historical method for enrichment culture of chemolithotrophs and other soil bacteria into the modern era of microbiology and demonstrates the potential of the Winogradsky column as a model system for investigating the effect of environmental variables on soil microbial communities.

Editor: Dionysios A. Antonopoulos, Argonne National Laboratory, United States of America

Funding: This work was supported by Vassar College start-up funds to DJE and Williams College divisional research funds to CDW and LMB. The funders had no role in study design, data collection and analysis, decision to publish, or preparation of the manuscript.

Competing Interests: The authors have declared that no competing interests exist.

* Email: daesteban@vassar.edu

Introduction

Sergei Winogradsky is a founder of modern microbiology and microbial ecology, and is credited with the discovery of chemolithotrophy [1,2]. His name is familiar to microbiologists and microbiology students as the originator of Winogradsky columns, commonly used in microbiology education to demonstrate microbial diversity and nutrient cycling. Traditionally, these enrichment cultures are created by culturing mud or sediment in a transparent column with a cellulose source and a source of sulfate and/or other nutrients. Over time, chemical gradients result in a vertical distribution of unique niches for microbial growth. In addition to an oxic niche at the top and anoxic sub-surface niches, the production of H_2S generated by sulfur reducing microbes results in high H_2S in anoxic layers, which diffuses up towards the oxic layer. This stratification in the column leads to the growth of different microbes at different depths, exemplified by the growth of pigmented microorganisms, including phototrophs, producing visible layers. Requiring only the input of light, the Winogradsky column establishes a structured microbial ecosystem, carrying out essential nutrient cycles including the carbon, nitrogen and sulfur cycles.

In nature, sediment and soil microbial communities contribute critically to biogeochemical cycling and the degradation of pollutants and toxins. The structure and function of these communities may be impacted by differences in oxygen concentration, water levels, nutrient levels, pH, and other factors [3–7] that make difficult the study of such communities and their responses to experimental manipulation. By contrast, Winogradsky columns are easy to create, replicate, and control. Thus they can be powerful models for discovering the impact of specific variables on stratified microbial communities as well as studying nutrient cycling and bioremediation.

Past studies have used Winogradsky columns in the study of soluble-reactive phosphate generation, bioremediation, and bio-hydrogen production by microbes [8–10]. Winogradsky columns (and similarly designed Winogradsky plates) have also been used as enrichment cultures for microbial groups including phototrophs and sulfur cycle microbes [11–13]. Novel microorganisms have been isolated and classified from columns [14–16]. Similarly, microcosms of rice paddy soils have been used to describe microbial communities along oxygen gradients [3], however these differ from Winogradsky columns in that they are incubated in the dark.

Soil microbial communities are exceptionally diverse and contain numerous species refractory to growth under laboratory culture conditions. Therefore, culture-based methods inaccurately represent natural microbial community composition [17]. Sequence-based surveys of environmental samples, using high throughput techniques, can reveal a more complete picture of microbial diversity than is possible through culture-based methods. Despite the fact that Winogradsky columns are widely used to demonstrate microbial diversity and have the potential for application to studies of microbial community dynamics in stratified ecosystems, to our knowledge they have never been evaluated using high-throughput sequencing methods.

Here we present the application of high-throughput sequencing to Winogradsky column microbial populations. We conducted a 16S rRNA gene survey of Winogradsky columns in order to investigate the diversity and structure of the communities present in these enrichment cultures, and the influence of environmental variables on these populations. We investigated the effects of depth, sediment source, and supplemental organic carbon source on microbial community structure. We demonstrate that Winogradsky column microbial communities are exceptionally diverse and that the community structure is determined by a founder effect from the sediment source used to create the column, which is further stratified by depth within the column.

Methods and Materials

Column Preparation

Twelve Winogradsky columns were created from two small ponds located near the campus of Williams College in Williamstown, Massachusetts: Buxton Pond (N 42.70° (42° 42′ 15″) W 73.2114° (73° 12′ 41″) and Eph's Pond (N 42.7201° (42° 43′ 12″) W 73.1975° (73° 11′ 51″). Buxton Pond is an artificial pond fed by ground water, and has an overflow that is piped to the east under Gale Road to Christmas Brook. It is in close proximity to a road and is entirely shaded. Eph's Pond is farther from roads and receives more direct sunlight than Buxton Pond. Prior to 1912, the present location of Eph's Pond was marsh and swampy land lacking in standing water. The pond was formed upon construction of a road that impeded the drainage from the center of the Williams College campus toward the Hoosic River. The hydrology of the pond has inputs from both nearby springs as well as storm sewers to the south, while the output is directed through a 24-inch pipe at the pond's northwest corner. Currently Eph's Pond has extensive cattail, sedge, and rush marsh bordering the declining extent of open water in a shallow pond (<1 meter depth) that is rapidly filling in with sediment. Both sampled sites are located on private land. The samples taken for this study qualify for the "minor activities" exemption to the Massachusetts Wetlands Protection Act (310 Code of Massachusetts Regulations 10.00, section 10.58(6)).

Sediment was collected from near the edge of each pond (under approximately 15–30 cm of water) in late October 2008; sticks and other large debris were removed. The pond sludge was mixed to homogeneity with 2% by weight $MgSO_4$, 2% by weight $CaCO_3$, and an equivalent volume of either oak and maple leaf litter or chopped vegetables (organically grown lettuce and red peppers from local farms) as supplemental organic carbon sources. Each mixture together with a small amount of pond water was packed into Plexiglas columns (5.5 cm diameter, 18 cm height, Carolina Biologicals), taking care to force out any trapped air. The packed sediment was overlaid with 2–3 cm of pond water, and the columns were covered tightly with saran wrap, which was secured with a rubber band. Three replicate columns were created for

each condition (Eph's Pond with vegetable scraps, Eph's Pond with leaf litter, Buxton Pond with vegetable scraps and Buxton Pond with leaf litter). Columns were incubated under mixed incandescent and fluorescent light (130 umoles/m2/sec) in a Conviron growth chamber at room temperature for 18 weeks. Samples were collected from each column at the soil water interface (SWI) and at 4, 8, and 12 cm below it. To collect samples, holes were drilled in to the sides of each column with a 0.25 inch drill bit, which was carefully washed between uses, and sediment was scooped with a small scoopula or drawn with a syringe. Additional samples from some columns were taken at other depths. Some samples were taken from the top surface by scraping the biofilm at the soil-water interface with a loop; these are referred to as "top surface" samples to distinguish them from SWI samples collected by drilling. Samples were pelleted by centrifugation at 13,000 rcf for 30 s. At least 0.1 g of soil was obtained after liquid removal, and frozen at −20°C until used. Samples from column replicates 1 and 2 were extracted at Williams College, and samples from column replicate 3 were extracted at Vassar College.

DNA Extraction, Amplification and Sequencing

DNA extraction was performed using a MoBio Powersoil DNA isolation kit (MoBio, CA) following the manufacturer's directions. The V4 region of the 16S rRNA gene was amplified using PCR following the protocol described in [18] and 515F and 806R primers targeting the V4 region [19]. Sequencing was performed using Illumina's MiSeq to generate paired-end reads. Sequences were deposited in MG-RAST (ID: mgp7374) and NCBI (BioProject ID: PRJNA234104).

Quality filtering and OTU picking

SpliceReads, a part of ALLPATHS software [20] was used for assembly of paired-end reads. Filtering, chimera-checking, OTU picking, and taxonomy assignment were conducted using Quantitiative Insights in Microbial Ecology (QIIME, v1.6.0) [21]. Briefly, de novo OTU clustering at 97% identity was performed using usearch (default settings) [22] and with greengenes OTUs for chimera detection. A representative sequence for each OTU was selected and an OTU table was generated. Taxonomic identities were assigned using the RDP Classifier [23] retrained with greengenes taxonomy in QIIME using default settings. Additional filtering for sequence errors was performed by removing OTUs appearing in fewer than 2 samples or containing fewer than 50 sequences across all samples using QIIME's filter_otus_from_otu_table.py script.

Diversity analysis

Alpha diversity was calculated using the Shannon index, OTU richness, and Berger-Parker Dominance index in QIIME. Rarefaction curves were generated by repeated (10 times) subsampling of 10 to 800 sequences, with steps of 79 sequences. We considered only those samples with four or more replicates; that is, samples scraped from the top surface, the SWI, and 4, 8, and 12 cm below the SWI.

To test for significant differences in alpha diversity, QIIME's compare_alpha_diversity.py script was used to run nonparametric two-sample t-tests. The default number of monte-carlo permutations (999) were used to calculate p-values in the nonparametric t-tests, and a significance threshold of $p < 0.05$ was used.

Phylogenetic analysis was performed by aligning representative sequences with PyNAST, followed by alignment filtering to remove non-informative positions, using default settings in QIIME. Phylogenetic beta diversity was calculated using both unweighted and weighted UNIFRAC [24,25]. To estimate the support for beta diversity results, rarefaction was used to generate 100 800-sequence subsamples from each sample. Principal coordinate plots and bootstrapped consensus UPGMA trees were generated from these rarefied distance matrices. Chi-square, Canberra, Pearson, and Spearman indices were also calculated in QIIME. To evaluate statistical significance of beta diversity results, we used nonparametric two-sample t-tests on weighted UNIFRAC distances calculated from the entire dataset (without rarefaction), with the default number of Monte Carlo permutations (999). A p value of less than 0.05 was considered a cutoff point for significant difference. All samples, including those with few or no replicates (that is, samples taken at 3 cm and 6 cm) were included in beta diversity analyses.

Comparison to other microbial communities

To compare Winogradsky column microbial communities to previously sequenced uncultured communities, we used MG-RAST [26]. In order to compare entire Winogradsky columns to uncultured communities we pooled data from the same columns. All steps, including quality filtering, assembly and normalization, used default settings. To minimize biases caused by the use of different sequencing techniques, we restricted our analysis to a comparison to paired-end sequences data obtained by Caporaso et al [18], which used identical primers but had shorter reads. We used the taxonomy-based Bray-Curtis beta diversity index in MG-RAST to generate principal coordinate plots.

Biomarker analysis

Linear discriminant effect size (LEfSe) [27] was used to identify microbial biomarkers for depth, sediment source, or supplemental organic carbon source. We included only those samples from the SWI, 4, 8, and 12 cm depths due to the lower number of replicates from other depths. Unclassified terminal features were removed from the data, leaving them in at higher, classified ranks. For example, OTUs unclassified at the genus level were removed from the genus rank, but were kept in the family and higher ranks. In this way, unclassified OTUs contribute to biomarker analysis at higher taxonomic ranks (where they are classified) but unclassified organisms (difficult to resolve and treat as taxonomic groups based on sequence information alone) are not classified as biomarkers. All settings were set to default with two exceptions. Depth biomarkers were identified using the less strict (one-against-all) setting, and sediment source biomarkers were identified with the requirement that only samples of the same depth be compared (thus to be considered a sediment source biomarker a microbe had to be consistently higher in abundance in one sediment source in each corresponding layer).

Results

Description of Columns and Samples

Winogradsky columns often have several bands of colors at different depths. All columns displayed bands of color at the soil water interface (SWI). At subsurface levels, most did not show the banding patterns commonly seen in Winogradsky columns, although patches of numerous different colors were seen throughout (Figure 1). Replicate columns did not necessarily look the same, but typically looked more like each other than columns with different sediment or supplemental organic carbon sources. Samples were taken by scraping the tops of some columns, and drilling holes at the SWI and at 4 cm intervals below in all columns. The local color of the sampling site was recorded (Table S1). The samples were quite loose, and the sediment was grey except for the 12 cm samples, which were black. Plating of samples on each of several standard bacteriological media yielded cultivable microbial populations that were strikingly similar across all fractions from a given column as well as across columns, with only a small number of distinct morphologies visible on each type of media (data not shown).

Sequence Data

Paired end reads of 175 bp were generated by Illumina sequencing. After assembly of paired end reads, there were 642,384 assembled sequences in 54 samples with an average read length of 253 bp. High throughput sequencing datasets are likely to contain sequencing errors and chimeric PCR products, so we applied chimera filtering in USEARCH and removed extremely rare sequences in QIIME. 135,155 sequences were removed in this process, leaving 507,229. The mean number of sequences per sample was 9393.

Alpha Diversity

In total, 31 phyla were present across all samples with an average of 23 phyla per sample (range: 12 to 30), indicating exceptional diversity (Figure S1). However, the column communities were dominated heavily by members of three abundant phyla (Proteobacteria, Bacteroidetes and Firmicutes) representing an average of 75% of each sample. 414 genera were identified across all columns with an average of 245 genera per sample (range: 102 to 323). The majority of these microbial genera (99%) appeared in three or more samples and 90% appeared in 10 or more.

A critical component of evaluating microbial community structure is the consideration of alpha diversity, or within-community diversity. The Shannon index measures the diversity of communities, taking into account their richness (or the number of distinct taxa that are present) and the evenness of taxon distribution. To investigate whether sequencing depth captured the diversity of our samples, and to assess the impact of soil depth, sediment source, and organic carbon source on Winogradsky column diversity, we generated rarefaction curves for the Shannon index.

The Shannon index rarefaction curves approached asymptotes, indicating that sampling depth was sufficient to capture the overall diversity of Winogradsky column microbial communities (Figure 2 A and D). Winogradsky column microbial diversity is exceptionally high (the average Shannon index value at a sampling depth of 800 sequences was 6.19) and both depth and sediment source impacted the diversity of Winogradsky column communities. Top surface samples were the least diverse (Figure 2 A). Significant differences were seen between top surface samples and samples taken from 4, 8, and 12 cm below the surface (non-parametric t-test, p = 0.021). Communities in Eph's Pond columns were significantly more diverse than those in Buxton Pond columns (Figure 2 D, nonparametric t-test, p = 0.001), but organic carbon source had no effect on diversity (p = 0.633, data not shown).

Richness is a count of the number of taxa present in a community. Rarefaction curves of OTU richness did not approach asymptotes (Figure 2 B and E), even when extended to maximum sampling depth (data not shown). Rarefaction of singles (OTUs appearing only once in a sample) also continued to rise with increased sampling depth (data not shown). Taken together, this suggests that either sequencing depth was insufficient to capture all rare organisms or that sequencing errors were contributing to apparent diversity. Top surface samples were significantly less rich than samples at 4, 8, and 12 cm below the SWI (p = 0.021 for surface vs. 4 cm, p = 0.021 surface vs. 8 cm, and p = 0.042 for

Figure 1. Winogradsky columns. Triplicate columns of each condition were prepared. Photographs were taken following sample collection. The teal plugs covering the holes drilled can be seen in some columns; the topmost plug is at the SWI. Column numbers are shown below the image.

surface vs. 12 cm) (Figure 2B). Samples taken from Eph's Pond columns were significantly more rich than those from Buxton Pond columns (Figure 2E) (p = 0.002). Organic carbon source did not have an effect on richness (p = 0.742, data not shown).

The Winogradsky column microbial communities had few abundant genera and numerous rare genera. Only 5–8% of genera were present in greater than 1% abundance, 23–34% of genera were present in 0.1–0.99% abundance, and the remaining 61–78% were very rare, present at less than 0.1% abundance (Table 1). This shows a fairly uneven distribution of taxa, so we applied the Berger-Parker dominance index, a measure of the relative abundance of the most abundant member of a microbial community. Dominance did not significantly differ by depth (Figure 2C), but Buxton Pond columns were significantly more dominated by single taxa than Eph's Pond columns (p = 0.015, Figure 2F). Organic carbon source did not have an effect on dominance (p = 0.467, data not shown).

Beta Diversity

To investigate the impact of column depth, sediment source and organic carbon source on beta diversity (between-sample diversity), we evaluated our samples using UNIFRAC, a metric evaluating the phylogenetic distance between pairs of communities based on the fraction of total phylogenetic tree branch length unique to both communities [25]. We used both unweighted UNIFRAC, which uses only OTU presence/absence data, and weighted UNIFRAC, which incorporates relative abundance information in to its calculation of phylogenetic distance [24].

Both depth and sediment source substantially impacted Winogradsky column microbial communities, while organic carbon source had a less pronounced effect (Figure 3). Samples scraped off the surface of columns and taken at the SWI were clustered separately from those taken at lower depths along a principal coordinate axis (PCoA) explaining 31.64% of the variation in the data when weighted UNIFRAC was considered (Figure 3A). Unweighted UNIFRAC also indicated separation based on depth; however it occurred along a principal coordinate axis explaining less of the variation in the data (14.94%) (Figure 3D). The compositions of the subsurface layers (4 cm, 8 cm, and 12 cm) were not distinguishable from one another in PCoA plots.

Samples clustered by sediment source in PCoA plots generated from both weighted and unweighted UNIFRAC, indicating a strong influence of sediment source on the column communities (Figure 3B and 3E). In weighted UNIFRAC analysis, samples clustered by sediment source along a second principal coordinate explaining 18.46% of the variation in the data. A third principal coordinate axis in both weighted and unweighted UNIFRAC less clearly separated samples by organic carbon source (Figure 3C and 3F). This axis explained less of the variation in the data (about 9% in weighted and unweighted UNIFRAC).

The substantial impact of both depth and sediment source on community structure in columns was further verified by nonparametric two-sample t-tests on within group vs. between group weighted UNIFRAC distances. These tests indicated that both depth and sediment source significantly impacted beta diversity

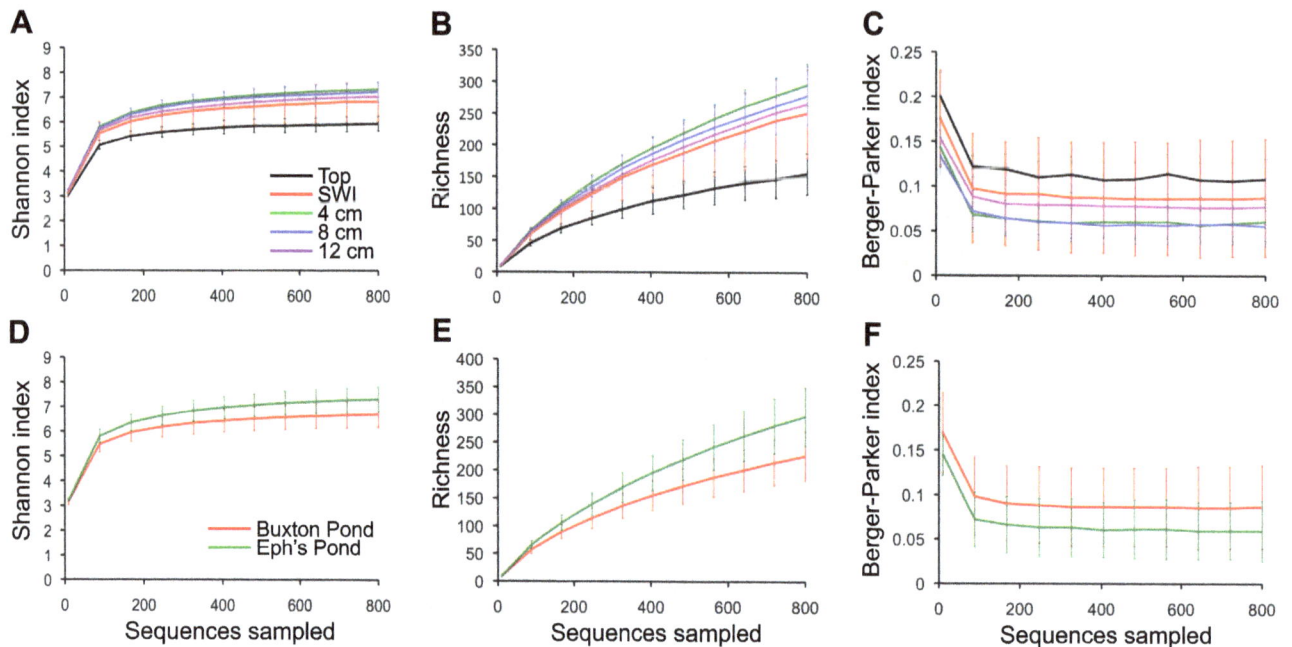

Figure 2. Alpha diversity of Winogradsky column samples. Alpha diversity indices were calculated on rarefied samples. Samples were pooled by layer (A,B,C) or by sediment source (D,E,F) and the average Shannon index (A,D), richness (B,F) and Berger-Parker dominance index (C,F) were calculated. Error bars represent standard error for each category. Significant differences were seen between the Shannon index of top surface samples and samples taken from 4, 8, and 12 cm below the surface (A, non-parametric t-test, $p = 0.021$) and between sediment sources (D, nonparametric t-test, $p = 0.001$). Top surface samples were significantly less rich than samples at 4, 8, and 12 cm below the SWI (B, non-parametric t-test, $p = 0.021$ for surface vs. 4 cm, $p = 0.021$ surface vs. 8 cm, and $p = 0.042$ for surface vs. 12 cm) and Eph's Pond columns are significantly more rich than Buxton Pond columns (E, non-parametric t-test, $p = 0.002$). Buxton Pond columns were significantly more dominated by single taxa than Eph's Pond columns (F, non-parametric t-test, $p = 0.015$).

(within vs. between depth nonparametric p value $= 0.001$ and within sediment source vs. between sediment source nonparametric p-value $= 0.001$). Organic carbon source also significantly impacted beta diversity, but less so (nonparametric p-value $= 0.01$), which is consistent with the less pronounced separation of the samples by organic carbon source in PCoA. The effect of both depth and sediment source on Winogradsky column beta diversity was also investigated by generating a consensus UPGMA tree from rarefied weighted UNIFRAC results. Two major clades were present, each comprised mainly of samples from a single sediment source (Figure 4A). Depth-based clades were not apparent; however, surface and SWI samples tended to branch separately from deeper level samples (Figure 4B).

Throughout the tree, we noted that different layers from the same column tend to be closely related, demonstrating similarity of the community within a whole column (Figure 4C). Replicates were seen to cluster together and we noted that column samples that were visibly similar had similar communities (Table S1). The 4 cm, 8 cm, and 12 cm samples of replicate columns 4.1, 4.2, and 4.3 (Buxton Pond, leaf litter) form a large clade along with several other Buxton Pond column samples, all of which were primarily black with flecks or patches of grey and white. The SWI, 4 cm, 6 cm, and 8 cm samples of column 6.3 (Eph's Pond, leaf litter) cluster together, and all are described as having various shades of green. The sample descriptions and overall appearance of columns 6.1 and 6.2 are quite different from that of 6.3, and this is also reflected in more divergent placement of these samples on the tree. Several triplicates with similar or identical descriptions cluster together. The 12 cm samples from columns 5.1, 5.2 and 5.3 (Eph's Pond, vegetable scraps) form a small clade, and the 8 cm samples form a clade along with two other column 5 replicate samples.

These are all described as being black with grey flecks or patches. In contrast, the 4 cm samples from same columns do not cluster closely and have differing descriptions. The 4 cm samples from columns 6.1, 6.2, and 6.3 also do not cluster together and have different sample descriptions.

There are numerous analysis methods available to compare microbial communities. Certain indices are well suited to detecting community differences in samples distributed along an environmental gradient, while others are more appropriate for identifying factors that contribute to sample clustering. We selected four taxonomy-based beta diversity metrics [28] based on their appropriateness for revealing microbial community differences along an environmental gradient, such as an oxygen gradient (Chi-Square and Pearson), or clusters, such as would be expected from sediment source effects (Spearman and Canberra). All analyses showed sample clustering by sediment source, while by depth, subsurface samples clustered separately from SWI and top surface samples (data not shown) supporting the results obtained with UNIFRAC analysis.

Finally, statistical analyses of weighted UNIFRAC distances by two-sample t-testing revealed that replicate samples tended to be more similar to one another than non-replicates ($p = 0.001$). Replicate samples tended to cluster together in principal coordinate analyses and UPGMA trees, though some heterogeneity was apparent, particularly in samples taken at the SWI (data not shown). The heterogeneity of SWI samples was confirmed by nonparametric two-sample t-tests, which indicated that replicate samples from the SWI were less similar to one another than other same-depth pairs of samples ($p = 0.001$). We also found no differences between samples extracted and prepared at Vassar College vs Williams College (data not shown).

Table 1. Distribution of abundant and rare genera.

Abundance range	Sample						
	All	Eph's	Buxton	SWI	4 cm	8 cm	12 cm
>10%	0 (0)	1 (0)	0 (0)	0 (0)	1 (0)	0 (0)	0 (0)
1–10%	24 (6)	23 (6)	30 (8)	25 (6)	19 (5)	24 (6)	25 (7)
0.1–0.99	135 (33)	135 (33)	94 (24)	126 (31)	134 (34)	104 (26)	87 (23)
<0.1%	255 (62)	246 (61)	273 (69)	257 (63)	239 (61)	266 (68)	296 (78)
Total	414 (100)	405 (100)	397 (100)	408 (100)	393 (100)	394 (100)	381 (100)

Samples were categorized by depth or sediment source and the average abundance of each genus was calculated. Number of genera (percent) is shown.

Microbial Biomarkers

Given the clear effects of both sediment source and depth on communities of Winogradsky columns, we next sought to determine the taxonomic groups that consistently and significantly differed in abundance in accordance with these variables. We used LEfSe, to identify biomarkers consistently varying in abundance according to depth or sediment source [27]. Our findings indicate that these Winogradsky columns are characterized by major phylum and class-level differences in microbial abundance associated with both depth and sediment source, and (to a lesser extent) organic carbon source (Figure 5). Across all taxonomic levels, 87 biomarkers were associated with depth, 194 were associated with sediment source, and 73 were associated with organic carbon source (Table S1).

The Proteobacteria, a biomarker for SWI, were highest in abundance at the tops of columns and decrease in abundance with increasing depth (Figure 6A and 6B). Interestingly, the degree to which Proteobacteria abundance decreased with depth differed substantially in a sediment-source dependent manner. In Buxton Pond columns Proteobacteria abundance dropped sharply below the SWI while in Eph's Pond columns abundance remained higher at greater depths. Based on this pattern we investigated the distribution patterns of the abundant classes in this diverse phylum (Figure 6C and 6D). Alphaproteobacteria and Betaproteobacteria, both biomarkers for SWI (Figure 5A), showed the same pattern in Buxton Pond columns as the phylum level Proteobacteria pattern, with a sharp decrease below the SWI (Figure 6C). In Eph's Pond columns, the Alphaproteobacteria were less abundant and concentrated in the upper layers, but the Betaproteobacteria and Deltaproteobacteria, while most abundant in the upper layers, were also abundant at greater depths (Figure 6D). Deltaproteobacteria were a biomarker for Eph's Pond (Figure 5B).

Firmicutes and class Bacteroidia were biomarkers for the lowest (12 cm) depth (Figure 5A). Both increased in abundance with increasing depth (Figure 6A and 6B). Below the SWI, Buxton Pond columns were dominated by Firmicutes, a biomarker for Buxton Pond, while Eph's Pond columns had abundant Proteobacteria, Firmicutes and Bacteroidetes. The Firmicutes population in both columns was largely dominated by the class Clostridia and order Clostridiales, while the Bacteroidetes were dominated by class Bacteroidia and order Bacteroidales (data not shown).

Other biomarkers included the Cyanobacteria, which only appeared in SWI samples, and the order Methylococcales and the Archaeal phylum Euryarchaeota, which were biomarkers for samples taken at 4 cm and 8 cm respectively (Figure 5A and Table S2). Interestingly, microbes often found in gut microbiota were identified as biomarkers for Buxton Pond – these were the genus *Bacteroides*, the order Enterobacteriales, and the genus *Ruminococcus* (Table S2). Additionally, several microbes associated with dehalogenation, degradation of aromatic compounds, and the degradation of industrial toxins were also identified including the genus *Dehalobacter* (higher in abundance in Buxton Pond), the genus *Desulfomonile* (higher in abundance in Ephs Pond), and the genus *Dechloromonas* (higher in abundance in Ephs Pond).

There were 73 biomarkers for organic carbon source, fewer than for depth or sediment source (Figure 5C and Table S2). These biomarkers were mainly at lower taxonomic levels than those seen for depth and sediment source. The few phylum-level biomarkers for organic carbon source included Tenericutes (higher in columns created with kitchen food scraps and dominated by the genus *Acholeplasma*), Synergistetes (low in abundance but marking kitchen food scraps columns), and Actinobacteria, a biomarker for leaf litter columns.

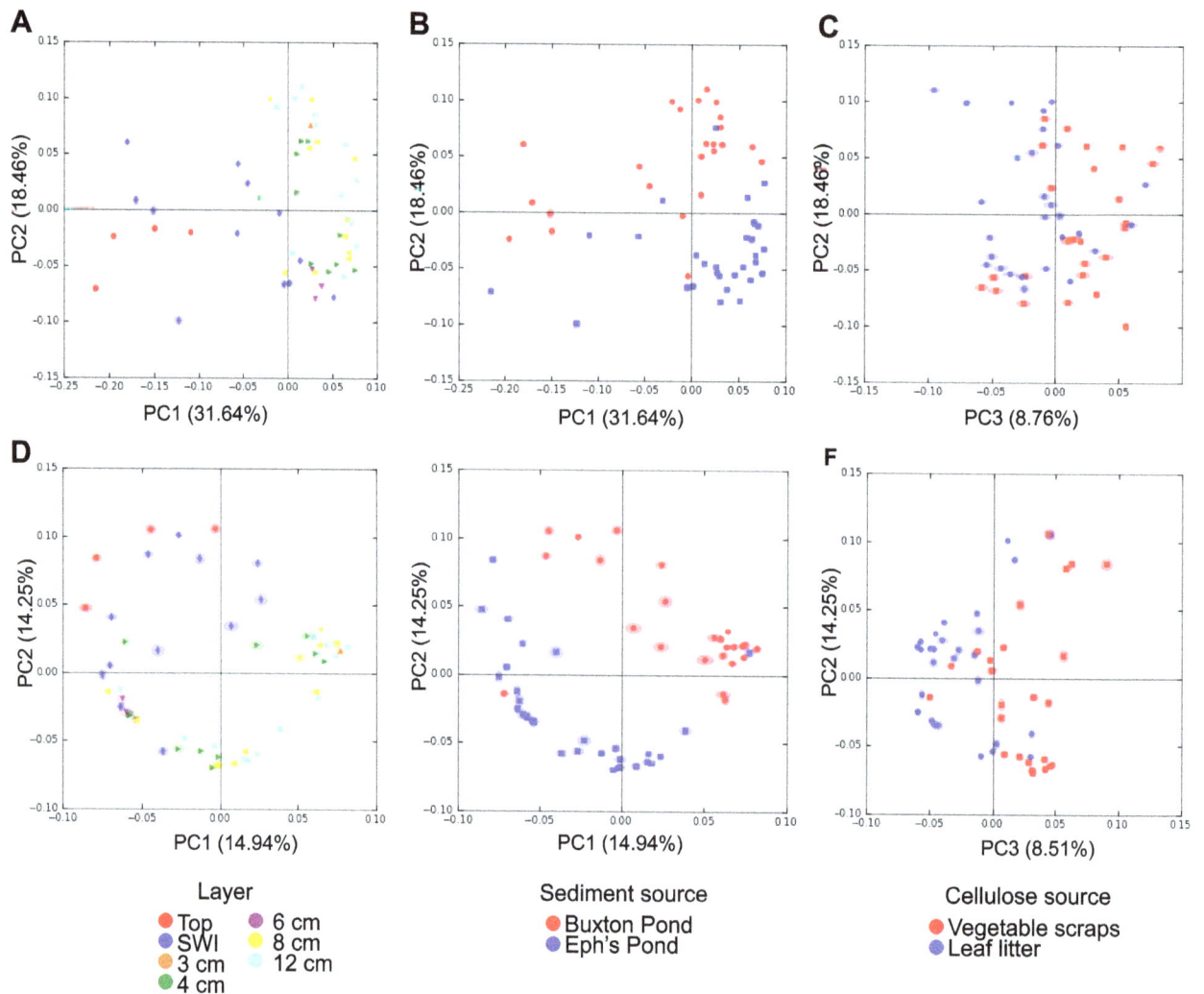

Figure 3. Principal coordinate analysis of Winogradsky column beta diversity. Principal Coordinate plots of weighted UNIFRAC (A,B,C) and unweighted UNIFRAC (D,E,F) results were generated and colored by depth (A,D), sediment source (B,E) or cellulose source (C,F). 100 rarefactions were conducted at a depth of 800 sequences per sample to estimate robustness of beta diversity patterns. Shading around each point represents interquartile range of that point's placement as calculated based on rarefied PCoA.

As an example of a nutrient cycle involving microbial processes, we investigated the abundance and distribution of sulfur-cycling organisms in the columns. Several different sulfur and sulfate reducing and oxidizing taxa were found throughout both Buxton Pond and Eph's Pond columns (Figure 7). In Buxton Pond columns, the Family *Peptococcaceae* of the phylum Firmicutes was the most abundant sulfur reducer, and was a biomarker. In Eph's Pond, several sulfur/sulfate reducing Deltaproteobacteria were identified as biomarkers. Sulfur and sulfide oxidizers in the Phyla Chlorobi and Proteobacteria were found in columns from both sediment sources, but were overall more abundant in Eph's pond. The chemolithotrophic *Hydrogenophilaceae* and phototrophic *Ignavibacteriaceae* were biomarkers for Eph's pond.

Comparison to Natural Communities

We wondered how Winogradsky column microbial populations compared to those seen in prior studies of uncultured microbial communities. We used MG-RAST to compare Winogradsky column communities to those found in several environments, including a freshwater lake, soil, and a creek [18]. Because our

sequencing reads were longer and paired-ends were be assembled, we compared the Winogradsky column samples independently to both the 5′ and 3′ reads from the Caporaso *et al* study. We found that Winogradsky column microbial communities clustered separately from other biomes, but tended to be closer to soil samples than to any other samples in principal coordinate plots (Figure 8 and data not shown).

Discussion

Winogradsky column community structure

Winogradsky column microbial populations can be exceptionally diverse. As many as 30 phyla and 323 genera were present in an individual column, and among all columns 31 phyla and 414 genera were identified. Similarly high diversity in taxonomic groups has been observed in past studies of sediments, including over 40 phyla identified in a study of salt marsh sediments, and 18 phyla identified in a suboxic freshwater pond [29,30]. The Shannon index also indicated an exceptionally diverse community, falling between 6 and 7 for SWI and deeper layers. This level of

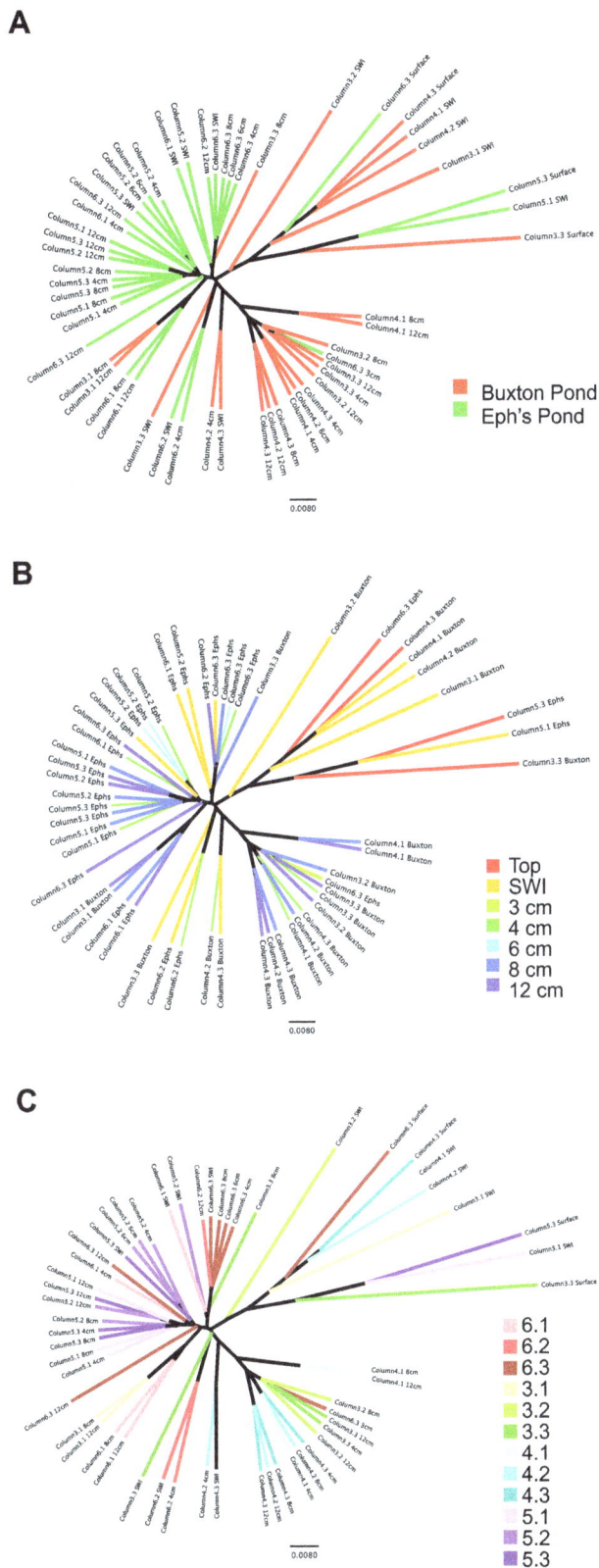

Figure 4. UPGMA trees of all Winogradsky column samples show separation by depth and sediment source. Rarefied weighted UNIFRAC results were used to generate a consensus UPGMA tree. Samples are colored by (A) sediment source, (B) depth, and (C) column.

diversity is not unprecedented but is notable; a prior study of salt marsh sediment microbial communities indicated Shannon index values of over 7 [31]. Furthermore, because the rarefaction curves reach a plateau, our sequencing effort was sufficient to effectively capture the diversity of the samples even at low sampling depth.

Our results show that these Winogradsky column communities contain few highly abundant taxa and a large number of more rare microbes. Most genera identified in the Winogradsky columns were present at less than 1% relative abundance, while between 19 and 30 genera were found at greater than 1% relative abundance. The Berger-Parker index demonstrated that community dominance was similar among all subsurface samples, and rarefaction curves of OTU richness continued to rise with additional sequences per sample (Figure 3). This distribution of few abundant and numerous rare microbes is typical of soil and other microbial communities studied using high throughput techniques [32–36]. In some cases, rare microbes have been cultured or shown to be functionally important in the ecosystem [37,38]. However, the concept that microbial communities contain a "rare biosphere" is controversial, as several studies have suggested that measurements of the rare biosphere, and therefore diversity, are inflated by sequencing errors [39–43]. Indeed, with increasing sampling depth and constant frequency of sequencing error, one would expect richness and singleton OTUs to continue to increase indefinitely due to error alone. The diversity of very low abundance OTUs therefore could be explained equally well by either rare microbes, sequencing artifacts, or a combination of both. We attempted to minimize the impact of sequencing errors on our data by clustering OTUs at 97% identity (rather than 100%), and quality filtering our sequences using both chimera detection and removal of extremely rare sequences. The remaining OTUs occurred in at least 2 samples and were represented by at least 50 sequences, which makes it less likely that these OTUs were the products of individual sequence errors or sequencing artifacts.

Effect of environmental variables on Winogradsky microbial community

Our results suggest that the composition of the community in a Winogradsky column is shaped by a founder effect followed by diversification in stratified niches. When sediment is collected from the pond site, a founding population is captured and poured into the column. The high diversity of soil and sediment communities ensures that there are microbes that can thrive in the variety of niches that result in the column, including gradients of oxygen and hydrogen sulfide along the depth of the column. It is likely that the chemical and physical properties of the sediment contribute to the establishment of these niches as well. Colonization of the human microbiome is also subject to a similar founder effect followed by diversification. Delivery mode (vaginal vs. Cesarean section) influences the structure of the founding population that colonizes the infant [44], and over time site specific communities develop on the skin, gut, and other niches on or in the human body [45].

In our study, we found evidence of the founder effect on Winogradsky communities, as sediment source was an important driver of the composition of the resulting community. Beta diversity measures showed that regardless of depth or organic carbon source, columns made from the same sediment source were more similar to each other than columns made from different sediment sources (Figures 3 and 4). Sediment source biomarkers indicated that the same niches in Winogradsky columns may be filled by phylogenetically distant microbes according to a founder effect. Overall, there were more biomarkers for Eph's Pond than for Buxton Pond columns (Figure 5B), consistent with our alpha

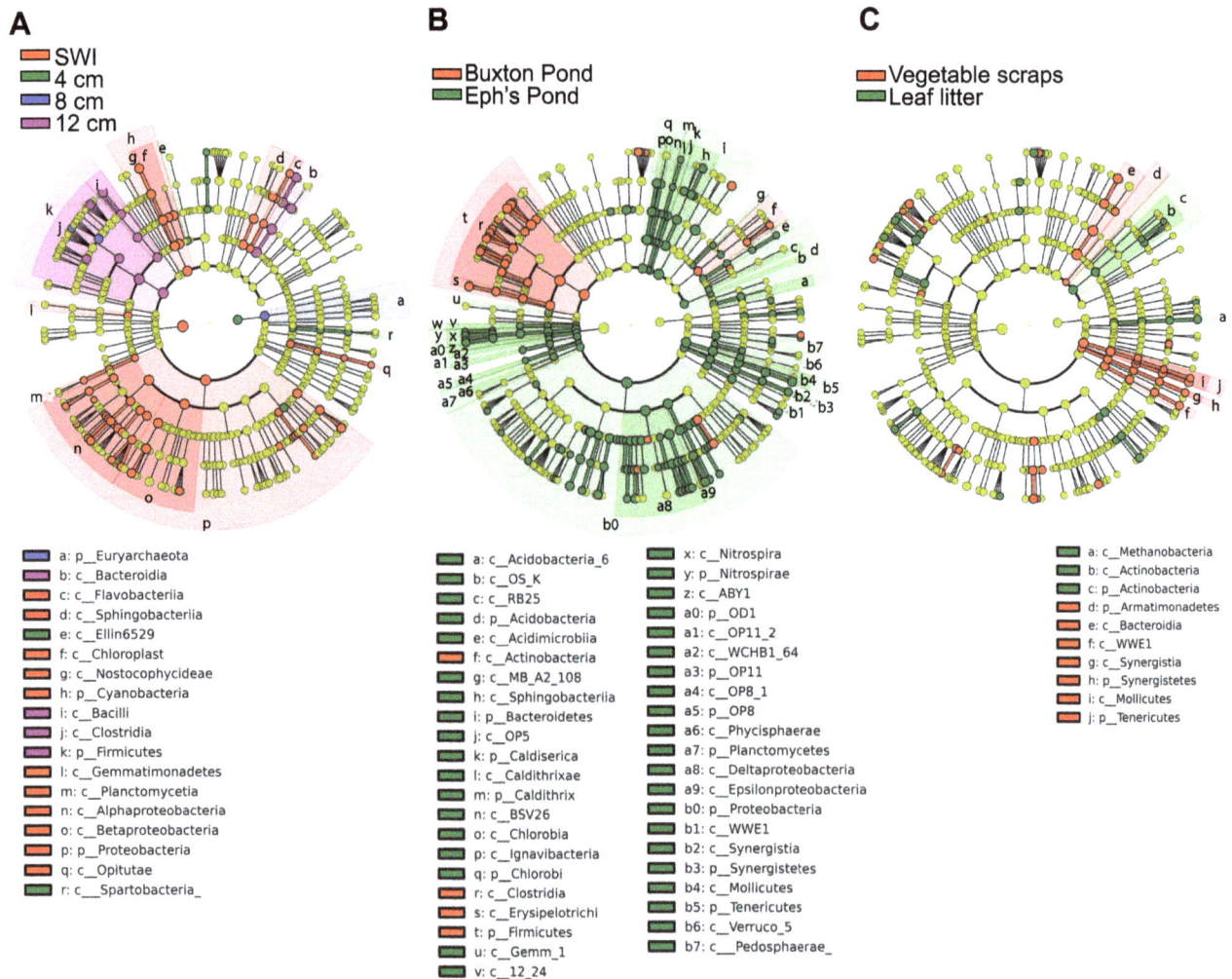

Figure 5. Cladograms of biomarkers for depth, sediment source, and cellulose source. LEfSe was used to identify biomarkers that discriminate (A) depth, (B) sediment source, and (C) cellulose source in Winogradsky columns. Concentric rings from outside in are genus, family, order, class and phylum, with the two central circles for bacteria and archaea. All taxa present are shown, with colored circles representing biomarkers and yellow circles representing non-discriminating taxa. Shaded areas show all taxa below phylum or class biomarkers. For clarity, only biomarkers at the phylum and class levels are labeled.

diversity results, which showed that Eph's Pond columns contained more diverse and rich microbial communities. Anaerobic members of the Proteobacteria, including sulfur cycle anaerobes within the Deltaproteobacteria, were higher in abundance in Eph's Pond columns at all layers than in the same layers of Buxton Pond columns. By contrast, the anaerobic phylum Firmicutes (including sulfate reducers within this phylum) was higher in abundance in Buxton Pond columns than Eph's Pond columns. These patterns suggest that both columns contained anaerobic and sulfur cycle niches dominated by different microbes according to a founder effect.

Once that founding population is added to the column, depth, and to a lesser extent, organic carbon source, allow the growth of specific bacteria in the different niches that are created. At the very top of the columns, the community, dominated by Cyanobacteria, was the least diverse and least rich (Figure 2). Samples collected by drilling into the SWI were the most variable but were no more diverse or rich than samples from greater depth. This indicates that, despite depth-based shifts in environmental conditions in Winogradsky columns, microbial diversity at all subsurface points

remains high. Because we don't have information on the structure of the pond microbial community or the organic matter content of the sediments, we are unable to explain why supplementation caused only a minor shift in the population. The added leaf litter or vegetable scraps may have provided too little supplemental carbon to make a sufficient difference, or both sources may have provided similar supplements, changing the population from the founding pond sediments but not from eachother.

Using beta diversity analyses, we found separation of the phylogenetic and taxonomic composition of the communities by depth (upper layers: top surfaces and SWI, and lower layers: 4 cm, 8 cm, 12 cm). The separation of surface and SWI samples from deeper samples was more pronounced in weighted UNIFRAC, where relative abundance is taken into account. Weighted UNIFRAC has been suggested to be more appropriate in highlighting community differences based on shifts in abundance, such as those associated with differences in metabolite concentration [24]. We therefore propose that the differences seen between upper-level and lower-level samples reflect the major shifts in microbial community composition and structure that occur as

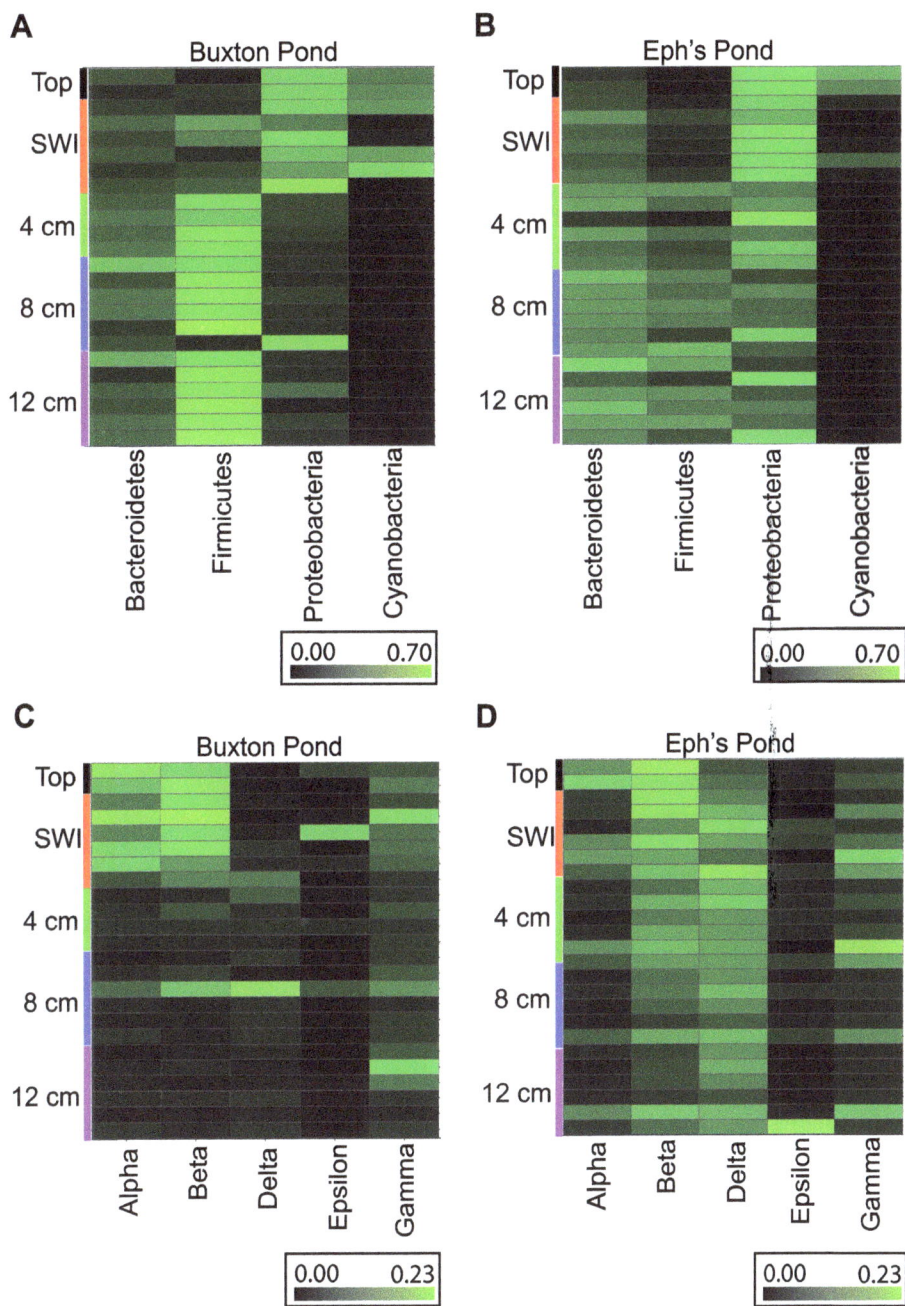

Figure 6. Heatmaps of abundant taxa in Winogradsky columns. Relative abundance of the six most abundant phyla is shown for (A) Buxton Pond and (B) Eph's Pond, and of Proteobacteria classes in (C) Buxton Pond and (D) Eph's Pond. Samples are organized by column depth from top to bottom.

conditions shift from oxic to anoxic. Similar patterns have seen in ponds receiving abundant organic matter and in flooded paddy soils, in which anoxic conditions rapidly develop close below the surface and anaerobic microbes produce methane and H_2S [3,29]. It will be interesting to determine if similarly high diversity as well as depth and sediment source effects are seen in Winogradsky columns prepared with materials from more distinct environments.

Composition of Winogradsky column communities

Proteobacteria, Firmicutes and Bacteroidetes made up more than 75% of the community of each sample. Proteobacteria were highest in abundance at the tops of columns and decreased in abundance with increasing depth. Proteobacteria are frequently identified as an abundant member of sediment microbial populations, and Alphaproteobacteria and Betaproteobacteria have been linked to oxic zones of vertical oxygen gradients in previous studies [3,29,46–48]. Our work shows that this association of members of the Proteobacteria with oxic zones is duplicated in Winogradsky columns. Interestingly, in both Eph's

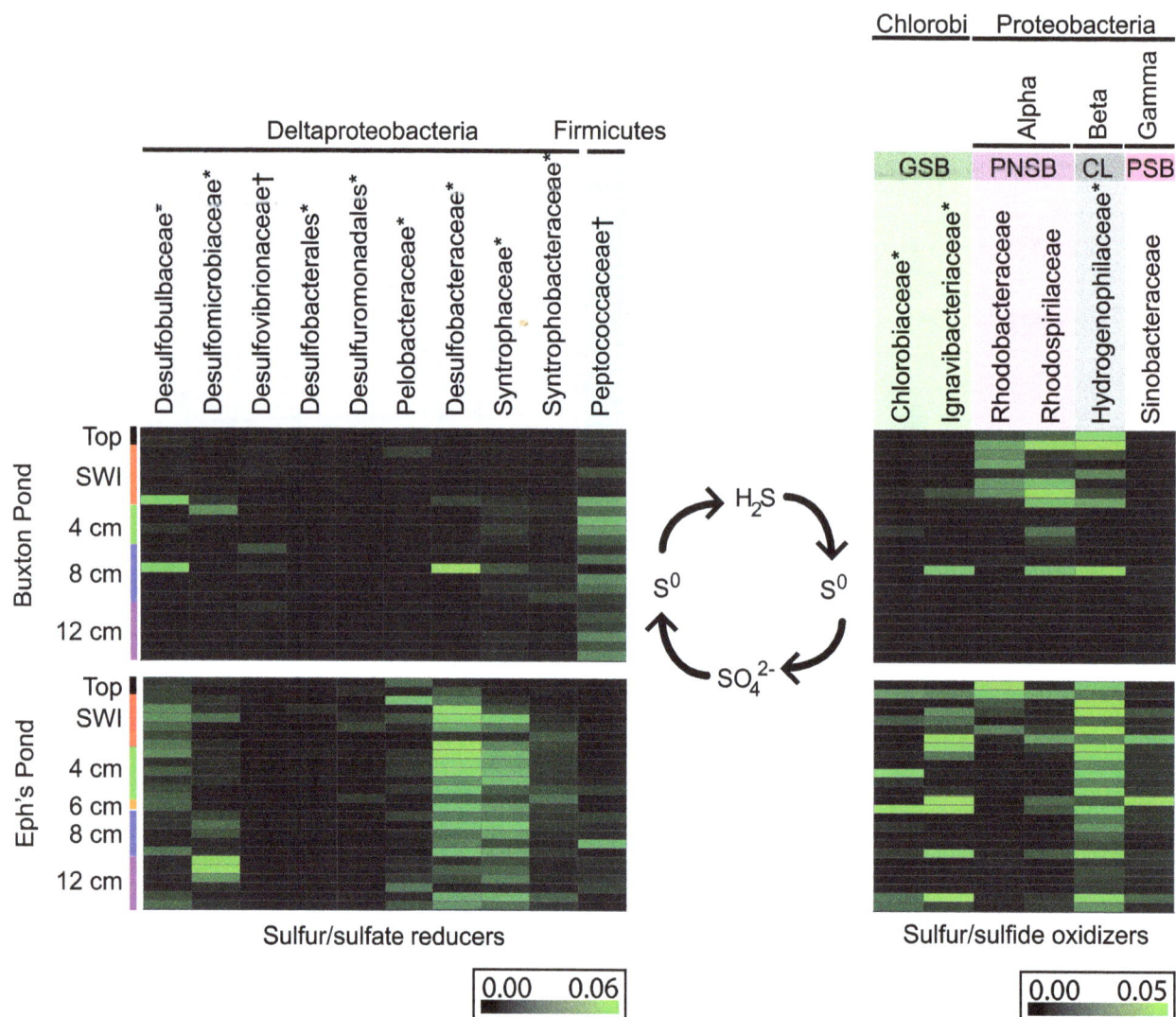

Figure 7. Sulfur cycling organisms identified in Buxton Pond and Eph's Pond Winogradsky column communities. Heat maps show the relative abundance of abundant sulfur and sulfate reducers (left) and sulfur or sulfide oxidizers (right). Samples are ordered by depth from top to bottom. *Eph's Pond column biomarkers, †Buxton Pond column biomarkers. GSB: green sulfur bacteria, PNSB: purple non-sulfur bacteria, CL: chemolithotroph, PSB: purple sulfur bacteria.

and Buxton Pond columns, in lower-depth samples, the communities were either abundant in Proteobacteria or Firmicutes, but not both, suggesting that local conditions favor growth of one or the other.

Firmicutes and Bacteroidetes were low in abundance at the tops of columns and increased in abundance with depth. Firmicutes, mainly Clostridium Cluster one, has previously been associated with anoxic zones in vertical oxygen gradients [3,46] and the genus *Clostridium* has long been described as an abundant member of the bottom layers of Winogradsky columns. The high abundance of the class Clostridia at lower layers supports past culture-based associations between Clostridia and anoxic zones of sediment.

Bacteroidetes is another anaerobic phylum and its increasing abundance with depth is likely a reflection of decreasing oxygen concentration with depth. Certain members of the Phylum Bacteroidetes are capable of degrading complex organic compounds like cellulose [49] and chitin and are likely key to the carbon cycle in the Winogradsky column. The simpler carbon

compounds produced by these decomposers can be utilized by other fermenting organisms. Overall the distribution of aerobic and anaerobic groups at different depths reinforces the implications of beta diversity analyses and strongly suggests a major impact of oxygen concentration on microbial communities in Winogradsky columns.

The sulfur cycle is a key biogeochemical cycle that is driven primarily by microbial processes. The Winogradsky column has been used to enrich for sulfur cycle organisms, and used as a teaching tool for demonstrating the sulfur cycle. High-throughput sequencing of Winogradsky columns allows for detailed analysis of the ecology and spatial distribution of sulfur cycle microbes and provides an example of taxonomically distant microbes filling the same niches in microbial communities. Dissimilative reduction of sulfur compounds is carried out by sulfur reducing bacteria for energy conservation. Anaerobic respiration of sulfate (SO_4^{2-}) generates hydrogen sulfide (H_2S), which has a distinctive odor and spontaneously forms ferrous sulfide with iron, visible as a black coloration of the soil. The sulfur reducing bacteria are not a

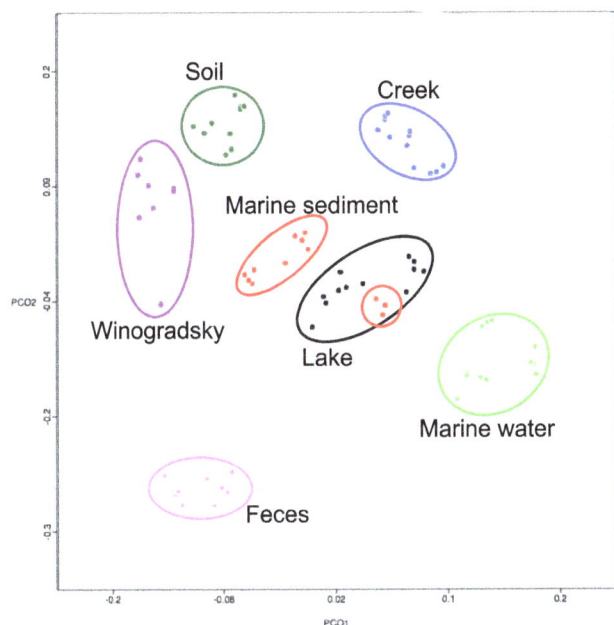

Figure 8. Comparison of Winogradsky column microbial communities to uncultured microbial communities. Samples from individual columns were pooled together to generate a single community for each column. MG-RAST was used to compare twelve samples from each of soil, marine sediment, creek, marine water, lake, and fecal biomes selected from previous metagenomics experiments (Caporaso et al.) targeting the V4 region of 16s rRNA (using single end sequencing of the 5′ end). These samples were compared to our data using the taxonomically-based Bray-Curtis index.

monophyletic group, but rather are defined physiologically. Many sulfur reducers are in the class Deltaproteobacteria, but some Firmicutes and Archaea are also capable of sulfate reduction [50]. In these columns, the particular sulfur/sulfate reducers found differed by sediment source used. The sulfate reducing Firmicutes in the family *Peptococcaceae* were identified as a biomarker for Buxton Pond columns while other sulfur/sulfate reducers were rare or absent (Figure 7). In Eph's Pond columns, sulfur/sulfate reducing organisms were more diverse and abundant and were almost exclusively Deltaproteobacteria, most of which were biomarkers for Eph's Pond columns (Figure 7). Deltaproteobacteria are commonly found in sediment communities and have been identified as an abundant member of black layer communities in wetlands [29,51,52]. The family *Desulfobacteraceae* was the most abundant sulfate reducer and was distributed throughout Eph's Pond columns. In the family *Syntrophaceae* about one third of the sequences classifiable to a genus belonged to *Syntrophus*, which cannot reduce sulfate; however other sequences may belong to sulfate reducing members of this family. The *Syntrophobacteraceae* family is also comprised of a mix of sulfur reducers and non-sulfur reducers. The differences in sulfur/sulfate reducing communities between the two sediment sources demonstrate the functional overlap between highly divergent microbial groups.

Sulfur oxidizers complete the cycle by oxidizing H_2S to elemental sulfur (S^0) and SO_4^{2-} through phototrophy or chemolithotrophy. The Purple Sulfur Bacteria (PSB) use H_2S, and sometimes other reduced sulfur compounds, as an electron donor in photosynthesis. They have red, orange, blue and yellow pigments, resulting in the red-violet zone of the Winogradsky column, in the upper-middle portion of the column. The process generates elemental sulfur that is stored in intracellular globules

and can be later oxidized to SO_4^{2-}. The PSB are all members of the class Gammaproteobacteria. Like the PSB, the Purple Non-sulfur Bacteria (PNSB) use H_2S in photosynthesis, but generally have lower H_2S concentration optima than the PSB. These are members of the Alphaproteobacteria and Betaproteobacteria. The Green Sulfur Bacteria (GSB) use H_2S as an electron donor in photosynthesis, generating S^0 and SO_4^{2-}. They are strict anaerobes, and are typically found in the "green zone" located in the lower-middle of the Winogradsky column. Neither the PSB nor the PNSB were found to be abundant in these columns (Figure 7). Although the GSB were more abundant, they were distributed unevenly in samples throughout the column rather than in the expected lower-middle zone. This is not surprising given that distinct green and red-violet zones were not apparent in these columns, although patches of color were present throughout at different intensities. Further, our sampling technique (drilling into the side of the column) may have excluded or destroyed surface-attached members of the microbial community.

Sulfur oxidation can also occur through chemolithotrophy, the use of a reduced sulfur compound as an electron donor in aerobic or anaerobic respiration. The chemolithotrophic sulfur oxidizer *Thiobacillus* was fairly abundant in the column, but especially in the upper-middle zones. They are shown in Figure 8 as *Hydrogenophilaceae*; the vast majority of the sequences identified in this family belong to the genus *Thiobacillus*. Another important chemolithotrophic sulfur oxidizer is the filamentous non-photosynthetic *Beggiatoa*, [53]. This species is strongly associated with Sergei Winogradsky, as its isolation and characterization by Winogradsky led to the concept of lithoautotrophy [54]. No *Beggiatoa* sequences were found in these columns.

We also identified a collection of microorganisms in the 4 cm and 8 cm layers involved in methane cycling. The phylum Euryarchaeota, one of the two Archaeal phyla detected in the columns, was a biomarker for samples at 8 cm. A prior study of microbial community composition in suboxic freshwater ponds identified Euryarchaeota as the only archaeal phylum in the sediment [29]. Euryarchaeota includes methanogenic microbes. Interestingly, the methanotrophic Methylococcales (a Gammaproteobacteria) was a biomarker for 4 cm samples, suggesting that methane produced in deeper layers is being used by methanotrophs above. The detection of Methylococcales in this layer is interesting, given that the black coloration of the sediment (indicative of metal sulfide precipitation) suggests an anoxic environment.

Natural history of Eph's and Buxton Ponds

Both Buxton Pond and Eph's Ponds are located near Williams College, Willamstown, MA. Buxton Pond is entirely shaded and receives a substantial amount of leaf litter, while Eph's pond receives more direct sunlight. Buxton Pond is also closer to small roads. Both ponds have in the past experienced exposure to human waste. A sewer rupture in 1994 allowed raw sewage to flow into Eph's Pond for several days; no remediation was performed. Until the mid-1990s Buxton Pond was the receiving water body for one of the Buxton boarding prep school's cesspools. Several fecal microbiome microbes were biomarkers for Buxton Pond (*Bacteroides*, Enterobacteriales, and *Ruminococcus*), perhaps reflecting past fecal contamination. However, these are also cellulose degraders, and may be present in greater abundance than in Eph's Pond due to the differences in the amount of leaf litter that falls on the ponds. Columns from both ponds have microbes associated with contamination with aromatic organic compounds (*Dehalobacter, Desulfomonile, and Dechloromonas*), consistent with their history and location in a lightly developed municipality.

Winogradsky columns as model communities

Microbial communities contribute critically to biogeochemical cycles, bioremediation, alternate fuel production, primary productivity, and numerous other processes critical in supporting ecosystems. They are also extraordinarily diverse and complex, containing numerous rare microbes, and are shaped by an enormous variety of factors including the presence or absence of other microbes, metabolic conditions, temperature, and pH. Studies of microbial communities in natural sediments are difficult to control for the impact of such variables. Here, we demonstrate the potential of Winogradsky columns for applying 16 s rRNA sequencing to the study of complex microbial communities. Winogradsky columns are easy to create, replicate, and manipulate, which allows for a degree of control not possible in field studies. We have demonstrated that Winogradsky column communities can be exceptionally diverse and display consistent patterns in microbial abundance based on depth, similar to natural ecosystems. Columns could also be manipulated to mimic the effects of changing temperatures, pollution, drought, or other effects relevant to current environmental challenges. Supplementing columns with different metabolites would offer a simple method for elucidating the effects of individual metabolites on stratified microbial communities. Time-course studies would offer insights in to the dynamics of communities over time as well as questions of succession and competition in sediment microbial communities. Winogradsky columns are self-contained and manipulatible ecosystems of diverse microbes and in combination with high-throughput sequencing could become a powerful tool for studies of microbial ecology.

Winogradsky columns in undergraduate education

Winogradsky columns are commonly used in undergraduate microbiology courses, for example as tools to teach principles of nutrient cycling [55]. Metagenomics is an increasingly important part of microbiology, and together, the two present a unique opportunity to introduce students to ground-breaking technology, laboratory techniques such as extraction of DNA and PCR, and analysis of large, complex datasets. Several successful programs have already integrated education and genomics research [56]. Some of the work presented in this paper was performed within introductory microbiology courses at Vassar College and Williams College, and we envision that short or more extensive units on Winogradsky metagenomics could be an effective component of undergraduate microbiology courses, or could be the focus of an advanced undergraduate course on microbial ecology or metagenomics. We have developed teaching materials to guide teachers on using wet-lab and bioinformatic techniques for Winogradsky metagenomics in their courses [57], and emphasize that the data generated in these studies are a publicly available resource that can be used to supplement laboratory instruction.

Supporting Information

Figure S1   Relative abundance of phyla in Winogradsky columns. A) Relative abundance in all samples, B–E) in samples at each depth, and F, G) in samples from the two mud sources. Vertical dashed lines demarcate phyla present at >1%, 1%–0.1%, and <0.1%.

Acknowledgments

Thanks to Elizabeth Collins (Vassar College) and Manuel Morales (Williams College) for helpful discussions, and to Jose Ruiz and Norm Bell (Williams College) for technical assistance. We are grateful to Henry Art (Williams College) for providing information about the natural history of the ponds. Thanks to Vassar College Microbiology (Biology 205) classes and interns (Spring 2009, Spring 2011 and Fall 2011) and the Spring 2008 and 2009 Williams College Microbiology (Biology 315) classes for assistance with sample preparation and data analysis. We thank David Fitzgerald (David.F.Fitzgerald@williams.edu), Horticulturalist and Grounds Supervisor at Williams College and Peter Smith, headmaster of Buxton School (petesmyth56@gmail.com) for permission to sample at Eph's Pond and Buxton Pond, respectively. We direct readers to them to obtain permission for future sampling.

Author Contributions

Conceived and designed the experiments: BB DJE LMB. Performed the experiments: DJE LMB CDW DW ER. Analyzed the data: ER DJE DW. Contributed reagents/materials/analysis tools: BB DW LMB DJE. Wrote the paper: DJE ER.

References

1. Winogradsky S (1888) Bietrage zur Morphologie und Physiologie de Bacterien. Zur Morphologie und Physiologie der Schwfelbacterien Heft 1.

2. Dworkin M (2011) Sergei Winogradsky: a founder of modern microbiology and the first microbial ecologist. FEMS Microbiol Rev 36: 364–379. doi:10.1111/j.1574-6976.2011.00299.x.

3. Lüdemann H, Arth I, Liesack W (2000) Spatial changes in the bacterial community structure along a vertical oxygen gradient in flooded paddy soil cores. Appl Environ Microbiol 66: 754–762. doi:10.1128/AEM.66.2.754-762.2000.

4. Fierer N, Schimel JP, Holden PA (2003) Variations in microbial community composition through two soil depth profiles. Soil Biol Biochem 35: 167–176.

5. Zhou J, Xia B, Huang H, Palumbo AV, Tiedje JM (2004) Microbial diversity and heterogeneity in sandy subsurface soils. Appl Environ Microbiol 70: 1723–1734.

6. Hansel CM, Fendorf S, Jardine PM, Francis CA (2008) Changes in bacterial and archaeal community structure and functional diversity along a geochemically variable soil profile. Appl Environ Microbiol 74: 1620–1633. doi:10.1128/AEM.01787-07.

7. Fierer N, Jackson RB (2006) The diversity and biogeography of soil bacterial communities. Proc Natl Acad Sci USA 103: 626–631. doi:10.1073/pnas.0507535103.

8. Guhathakurta H, Biswas R, Dey P, Mahapatra PG, Mondal B (2007) Effect of organic forms of phosphorus and variable concentrations of sulfide on the metabolic generation of soluble-reactive phosphate by sulfur chemolithoauto-trophs: a laboratory study. ISME J 1: 545–550. doi:10.1038/ismej.2007.61.

9. de Sousa ML, de Moraes PB, Lopes PRM, Montagnolli RN, de Angelis D de F, et al. (2012) Textile Dye Treated Photoelectrolytically and Monitored by Winogradsky Columns. Environ Eng Sci 29: 180–185. doi:10.1089/ees.2010.0259.

10. Loss RA, Fontes ML, Reginatto V, Antônio RV (2013) Biohydrogen production by a mixed photoheterotrophic culture obtained from a Winogradsky column prepared from the sediment of a southern Brazilian lagoon. Renewable Energy 50: 648–654. doi:10.1016/j.renene.2012.07.011.

11. van Niel CB (1971) Techniques for the enrichment, isolation, and maintenance of the photosynthetic bacteria. Method Enzymol 23: 3–28. doi:10.1016/S0076-6879(71)23077-9.

12. Charlton PJ, McGrath JE, Harfoot CG (1997) The Winogradsky plate, a convenient and efficient method for the enrichment of anoxygenic phototrophic bacteria. J Microbiol Method 30: 161–163.

13. Postgate JR (1959) Differential media for sulphur bacteria. J Sci Food Agric 10: 669–674. doi:10.1002/jsfa.2740101206.

14. McInerney MJ, Bryant MP, Hespell RB, Costerton JW (1981) Syntrophomonas wolfei gen. nov. sp. nov., an anaerobic, syntrophic, fatty acid-oxidizing bacterium. Appl Environ Microbiol 41: 1029–1039.

15. Neutzling O, Imhoff JF, Truper HG (1984) Rhodopseudomonas adriatica sp. nov., a new species of the Rhodospirillaceae, dependent on reduced sulfur compounds. Arch Microbiol 137: 256–261.

16. Janssen PH, Harfoot CG (1991) Rhodopeudomonas rosca sp. nov., a new purple nonsulfur bacterium. Int J Sys Bacteriol 41: 26–30.

17. Daniel R (2005) The metagenomics of soil. Nat Rev Microbiol 3: 470–478.

18. Caporaso JG, Lauber CL, Walters WA, Berg-Lyons D, Lozupone CA, et al. (2011) Global patterns of 16S rRNA diversity at a depth of millions of sequences per sample. Proc Natl Acad Sci USA 108: 4516–4522.

19. Walters WA, Caporaso JG, Lauber CL, Berg-Lyons D, Fierer N, et al. (2011) PrimerProspector: de novo design and taxonomic analysis of barcoded polymerase chain reaction primers. Bioinformatics 27: 1159–1161. doi:10.1093/bioinformatics/btr087.

20. Butler J, MacCallum I, Kleber M, Shlyakhter IA, Belmonte MK, et al. (2008) ALLPATHS: De novo assembly of whole-genome shotgun microreads. Genome Res 18: 810–820.

21. Caporaso JG, Kuczynski J, Stombaugh J, Bittinger K, Bushman FD, et al. (2010) QIIME allows analysis of high-throughput community sequencing data. Nat Method 7: 335–336.

22. Edgar RC (2010) Search and clustering orders of magnitude faster than BLAST. Bioinformatics 26: 2460–2461. doi:10.1093/bioinformatics/btq461.

23. Wang Q, Garrity GM, Tiedje JM, Cole JR (2007) Naive Bayesian classifier for rapid assignment of rRNA sequences into the new bacterial taxonomy. Appl Environ Microbiol 73: 5261–5267. doi:10.1128/AEM.00062-07.

24. Lozupone CA, Hamady M, Kelley ST, Knight R (2007) Quantitative and Qualitative Diversity Measures Lead to Different Insights into Factors That Structure Microbial Communities. Appl Environ Microbiol 73: 1576–1585. doi:10.1128/AEM.01996-06.

25. Lozupone C, Knight R (2005) UniFrac: a new phylogenetic method for comparing microbial communities. Appl Environ Microbiol 71: 8228–8235. doi:10.1128/AEM.71.12.8228-8235.2005.

26. Meyer F, Paarmann D, D'Souza M, Olson R, Glass EM, et al. (2008) The metagenomics RAST server – a public resource for the automatic phylogenetic and functional analysis of metagenomes. BMC Bioinformatics 9: 386. doi:10.1186/1471-2105-9-386.

27. Haas BJ, Gevers D, Earl AM, Feldgarden M, Ward DV, et al. (2011). Genome Biol 12: 494–504. Available: http://genome.cshlp.org/cgi/doi/10.1101/gr.112730.110.

28. Kuczynski J, Liu Z, Lozupone C, McDonald D, Fierer N, et al. (2010) Microbial community resemblance methods differ in their ability to detect biologically relevant patterns. Nat Method 7: 813–819. doi:10.1038/nmeth.1499.

29. Briée C, Moreira D, López-García P (2007) Archaeal and bacterial community composition of sediment and plankton from a suboxic freshwater pond. Res Microbiol 158: 213–227. doi:10.1016/j.resmic.2006.12.012.

30. Bowen JL, Ward BB, Morrison HG, Hobbie JE, Valiela I, et al. (2011) Microbial community composition in sediments resists perturbation by nutrient enrichment. ISME J 5: 1540–1548. doi:10.1038/ismej.2011.22.

31. Bowen JL, Crump BC, Deegan LA, Hobbie JE (2009) Salt marsh sediment bacteria: their distribution and response to external nutrient inputs. ISME J 3: 924–934. doi:10.1038/ismej.2009.44.

32. Neufeld JD, Mohn WW (2005) Unexpectedly high bacterial diversity in arctic tundra relative to boreal forest soils, revealed by serial analysis of ribosomal sequence tags. Appl Environ Microbiol 71: 5710–5718. doi:10.1128/AEM.71.10.5710-5718.2005.

33. Youssef NH, Couger MB, Elshahed MS (2010) Fine-Scale Bacterial Beta Diversity within a Complex Ecosystem (Zodletone Spring, OK, USA): The Role of the Rare Biosphere. PLoS ONE 5: e12414. doi:10.1371/journal.pone.0012414.

34. Gobet A, Boer SI, Huse SM, van Beusekom JEE, Quince C, et al. (2011) Diversity and dynamics of rare and of resident bacterial populations in coastal sands. ISME J 6: 542–553. doi:doi:10.1038/ismej.2011.132.

35. Bowen JL, Morrison HG, Hobbie JE, Sogin ML (2012) Salt marsh sediment diversity: a test of the variability of the rare biosphere among environmental replicates. ISME J 6: 2014–2023. doi:10.1038/ismej.2012.47.

36. Sogin ML, Morrison HG, Huber JA, Welch DM, Huse SM, et al. (2006) Microbial diversity in the deep sea and the underexplored "rare biosphere." Proc Natl Acad Sci USA 103: 12115–12120. doi:10.1073/pnas.0605127103.

37. Pester M, Bittner N, Deevong P, Wagner M, Loy A (2010) A "rare biosphere"microorganism contributes to sulfate reduction in a peatland. ISME J 4: 1591–1602.

38. Shade A, Hogan CS, Klimowicz AK, Linske M, McManus PS, et al. (2012) Culturing captures members of the soil rare biosphere. Environ Microbiol 14: 2247–2252. doi:10.1111/j.1462-2920.2012.02817.x.

39. Huse SM, Welch DM, Morrison HG, Sogin ML (2010) Ironing out the wrinkles in the rare biosphere through improved OTU clustering. Environ Microbiol 12: 1889–1898. doi:10.1111/j.1462-2920.2010.02193.x.

40. Wu J-Y, Jiang X-T, Jiang Y-X, Lu S-Y, Zou F, et al. (2010) Effects of polymerase, template dilution and cycle number on PCR based 16 S rRNA diversity analysis using the deep sequencing method. BMC Microbiol 10: 255. doi:10.1186/1471-2180-10-255.

41. Lee CK, Herbold CW, Polson SW, Wommack KE, Williamson SJ, et al. (2012) Groundtruthing next-gen sequencing for microbial ecology-biases and errors in community structure estimates from PCR amplicon pyrosequencing. PLoS ONE 7: e44224. doi:10.1371/journal.pone.0044224.

42. Schloss PD, Gevers D, Westcott SL (2011) Reducing the Effects of PCR Amplification and Sequencing Artifacts on 16S rRNA-Based Studies. PLoS ONE 6: e27310. doi:10.1371/journal.pone.0027310.

43. Dickie IA (2010) Insidious effects of sequencing errors on perceived diversity in molecular surveys. New Phytol 188: 916–918. doi:10.1111/j.1469-8137.2010.03473.x.

44. Dominguez-Bello MG, Costello EK, Contreras M, Magris M, Hidalgo G, et al. (2010) Delivery mode shapes the acquisition and structure of the initial microbiota across multiple body habitats in newborns. Proc Natl Acad Sci USA 107: 11971–11975. doi:10.1073/pnas.1002601107/-/DCSupplemental.

45. Costello EK, Lauber CL, Hamady M, Fierer N, Gordon JI, et al. (2009) Bacterial Community Variation in Human Body Habitats Across Space and Time. Science 326: 1694–1697. doi:10.1126/science.1177486.

46. Noll M, Matthies D, Frenzel P, Derakshani M, Liesack W (2005) Succession of bacterial community structure and diversity in a paddy soil oxygen gradient. Environ Microbiol 7: 382–395. doi:10.1111/j.1462-2920.2005.00700.x.

47. Shivaji S, Kumari K, Kishore KH, Pindi PK, Rao PS, et al. (2011) Vertical distribution of bacteria in a lake sediment from Antarctica by culture-independent and culture-dependent approaches. Res Microbiol 162: 191–203.

48. Xiong J, He Z, Van Nostrand JD, Luo G, Tu S, et al. (2012) Assessing the Microbial Community and Functional Genes in a Vertical Soil Profile with Long-Term Arsenic Contamination. PLoS ONE 7: e50507. doi:10.1371/journal.pone.0050507.

49. Bayer EA, Shoham Y, Lamed R (2006) Cellulose-Decomposing Bacteria and Their Enzyme Systems. In: Dworkin M, Falkow S, Rosenberg E, Schleifer K-H, Stackebrandt E, editors. THe Prokaryotes. Springer. pp.578–617. doi:10.1007/0-387-30742-7.

50. Rabus R, Hansen TA, Widdel F (2006) Dissimilitory Sulfate and Sulfur Reducing Prokaryotes. In: Dworkin M, Falkow S, Rosenberg E, Schleifer K-H, Stackebrandt E, editors. The Prokaryotes. Springer. pp.659–768. doi:10.1007/0-387-30742-7.

51. Llobet-Brossa E, Rabus R, Böttcher ME, Könneke M, Finke N, et al. (2002) Community structure and activity of sulfate-reducing bacteria in an intertidal surface sediment: a multi-method approach. Aquat Microb Ecol 29: 211–226. doi:10.3354/ame029211.

52. Mussmann M, Ishii K, Rabus R, Amann R (2005) Diversity and vertical distribution of cultured and uncultured Deltaproteobacteria in an intertidal mud flat of the Wadden Sea. Environ Microbiol 7: 405–418. doi:10.1111/j.1462-2920.2005.00708.x.

53. Teske A, Nelson DC (2006) The Genera Beggiatoa and Thioploca. In: Dworkin M, Falkow S, Rosenberg E, Schleifer K-H, Stackebrandt E, editors. The Prokaryotes. Springer. pp.784–810. doi:10.1007/0-387-30746-X.

54. Winogradsky S (1887) Uber Schwefelbakterien. Bot Zeitung 45: 489–610.

55. Rogan B, Lemke M, Levandowsky M, Gorrell T (2005) Exploring the sulfur nutrient cycle using the Winogradsky column. Am Biol Teach 67: 348–356.

56. Jurkowski A, Reid AH, Labov JB (2007) Metagenomics: a call for bringing a new science into the classroom (while it's still new). CBE Life Sci Educ 6: 260–265. doi:10.1187/cbe.07-09-0075.

57. Banta LM, Crespi EJ, Nehm RH, Schwarz JA, Singer S, et al. (2012) Integrating Genomics Research throughout the Undergraduate Curriculum: A Collection of Inquiry-Based Genomics Lab Modules. CBE Life Sci Educ 11: 203–208. Available: http://www.lifescied.org/content/11/3/203.full.

Permissions

List of Contributors

Simon Thorn and Claus Bässler
Sachgebiet Forschungund Dokumentation, Nationalparkverwaltung Bayerischer Wald, Grafenau, Germany

Thomas Gottschalk
Hochschule für Forstwirtschaft Rottenburg, Rottenburg am Neckar, Germany

Torsten Hothorn
Abteilung Biostatistik, Universität Zürich, Zürich, Switzerland

Heinz Bussler
Bavarian State Institute for Forestry, Freising, Germany

Kenneth Raffa
Department of Entomology, University of Wisconsin-Madison, Madison, United States of America

Jörg Müller
Sachgebiet Forschung und Dokumentation, Nationalparkverwaltung Bayerischer Wald, Grafenau, Germany
Chair for Terrestrial Ecology, Department of Ecology and Ecosystem Management, Technische Universität München, Freising, Germany

Sara A. O. Cousins and Regina Lindborg
Landscape Ecology, Department of Physical Geography and Quaternary Geology, Stockholm University, Stockholm, Sweden

Mitja Kaligarič
University of Maribor, Biology Department, Faculty of Natural Sciences and Mathematics, Maribor, Slovenia
Faculty of Agriculture and Life Sciences, University of Maribor, Pivola 10, Hoče, Slovenia

Branko Bakan
University of Maribor, Biology Department, Faculty of Natural Sciences and Mathematics, Maribor, Slovenia

Peter O. Alele
Key Laboratory of Tropical Forest Ecology, Xishuangbanna Tropical Botanical Garden (XTBG), Chinese Academy of Sciences, Kunming, Yunnan, P. R. China
University of the Chinese Academy of Sciences, Beijing, P. R. China
Great Nile Conservation Centre (GNCC), Lira, Uganda Kabale, Uganda

Institute of Tropical Forest Conservation (ITFC), Mbarara University of Science and Technology (MUST), Engineering, Southern Cross University, Lismore, New South Wales, Australia

Douglas Sheil
Department of Ecology and Natural Resource Management, Norwegian University of Life Sciences, Ås, Norway
Center for International Forestry Research (CIFOR), Bogor, Indonesia
Department of Ecology and Natural Resource Management, School of Environment, Science and nstitute of Tropical Forest Conservation (ITFC), Mbarara University of Science and Technology (MUST), Engineering, Southern Cross University, Lismore, New South Wales, Australia

Yann Surget-Groba
Key Laboratory of Tropical Forest Ecology, Xishuangbanna Tropical Botanical Garden (XTBG), Chinese Academy of Sciences, Kunming, Yunnan, P. R. China

Shi Lingling
Key Laboratory of Tropical Forest Ecology, Xishuangbanna Tropical Botanical Garden (XTBG), Chinese Academy of Sciences, Kunming, Yunnan, P. R. China
University of the Chinese Academy of Sciences, Beijing, P. R. China

Charles H. Cannon
Key Laboratory of Tropical Forest Ecology, Xishuangbanna Tropical Botanical Garden (XTBG), Chinese Academy of Sciences, Kunming, Yunnan, P. R. China
Texas Tech University, Lubbock, Texas, United States of America

Jacqueline Loos, Ine Dorresteijn, Jan Hanspach and Joern Fischer
Institute of Ecology, Leuphana University, Lueneburg, Germany

Pascal Fust
Organic Agricultural Science Group, University Kassel, Witzenhausen, Germany

Lászlo´ Rakosy
Department Taxonomy and Ecology, Babes-Bolay University, Cluj-Napoca, Romania

Liming Lai and Steven X. Ge
Department of Mathematics and Statistics, South
Dakota State University, Brookings, South Dakota,
United States of America

Dana H. Ikeda, Stephen M. Shuster and Thomas G. Whitham
Department of Biological Science, Northern Arizona
University, Flagstaff, Arizona, United States of America
Merriam-Powell Center for Environmental Research,
Northern Arizona University, Flagstaff, Arizona,
United States of America

Kevin C. Grady
Merriam-Powell Center for Environmental Research,
Northern Arizona University, Flagstaff, Arizona, United
States of America
School of Forestry, Northern Arizona University,
Flagstaff, Arizona, United States of America

Daniel P. Bruschi and Shirlei M. Recco-Pimentel
Departamento de Biologia Estrutural e Funcional,
Instituto de Biologia, Universidade Estadual de
Campinas - UNICAMP, Campinas, São Paulo, Brazil

Elaine M. Lucas
Área de Ciências Exatas e Ambientais/Mestrado em
Ciências Ambientais, Universidade Comunitária da
Região de Chapecó - UNOCHAPECÓ, Chapecó, Santa
Catarina, Brazil

Paulo C. A. Garcia
Departamento de Zoologia, Instituto de Ciências
Biológicas, Universidade Federal de Minas Gerais -
UFMG, Belo Horizonte, Minas Gerais, Brazil

Ramón Silva-Flores
Universidad Juárez del Estado de Durango, Ciudad
Universitaria, Durango, México

Gustavo Pérez-Verdín
Instituto Politécnico Nacional, CIIDIR Durango,
Durango, México

Christian Wehenkel
Instituto de Silvicultura e Industria de la Madera,
Universidad Juárez del Estado de Durango, Ciudad
Universitaria, Durango, México

Tomoharu Inoue, Shin Nagai, Hadi Fadaei, Reiichiro Ishii and Rikie Suzuki
Department of Environmental Geochemical Cycle
Research, Japan Agency for Marine-Earth Science and
Technology (JAMSTEC), Yokohama, Japan

Satoshi Yamashita, Kimiko Okabe and Hisatomo Taki
Department of Forest Entomology, Forestry and Forest
Products Research Institute (FFPRI), Tsukuba, Ibaraki,
Japan

Yoshiaki Honda and Koji Kajiwara
Center of Environmental Remote Sensing, Chiba
University, Chiba, Japan

Elisa Baldrighi and Elena Manini
Institute of Marine Sciences, National Research Council
(ISMAR-CNR), Ancona, Italy

Marc Lavaleye
Department of Marine Ecology, Royal Netherlands
Institute for Sea Research (NIOZ), Texel, The
Netherlands

Stefano Aliani
Institute of Marine Sciences, National Research Council
(ISMAR-CNR), La Spezia, Italy

Alessandra Conversi
Institute of Marine Sciences, National Research Council
(ISMAR-CNR), La Spezia, Italy
Marine Institute, Plymouth University, Plymouth,
United Kingdom

Tessa Mazor and Salit Kark
ARC Centre of Excellence for Environmental Decisions,
School of Biological Sciences, The University of
Queensland, Brisbane, Queensland, Australia

Hugh P. Possingham
ARC Centre of Excellence for Environmental Decisions,
School of Biological Sciences, The University of
Queensland, Brisbane, Queensland, Australia
Grand Challenges in Ecosystems and the Environment,
Silwood Park, Imperial College, London, United
Kingdom

Dori Edelist
Leon Recanati Institute for Maritime Studies,
Department of Maritime Civilizations, University of
Haifa, Mount Carmel, Haifa, Israel

Eran Brokovich
Department of Geography, The Hebrew University of
Jerusalem, Mount Scopus, Jerusalem, Israel

Xuemei Han
NatureServe, Arlington, Virginia, United States of
America
Department of Environmental Science and Policy,
George Mason University, Fairfax, Virginia, United
States of America

Regan L. Smyth, Bruce E. Young, Alexandra Sánchez de Lozada and Healy Hamilton
NatureServe, Arlington, Virginia, United States of America

Thomas M. Brooks
NatureServe, Arlington, Virginia, United States of America
International Union for Conservation of Nature, Gland, Switzerland
World Agroforestry Center, International Center for Research in Agroforestry, University of Philippines, Los Baños, Laguna, Philippines
School of Geography and Environmental Studies, University of Tasmania, Hobart, Australia

Philip Bubb
United Nations Environment Programme World Conservation Monitoring Centre, Cambridge, United Kingdom

Stuart H. M. Butchart
BirdLife International, Cambridge, United Kingdom
Conservation International, Arlington, Virginia, United States of America

Frank W. Larsen
European Environment Agency, Copenhagen, Denmark

Matthew C. Hansen
Department of Geographical Sciences, University of Maryland, College Park, Maryland, United States of America

Will R. Turner
Conservation International, Arlington, Virginia, United States of America

Matthias Schröter
Environmental Systems Analysis Group, Wageningen University, Wageningen, The Netherlands
Norwegian Institute for Nature Research (NINA), Trondheim/Oslo, Norway

Graciela M. Rusch, David N. Barton, Stefan Blumentrath and Björn Nordén
Norwegian Institute for Nature Research (NINA), Trondheim/Oslo, Norway

Estefany Goncalves
Centro de Ecología, Instituto Venezolano de Investigaciones Científicas, Caracas, Venezuela
Departamento de Estudios Ambientales, Universidad Simón Bolívar, Caracas, Venezuela

Ileana Herrera, Margarita Lampo and Grisel Velásquez
Centro de Ecología, Instituto Venezolano de Investigaciones Científicas, Caracas, Venezuela

Mile´n Duarte and Ramiro O. Bustamante
Departamento Cs. Ecoló gicas, Facultad de Ciencias, Universidad de Chile, Santiago, Chile
Instituto de Ecología y Biodiversidad, Facultad de Ciencias, Universidad de Chile, Santiago, Chile

Gyan P. Sharma
Department of Environmental Studies, University of Delhi, Delhi, India

Shaenandhoa García-Rangel
Departamento de Estudios Ambientales, Universidad Simón Bolívar, Caracas, Venezuela

Sandrine Baillon and Annie Mercier
Department of Ocean Sciences, Memorial University, St. John's, Newfoundland and Labrador, Canada,

Jean-François Hamel
Society for the Exploration & Valuing of the Environment (SEVE), St. Philips, Newfoundland and Labrador, Canada

Maria H. Hällfors, Marko Hyvärinen and Leif E. Schulman
Botany Unit, Finnish Museum of Natural History, University of Helsinki, Helsinki, Finland

Elina M. Vaara
Botany Unit, Finnish Museum of Natural History, University of Helsinki, Helsinki, Finland
Faculty of Law, University of Lapland, Rovaniemi, Finland

Markku Oksanen
Department of Behavioural Sciences and Philosophy, University of Turku, Turku, Finland

Helena Siipi
Department of Behavioural Sciences and Philosophy, University of Turku, Turku, Finland
Turku Institute for Advanced Studies, University of Turku, Turku, Finland

Susanna Lehvävirta
Botany Unit, Finnish Museum of Natural History, University of Helsinki, Helsinki, Finland
Department of Environmental Sciences, University of Helsinki, Helsinki, Finland

Mijoro Rakotoarinivo
Kew Madagascar Conservation Centre, Ambodivoanjo Ivandry, Antananarivo, Madagascar

John Dransfield, Steven P. Bachman, Justin Moat and William J. Baker
Royal Botanic Gardens, Kew, Richmond, Surrey, United Kingdom

Ethan A. Rundell and David J. Esteban
Department of Biology, Vassar College, Poughkeepsie, New York, United States of America

Lois M. Banta and Corey D. Watts
Department of Biology, Williams College, Williamstown, Massachusetts, United States of America

Doyle V. Ward and Bruce Birren
Genome Sequencing Center, Broad Institute, Cambridge, Massachusetts, United States of America

Index